Composite Structures for Civil and Architectural Engineering

Composite Structures for Civil and Architectural Engineering

D.-H. Kim
President, Korea Composites, Seoul, Korea

E & FN SPON
An Imprint of Chapman & Hall

London · Glasgow · Weinheim · New York · Tokyo · Melbourne · Madras

Published by E & FN Spon, an imprint of Chapman & Hall, 2–6 Boundary Row, London SE1 8HN, UK

Chapman & Hall, 2–6 Boundary Row, London SE1 8HN, UK

Blackie Academic & Professional, Wester Cleddens Road, Bishopbriggs, Glasgow G64 2NZ, UK

Chapman & Hall GmbH, Pappelallee 3, 69469 Weinheim, Germany

Chapman & Hall USA, One Penn Plaza, 41st Floor, New York NY 10119, USA

Chapman & Hall Japan, ITP-Japan, Kyowa Building, 3F, 2-2-1 Hirakawacho, Chiyoda-ku, Tokyo 102, Japan

Chapman & Hall Australia, Thomas Nelson Australia, 102 Dodds Street, South Melbourne, Victoria 3205, Australia

Chapman & Hall India, R. Seshadri, 32 Second Main Road, CIT East, Madras 600 035, India

First edition 1995

© 1995 E & FN Spon

Typeset in Great Britain by Techset Composition Ltd., Salisbury, Wilts. Printed in Great Britain at the University Printing House, Cambridge

ISBN 0 419 19170 4

A catalogue record for this book is available from the British Library

Library of Congress Catalog Card Number: 94-68234

∞ Printed on acid-free text paper, manufactured in accordance with ANSI/NISO Z39.48-1992 (Permanence of Paper).

And openly
I pledged my heart to the grave and
suffering land, and often in the consecrated
night, I promised to love her faithfully until
death, unafraid, with her heavy burden of
fatality, and never to despise a single one of
her enigmas. Thus did I join myself to her
with a mortal cord.

— Hölderlin —

Contents

Preface

During the long history of human civilization, four basic construction and structural components/systems have evolved. They are beams and columns, stone arches, wooden trusses and modern reinforced concrete and steel structures. The development of these basic construction systems is consistent with the materials available at the time and the theories developed that were applicable in the utilization of such materials. Historical ages are even referred to in terms of materials of construction: stone age, bronze age, iron age, and so on. The present age is the age of flexibility of choice. Modern science and technology have produced, and will continue to produce, numerous advanced new materials, applicable theories and methods for their utilization. It is therefore not possible nor appropriate to characterize the present age by any one single material. The right material could be often selected and used at the right time in the right place. This situation, however, requires optimal selection of materials, structural forms and configurations, manufacturing and fabrication methods, compared to situations where no material choices were possible. Because of the diversity in modern material properties, the engineer or designer must now be able to select and design the materials. The new structural concept includes the determination of a structural system and the selection of a material best suited for this system in consideration of all relevant conditions.

The author calls the above concept the fifth basic concept for structural construction. This concept, without being recognized by many, has been implemented by engineering professionals of various disciplinary branches to a considerable extent. Today, a civil engineer or architectural designer, without that knowledge and ability to apply such concepts in practice, will be in as helpless a situation as a bank teller without the ability to handle a calculator or personal computer.

When the concept is realized as the product, it can be identified as a composite or composite structural system. The composite is made from two or more kinds of different materials combined together to obtain the properties that are not available in the original material systems alone. The characterization of materials and theories for composites are very well established, mostly through advances in material sciences and related engineering applications, particularly aeronautics and aerospace engineering. However, present day technologies are too far diversified, and there is little communication between the different branches of engineering. The lack of communication even exists quite seriously between different specializations within the same engineering branch. Application of composites requires a certain degree of knowledge of all branches of engineering and technological areas concerned. This book is intended to bridge such communication gaps by integrating current knowledge on composites from various disciplinary sources into a coherent volume to facilitate practical applications.

When a war is over and peace comes, with ensuing prosperity, those belonging to the new generation might never know how the battles were fought. The pioneering generation, that faced the grim task of building a nation, had to do everything, from basic planning, acquiring financing sources, managing, designing and constructing all kinds of projects in civil works. On the other hand, the generation of a peaceful prosperity, in general, does routine work within the domain of his/her professional specialization. However, the pace of development of science and technology requires an awakening of this generation's focus. Furthermore, as additional advances are made, engineers will be in charge of responsibilities which require diversified knowledge, as well as the courage to make critical judgement and to take responsibility. Composite designs require such a pioneering spirit, with a boldness to face all challenges that are brought forth by the scope and complexity of the rapidly advancing technology. Any engineers without such an ability may become obsolete in the near future. This book provides the necessary baseline technology to assist the new generation of engineers as well as senior engineers who once belonged to the pioneering frontiers, to meet the challenges they face in composite design and construction.

In order to utilize the benefits of new materials to the maximum, the modern engineer is not just a material user, but a material designer as well, who must select and design the material and manufacturing method best suited for their project. The engineer's specifications for the material and the manufacturing method must include the standards for quality control. However, regardless of all complexities involved, understanding and applying composites is not difficult if certain rules are understood. It is the aim of this book to translate the difficult existing composite theories into terms of easy and familiar language for use by civil and architectural engineers.

In Chapter 1, the basic concept of composites is given in detail. This would give the reader the technical insight and necessary background to the rest of the volume. In Chapter 2, the properties of the constituent materials, both reinforcements and matrices, are given. Some materials of particular interest to the civil engineers are introduced. The mechanical properties of several kinds of composites based on polymer, metal, ceramic and cement matrices are also given. An extensive review on classical elasticity and structural mechanics is presented in Chapter 3. Emphasis is placed on the structural forms which may be useful for composite designs. Eigenvalue problems to both buckling and vibration of structures are introduced in Chapter 4. A method which can be used to solve eigenvalue problems for structural elements with variable cross-sections and arbitrary boundary conditions is also given. Anisotropic elasticity including stress–strain relations, engineering constants, on-axis to off-axis (or *vice versa*) transformation equations, invariants, laminates and micromechanics are presented in Chapter 5. After explaining one-dimensional composite mechanics in Chapter 6, composite plates are discussed in Chapter 7 in detail, including plate boundary values as well as eigenvalue problems. After obtaining the maximum stresses and strains by the methods given in the previous chapters, the element maximum strength and strain must be compared with such quantities in the design procedure. Extensive discussions on failure and strength theories of composites are described in Chapter 8. Connecting methods for both mechanical and adhesive bond joints for composites are elaborated in Chapter 9. With methods of analysis, the methods of joining, and strength theory of composites given, a practical design can be executed. The procedures and methods of designing composite structural elements are presented in Chapter 10.

The author is greatly indebted to his mentors, the eminent engineer and professor, Dr John E. Goldberg, and the eminent mathematician and physicist, Dr S. S. Shu, both formerly with Purdue University. Their wide spectrum of knowledge on structures, physics and other subjects, gave the author the ability to grasp and develop the ideas and methods covered in this text in terms of the technical language common to engineers of civil construction.

The author extends special thanks to Dr S. W. Tsai, professor of aeronautics and astronautics of Stanford University and the former chairman of the International Conference on Composite Materials, Dr S. J. Dastin of Grumman Aerospace, and Dr B. A. Wilson, the former president of the Society for the Advancement of Material and Process Engineering, and others for willingly providing reference materials. Special appreciation goes to those dedicated people, especially Ms H. O. Whang, Mr K. J. Kim, D. S. Shim, C. M. Won, K. S. Kim and other graduate students of Kangwon National University, who have worked day and night to assist the author to prepare the manuscript.

It must be recorded that the publication of this book was made possible by Professor K. P. Chong, Dr S. C. Liu and Professor H. J. Lagorio, all of the United States National Science Foundation.

Duk-Hyun Kim

1 Introduction

1.1 Historical necessity

Throughout the long history of human civilization, four basic concepts of building construction have evolved and developed. These concepts, namely, **beam and column**, **masonry arch**, **wooden truss** and the modern **steel truss and frame**, were made possible by available construction materials and applicable technical knowledge of each age.

The ruined structures at Chaldea were built in 5000 BC. However, before the Greeks, the best structures were built by the great Egyptians and the concept was of the beam and column. The best of the Egyptian temples is the great Ammon temple in Karnak. This structure, built around 1500 BC, has a length of 360 m and width of 108 m. The roof made of great masonry slabs is supported by a beam–column system. The central columns have diameters of 3.5 m and lengths of 20.7 m.

The Romans seem to be the first people to have used the arches in structural construction. They built viaducts using masonry double arches, and some of these are still in use. The Pantheon in Rome is a masonry dome with a diameter of 44 m and was built in AD 120.

Another important construction concept, the wooden truss, was used by Andrea Palladio (1518–1580) of Italy. However, the importance of his invention was not recognized until the the 18th century.

The development and application of steel in building construction (Bessemer, 1856) are the most memorable human achievements in this field after the Romans. Application of structural steel with high strength made the construction of multistory buildings, long span bridges, high rise towers, large ships and so on, possible, and helped to make human civilization come to its full bloom.

Modern metallurgy and chemical engineering have produced numerous new structural materials day by day, ready to make existing materials such as steel obsolete. As human beings have developed structural concepts with usable materials and applicable technologies, it is necessary for us to develop a new structural concept (or concepts) suitable for new materials.

The author wishes to call this **the fifth basic concept of structures.**

From the dawn of human history, materials have been the weapons of progress. Historical ages were even named after them: the stone age, bronze age, iron age and so on. Today, we cannot qualify our age after just one material, for our era is the era of choice. We have the possibility of using the right material at the right place. 'Era of choice' means there is strong competition. In this fight, composites are very well armed and are beginning to show their strengths. One of their main strengths is their wide range of properties.

The competitors also have a strong weapon by virtue of low cost. Composites have only been able to conquer the 'high performance/low volume' industry such as aerospace where cost has been a secondary factor. Now that the price structure of materials shows a sign of decreasing, and efficient manufacturing methods are evolving day by day, engineers must look into these materials very seriously. The design method plays a great role in cutting the price. There are several civil structures, designed with composites, that have proven more economical than with other material construction.

Simply stated, a **composite** is a combination of materials joined into a whole to create an end product for a specific purpose. This concept is not new. Ancient people, eastern and western, built houses with mud and chopped straw. Wood, the most widely used structural material from earliest times, is a supreme example of a naturally occurring composite material.

Other composites have been around for more than a hundred years. Reinforced concrete is an excellent example.

From glass fiber reinforced plastics to advanced carbon fiber reinforced epoxy, composite materials are very much diversified. The exact composition of some composites such as metal–metal and carbon–carbon is subject to national security. Several kinds of composites such as aramid reinforced ones can be found in applications for automobiles, sporting goods, armor and others.

Composites are made of **reinforcement** and other material which binds the reinforcement. This binder is called a **matrix**. From polyester to super metals such as titanium or magnesium, the matrices used today are very much diversified. In all cases, however, matrices only support and bind the reinforcement, and all composite mechanical properties in the longitudinal direction are taken care of by the reinforcement. However, the matrix controls most of the properties in the transverse direction.

Compared with metals, composites have higher specific strength and higher specific modulus. Generally speaking, composites are **corrosionless**. Many engineering professions such as aerospace, shipbuilding and even automobile builders have used composites for several decades. Civil and architectural engineers, however, have been very slow in using them and have had, to a certain degree, prejudices against them.

For large scale civil and architectural structures, mainly steel, concrete and aluminum have been used. Advancement in structural technology made large size structures possible, and weight and corrosion of materials became a major concern. It is natural that engineers want to have new materials which can reduce the weight and improve the life span of structures, with reduced maintenance cost. It is out of this necessity that engineers should pay attention to composites using materials such as carbon fibers (graphite) with high tensile strength (sometimes over $5.5\,GPa = 800\,000\,psi$) and **light weight** (about a quarter of steel).

In addition to very light weight and freedom from corrosion, structural engineers will find the following advantages against conventional materials, when composites are used in designs:

- superior toughness;
- superior thermal properties;
- superior electromagnetic properties;
- excellent damping effect;
- reduced number of parts;
- possibility of reinforcing to any direction;

- possibility of strengthening the members without geometric reinforcements, such as stiffeners;
- possibility of producing optimum members/structures depending on structural requirements.

Application of composites in the field of large structures is still in its infancy. However, modern science and technology are advancing so fast that by the time students finish education and go to engineering practice, they will find that knowledge of and capability to design with composites are immensely important.

1.2 Basic concept of composites

In composites, a second material is added to obtain specific performance properties not available in the unmodified material. Many engineers tend to have a misconception about composites, such that they consider only polymer materials are composites. However, the range of composites is enormously diversified and the list of candidate materials is almost endless. Carbon, glass, aramid, or other fibers are placed in thermoset or thermoplastic polymer resins to obtain higher strength and stiffness. Ceramic whiskers are added in ceramic composites to improve toughness. Silicon carbide particles are put in metal matrix composites to control thermal expansion. Nearly all conventional engineering materials can be used as a matrix and a reinforcement. According to a SAMPE (Society for the Advancement of Material and Process Engineering) journal, several scores of new materials are reported every month. However, regardless of the dizzying pace of development of composites, basic concepts are not difficult to grasp. If certain rules and principles are clearly understood and are thoroughly followed, practicing engineers can have maximum use of these wonderful materials.

Needless to say, young engineering students should be taught composite theory from the beginning of engineering study. The mechanics of conventional materials can be considered as a special case of composite mechanics.

Composite materials are made by controlled distribution of one or more materials, the reinforcement (first phase), in a continuous phase of a second, the matrix. The boundary between the matrix and the reinforcement, the **interface** (third phase), is controlled to obtain the desired properties from a given pair of materials. This interface may be intentionally made weak by minimizing the chemical coupling of reinforcement to matrix such as the cases of some ceramic composites, and of when dissipation of impact energy is required. However, it is much more common to try to maximize the coupling between two phases. This interfacial coupling allows stresses dispersed through the matrix to be transferred to the reinforcements. Coupling is provided by **wetting** the reinforcement by the matrix in molten or low viscosity state. Wetting can be done in several ways depending on the manufacturing method.

Composite raw stocks come in several different forms. Some of them are continuous fibers, tapes and broad goods. The most basic is a 'tow' which is a quantity of continuous fibers wound on a reel. In the **pultrusion process**, fibers are drawn through a resin mix, to provide intimate fiber–resin wetting, onto a mandrel, into a mold, around a form, or whatever, then put through the curing and fabrication process. In **filament winding**, fibers are drawn through a resin, as in pultrusion, to provide wetting, then onto a rotating mandrel. Tapes usually come in widths of 7.6–15.2 cm and comprise unidirectional continuous fibers with binder already applied. Broad goods are sheets of fibers (usually unidirectional) impregnated with the binder material. Both tapes and broad goods, already impregnated with binder, are called 'prepregs'. Reactive resins are usually partially cured (B-staged) for easier handling. In metal matrix composites, squeeze casting may be used to apply high pressure to obtain good wetting of fibrous preforms.

Sometimes, reinforcements of polymer matrix composites are treated with coupling agents to improve interface compatibility.

There are many occasions when engineers mention fiber reinforced plastics (FRP). This terminology is improper because, in composites, it is the fiber which produces the mechanical properties required. The plastic acts as a matrix to bind the reinforcements.

1.3 Matrix

In addition to just binding reinforcements, the matrix plays vital roles for overall composite characteristics. In general, reinforcements have high strength and stiffness, but tend to be brittle. It is the matrix which protects reinforcements against abrasion or environmental corrosion which can initiate fracture. The load carried by longitudinal reinforcements is distributed by the matrix. In order to transfer the loads and to reduce the chance of failure in the matrix, adhesion to reinforcements must be coupled with sufficient matrix shear strength, which is, in general, proportional to the tensile strength. However, high strength matrices tend to be brittle. Civil engineers should recall the principles used in designing reinforced concrete. In composite laminates, forces in the transverse direction are carried by the matrix. In a reinforced concrete beam, transverse shear, exceeding the concrete shear strength, is carried by stirrups. When a composite is used for a large structural member, a similar concept should be borne in mind. If fiber breakage occurs, the matrix re-distributes the load among neighboring reinforcements and both halves of broken reinforcement. While high tensile and flexural properties are provided by strong and stiff, but brittle, reinforcements, fracture toughness is provided by plastic flow at crack tips in the matrix, which absorbs energy and reduces stress concentration. Plastically deforming matrices also deflect cracks parallel to fibers, and serve to prevent the failure of fibers all in one plane. For a crack to extend through ductile matrices, it is necessary for the fibers to pull out of the matrix as they break. Engineers with steel design experience should recall the importance of ductility of steel. These functions of the matrix must be performed across the anticipated temperature range (Figure 1.1). The matrix must resist expected chemical and environmental stresses. In polymer matrices, temperature combined with moisture plays a critical role. This phenomenon is called the **hygrothermal effect**.

Figure 1.1 Usable temperatures of high temperature resistant materials. (Courtesy of Textron Specialty Materials.)

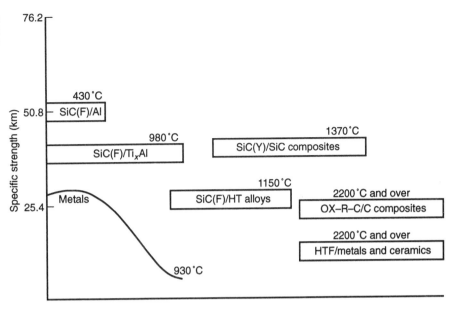

In metal and ceramic matrix composites, thermal and environmental stability are less of a problem than in polymer composites. Aluminum is one of the leading metal matrices because of its light weight, relative ease of processing, and excellent property improvement with reinforcement. Copper is heavier, but has higher shear strength than aluminum above 700°F. Titanium is lightweight and resistant to elevated temperatures (up to 1200°F) but is more costly and difficult to reinforce. Magnesium has good compatibility with reinforcement. As far as high temperature is concerned, carbon–carbon composite is best suited (up to 6000°F in inert atmosphere or vacuum).

There are several scores of different **polymer matrices** used in 'advanced' composites. The permutations of different fibers and epoxy binders alone could run into the hundreds for advanced composites. However, within such a huge advanced composite family there are two major types which determine the methods of manufacturing. One uses a **thermosetting** binder and the other uses a **thermoplastic** (Figure 1.2).

A polymer is a long chain molecule made by connecting many smaller molecules. A thermoplastic polymer can be considered as long chains of molecules lying next to each other. Various types of electrostatic attraction hold the molecules in the same position relative to each other. At a certain temperature, the molecular motion becomes so great that the attractions break down and the molecules begin to slide past one another. This temperature is called the melt temperature, T_m. At or above T_m, the polymer can be molded or formed into a shape. Upon cooling the polymer will retain its new form. This heating, forming and cooling can be repeated a number of times.

A thermosetting polymer can be thought of as one large molecule, since each molecular chain is chemically bound to its neighbor. The bonds between chains are called crosslinks. The thermosetting polymer is crosslinked during the manufacture of the composite, usually by heating in an oven, mold or autoclave.

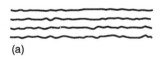

(a)

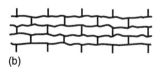

(b)

Figure 1.2. (a) Thermoplastic and (b) thermoset polymers.

A minimum curing time is required, depending on the particular thermoset formulation to be processed. During the curing process, anywhere from ten minutes to six hours, the thermoset changes from liquid to solid, crosslinked polymer. This change is irreversible; the polymer cannot be remelted or liquefied.

Because of these basic differences, there are advantages for both types of composites. Thermosets have the advantages of lower raw material costs (with some exceptions), better chemical resistance, high application temperatures and better resistance to creep under load. Thermoplastics allow faster processing speeds, postformability, versatile bonding methods, such as ultrasonic or heat welding and solvent bonding, high damage tolerance and the possibility to reprocess scrap material.

Thermosetting binders are cured once and can never be reformed, recured or reused. Unlike metals, which can be remelted or recast, the thermosets have a single use. It is important to note that the curing process starts the moment the epoxy is formulated. At a high enough temperature it is hours or minutes. At room temperature, the curing is usually about two weeks, and at 0°F it will take two years or longer. Thermoset inventories must be stored at a subzero temperature for this reason.

Recently, both families of binders are seeing a dizzying pace of development. A company is ready to report an epoxy which can be cured in a second. Epoxies which can be stored at room temperature up to six months were reported. The recently developed thermoplastic family of binders is causing a serious re-evaluation of this type of binder. When the thermoplastics first appeared, they started to lose their strength at 300°F. Now, some must be raised to 600°F before the reforming and recuring process starts.

It is necessary for civil engineers to give attention to these new materials so that the maximum benefit of composite materials for civil structure applications can be achieved. Some of the polymer matrices both thermosetting and thermoplastic, are shown in Table 1.1.

Table 1.1   Some thermosetting and thermoplastic polymer matrices

Thermosets	Thermoplastics
Acrylamate polymers	Acrylics (from MMA monomer)
Alkyd and diallyl phthalate (DAP)	Nylons (NY)
Bismaleimides (BMI)	Polyamide-imide (PAI)
Epoxies (EP)	Polyarylene sulfide (PAS)
Melamines	Polycarbonates (PC)
Phenolics	Polyesters
Polyesters	Polyetheretherketone (PEEK)
Polyimides (PI)	Polyetherketone (PEK)
Polyurethanes (PUR)	Polyetherimides (PEI)
Silicone (SI)	Polyether sulfone (PES)
Vinylesters	Polyethylenes (high density) (HDPE)
	Polyethylene terephthalate (PET)
	Polyphenylene sulfide (PPS)
	Polypropylenes (PP)
	Polystyrene (PS)
	Polysulfone (PSU)
	Styrenic copolymers (ABS, ACS, etc.)
	Thermoplastic polyimide (TPI)

1.4 Reinforcements

The transfer of loads and improved toughness provided by the matrix and interface are necessary prerequisites for the excellent properties of composites. But it is the reinforcement that is primarily responsible for these properties.

Reinforcements fall into three main categories: **fibers**, **flakes** and **spheres or particulates**.

A fiber is a long fine filament of matter with a diameter generally of the order of 10 μm and an aspect ratio of length to diameter between a thousand and virtually infinity for continuous fibers.

Reinforcements may be called by different names according to sizes, such as whisker (<0.025 mm), fiber (0.025–0.8 mm), wire (0.8–6.4 mm), rod (6.4–50 mm) and bar (>50 mm). Glass, carbon (sometimes called graphite in the USA until 1988), boron, aramid (called Kevlar by Du Pont) and other organics dominate fiber usage.

Ceramics and metals are generally in whisker form but sometimes in fiber form.

Glass, mica, metal and some other materials are used in flake form. Many plastics, glass, carbon and ceramic materials are used as microspheres.

Fillers such as calcium carbonate, clay, talc and related products reduce cost, improve surface quality and provide other nonstructural benefits. Sometimes sand is used as a filler for some large size structures, reducing significant amounts of the resin matrix required. Microspheres improve thermal insulation and shock resistance, and add dimensional stability and compressive strength (recall the function of aggregates in concrete). Flake reinforcement is used where a good barrier against moisture, gas or chemicals, or provision for high thermal and electrical resistance or conductivity is needed.

Most structural composites are reinforced with glass fibers, used as rovings (collections of continuous filaments), yarns (collected filaments with a twist applied), woven rovings, fabrics, mats, stitched or chopped staples. Depending on coupling agents used for enhanced strength and chemical/moisture resistance with specific matrix systems, fiber glass shows different properties and has several different names such as E, S, S2 and T.

Some of the materials for reinforcement are as follows: Alumina, aluminum, boron, boron nitride, beryllium, carbon (graphite), glass, aramid (Kevlar), silicon carbide, silicon nitride, steel, titanium, tungsten.

1.5 Filamentary type composites

Among many different types of reinforcements, it is the filamentary type which is most important from a structural applications point of view. In general, **continuous filament reinforced composites** exhibit high specific strength and stiffness.

Matrices and reinforcements used in this type of composites are numerous and vastly diversified, and there is much related literature. In Chapter 2, properties of some materials and composites with potential civil and architectural application are described.

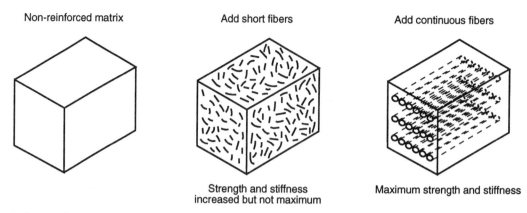

Non-reinforced matrix   Add short fibers   Add continuous fibers

Strength and stiffness
increased but not maximum

Maximum strength and stiffness

Figure 1.3 Concept of composite.

In this section, terminologies related to filamentary type composites are explained. As a starting point, a single ply **lamina** is considered. When reinforcements are discontinuous, fibers can be either 'random' oriented or 'preferred' oriented. Single ply sheet with random orientated discontinuous fibers, such as chopped fiber (Figure 1.3), may be considered isotropic in the plane of the lamina, provided that special care is taken. The maximum strength (as well as

Figure 1.4 Continuous fiber reinforced composite.

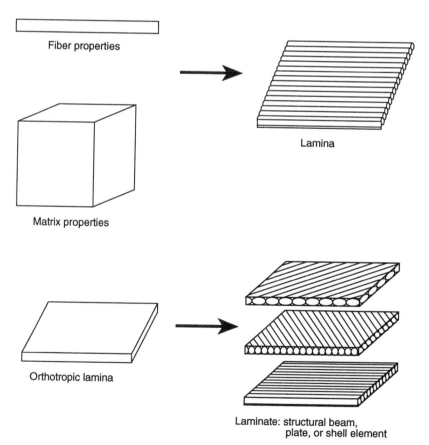

Fiber properties

Lamina

Matrix properties

Orthotropic lamina

Laminate: structural beam,
plate, or shell element

stiffness) of composites can be obtained when the reinforcements are in the form of continuous fibers (Figure 1.3). These fibers can be arranged either to the same direction (**unidirectional**) or at certain angles (**angle ply**) (Figure 1.4).

Most of the structural members require a certain thickness with several layers of plies. These are called **laminates**. Sometimes, layers with different material constituents may be used. This is called **hybrid laminate** (Figure 1.5). It should be noted that reinforcement fibers can also be of hybrid type.

There are certain fundamental, though trivial, principles regarding fibers in filamentary type composites.

1. The elongation of the fiber in a composite must be less than and its stiffness higher than that of the matrix to obtain an effective composite.
2. Mechanical properties of a laminate are determined more by the amount and form, i.e. length and orientation, than by the type of reinforcement.
3. The strength of a composite structural member increases in proportion to increasing fiber content. The higher the content, the stronger the member.
4. The longer the 'straight' fiber length to a given direction, the greater the continuity of stress transfer to that direction, thus making the member exhibit higher load carrying capacity in that direction.
5. The fiber orientation determines the direction of strength as well as the fiber content which determines the strength level achieved:
    (a) **Unidirectional**. The fiber alignment is greater and the maximum fiber content can be obtained. Note that the maximum strength is achieved in the 'fiber direction'. Up to 85% (by weight) continuous fiber content can be obtained by filament winding, pultrusion and 'prepregging'.
    (b) **Bidirectional**. Continuous fibers are at right angles (woven roving and cloth). Strengths are higher in those two directions. Up to 65% (by weight) woven fiber content is possible. The practice is, however, to use fiber mat for shear strength, thus decreasing the fiber content to a certain degree.
    (c) **Multidirectional**. As mentioned previously in this section, chopped fiber strands can be placed randomly to obtain isotropic strengths, i.e. equal strengths in all directions. Mat or chopped strands are used in many processes with fiber content up to 65% (by weight).

Figure 1.5 Hybrid laminate. Cost can be reduced by mixing carbon fibers with cheaper glass continuous fibers.

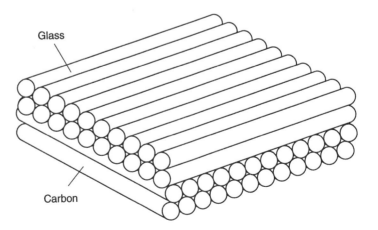

Glass

Carbon

From the above principles, the following facts can be understood easily. Continuous parallel (unidirectional) fibers provide high strength to the direction of the fibers, but very low transverse strength. Adding a 90° cross-ply (bidirectional) will provide fairly high strength in the 0° and 90° directions, but the composite is weak in the 45° direction. If fibers are laid in 120° triple plies, the composite will have 'moderate' strength in 'all' directions. If discontinuous fibers are laid, the composite strength may be reduced by an order of magnitude. As an example, with fibers of 3.45 GPa tensile strength, a unidirectional fiber composite provides around 1.38 GPa tensile strength in the fiber direction. A bidirectional arrangement (cross-ply) can provide about 690 MPa to the 0° and 90° directions. A 120°, three ply composite may exhibit approximately 480 MPa in all directions.

It should be kept in mind that the above facts explain only the general behavior of composites, and that detailed mechanical properties are dependent on several factors, such as material properties and the method of manufacturing. Mechanical properties of some of the composite systems are explained in Chapter 2.

1.6 Composite manufacture

As explained in the previous section, selecting proper matrix and reinforcement is important to composite properties, but more important is the way in which these constituent materials are arranged. The methods of combining these constituents have developed rapidly from predominantly manual placement, and other processes, resulting in great advances in precision, quality control, and reproducibility.

Some of the composite fabrication processes and resulting mechanical properties are shown in Table 1.2.

Reinforcement types and precombining methods or compound types are also given. There are other methods not shown in this table. Some of these are vacuum pressure bag molding, cold press molding, stamping, centrifugal casting, continuous laminating, pressure and roll bonding, plasma spraying techniques, powder metallurgy, controlled solidification and pneumatic impaction.

Two process methods are of particular interest to civil engineers; these are filament winding and pultrusion.

Filament winding. Of all the manufacturing processes available to the composites industry, filament winding is considered to be the oldest mechanical process and has benefited the most from the introduction of computer technology. Filament winding machine is not really an adequate description. 'Fiber placement machine' might be more proper as it describes more accurately the ability to lay fibers at any angle, in any direction and to employ as many permutations of these movements as required by the structural member's design.

The basic components of a filament winding system are shown in Figure 1.6. The mandrel on which the fibers are to be wound is rotated at a constant speed. Fibers come from a creel through a resin impregnation stage and a 'pay-out eye' which is mounted on a moving transverse carriage. The carriage motion is controlled relative to the mandrel rotation to give the required fiber

Table 1.2 Manufacturing methods and related mechanical properties

	Hand lay-up	Spray-up	Structure RTM/RRIM	Bulk molding compound	Preform	Sheet molding compound	Injection molding FGRTP	Filament winding	Pultrusion (profile-rod)
Tensile strength (ksi (MPa))	9–50 (62–344)	5–18 (35–124)	20–28 (138–193)	3–12.5 (21–86)	10–30 (69–207)	8–39 (55–269)	6–30 (41–207)	80–200 (550–1380)	40–180 (275–1240)
Tensile modulus (Msi (GPa))	0.6–4.5 (4–31)	0.8–1.8 (6–12)	0.5–1.5 (3–10)	1–2.5 (7–9)	0.8–2.0 (6–14)	1–2.8 (7–17)	0.5–1.8 (3–12)	4.4–7.5 (30–50)	3–6 (21–41)
Flexural strength (ksi (MPa))	16–80 (110–550)	12–28 (83–190)	30–45 (207–310)	7–20 (48–138)	24–40 (165–276)	15–70 (103–483)	8–45 (55–310)	100–250 (690–1725)	75–210 (517–1448)
Flexural modulus (Msi (GPa))	0.9–4.0 (6–28)	0.7–1.3 (5–9)	1.2–2.2 (8–15)	1–2.5 (7–17)	1–2.5 (7–17)	1–2.61 (7–18)	0.4–2.2 (3–15)	5–7 (34–48)	3–6 (21–41)
Compression strength (edgewise) (ksi (MPa))	18–50 (124–344)	15–30 (103–207)		15–30 (103–207)	15–30 (103–207)	15–35 (103–240)	6–26 (41–180)	50–80 (345–550)	40–100 (276–690)
Shear strength (ksi (MPa))	4–6 (28–41)					3–5 (21–34)		7–10 (48–69)	5–10 (35–69)

All composites are reinforced by glass fiber.
RTM: resin transfer molding; RRIM: reinforced reaction injection molding; FGRTP: fiber glass reinforced thermoplastic.

Figure 1.6 Filament winding system.

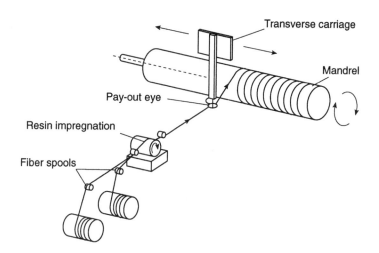

orientation. The transverse carriage moves back and forth until the required amount of fiber is applied.

Most of the mechanically operated winding equipment is capable of handling two or three axes, but with the introduction of computer control, it is possible to develop equipment with up to nine axes.

Many types of fibers can be used; for large structures, glass fibers are most suited. Tensions in the range 300–3000 gf are applied to fibers depending on the design.

The most popular resin types for filament windings are polyester, vinylester and epoxy. Resins can be applied during the winding process (wet winding), or to the fibers as prepreg. In wet winding, lower viscosity of resin improves the wetting of the fibers. The dominant resins for filament winding have been thermosets. However, considerable development work is being done with thermoplastics such as nylon, polypropylene, polyetheretherketone (PEEK), and polyphenylene sulfide.

There are several types of filament winding, such as hoop or circumferential, helical, multi-directional and polar. Many types of machines are capable of doing all of these.

Filament winding accounts for up to 10% of total market consumption of composites. Pipe and tank production is the largest single application for filament wound products satisfying both small and large diameter ones. Rocket cases, aircraft fuselages, automobile bumpers and leaf springs, and many structural members have been manufactured by this method. A series of filament wound tubes can be joined to make a space frame structure. Several new structural concepts for civil engineers can be realized by this process.

Pultrusion. Pultrusion is a continuous molding process. Continuous fiberous reinforcement rovings and strand mat (or other reinforcements) are pulled through tanks of resin, a heated forming and curing die to form a completed composite structural shape, a pulling station and finally reach a cut-off unit (Figure 1.7).

Fiber placement, resin formulation, catalyst level, die temperature and pull speed are critical process variables. These variables must be established during product design, and monitored during the manufacturing process to assure that

Figure 1.7 Pultrusion schematic.

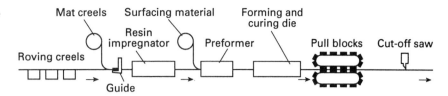

the finished pultruded structural member has the proper appearance and specified physical and chemical properties.

Glass fiber is the most common reinforcement for pultruded shapes, even though other reinforcements such as Kevlar, boron, carbon fibers and even metal materials are used in some cases. The combinations of reinforcements for a given shape are engineered to provide the strength required in both longitudinal and transverse directions.

The resin systems used for structural shapes are designed to meet required service conditions such as chemical and corrosion resistance, good electrical insulating properties, as well as temperature, weather and water resistance. Sometimes, some additives are used to meet fire retardance requirements. The most common resin is a thermosetting polyester; vinylesters and epoxies are also used.

Recently, there has been intensive development in the use of thermoplastics, such as polyetheretherketone (PEEK), polyethylene terephthalate, polysulfone and polyphenylene sulfide.

The first patent issued strictly for pultrusion was in 1951. The pultruded shapes began to be used for sporting goods; the usage of pultruded shapes has recently expanded to almost all sectors of industry. One of the merits of using the pultrusion process is that most of the standard structural shapes can be manufactured by this method. The engineer has very specific design possibilities with these shapes. For instance, flexural modulus, both longitudinal and transverse, can be changed by simply altering fiber orientation without changing the cross-section.

There are some other means of composite processing, which may become significantly important for structural engineers in the near future.

1. **Drawing**. A thermoplastic formable sheet is reinforced with unidirectional but discontinuous fibers. During drawing of the sheet, the fibers slide past one another, providing extensibility not available with continuous fiber reinforcement. When this method is fully developed, some design problems which require high ductility might be solved.

2. **ARALL**. The ARALL laminate concept was developed by Delft University and Fokker Aircraft during the 1970s. ARALL, which stands for aramid aluminum laminate, is a hybrid composite material system based on the concept of bonding high strength aramid fibers embedded in an epoxy resin between thin sheets of high strength aluminum alloys. The resulting laminate offers significant structural and weight saving advantages over monolithic construction, especially in tension dominated, fatigue and fracture critical applications. Some composites have relatively weak stiffness compared with tensile strength. The high stiffness of metals and high

strength of composites can be combined by this scheme to obtain a structural material with superb properties. This concept can be further extended to hybrid composites using steel and other materials.

3. **Braiding**. Braiding has become increasingly attractive for composite manufacture because of its high rates of fiber deposition, adaptability to automation and ability to produce strong, complex shapes. In general, composite laminates are weak against transverse shear and subject to delamination. Braiding is a method of three-dimensional placement of reinforcements and is capable of solving such problems. There are two- and four-step braiding processes. The two-step braiding is of interest to civil engineers since manufacture of standard members may be possible. Three-dimensional fabrics are also considered to solve those problems associated with large structural members.

One of the most important steps in composite manufacturing is **flaw detection**. Because of the number of processing variables involved, manufacturing defects are quite likely. Design engineers should specify how to differentiate between noncritical anomalies and defects with potentially destructive results, and the acceptance criteria. There are many types of flaws in composites such as delamination, voids, porosity, broken or misoriented fibers, improperly cured resin, improper resin content and moisture in the laminate. However, **delamination** and **voids** are of most concern. Improper surface preparation, foreign matter or impact damage during handling may cause delamination. If a resin with high volatile content is cured with a short curing cycle, the resulting composite may have an unacceptable number of voids.

Ultrasonic testing and **radiography** are the most widely used methods. Many manufacturers prefer a combination of the two, using the former for general inspection and the latter for a closer look at flaws or questionable areas. **Computer tomography** applied to radiography allows three-dimensional X-ray images. **Acoustic microscopy**, computer controlled ultrasonic systems, and laser based ultrasonics are beginning to be used. However, future systems will combine more than one technology.

The most troubling drawback in using composites is **high cost**. The manufacturing of high performance composites used to be labor intensive, resulting in costly processing. Because of such problems, a lot of work has been done aimed at developing process shortcuts and increasing process automation. Computer aided design, engineering and manufacturing (CAD, CAE, CAM) are becoming popular among several composite manufacturers, with CAM playing a significant role in automation, thus obtaining the goal of reducing the cost of composites.

1.7 Applications, present and future

Although manufactured composites have been used for thousands of years, the high technology era of composites began in 1964 when carbon fiber exhibiting sufficiently high strength and stiffness was produced at the Royal Aircraft Establishment in England. Soon, composites found their applications in sporting goods such as fishing rods, tennis rackets, boats and bicycles, followed by the automobile industry. Application in civil structures is now underway. If the slow pace of application in the past is considered, one may think that it may take years to achieve wide application of composites in civil engineering. However, once successful applications bring confidence to the engineers and users, further applications will be brought, stimulating new development by material suppliers, designers and manufacturers, making composites the major structural materials in the near future.

Some developments and applications of composites are briefly reviewed as follows.

1.7.1 Aerospace

Early aeroplanes were made of wood, fabrics and piano wires. Aeroplanes made of aluminum appeared in early 1920s, and most modern aeroplanes still employ an aluminum structure. In 1944, the first fiberglass reinforced thermoset resin laminate airplane made its virgin flight at Wright Patterson Air Force Base.

In the late 1960s, the first flight-worthy composite component appeared in the form of the F-111 horizontal stabilizer, made of carbon and boron reinforced epoxy. In 1969, the US Navy with Grumman and General Dynamics, manufactured the F-14 horizontal stabilizer by boron–epoxy skin covered aluminum honeycomb sandwiches. Carbon (graphite)–epoxy composite appeared in the early 1970s and was cheaper ($40–70 per unit) than boron–epoxy composite ($350 per unit) and had more manufacturing flexibility. The F-15, F-16, F-18, X-29A and AV-8B all employed some composite parts.

The Boeing 767 used almost two tons of composites for its floor beams and surfaces, the Russian Antonov 124 five and a half tons of composites. Recently, there have been all-composite structured airplanes, the Rutan designed Voyager and Beechaircraft's Starship 1. The Chinese Dauphine 2 helicopter has 87% of its structure made of composite. The Japanese next-generation fighter FSX will have 80% of its structure made of single mold composite, using advanced material technology. The B-1B lower wing skin measures 8 ft by 49 ft and is tapered from 2.5 in inboard to 0.2 in at the outboard end. This panel weighs 2100 lb and required more than 18 miles of tapes.

1.7.2 Automotive industry

A car which never rusts even if it is scratched, and which is cheaper than a conventional one, may be realized if composites are used fully in the automotive industry. Composite materials entered the automotive sector in the early 1970s. In 1971, Renault decided to manufacture bumpers of small cars using composites, and this application spread to side protections of the same car and then to other models.

According to a French study in 1986, the automotive industry accounted for 25% of new materials outlets.

For bumpers and side protections, sheet molding compound (SMC), usually polyester reinforced with chopped fiber glass, has been used. However, use of SMC type composites cannot achieve smooth surface and body-colored bumpers at the right cost. The trend to overcome such drawbacks is to use either new thermoplastics such as reinforced reaction injection molding (RRIM) polyurethanes or bulk molding compound (BMC), possibly combined with SMC. Among many advantages of using composites for bumpers and side protections, the nature of a truly three-dimensionally reinforced material and ease of repair are most interesting.

Studies on using composites for leaf springs and torsion bars are underway, and several models of car with composite leaf springs have been on the road since 1986. One of the advantages is, if failure occurs, the leaf spring will just lose its properties rather than breaking, as will other materials.

Studies on frames and their constituent parts such as chassis system, floorpan, front structure and front end, are still at an early stage. One of the advantages of composites for these applications is that of consolidation of parts, which may cut costs drastically. In general, a car body has 300 or more stamped components; some designers believe that use of composites may reduce this number to only 25 different parts.

Additional advantages of composites over steel are noise reduction, weight saving and good resistance to impact. In order to exploit to the full such advantages, more studies on design methods such as single piece design, and connection methods may be necessary.

1.7.3   Civil and architectural structures

In 1967, the American Society of Civil Engineers Task Committee made a brief but comprehensive report on properties and problems of structural plastics (the word composite with its present meaning came later). The report mentioned that the composites would excel in services requiring high tensile strength, minimum weight and high resistance to chemical attack, but the performance would not be acceptable if high temperature or low deformations were required. After more than 20 years continuous intensive development, the properties of modern composites far excel those of other materials. The problem of concern is the price structure acceptable to civil construction.

Assuming that material price and manufacturing cost are fixed (even though these are coming down day by day), price structure can be reduced by a good design scheme.

Since 1949, one American company has manufactured over 20 million feet of fiber wound tubings. A European company has manufactured fiber reinforced pipes for over 20 years. These pipes are lightweight, corrosion resistant, easy to install and have many other advantages over those made in other materials, so that application to water supply and sewerage systems, desalination plants, piping in chemical plants, long distance pipe-lines for oil and gas, and so on, may explode provided that the cost limit is met.

According to an international publication on composites, the manufacturing cost of small diameter pipes (less than 24 in) is highly competitive in both metal and thermoplastic materials. Pipes with diameters in excess of this size, up to 60 in, are produced in both metal and composites, but beyond this size, composites come into their own. However, the author designed some projects involving high pressure composite pipes with diameters less than 10 in which were cost effective. The 682 ft high reinforced concrete chimney at the Inter-mount Power Plant in Utah has an inside liner made of composite material. The diameter is 28 ft.

Soil and foundation engineers have used geotextiles for strengthening foundations, embankments, slopes, and so on. Piles made in plastic are already in use in several countries. Because of the nature of corrosion resistance, use of composites for constructions of off-shore oil platforms, harbors, waterways and underground facilities may become extensive.

Several companies around the world are manufacturing standard composite structural members. These members are used for decking and walkway systems in waste water treatment plants, paper processing plants, and so on. Computer centers and hospital magnetic resonance imaging rooms use composites because they do not conduct electromagnetic waves. Rebar made of composite material has an ultimate tensile strength of 100 000 psi.

The use of composites in bridge construction has begun, though at a slow pace. The city of Düsseldorf, Germany, is the site of the world's first road bridge using composite material. The 16 m wide, 47 m span bridge, opened in 1986, is the first prestressed concrete structure using composite rods made of glass fiber and polyester resin. In Austria, a jetty and boat landing stage was built by blow molded composite pontoons held together by bolts. The Chinese are known to have built a vehicular bridge completely from synthetic composites. Though the performance result is not known, the bridge has carried 9 ton trucks. The Chinese also have built a cable-stayed, 13 ft wide and 89 ft span pedestrian bridge in Chongqing. The Heavy Assault Bridge made for the US Army is to be carried in three jointed sections on an armored vehicle, unfolds hydraulically to create a 106 ft span, and supports a 70 ton load in the unsupported center section. Twelve carbon–epoxy chords, about 38 ft long and 4×5 in in cross-section, support the structure. The idea of joining Europe and Africa across the strait of Gibraltar is under serious consideration. Two methods are noteworthy. The first is to use a carbon fiber composite cable-stayed bridge. The other is to lay composite tubes under water.

A Swiss federal laboratory has studied the future use of advanced composites in bridge construction. Among others, the report on the method of strengthening existing structures is interesting. When carbon–epoxy laminates are used to strengthen the existing bridge, the cost required could be reduced by about 20% compared to the method with steel plates.

In several countries, flat and corrugated composite sheets are used for domes, parts of internal walls and ceilings, and roofs. Composite made molds for concrete construction and built-up houses are often seen. Many transportation schemes such as ships and containers have used composites.

According to a report in 1987, about 60% of 560 000 bridges in the United States are either structurally defective or functionally obsolete. Steel girders become rusty; the reinforcing bars embedded in concrete beams or slabs are subject to corrosion caused by electro-chemical action. Underground fuel tanks are under similar conditions. In 1979, the US Bureau of Standards study showed that yearly loss caused by corrosion related dâmages amounted to 82 billion dollars, about 4.9% of GNP. About 32 billion dollars could be saved if existing technologies were used to prevent such losses.

In general, the sizes in civil construction are huge and economy is one of the major factors to be considered. In this connection, combining conventional materials with new materials is considered as the first step to be taken. A French company built retaining walls using polyester thread reinforced sand. The filaments provided a high internal angle of friction and cohesion, and the cost was about half that of concrete walls. There are many examples of such concepts: polypropylene reinforced asphalt runway surfaces, polyester styrene polymer road overlays, fiber reinforced concrete slabs, pavements, and others. Polymer bound sand is used as a filler for large size structural members.

When composites are applied to civil and architectural structures, design efforts must be made to utilize to the maximum the advantages of such materials, including developing new basic **concepts on structural forms**. Practicing engineers must be re-educated, students must study new principles involved in composite design. The engineers of the past were materials selectors. The new generation engineers will be material designers.

Further reading

Books

Finch, J. K. (1951) *Engineering and Western Civilization*, McGraw-Hill, New York.
McGuire, J. G. and Barlow, H. W. (1956) *The Engineering Profession*, Addison-Wesley, Reading, MA.
Smith, R. J. (1956) *Engineering as a Career*, McGraw-Hill, New York.
Timoshenko, S. P. (1953) *History of Strength of Materials*, McGraw-Hill, New York.
Tsai, S. W. (1988) *Composites Design*, Think Composites, Dayton, OH.
Wells, H. G. (1956) *The Outline of History*, Garden City Books, Garden City, New York.
Wickenden, W. E. (1949) *A Professional Guide for Young Engineers*, Engineers Council for Professional Development, New York.

Articles and reports

Note: ICCM = International Conference on Composite Materials; SAMPE = Society for the Advancement of Material and Process Engineering.
Anon. (1985) Filament winding. *Int. Reinf. Plast. Ind.*, Mar./Apr. 4–7.
Anon. (1985) Filament wound chimney liner. *Int. Reinf. Plast. Ind.*, Mar./Apr., 14.
Anon. (1986) Blow molded pontoons replace wooden jetties. *Mod. Plast. Int.*, Apr., **16**(4), 34.
Anon. (1986) Bridge-building composite – literally. *Int. Reinf. Plast. Ind.*, Nov/Dec., **5**(8), 16.
Anon. (1987) Confidence in plastic builds. *Engng News Rec.*, 17 Sep., **219**(12), 29.
Anon. (1987) Glass bridges shimmer in future. *Engng News Rec.*, 17 Sep., **219**(12), 44.
Anon. (1987) New Lockheed composite materials. *Aircraft Engng*, **58–59**, Jun., 31.
Anon. (1987) New structural ceramic composite for oxidation environments. *Mater. Process. Rep.*, June.
ASCE (1967) Structural plastics: properties and problems. Report of the Task Committee on Properties of Selected Structural Plastics and Systems, STI, American Society of Civil Engineers, Feb.
Bak, D. J. (1986) Vapor deposition improves metal matrix composites. *Des. News*, **42**, 16 Jun., 114–5.
Bakale, K. M. (1994) America shows its age … pultruded composites offer successful solutions for the decaying infrastructure, in *Preprints, The Second Annual Wilson Forum on Existing and Potential·Applications of Composite Materials in the Infrastructure* (ed. Wilson, B.), Santa Ana, CA, 18–19 Apr.
Brady, D. G. (1986) Aerospace discovers thermoplastic composites. *Mater. Engng*, **103**(9), 41–4.

Bunsell, A. R. (1987) Fibre reinforcements, past, present and future, in *Proceedings ICCM 6* (ed. Matthews, F. L.), Vol. 5, London, 20–24 July, p. 1.

Chang, F. K. (1987) Damage in and residual strength of bolted composite joints in tension or shear-out mode failure, in *Proceedings ICCM 6* (ed. Matthews, F. L.), Vol. 5, pp. 173–82.

Curtis, P. T. (1987) A comparison of the fatigue performance of woven and nonwoven CFRP laminates in reversed axial loading. *Int. J. Fatigue*, **9**(2), 67–78.

Dastin, S. J. (1986) Advanced composites for future aerospace systems. *Horizons*, **22**(2), 10.

Dorey, G. (1987) Impact damage in composites, in *Proceedings ICCM 6* (ed. Matthews, F. L.), Vol. 3, pp. 1–26.

Garza, R. and Sharp, B. (1986) Holography for noncontact structural analysis. *Sensors*, Sep.

Gottesman, T. (1987) Criticality of impact damage in composites sandwich structures, in *Proceedings ICCM 6* (ed. Matthews, F. L.), Vol. 3, pp. 27–35.

Harosceugh, R. I. (1987) Composites – the way ahead, in *Proceedings ICCM 6* (ed. Matthews, F. L. *et al.*), Vol. 5, 20–24 July, London, p. 15.

Harris, B. J. (1969) Local failure of plastic-foam sandwich panels. *J. Struct. Div., Proc. ASCE*, Apr. Vol. 95, ST4, pp. 585–609.

Hashemi, S. (1987) Interlaminar fracture of composite materials, in *Proceedings ICCM 6* (ed. Matthews, F. L.), Vol. 3, pp. 254–64.

Hasson, D. F. (1987) Impact behavior of ceramic matrix composite materials, in *Proceedings ICCM 6* (ed. Matthews, F. L.), Vol. 2, pp. 40–47.

Hiel, C. C. (1987) Durability of composites, in *Proceedings ICCM 6* (ed. Matthews, F. L.), Vol. 4, pp. 172–9.

Hogg, P. J. (1987) The impact properties of metal–composite laminates, in *Proceedings ICCM 6* (ed. Matthews, F. L.), Vol. 3, pp. 46–56.

Hurwitz, F. (1987) Ceramix matrix and resin matrix composites: a comparison, in *Proceedings 32nd International SAMPE Symposium and Exhibition* (eds Carson, R. *et al.*) Anaheim, CA, 6–9 Apr., p. 433.

Isley, F. (1994) Application of composites to infrastructure problems: repair of concrete and masonry structures with composite materials, in *Preprints, The Second Annual Wilson Forum on Existing and Potential Applications of Composite Materials in the Infrastructure* (ed. Wilson, B.), Santa Ana, CA, 18–19 Apr.

Jardon, A. (1987) Mass production composites, in *Proceedings ICCM 6* (ed. Matthews, F. L.), Vol. 1, pp. 1–4.

Kharbhari, V. M. (1993) Composites and infrastructure: a perspective influenced by a panel study in Japan. *SAMPE J.*, **29**(3), 7.

Kim, D. H. (1973) Definition and fields of civil engineering with stresses on professionalism. *J. Korean Soc. Civ. Engrs*, **21**(2), 38.

Kim, D. H. (1993) The importance of concept optimization of composite structures, Keynote lecture note, in *Third Pacific Rim Forum on Composite Materials* (ed. Tsai, S.), Honolulu, Hawaii, 2–4 Nov.

Kim, D. H. (1994) Cement problems: applications of composite materials for the infrastructure, in *Preprints, The Second Annual Wilson Forum on Existing and Potential Applications of Composite Materials in the Infrastructure* (ed. Wilson, B.), Santa Ana, CA, 18–19 Apr.

Kliger, H. S. and Segal, C. L. (1993) Opportunities for structural composites in high speed and light rail passenger cars, in *Proceedings 38th International SAMPE Symposium and Exhibition* (eds Bailey, V. *et al.*), Anaheim, CA, May, p. 1071.

Ling, Y. (1987) Contact stress analysis of mechanical fastened joint in composite, in *Proceedings ICCM 6* (ed. Matthews, F. L.), Vol. 5, pp. 160–72.

Lockett, F. J. (1987) The provision of adequate materials property data, in *Proceedings ICCM 6* (ed. Matthews, F. L.), Vol. 1, pp. 5–27.

Obraztsov, I. F. (1987) First approximation geometrically nonlinear composite shell theory, in *Proceedings ICCM 6* (ed. Matthews, F. L.), Vol. 5, pp. 113–22.

Phillips, D. C. (1987) High temperature fiber composites, in *Proceedings ICCM 6* (ed. Matthews, F. L.), Vol. 2, pp. 40–47.

Reifsnider, K. L. (1987) Life prediction analysis: directions and divagations, in *Proceedings ICCM 6* (ed. Matthews, F. L.), Vol. 4, pp. 1–31.

Singh, G. (1987) Large deflection behavior of thick composite plates. *Composite Struct.*, **8**(1), 13.

Stone, D. E. W. (1987) Non-destructive evaluation of composite structures – an overview, in *Proceedings ICCM 6* (ed. Matthews, F. L.), Vol. 1, pp. 28–59.

Tooth, A. S. (1987) A design procedure for horizontal cylindrical GRP vessels supported on twin saddles, in *Proceedings ICCM 6* (ed. Matthews, F. L.), Vol. 5, pp. 123–33.

Wilkison, A. R. and Ganga Rao, V. S. (1993) Fiber reinforced polymer matrix composites in construction, in *Proceedings 38th International SAMPE Symposium and Exhibition* (eds Bailey, V. *et al.*), Anaheim, CA, May, p. 1061.

Wood, A. S. (1987) Advanced thermoplastic composites get the full automation treatment. *Mod. Plast. Int.*, **17**(4), 44–7.

Zhang Xigong (1987) The failure of composite cylindrical shells with circular holes under internal pressure, in *Proceedings ICCM 6* (ed. Matthews, F. L.), Vol. 5, pp. 153–9.

Zhou, C. T. (1987) Some basic problems in the analysis of instability to composite cylindrical shells, in *Proceedings ICCM 6* (ed. Matthews, F. L.), Vol. 5, pp. 101–12.

Manufacturers' data

Advanced thermoplastic composites. Technical Pamphlets, Phillips Petroleum Company, 1987.

Carbon–carbon for advanced applications. LTV Missiles and Electronics Group, USA, Apr. 1987.

Chemical control of rebar corrosion. Grace Corp. 1987.

Fiber glass reinforced plastics by design. Engineering data book, PPG Industries, Inc.

2 Properties of composites

2.1 Introduction

The design and analysis of any structural member require a detailed knowledge of material properties, which are dependent on the manufacturing and fabrication methods as well as the nature of the constituent materials.

In the cases of conventional materials such as concrete or steel, design engineers tend to pay less attention to such details probably because so much research and construction experience has resulted in fairly complete specifications and because some properties such as ductility of steel permit less rigorous details. However, there are many cases of problems in practice even with steel structures and the steel codes have been revised continuously. As for the structures made of composites, there are numerous excellent candidate constituent materials, but available standards and specifications are very few. This situation forces the structural engineer to have thorough knowledge on the candidate materials' properties, as well as manufacturing and fabrication procedures. Nevertheless, it should be borne in mind that the list of constituent material properties of composites is not much different from that of conventional materials except in the anisotropic nature of composites, that is, the properties have directional characteristics. The list of candidate materials of both reinforcements and matrices used for composites is huge and the range of properties of each material is very much diversified. Some materials are supplied by many manufacturers and these ranges differ from one supplier to another.

The author intends to avoid any specific supplier and to give a general idea to the reader so that a formula of specific materials can be decided for each structure he designs.

2.2 Reinforcements

2.2.1 Glass fiber

On a specific strength, i.e. strength to weight, basis, glass fiber is one of the strongest and most commonly used structural materials. Some laboratory made fibers can reach strengths of over 10^6 psi (6896 MPa) and commercial grades range from 500 000 to 700 000 psi (3448–4830 MPa) [1–3].

There are two basic processes to manufacture continuous glass filament. The first is the **marble melt process** in which an appropriate mixture of raw materials is melted and glass marbles with diameters of 0.8–1.2 in (2–3 cm) are formed, and these marbles are remelted and formed into the glass fiber product. In the second, the **direct melt process**, the raw materials are melted and formed directly into the glass fiber product.

Glass filaments are highly abrasive to each other. In order to minimize abrasion-related degradation of glass fibers, surface treatments, or **sizings** are applied before the fibers are gathered into strands. There are two types: the

21

first, sometimes called **textile sizings**, may be temporary, as in the form of a starch–oil emulsion which will be subsequently removed by heating and replaced with a glass to matrix coupling agent called a **finish**. The second, often called **reinforcement sizings**, may be a compatible treatment in which complexes of film formers combined with wetting agents and surface active ingredients perform several necessary functions during the subsequent forming operation and act as a coupling agent to the matrix during impregnation. In order to keep fibers in position during fabrication, 'binders' may be applied to the virgin blown fibers or 'sized' strands.

In the fiber glass industry, a specific filament diameter is referred to by a specific **alphabet designation**. For conventional composite reinforcement, filament diameters that range from G (9–10.2 μm (3.5–4.0 $\times$ 10^{-4} in)) to T (22.9–24.1 μm (9.0–9.5 $\times$ 10^{-4} in)) are used.

The major finished glass **fiber forms** are as follows.

- **Continuous strands**. Fiber glass roving is produced by collecting a bundle of strands into a single large strand. In composite manufacture, fiber roving designation is often made by the **yield** (yd/lb). Rovings are used in many applications. When continuous strands are chopped into short lengths, usually 3.2–12.7 mm (1/8–1/2 in), uses can be found in spray-up, sheet molding compound (SMC), injection molding, bulk molding compound (BMC), resin transfer molding (RTM), and others. Continuous single-end rovings are used in filament winding and pultrusion processes to manufacture pipes, tanks, and many other conventional structural forms.
- **Woven roving**. Continuous strands are woven into many weave configurations to be used in numerous hand lay-up and panel molding processes. Plain or twill weaves provide strength in both directions, while unidirectionally stitched or knitted fabrics provide primary strength in one direction. There are many other types of weaves, such as basket, satin, leno, biaxial and double bias.
- **Surfacing mats**. These may be produced in two types. A **chopped-strand mat** is formed by chopping strands to various lengths and depositing in a random pattern, and are held in place by a chemical binder, usually a thermoplastic resin, or by 'needling'. It is used mostly for contact molding. A **continuous-strand mat** is formed by depositing continuous strands in a swirl pattern to provide random distribution. Both chopped and continuous strands are used to form a **combined mat**.

Reinforcing fibers can be woven and knitted into three-dimensional (3-D) [4, 5] and continuous shapes for reinforcing some structural parts. Other reinforcement forms include **milled fibers**, in which fiber lengths are made 0.79–6.4 mm (1/32–1/4 in), and **fiberglass paper**. A combination of mat and woven roving is used in many lay-up processes to save fabrication labor.

Fabric with three primary reinforcement directions is called **triaxial**. Quadraxial fabric has four primary reinforcement directions. Quadraxial fabrics with the ply orientations evenly spaced, for example, 0°, +45°, 90°, −45°, are often referred to as quasi-isotropic. **Multidirectional woven products** such as 5-D, 7-D and 11-D fabrics approach isotropy in construction and resulting physical and mechanical properties.

Some advantages of 3-D fabrics are:

1. Integrated structure, eliminating trimming of composites, is possible.
2. Delamination, which is one of the weak points of 2-D laminated composites, can be avoided.
3. Impact resistance can be improved.
4. Compressive strength can be improved.

In 3-D fabrics, orthogonal orientation is the most common but variations are available depending on structural requirement. 5-D fabrics are woven with yarn reinforcement in the V, W, X, Y and Z directions. The V-W-X-Y plane demonstrates a reinforcement direction every 45° with an axial Z direction plane perpendicular to V, W, X, Y reinforcement. In 7-D or 11-D 'isotropic' woven fabrics, blocks and plates may be woven with yarns oriented in seven or eleven directions. The 7-D structures may be of two types:

1. Three orthogonal directions and four diagonal reinforcement directions across the corners.
2. Three orthogonal directions and four diagonal reinforcement directions across the sides.

The 11-D structures have both of the above diagonal types plus the three orthogonal reinforcement directions. The 7-D structures have a more isotropic substrate than the 3-D but with less fiber concentration in the three orthogonal directions.

There are several **glass fiber types** with different chemical compositions, providing the specific physical/chemical properties [5–9].

E-glass is a family of glasses with a calcium aluminoborosilicate composition and a maximum alkali content of 2.0%. E-glass, or electrical grade, is best for general purpose structural uses, as well as for good heat and electricity resistance. This type of fiber may become the most important material for civil and architectural structures.

S-glass, which has a magnesium aluminosilicate composition, is a special glass with higher tensile strength and modulus with good heat resistance. S-2 glass fibers have the same glass composition as S-glass but different coating [1]. The additional benefit of S-2 glass over E-glass is strong resistance to acids such as H_2SO_4, HCl and HNO_3. Some property comparisons with E-glass are shown in Figure 2.1.

C-glass has a soda–lime–borosilicate composition. C-glass has good chemical stability in chemical corrosive environments.

N-VARG is alkali-resistant glass fiber and is used in composites with a cement based matrix.

T-glass fiber, in comparison to E-glass, is claimed to have a remarkably improved performance, with a 36% increase in tensile strength, a 16% increase in the tensile modulus, an increased heat resistance of 135°C and a 40% decrease in the coefficient of thermal expansion. T-glass has additional advantages over E-glass such as improved elongation, improved impact strength and electrical properties, and improved thermal and chemical resistance.

R-glass has a magnesium–lime–aluminosilicate composition. R-glass has higher tensile strength and modulus relative to E-glass, and gives beneficial

Figure 2.1 Comparative specific strength and modulus.

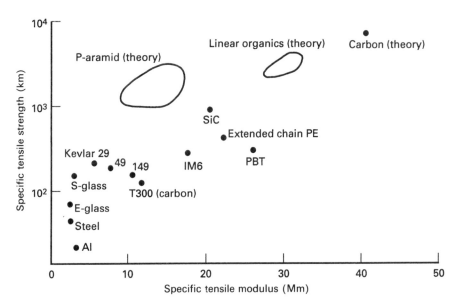

results with higher resistance to fatigue, aging, temperature and corrosion. After aging for 100 days in distilled water, the residual strength of R-glass is 49% higher than that of E-glass.

A-glass is high alkali glass, and consists of soda–lime–silica. The fiber form of this composition is used for applications requiring good reinforcing properties with good chemical resistance.

ECR-glass is chemically resistant E-glass. Applications are made when good electrical properties with better chemical resistance are required.

AR-glass fiber is alkali-resistant and is used for glass fiber reinforced cement (GFRC). When glass fibers are used to reinforce cement, degradation in strength and toughness occurs when exposed to outdoor weathering, especially in humid conditions. This process takes place even with AR-glass fibers, although at a slow rate. Studies have been carried out to improve AR properties and results are promising.

Some of the noncommercialized glass fibers are **high-modulus glass** (16×10^6 psi (110 GPa)) based on beryllium oxide, **lead glass** for radiation protection, **D-glass** for low dielectric constant, and **lithium-oxide based glass** for X-ray transparency.

The physical properties of some fibers are given in Table 2.1.

The tensile strength of a glass fiber strand may be 20–40% lower than the values measured with a pristine single filament because of surface defects incurred during the strand forming process. Moisture has a detrimental effect on the fiber strength as does increasing temperature. The loss in strength of glass exposed to moisture, and under an applied load, is known as static fatigue.

When the glass fiber is heated, the modulus of elasticity gradually increases. For most glasses, Poisson's ratio is between 0.15 and 0.26. The ratio for E-glass is 0.22 up to 510°C.

There are many detailed properties of fibers not given here, but can be found in the further reading section at the end of this chapter.

Table 2.1 Comparison of fiber properties

Fiber type	Specific gravity	Young's modulus, E		Tensile strength		Strain to failure (%)	Highest usable temp. (°C)
		(GPa)	(10^6 psi)	(GPa)	(10^3 psi)		
E-glass	2.5–2.6	69–72	10–10.4	1.7–3.5	246.5–507.5	3–4.8	350
S,S$_2$-glass	2.49	86.9–93.1	12.6–13.5	4.58	665	5.4	350
A-glass	2.5	69	10.0	3.04	440.8		
ERC-glass	2.62	72.5	10.5	3.63	526.4		
R-glass	2.55	86	12.5	4.4	638.0		780
Boron	2.4–2.6	365–440	52.9–63.8	2.3–3.5	333.5–510	1.0	2000
Carbon HM[a]	2.15	725	105	2.2	325.0	0.30	600
Carbon HS[b]	1.77	295	42.8	5.66	820	1.91	500
Al$_2$O$_3$ (Du Pont)	3.95	379	55	1.38–2.1	200.1–304.5	0.4	1000
Nicalon SiC	2.8	45–480	6.5–69.6	0.3–4.9	43.5–710.5	0.6	1300
Avco SiC	2.7–3.3	427	61.9	3.4–4.0	493.0–580.0	1	
Tyranno UBE[c]	2.4	120	17.4	2.5	362.5	2.2	1300
Nextel-3M[d]	2.5	152	22.0	1.72	249.4	1.95	1200
Al$_2$O$_3$ (ICI)	3.3	330	43.5	2	290.0	1.5	1000
Quartz (SiO$_2$)	2.2–2.5	75	10.9	5.9	855.5	1.5–1.8	1100
Nylon 66	1.2	<5	<0.7	1	145.0	20	150
Polyester	1.38	<8	<2.6	0.8	116.0	15	150
Technora-HM50[e]	1.39	70	10.2	3.04	440	4.3	250
Spectra 900[f]	0.97	117	17.0	3	435.0	3.5	120
Kevlar 29	1.44	82	12	3.61	525	4.4	
Kevlar 49	1.45	131	19.0	3.61	525	2.9	250
Kevlar 149	1.47	186	27	3.45	501	1.8–1.9	

[a] High modulus. [b] High strength. [c] SiC. [d] Al-boria-silica. [e] Aramid. [f] UHMWPE.

2.2.2 High-performance fibers

Although glass fibers have excellent high strength with low cost and provide the overwhelming majority of structural composites, special structural requirements, such as limitation on deformation under repeated loading, call for higher elastic moduli/higher fatigue strength than can be provided by glass fibers. For such cases, high-performance fibers can be used, however, at increased cost. Typical high-performance fibers are as follows.

Carbon fibers

Carbon fibers made by carbonizing cotton fibers and, later, bamboo, were the first filaments used in Edison's incandescent electric lamps [10]. Many fibers can be converted into carbon fibers as long as the precursor fiber carbonizes instead of melting when heated.

There are three different precursor materials used, at present, to produce carbon fibers.

Rayon precursors, derived from cellulosic materials, were among the earliest precursors used to make carbon fibers. The significant disadvantage is that only a small part (typically 25%) of the initial fiber mass remains after carbonization, resulting in expensive fibers.

Polyacrylonitrile (PAN) precursors are the basis for the majority of carbon fibers commercially available. The carbon fiber conversion yield is 50 to 55%.

PAN precursor based carbon fiber generally has a higher tensile strength than a fiber based on any other precursor, due to a lack of surface defects.

Pitch precursors based on petroleum asphalt, coal tar, and PVC can also be used to produce carbon fiber. Pitches are relatively low in cost and high in carbon yield. However, nonuniformity from batch to batch is a serious problem when pitch precursors are used. Some efforts have been made to alleviate the problem.

Carbon fiber conversion processes include stabilization at temperatures up to 400°C (750°F), carbonization at temperatures from 800° to 1200°C (1470° to 2190°F), and graphitization in excess of 2000°C (3630°F), and surface treatments, sizings and spooling as with glass fibers.

New processes have been developed and some may be ready for commercial production. All commercial production of PAN precursor carbon fibers is by spinning, and the fiber cross-section is round. One new area for acrylic based precursor fibers involves a melt assisted extrusion as a key part of the spinning process. An advantage of this method is its ability to produce different precursor shapes, including rectangular, I-type and cross. The advantage of nonround shapes is that closer fiber packing in composite is possible, permitting higher load carrying capacity than round fibers.

Another novel method, recently reported, involves growing carbon fibers from the vapor phase of a hydrocarbon, such as natural gas or benzene, using metal catalytic particles. The fibers are not continuous as are PAN based fibers but they offer possibilities of application for many products where chopped or milled carbon fibers are used. The cross-sectional structure of **vapor-grown carbon fibers** resembles the annular rings of a tree. This annular ring structure gives the vapor-grown fibers an unusual fracture behavior. Instead of a sudden brittle failure in tension as seen in the case of PAN or pitch based fibers, vapor grown fibers yield in a mode described as a 'sword-in-sheath noncatastrophic failure'. The production costs of this process may be significantly less than that of conventional PAN based fibers, suggesting application in automotive and civil structural composites.

Note that a **graphite** fiber is defined as consisting of graphene-layer planes stacked with three-dimensional ordering. Carbon fibers have two-dimensional ordering. However, both carbon and graphite fibers are carbon fibers in a broad sense. An ISO subcommittee on carbon fibers recommends that the word graphite, as applied to fiber, be abandoned. The editorial of the 1988 January/February issue of *Advanced Composites*, **3**(1) recommended to do the same. The author will follow this decision and the term carbon fibers will include graphite fibers.

The range of properties of carbon fibers used in composites is shown in Table 2.2. The values given show only the typical ranges and there are considerable ranges of carbon fibers depending on manufacture.

Rayon and isotropic pitch precursors are used to produce low modulus carbon fibers (≤ 50 GPa, 7×10^6 psi) [11]. High modulus carbon fibers (≥ 200 GPa, 30×10^6 psi) are made from PAN or liquid crystalline (mesophase) pitch precursors.

A pitch based carbon fiber with an elongation to break of 2% has been reported [12]. It is claimed to have a strength of 5.16 GPa (748 000 psi) and a modulus of 252 GPa (36.5×10^6 psi). This is comparable to respectable PAN

Table 2.2 Carbon fiber properties

Grade	Modulus (GPa)	Modulus (10^6 psi)	Strength (GPa)	Strength (10^3 psi)	Elongation (%)
PAN based fibers					
Hysol Grafil Apollo IM[a]	300	43.5	5.20	754	1.73
HM[b]	400	58	3.5	507.5	0.88
43–750	296	43	5.66	820	1.91
HS[c]	260	37.7	5.02	728	2.00
BASF Celion G40–700	300	43.5	4.96	720	1.66
Gy80	572	83	1.86	270	1.66
Hercules AS6	241	35.0	4.14	600.3	1.77
Torayca T300	234	33.9	3.53	511.9	1.51
Thornel (Amoco) T40	295	42.8	5.66	820	1.8
T50	390	57	2.42	350	
Pitch based fibers					
Union Carbide P120	827	120.0	2.20	325	0.27
P100	724	105	2.2	325	0.31
P75-S	520	75.0	2.10	300	0.40
P55-S	380	55.0	1.90	275	0.50
P25-W	160	23.0	1.40	200	0.90
Teijin (Patent abstract)	252	36.5	5.16	748.2	2.00

[a] Intermediate modulus. [b] High modulus. [c] High strain.

based fibers. Because, with pitch, processing can be made cheap and carbon yield is high, 50% higher than PAN, production of cheaper carbon fiber may be possible.

Pitch based short fibers are used for metal matrix composites, carbon–carbon composites, and for reinforcing thermosetting as well as thermoplastic polymer matrices, and cement. Extensive use in general civil and architectural structures is expected in the near future. The basis of this expectation is its possibility of lower cost, maintaining good resistance properties against high temperature and the environment.

Carbon fibers, in general, are not affected by moisture, atmosphere, solvents, bases or weak acids at room temperature [13]. The threshold for oxidation for extended operating time is 350°C for low-modulus PAN based fibers and 450°C for high-modulus PAN or pitch based fibers. Improving oxidation resistance may be achieved by using higher-purity fibers and special resins in the future.

Aramid fibers

Aramid fiber (aromatic polyamide) was introduced in 1972 by Du Pont under the name Kevlar [14, 15]. Recently, related aramid fibers have become available from several other manufacturers. The structure of aramid fiber is anisotropic and gives higher strength and modulus in the fiber longitudinal direction. The ranges of tensile strength and modulus are given in Table 2.1.

Aramid is resistant to fatigue, both static and dynamic. Aramid fiber responds elastically in tension but it exhibits nonlinear and ductile behavior under compression. Special attention is necessary when application of aramid fiber involves high strain compression or flexural loads. Aramid fiber exhibits good toughness and general damage tolerance characteristics. Because of its high toughness, aramid is used for impact resistance and ballistic resistance armor.

Applications of aramid fiber in civil structures include ropes, cables, curtain walls, floors and ceilings, pipes and prestressing tendons.

Linear organic fibers

Linear organic fibers may become one of the major reinforcements for civil and building structures in the future. A high strength and high modulus organic fiber can be produced by arranging the molecular structure of simple polymers to become straight, during their manufacture. Theoretically, a maximum rigidity of 240 GPa (34.6×10^6 psi) can be obtained. With a lower density of 0.97, high modulus polyethylene fibers produced both in USA and Holland have properties comparable to those of aramid fibers: tensile modulus of 117 GPa (25×10^6 psi) and tensile strength of 2.9–3.3 GPa (430–480×10^3 psi). These properties were obtained at ambient temperatures and decrease rapidly with increasing temperature. Creep is another problem. However, recently, it has been claimed that this could be solved by cross-linking using radiation. In view of the present rate of advances in composite technology, its extensive usage may be expected in the near future. Molecular tailored organic fibers will then have an important role.

A general outline of specific tensile properties of reinforcements under both commercial and experimental stages, compiled from several suppliers of technical materials, is shown in Figure 2.1.

Other noble fibers

Both **boron** and **silicon carbide** fibers are produced by chemical vapor deposition (CVD) on substrate wires such as tungsten wire or a carbon filament. Boron fiber was the first to be produced which had a modulus comparable to glass fibers, and began to be used for the F111 horizontal stabilizer as carbon and boron reinforced epoxy in the late 1960s. The US space shuttle is made up of more than 200 boron fiber reinforced aluminum tubes. Boron fibers have high strength and modulus (Table 2.1), but the high cost of fibers prohibits wide application in civil structures.

Silicon carbide (SiC) fibers can be used to reinforce metal matrices such as aluminum, titanium, magnesium and copper, ceramics, and both thermoset and thermoplastic polymers. **SiC reinforced glass** composite has the flexural strengths of 296 MPa (43 000 psi) at 22°C, and 489.7 MPa (71 000 psi) at 600°C, and flexural modulus of 117 GPa (17×10^6 psi). The properties of such composites are available from the manufacturers' technical bulletins given in the further reading section. With the price about one third of that of boron, silicon carbide fibers may be applied in some civil structures in the future.

Many types of fine **ceramic fibers** have been developed since 1965. The term has been used since a new fiber was derived from kaolin and other mineral sources in the early 1950s. Kaolin is a form of clay used for producing ceramics. Ceramic fibers are quite diversified in both composition and properties, and in application. Ceramic fiber types are defined on the basis of composition as follows: **Oxide fibers** include aluminum oxide fibers, alumina–silica fibers (alumina fibers), alumina–boria–silica fibers, zirconia–silica and fused silica fibers (quartz fibers). The fused silica fiber (99.9% SiO_2) has higher strength-to-weight ratio than any other ceramic fiber. **Nonoxide fibers** include silicon carbide (SiC) fibers and nitride fibers.

Other **organic fibers** include acrylic, nylon, polybenzimidazole (PBI), polyester, polypropylene and Teflon.

Some of the **metal reinforcements** are aluminum, beryllium, copper, tungsten and stainless steel, and are used for metal matrix composites. Steel fiber is used for reinforced concrete.

2.3 Matrices

A brief outline of matrices used for composites is given in Chapter 1. Of metal and ceramic matrices, cement is of most importance to civil engineers. Cement matrix based composites will be discussed later in this chapter.

A polymer is an organic material composed of molecules made of many(poly) repeats of some simpler unit called the mer, or the monomer. A polymer may be classified by various schemes, but for engineers, classification by either properties or end-use purposes, such as high temperature resistance, may be of interest. Engineered forms such as fibers and foams may be another means of classification.

The chemical structures of polymers are very involved and beyond the scope of this book.

2.3.1 Properties of polymeric matrices
In this section, important properties of some of the polymer matrices which may be of importance to civil and architectural structures are briefly presented.

Thermal properties
Dimensional stability is the most important thermal property for polymers because no polymer can be used at a temperature above which it loses dimensional stability. This temperature is called **glass transition temperature** (T_g) of the polymer, below which amorphous and partially crystalline polymers become glassy. Structural components cannot be permitted to reach this temperature. The change in properties at T_g occurs not at a distinct temperature but over a range of temperatures. The behavior and magnitude of T_g of each polymer are different from those of other polymers. In general, thermosets have higher T_gs than thermoplastics. At a **thermodynamic melting point** (T_m), crystalline polymers abruptly change into mobile liquids, losing all mechanical properties. If the crystallinity of a polymer is high, the region of acceptable dimensional stability above T_g can be extended. At the **decomposition temperature** (range) (T_d), a polymer may become flammable. At the **flash ignition** and **autoignition temperatures**, polymers react with oxygen to start burning. The decomposition temperature and the chemical resistance of a polymer are increased by stronger bonds and by the inclusion, in the mer, of elements and bonds which are not easily attacked by chemicals or other agents.

Table 2.3 Typical values of coefficient of thermal expansion (CTE) of some materials

Material	CTE, 10^{-6}/K
Epoxy	48 to 85
ABS	53
PMMA	34
Liquid crystal polymer	5
Nylon	25 to 40
Unsaturated polyester	16
PPS	30
Carbon fiber	−0.5 to 1.6
E-glass	5.4
S-glass	1.6
Aluminum	23.4
Steel	10.8
Titanium	10.1
Stainless steel	18
Carbon–epoxy (0°)	0.43
Carbon–epoxy (0°/±45°/90°)	3.4
E-glass–epoxy (0°)	8.6
Aramid–epoxy (0°)	−5.4
Concrete	10.8

Polymers generally have higher **coefficients of thermal expansion** (CTE), than conventional materials such as metals and concrete (Table 2.3). This characteristic is an important consideration in structural design. By definition, CTE is the change in length or volume per unit length or volume produced by a 1° rise in temperature. These coefficients vary significantly with temperature ranges. Thus, CTE is calculated as the slope of the secant line of the thermal expansion curve between the reference temperature (generally room temperature) and the temperature of interest. In the case of a thermoset, the rate of thermal expansion is influenced mostly by the degree of cross-linking and the overall stiffness of the units between cross-links. In a thermoplastic, thermal expansion is controlled more by the strength of the secondary bond between molecules than by the stiffness of the chains. Thermal expansion is also greatly reduced by crystallinity.

Thermal conductivity of polymers is generally lower than that of metals. By definition, thermal conductivity is the ability of a material to conduct heat and is measured by the quantity of heat that passes through a unit cube of the material in unit time when the difference in temperature of two faces is 1°. The thermal conductivity of a polymer can be increased by adding metallic fillers, and can be decreased by foaming with air or some other gas.

Mechanical properties

1. **Stiffness**. As with thermal expansion, the degree of cross-linking and the overall flexibility of a thermoset are important for stiffness. In the case of a thermoplastic, crystallinity and secondary bond strength dictate stiffness.
2. **Strength**. The concept of strength is rather more complex than that of stiffness. The tensile, compressive and flexural strengths are, in general, different from one another. Types of strength are diversified, such as

short- and long-term strengths, static and dynamic strengths, and impact
strength. Some characteristics of strength are related to those of toughness.
The short-term yield strength of a polymer is largely dependent on the
bonding that holds the polymer together. Increased cross-linking promotes
short-term yield strength, but has an adverse effect on toughness. Long-term
rupture strengths in thermoplastics are increased by increased secondary
bond strength and crystallinity. Fatigue strength is similarly influenced.
Because substantial heating is often encountered in fatigue, all factors that
influence thermal dimensional stability also influence fatigue strength.
Short-term failure strengths and impact strengths are determined by the
same factors which control toughness.

3. **Toughness**. This is as complex as strength. It represents the ability of a
    material to absorb energy and is defined as the work required to rupture
    a unit volume of the material. It is proportional to the area under the
    load–deflection curve from the origin to the rupture point. In general,
    increased stiffness results in decreased toughness. A method to increase
    toughness is to blend, fill or copolymerize a brittle (but with high stiffness)
    polymer with a tough one, but with a resulting sacrifice in strength. One
    must compromise between strength and toughness.

Chemical properties

1. **Solubility**. A polymer may be dissolved in various solvents. A solvent may
    diffuse into the polymer, or it may make the polymer swell. In selecting any
    polymer matrix for an application, attention should be given to study if any
    chemical with such a nature may come into contact. The general guideline
    to start is that a polymer will not dissolve in a solvent unless the chemical
    structure of its mer is similar to that of the solvent. Plasticization is one of
    the useful aspects of solubility; a plasticizer may be added to a polymer to
    improve its processing workability or to alter its properties. However, the
    plasticizer may lower the polymer's temperature resistance, hardness,
    stiffness and tensile strength, though its toughness may be increased. In
    some polymers, plasticizers are used in processing to decrease viscosity
    without increasing temperature.

2. **Permeability**. A polymer may be permeable to gases or other small mol-
    ecules. A polymer with higher crystallinity/density may have lower perme-
    ability. Cross-linking also reduces permeability.

3. **Chemical resistance**. The ability of a polymer to resist attack by chemicals,
    environment and radiation depends on the chemical nature and bonding
    in the mer and the shape of links in the polymer chain. Detailed discussion
    on this subject is beyond the scope of this book. Further discussion on
    degradation in general is given later in this section.

Electrical properties

When high electrical resistivity is important in structural design, a polymer
matrix is a suitable choice. Matrices, fillers and reinforcements can be chosen
to meet electrical requirements. Electrical properties of polymers of interest to
structural engineers are as follows.

1. **Dielectric properties**. In general, polymers are good insulators. These polymers can be used as good dielectrics because of their ability to store electrical charge effectively. A dielectric is defined as a material able to resist the flow of an electric current.
2. **Conductivity**. This can be produced by adding a conductive material to the polymer. The carbon fiber reinforcement in a polymer can provide electrical conductivity.

Optical properties
Most polymers are colorless and coloring may be possible.

Properties of a polymer in a molten state
Viscosity and pot life are important in selecting a processing method. Pot life, or working life, is the length of time a catalyzed thermoset matrix maintains a viscosity low enough for processing.

Degradation
Degradation of polymers is caused by oxidation, radiation of wavelengths lower than 300 nm (e.g. ultraviolet), photo-oxidation and ionization, mechanochemical reaction, micro-organisms, chemical attacks, and so on [16]. To cope with such degradation problems, additives such as flame retardants, heat stabilizers, antioxidants, antiozonants and ultraviolet stabilizers, are used. Environmental problems will be discussed in more detail in design considerations.

2.3.2   Some polymer matrices of interest to structural engineers
There are many types of both thermoplastics and thermosets which can be used as matrices for composites. Some of the polymer matrices of interest to civil structural engineers are given as follows.

Unsaturated polyesters
For civil and architectural structures, unsaturated polyesters are and will be the most commonly used because of their low cost and ease of processing, coupled with their ability to make good quality composites. Unsaturated polyesters are already widely used in several applications, and are soon likely to replace conventional materials including concrete and steel.

Many source materials can be combined to produce unsaturated polyesters (commonly known just as polyesters) resulting in many varieties of properties. **Orthophthalic resins** are widely used and referred to as general purpose. The weakness of these resins lies in their lack of thermal stability, chemical resistance and processability. **Isophthalic** or **terephthalic polyester resins** have higher quality with better thermal resistance. **Bisphenol A(BPA) fumarates** show a higher degree of hardness and rigidity, and good thermal performance and chemical resistance. **Chlorendics** have excellent chemical resistance and some degree of flame resistance. **Dicyclopentadiene**-containing unsaturated polyesters show resistance to thermal oxidative decomposition at high temperature.

The mechanical properties of an unsaturated polyester composite are greatly improved by the use of fiber reinforcements. The properties which can be obtained depend on the amount and type of reinforcement used.

Polyester resins have good resistance to chemical attack and have been used for many corrosion-resistant structures. Depending on the chemical to be contained, a specific formula (or system) has to be designed. Most of the material suppliers provide tables for various anti-corrosion properties. Care must be taken when glass fiber is used since it may be dissolved if attacked by some chemicals.

If unsaturated polyesters are mixed with a polymerizable unsaturated monomer, such as styrene, the two react and cross-link to form a three-dimensional structure. This hardening or curing process is initiated and promoted by hardeners, or catalysts, such as an organic peroxide. If this organic peroxide is chosen properly, cross-linking (curing) can be accomplished from room temperature (15°C (60°F)) to over 150°C (300°F). Heat accelerates the reaction. Polymerization (curing) is also carried out by radiation (e.g. ultraviolet). The cross-linking can be accelerated by the addition of promoters, such as cobalt naphthenate, and, if necessary, can be retarded by the addition of inhibitors such as hydroquinone.

When some degree of resistance to burning is required, a flame-resistant additive such as bromine may be used. Carbon black screens out ultraviolet radiation. Some additives such as hydroxybenzophenones absorb ultraviolet radiation and re-emit it later as thermal energy.

Epoxies

Epoxies are a broad range of products with a common epoxy ring consisting of two carbon atoms single bonded to an oxygen atom. Almost all known phenols, bisphenols, alcohols, glycols and polyols are readily converted to epoxies, and there are hundreds of different systems. However, there are two basic types of epoxies. The first, known as **glycidals**, are made by a reaction with epichlorohydrin, and the others, known as **cycloaliphatics**, are made by peroxidizing olefins.

Epoxies, with many commercial forms, have been the most popular resins in the composite industry.

Bisphenol A epoxies are based on the diglycidyl ether of bisphenol A (DGEBA), produced by reacting bisphenol A and epichlorohydrin. DGEBA provides an excellent balance of physical, chemical and electrical properties, and has been used for general purpose areas.

Aliphatic epoxies or aliphatic glycidyl ethers are made by epoxidizing aliphatic backbones containing hydroxyl (OH) units, such as alcohols, glycols and polyols, by reaction with epichlorohydrin. These groups are used as diluents or flexibilizers for DGEBA resins. Cycloaliphatic resins are used when stability to ultraviolet light and good electrical properties are needed.

Novolac epoxies are made by reacting phenol with formaldehyde and reacting this product with epichlorohydrin. These epoxies can be cured to a high cross-link density resulting in high temperature properties and improved chemical resistance.

Multifunctional epoxies contain two or more epoxy groups in the same molecule. A specific system can be developed to meet certain requirements.

Brominated epoxies are produced by reacting tetrabromobisphenol A with a DGEBA type resin. These resins, with from 15 to 50% bromine, are used for their flame retarding properties.

In general, epoxies have good high temperature resistance and are normally used at temperatures up to 177°C (350°F), sometimes as high as 316°C (600°F). Compared with polyesters, epoxies have higher specific strength and dimensional stability, and better resistance to solvents and alkalis but poor resistance to acids. Poor ultraviolet resistance and weathering are weak points of epoxies.

Epoxies are used with coreactant, or curing agent and the composite properties are influenced by the choice of curing agent; these include amines, anhydrides, acids, phenolics and amides. Cross-linking can be accomplished at room temperatures with the aid of accelerators, but it is common to use heat, which results in superior properties. Depending on the system employed, this curing may require a multiple step schedule including post-cure.

Applications include various types of coatings and adhesives, and a wide range of structural purposes. Glass, carbon, aramid and other fibers may be used as reinforcements. For a civil engineering structure, glass–epoxy composites may be of higher quality than glass–polyester composites. Unlike polyester, epoxy does not exhibit styrene-related environmental problems. However, higher cost and difficulty in processing of epoxy may offset such advantages.

Vinylesters

Vinylesters are unsaturated esters of epoxy resins. A typical chemical resistant resin is produced by reacting bisphenol A (BPA) epoxy with methacrylic acid. This product is diluted with 30 to 50% styrene monomer.

Vinylester resins have similar properties as epoxies and processibility of a polyester. These resins are often identified as a class of unsaturated polyester thermosetting resins because of the curing and processing similarities.

A multifunctional novolac epoxy may be used as the starting material to improve resistance to solvents, high temperatures and corrosion. The T_g of such vinylesters are 30°–50°C (50°–100°F) higher than BPA epoxy base resins and are classified as higher temperature resins. Flame retardant properties can be obtained if tetrabromobisphenol A is used instead of bisphenol A.

In general, vinylester resins

1. wet out and bond well to glass fibers;
2. provide cured composites with resistance to chemicals from strong acids to strong alkalis;
3. have easy processibility with toxicity as low as polyesters.

Because of such properties, both polyesters and vinylesters may become the major polymer matrices for composites used by civil engineers. The wide range of applications includes pipes, tanks, underground storage tanks, ducts, pumps, linings, scrubbers and chimneys.

Like polyesters, processing can be done at either ambient temperature or elevated temperatures. The range of process methods is as diversified as with polyesters.

The terms **basis** and **stage**, used by material engineers, are defined as follows.

- The '**A**' **basis** mechanical property is the value above which at least 99% of the population of the values is expected to fall, with a confidence of 95%, and the '**B**' **basis**, 90% of the population, with the same percentage of confidence. The '**S**' **basis** property is the minimum value specified by the appropriate authorities or specifications, while 'typical' property value is an average value without statistical assurance.

- The **A-stage**, also called **resole**, is an early stage in the polymerization process of thermosetting resins. The material is still soluble in certain liquids and is fusible.
- The **B-stage**, or **resitol**, is an intermediate stage. The material, such as the resin in an uncured prepreg or premix, softens when heated and is plastic and fusible. But it may not entirely dissolve or fuse.
- The **C-stage**, or **resite**, is the final stage of the same process. The material is insoluble and infusible.

As with fibers there are many kinds of noble polymer matrices which are not presented here. Matrix materials which have been extensively used by other branches of industry are **phenolics**. Phenolic thermosetting resins are products of the condensation reaction of phenol and formaldehyde. They have good flame retardant properties, low smoke generation and high heat resistance (up to 205–316°C (400–600°F)). They also have good dimensional stability and resistance to acids. Glass filled grades can have thermal expansion equal to metals. Phenolics are available, in general, in black or brown, and are not stable to ultraviolet radiation, which causes the color to darken.

2.4 Particulate composites

In particulate composites, particles of one or more materials are embedded in a continuous matrix of another material. The particles as well as matrices may be either metals or nonmetals. Numerous combinations are possible to form a composite for a specific purpose. The most important particulate structural composite for civil engineers is portland cement concrete. Concrete is made of particulates (gravels and sands) suspended in cement matrix. Reinforced concrete is both particulates and fibrous since steel reinforcement (fibrous) is embedded in concrete. Polymers may improve properties of portland cement concrete significantly. In polymerous concretes (or concrete polymer composites), polymers are used together, or in place of portland cement to improve properties. Depending on manufacturing processes one may classify them in more detail as follows (Figure 2.2).

2.4.1 Polymer impregnated concrete (PIC)

Polymer impregnated concrete (PIC) is ordinary cured reinforced concrete whose voids are impregnated by monomers, which eventually polymerizes to form a polymer–concrete 'unified' body with improved properties [17–20]. In general, ordinary cured concrete can be dried by any method such as autoclave, hot blown air, or steam. After the air is taken away from the voids in the structure by vacuum, surface impregnants made of low viscosity monomer, initiator, promoter and cross-linking agent (if necessary), which will be chosen by the engineer, are applied to the surface of the existing concrete. The monomers frequently used are methyl methacrylate (MMA) and styrene (S). Acrylonitrile and epoxy are also used. Use of polymethyl methacrylate (PMMA) increases strength as well as brittleness. Adding butylacrylate (BA) or styrene

Figure 2.2 Concrete composite system.

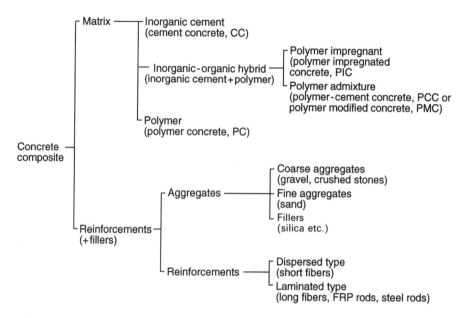

(S) increases ductility. Use of 50% MMA and 50% BA increases strength as well as ductility suitable for construction materials. Benzoyl peroxide (BPO) and azobisiso butylonitrile (AIBN) are used as initiators. Dimethylaniline (DMA) is one of the promoters. For increased thermal stability, cross-linking agents such as trimethylolpropane trimethacrylate (TMPTMA) or thermosetting monomers such as polyester–styrene may be used. Impregnation may be carried out in full depth or partial depth, the latter being used for on-site constructions or for large structural members. Polymerization may be done by radiation or heat. Use of heat is more common in industry. For impregnation, pressure may be applied by the use of such as autoclaves. For on-site large structures such as bridge decks, soaking may be used by flooding the surface with impregnants.

Polymer impregnated concrete has been developed since the 1960s; application examples can be found in dams, atomic energy facilities and highway and marine structures. Recently, concrete structure deterioration caused by salt and freezing is reported to be serious and PIC began to attract more attention. By using thin PIC precast panels as both surface protection and formwork panel of concrete structural members, improved properties and mass production of the product can be obtained resulting in decreased cost. Properties of PIC depend primarily on impregnation rate and depth. Better impregnation rate and increased depth increases cost, and judgement is necessary based on consideration of all factors such as improved properties and cost.

Mechanical properties of PIC are greatly increased compared with untreated concrete. In general, compressive strength of 17 000–28 000 psi (117–193 MPa), bending strength of 2560–4270 psi (17.6–29 MPa) and Young's modulus of 6.4×10^6 psi (44 GPa) can be obtained. Poisson's ratio is 0.18–0.20. It was reported that a compressive strength of 39 800 psi (275 MPa) could be obtained. In general, the concrete becomes elastic to failure and creep is greatly reduced and becomes negligible. One of the outstanding properties of PIC is its ability to withstand freeze–thaw action. After 1000 freeze–thaw cycles, ordinary concrete loses about 30% of its Young's modulus. PIC maintains its initial value. Abrasion loss is less than one third that of ordinary concrete. It has excellent anti-chemical and anti-weathering properties. After 3000 h exposure to ultraviolet light, decrease in strength is zero. The fatigue characteristic is similar to that of ordinary concrete, maintaining a higher strength for the same ratio. PIC is a good candidate material for anti-corrosion and a fabrication method study is in progress.

2.4.2 Polymer cement concrete

Polymer cement concrete (PCC) or polymer modified concrete (PMC) has a hybrid matrix made of ordinary portland cement and polymers [21]. However, unlike polymer impregnated concrete, monomers and polymers are directly incorporated into the standard cement concrete mixes. After standard portland cement concrete is mixed, a monomer or a polymer latex (polymer emulsified in water) is added to the mix, and the whole mix is deposited and cured. ACI defines PMC (PCC) as follows [22]: 'Polymer modified concrete is a premixed material in which either a monomer or polymer is added to a fresh concrete mixture in a liquid, powder or dispersed phase, and subsequently allowed to cure, and, if needed, polymerized in place'.

Polymer admixtures used for polymer cement concrete (or mortar) are diversified as shown in Figure 2.3. Styrene-butadiene rubber (SBR) latex and polymer dispersions such as ethylene vinyl acetate (EVA) and polyacrylic acid ester (PAE) emulsions are widely used. In order to improve the workability and bond stress of the fresh mortar, powdered emissions of EVA or others, and water soluble polymers such as methyl cellulose (MC), polyvinyl alcohol and others are widely used. Epoxy resins were widely used in Europe and America, but there is an imbalance of function versus cost since the polymer cement ratio has to be over 30% in order to obtain a good result. Polymethyl methacrylate (PMMA), polyester–styrene, and vinylidene chloride have also been used. Recently, the annual use of polymer dispersions in Japan alone is over 100 000 tons.

Addition of polymer latex materials to ordinary portland cement concrete increases the mechanical properties remarkably. One percent of PMMA added to ordinary concrete mix produces twice the compressive strength. By the use of vinylidene chloride latex, the bond stress could be increased up to five times that of ordinary cement concrete. Unlike ordinary cement concrete, the mix design of polymer cement concrete (mortar) must be based on consideration of the whole set of factors such as compression, tension and bending strengths, ductility, bond stress, watertightness and anti-chemical properties. Since such properties depend on the polymer–cement rather than the water–cement ratio, the former ratio must be determined according to the overall requirement, accompanied by actual tests. In general, the polymer–cement ratio is between 5 and 30%. In order to improve bending strength and toughness of polymer cement concrete, short fibers of steel, glass, organic, or carbon may be added.

Polymer cement concrete (or mortar) has already become a popular construction material. It is used extensively in restoring and resurfacing deteriorated bridges and road surfaces. It shows excellent resistance to freezing and thawing,

Figure 2.3 Types of polymer admixtures.

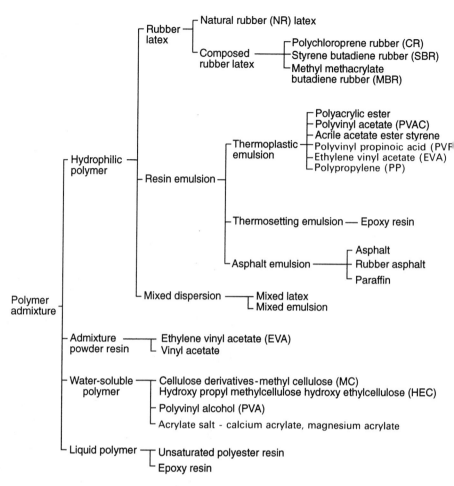

penetration by de-icing salts, and abrasion. Because of its excellent bond strength to existing concrete, polymer cement concrete (mortar) is an excellent material for repairing reinforced concrete structures in general.

Recently, some types of ultrarapid hardening polymer modified concretes have been developed [23]. One method uses the mixture of ordinary portland cement and monomers which polymerize immediately at room temperature, such as magnesium acrylate. Since the curing time can be made less than one second, this is used for shotcrete. In another method, high early strength cement, instead of ordinary portland cement, is mixed at the site with polymer dispersion, such as SBR latex, and is used for emergency repair and overlay of roads. For underwater work such as waterfronts, water soluble polymers such as hydroxy propyl methylcellulose and hydroxy ethylcellulose are used at a polymer–cement ratio of 0.2–2.0%.

Macro-defect free (MDF) cement has been developed since 1981. Water soluble polymers such as polyvinyl alcohol and polyacryl amide are used at a polymer–cement ratio of 2–8%. This composite has flexural strength of over 14 000 psi (98 MPa) but is subject to unfavorable hygrothermal effects. Extensive studies are underway to cope with such problems.

High-strength concrete using a water absorbant polymer such as polyacrylate salt was reported in 1989. It is noted that several kinds of precast concrete products of such concrete can be made by injection molding or compression molding. The advantage of polymer cement concrete is that the technology used for ordinary portland cement concrete can be adopted without difficulty. Special attention should be given to MSC meaning 'containing micro-silica' and DSP (densified with small particles) concretes. Very high compressive strength can be obtained by DSP concrete (up to ten times that of standard concrete). Engineers should be aware of the new materials that are being obtained by modifying conventional materials.

2.4.3 Polymer concrete

Polymer concrete (PC), sometimes called 'resin concrete', is a composite material formed by polymerizing a monomer and aggregate mixture [22]. The polymer matrix chosen completely replaces the portland cement or other standard concrete binders.

Generally, the same aggregates used for portland cement concrete are good for polymer concrete. One major difference is that these aggregates must be **well graded** for minimum void ratio to minimize the resin required (and thus cost), since the polymers employed are considerably more expensive than portland cement. If the mix is well graded the percentage of resin required is approximately 6%, similar to that of polymer impregnated concrete. However, it is common to use 9–25% by weight of resins. Liquid resins used for polymer concrete are as shown in Figure 2.4. Among these resins, unsatured polyesters, epoxies, furans, methyl methacrylate (monomer), urethane, vinylester and styrene are commonly used. It may be possible that resins made from recycled plastics will be available at lower cost. The curing time needed for polymer concrete is less than one day and can be made less than a minute by proper choice of curing temperatures and chemical additives.

Recently, methyl methacrylate has attracted some attention for improving workability, low temperature curing characteristics and environmental resistance of polymer concrete (mortar). Glycerine methyl methacrylate–styrene system matrices have also been developed for the same purposes. The properties of polymer concrete (mortar) greatly depend on the type, property and quantity

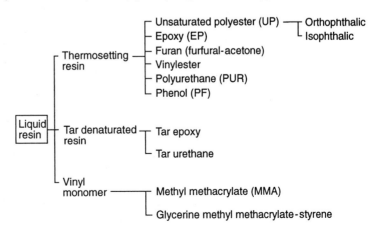

Figure 2.4 Types of liquid resins.

of the matrix resin, and the property of aggregates. In any case, engineers should be able to design the mixes and to test the specimen as with the standard portland cement concrete. Some guidelines for optimum mix design can be found in references [24, 25]. Short fibers such as those of steel, glass and sometimes pitch based carbon, are used to reinforce polymer concrete itself. When polymer concrete is used to form a structural member, steel bars for ordinary reinforced concrete, steel tendons or bars for prestressed concrete, or composite (usually glass fiber plus polymer matrix) rods may be used for reinforcement.

The manufacturing process is similar to that for portland cement concrete, i.e. mixing the catalyzed polymer with the aggregate and depositing the mix in the form or mold. Almost the same equipment for standard cement concrete can be used for polymer concrete. Polymer concrete has been used in a variety of applications. In building construction, panels, skins of sandwich panels, and floors made of polymer concrete are abundant. It is used for surfaces of roads and bridges, for anti-corrosion applications in marine structures, chemical factories, for maintenance of existing civil and architectural concrete structures and for waterproofing purposes.

Artificial marble and cores of fiber reinforced polymer (FRP) sandwich pipes are some of the other applications.

A general outline of properties of standard concrete, polymer concrete and polymer cement concrete (polymer modified concrete) can be found in the paper of Fowler [26] and in Table 2.4. The reader is reminded that properties depend on many factors including polymer types and quantities.

2.4.4 Flakes

Flakes of nonmetallic materials such as mica or glass suspended in a polymer or a glass produce an effective composite material. Since flakes have a two-dimensional geometry, a flake composite, when the flakes are packed parallel to each other, has lower permeability to liquids than an ordinary composite. Mica flakes in glass composites have good insulating qualities and are used in electrical applications. Metal flakes such as silver or copper in a polymer increase the electrical conductivity greatly. Frequently, metallic particles are suspended in metallic matrices. A metallic particle in a metallic matrix does not dissolve as does an alloy. Lead particles in copper increase machinability and act as a natural lubricant in copper alloy bearings.

Table 2.4 Relative properties of concrete-polymer materials compared to normal concrete

	Conventional portland cement concrete	PC	PCC (PMC)
Compressive strength	1	1.5–5	1–2
Tensile strength	1	3–6	2–3
Modulus of elasticity	1	0.05–2	0.5–0.75
Water absorption	1	0.05–0.2	–
Freeze–thaw resistance			–
(No. cycles/% weight loss)	700/25	1500/0–1	
Acid resistance	1	8–10	1–6
Abrasion resistance	1	5–10	10

Nonmetallic particles such as ceramics are suspended in metal matrix. The resulting composite is called a **cermet**. Cermet is widely used in applications requiring high temperature stability, high hardness, and high corrosion and abrasion resistance.

2.4.5 Main reinforcements

The most common reason for early degradation of a concrete structural member is the corrosion of the reinforcing steel. By employing reinforcements which do not corrode, the service life of a concrete structure can be greatly improved.

The most frequently used noncorrosive reinforcement is epoxy coated steel bars. The epoxy, with thickness 180–200 μm, is coated over the surface of the deformed steel bars and cured. Examples of application of such bars for marine structures and for cold region highway structures are abundant.

Reinforcing bars or cables of composites with several types of long fibers have been reported.

Kajima-FRC reported a type of composite concrete called '3R-FRC' in which three-dimensional fabric, made by weaving the fiber rovings in three directions, is impregnated by epoxy and cured, and is employed as the main reinforcement. The fiber is a hybrid of PAN based carbon, aramid and vinylon fibers. In addition to the long fibers, aramid and vinylon short fibers (volume fraction $V_f = 1.0$–1.5%) are mixed into a mortar matrix to improve the flexural characteristics. In Japan, 3D-FRC was applied to the partition panels of a hydrogen storehouse and as the granite-facing parapet of a building. Studies are underway to apply 3D-FRC to major structural members such as beams, walls and floors.

NEFMAC is a kind of composite reinforcement for concrete. A hybrid of continuous carbon, glass and aramid fibers is impregnated with resin and formed into a mesh. This type of reinforcement does not need to be fully encapsulated in the concrete, enabling thinner concrete sections to be used.

ARAPREE is a composite prestressing tendon consisting of aramid filaments and epoxy matrix. While glass fiber is susceptible to corrosion by alkalis, aramid is alkali resistant. The parallel aramid filaments are simultaneously impregnated and coated by epoxy resin. The fiber volume fraction is about 50%. ARAPREE has both circular and rectangular cross-sections, and is delivered in coils. The necessary anchorage length of a rectangular strand (strip) with 100 000 filaments (20 mm × 1.5 mm, fiber cross-section 11.2 mm^2) in a C 45 grade concrete for full prestressing force is 1.97 in (50 mm). The breaking load of this strand is 7.87 kips (35 kN). Prestressing is done up to 60% of typical tensile strength. It is claimed that the typical uniaxial tensile strength of a rectangular strip is 400 000 psi (2800 MPa), Young's modulus: (18–19) × 10^6 psi (125–130 GPa), failure strain 2.4%, density 2.4. Relaxation is 15% in a dry environment and 20% in wet conditions.

AFRP Rod from Teijin is another concrete reinforcing tendon using Technora aramid fibers. The matrix used is vinylester. Both circular and rectangular sections, and deformed bars to improve concrete bond, are available. Mechanical properties of a typical 6 mm diameter rod with 65% fiber volume are: tensile strength 270 000 psi (1862 MPa); tensile modulus 7.67 × 10^6 psi (52.9 GPa); failure strain 3.7%; relaxation 7–14%. The **relaxation of tendons** for prestressing is defined as the rate of stress loss due to relaxation at a time approaching infinity, over initial stress in tendon cross-section after release of prestressing anchors (55% of short-term strength or 65% of long-term strength of the tendons).

CFCC is the carbon fiber composite cable from Tokyo Rope Manufacturing Company. PAN based carbon fiber is preimpregnated with ordinary epoxy for use at up to 130°C, heat-resisting epoxy for use up to 180°C, or bismaleic amid resin for use up to 240°C. The prepreg is turned into strand prepreg coated with another resin, made into strands and cured. The cables are supplied in several sizes with either single or several strands in coils. The mechanical properties of a typical cable (seven strands, dia. 0.49 in) are: tensile strength 307 000 psi (2118 MPa); tensile modulus 19.9×10^6 psi (137 GPa); specific weight 1.5; elongation at break 1.57%. Since the tensile strength of a typical PC steel cable with similar diameter is 277 000 psi (1912 MPa), the service range of applied stress is similar to steel PC strands. The coefficient of thermal expansion is $0.6 \times 10^{-6}/°C$ (steel, $12 \times 10^{-6}/°C$). Relaxation loss at 20°C (0.8 Pu, 10 hours) is 0.66% (steel, 1.4%) and at 60°C (0.8 Pu, 16 hours) is 2.46% (steel, 5.8%). This property is important when the tendons are used for structural purposes such as concrete reinforcement. The creep at 130°C (0.6 Pu, 1000 hours) is 0.04% (steel, 0.07%). The bond stress to concrete (embedding 6 in) is 1048 psi (7.2 MPa) (steel, 413 psi (2.8 MPa)). The fatigue performance is better than steel. The efficiency of impact loading is 88% (steel, 90%).

In general, composite products for main reinforcement of concrete are more expensive than steel. However, their corrosion resistance, nonmagnetic properties, low electrical conductivity, weather durability, light weight, and other properties may play an important role for engineers to select composite reinforcement instead of conventional materials.

Some of these products have been applied to bridges and other structures with success.

2.5 Other matrix based composites

Some of the new very promising materials are carbon, metal and ceramic matrix composites. Price structures of constituent materials for such composites have been out of reach of civil engineers but some of these are coming down dramatically. Figure 2.5 shows the general tendency of price changes of silicon carbide reinforced aluminum composite. A manufacturer has projected the price of SiC–Al close to those of both steel and aluminum. At present, carbon–carbon composites are relatively higher in cost and manufacturing processes are under further development. Whisker reinforced ceramics have high-temperature stability as well as high toughness properties. Fiber reinforced glass matrix composites show promise up to 1000°C. Metal matrix composites have a metal base which is reinforced by one or more constituent materials such as continuous carbon, alumina, silicon carbide, or boron fibers, and discontinuous carbon or ceramic whisker materials. Optimum composite properties are obtained by using oriented continuous fibers. Composites reinforced by whisker or particulates exhibit isotropic characteristics, as with other matrix based composites. General outlines of mechanical properties of ceramic matrix composites are

Figure 2.5 Tendency of price changes of SiC–Al composite billet.

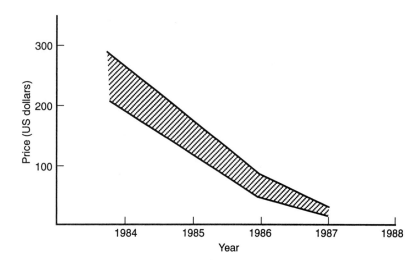

shown in Tables 2.5 and 2.6, which are based on Du Pont products. The mechanical properties of some metal matrix composites are shown in Table 2.7.

2.6 Mechanical properties of fibrous composite

Even though composite members with long, oriented fibers, frequently laminated, have the maximum available strengths, those with randomly distributed short fibers may have wide application in civil and architectural engineering. Particulate composites such as cement based ones can also be reinforced by short fibers. Short fiber reinforced composites have an isotropic nature in general, unless random distribution processes of fibers are 'disturbed' by the

Table 2.5 Mechanical properties of alumina (Al_2O_3) ceramic matrix laminates (Du Pont – Lanxide)

Reinforcement:	Silicon carbide (SiC)	
V_f (%):	55	35
Type:	5–20 μm particles	Continuous
Density at 25°C	3.4–3.5	2.9
Flexural strength (ksi (MPa))		
25°C	58–73(400–500)	72(500)
1000°C	29–36(200–250)	
1200°C	26–33(180–230)	43(300)
Fracture toughness (ksi in$^{1/2}$ (MPa m$^{1/2}$))		
25°C	5.4–6.3(6.0–7.0)	14(15)
1000°C	2.3–3.2(2.5–3.5)	8(9)
1200°C	2.3–3.2(2.5–3.5)	6(6)
Modulus of elasticity at 25°C (Msi (GPa))	43–51(300–350)	29.1(200)
Shear modulus at 25°C (Msi (GPa))	14–22(100–150)	13.1(90)

Table 2.6 Mechanical properties of silicon carbide (SiC) ceramic matrix laminates (Du Pont)

Reinforcement:	Carbon	SiC
V_f (%):	45	40
Type:	T300 fabric	0°/90° Nicalon fabric
Density	2.1	2.5
Tensile strength (ksi (MPa))		
23°C	51(350)	29(200)
1000°C	51(350)	29(200)
1400°C	48(330)	22(150)
Modulus (Msi(GPa))		
23°C	13(90)	33(230)
1000°C	15(100)	29(200)
1400°C	15(100)	25(170)
Elongation		
23°C	0.9	0.3
1000°C	NA	0.4
1400°C	NA	0.5
Flexural strength (Msi(MPa))		
23°C	73(500)	44(300)
1000°C	102(700)	58(400)
1400°C	102(700)	41(280)
Fracture toughness (ksi in$^{1/2}$ (MPa m$^{1/2}$))		
23°C	32(35)	27(30)
1000°C	32(35)	27(30)
1400°C	32(35)	27(30)

NA: not applicable.

manufacturing processes, such as injection molding, or by any other means, resulting in oriented or directional properties.

2.6.1 Cement based composites

Steel fibers have been used to reinforce concrete structural members, including polymer concrete, which is not a cement based matrix composite, in order to improve toughness, through enhanced ductility and energy absorption capacity, strength and stiffness.

Addition of 2% volume fraction of straight round steel fibers with length to diameter (aspect) ratio of 60 increases the tensile strength by 60–80%, the ultimate bond resistance by 50% and the post-peak frictional bond resistance by 140%. Noble fibers such as carbon fibers (mainly pitch based), aramid fibers, glass fibers (alkaline resistant) and other polymer based fibers are developed as secondary reinforcements for concretes, namely as temperature and shrinkage reinforcements as well as improving the properties given above. Some of the drawbacks to using such fibers are difficulties of uniform mixing and either floating to the surface or sinking to the bottom after distribution in the concrete mix.

Instead of using monofilaments, adapting three-dimensional fibrillation may solve such problems. For example, Forta CR is a collated, fibrillated polypropylene fiber reinforcement [27]. The synthetic polymer is drawn or stretched

Table 2.7 Mechanical properties of some metal matrix composites

Matrix:	Aluminum								Copper	Titanium
Reinforcement:		SiC	Carbon	Carbon P55	Carbon P100	Avco SiC	Avco SiC	Avco SiC	Avco SiC	Avco SiC
V_f (%):	0	20	47	50–60	50–60	47	47	47	48	35
Orientation:		Whisker	Continuous 0°	Fibril	Fibril	Continuous 0°	Continuous 0°/90°	Continuous ±45°	Continuous 0°	Continuous 0°
Density	2.77	2.82	2.49	–	–	2.85	2.85	2.85	6.37	3.87
Tensile strength (ksi (MPa))	40–80 (275)–(550)	75–90 (517)–(620)	100 (689)	75–90 (517)–(620)	80–121 (550)–(834)	230 (1586)	110 (759)	45 (310)	130–200 (897)–(1380)	240± (1655)±
Tensile modulus (Msi (GPa))	10 (69)	15–20 (104)–(138)	30 (207)	28–32 (193)–(220)	55–60 (380)–(414)	≥30 ≥(207)	19 (131)	17 (117)	25–35 (172)–(240)	32± (220)±
Elongation (%)	14–18	2	0.5	–	–	0.9	0.9	–	–	1.1
Max. use temp. (°C)	260	260	260	–	–	260	260	260	–	538

Matrix:	Aluminum		
Reinforcement:	Avco SiC	Nicalon SiC	Alumina
V_f (%):	48	35	48
Orientation:	Quasi isotropic 0°/±45°/90°	Continuous 0°	Continuous 0°
Density	2.85	2.60	3.6
Tensile strength (ksi (MPa))	83 (572)	114–129 (786)–(890)	180 (1240)
Tensile modulus (Msi (GPa))	18 (124)	14–16 (97)–(110)	30 (207)
Elongation (%)	1.0	–	0.7
Max. use temp. (°C)	260	260	260

axially into thin film sheets instead of circular cross-section monofilaments. The film sheets are slit longitudinally into tapes and then further worked on to produce fine fibers which are collated or held together by cross-linking along their length. The stretched fiber has a tensile strength of 70–100 ksi (483–690 MPa), a modulus of elasticity of 700 ksi (4.83 GPa), specific gravity of 0.9, and the elongation at break of 8% minimum. When this fiber is put into an ordinary concrete mix, usually 0.1% by volume, the mixing action distributes and separates the fiber bundle to form a three-dimensional network inside the whole mass of concrete. Compared with conventional wire mesh construction, even the overall short-term cost is less, in addition to the long-term benefits with the added durability which results in lower maintenance cost. The wire mesh used as secondary reinforcement has single plane distribution and holds the concrete together after it cracks. The three-dimensional reinforcing benefits include reduction in both plastic and hardened concrete shrinkage (about half that of plain concrete), and improved impact resistance, ductility and durability. Permeability and floor abrasion are reduced. As for fatigue strength tests of beams, the plain concrete specimens failed at 55 300 cycles with a load equal to 57.5% of 720 psi. According to the manufacturer's test results, the three-dimensional fiber reinforced beams survived 10^7 cycles of repetitive loading. Two million cycles were at 70.6% of 775 psi. When used for shotcrete, either wet mixed or dry mixed, the clogging in the shotcrete lines or at the nozzle, which is one of the typical problems with either steel or glass fibers, does not occur. Organic synthetic fibers alleviate problems such as corrosion, chemical and alkali attack. Such fibers have been successfully used for many kinds of projects.

Du Pont reported test results on oriented polyethylene pulp, with trade name Pulplus, for asbestos replacement in cement sheet and pipe building products [28]. Some substitutes for asbestos such as glass and cellulosics have had durability problems because of their sensitivity to alkalinity. Another substitute, synthetic staple fibers, do not contribute to cement retention. It is claimed that Pulplus overcomes such problems and brings enhanced toughness, good dimensional stability and crack-free performance. When 6% by weight of this fiber is added to type 1 portland cement concrete, the maximum flexural strength is 3770 psi (26 MPa), modulus 2.25 Msi (15.5 GPa) and toughness 2.3 kJ/m². Chen and Chung [29] report that unsized pitch based short carbon fibers in the amount of 0.5% by weight of cement increase the toughness of concrete by up to 56%, the flexural strength by up to 44% and the compressive strength by up to 22% at 14 days of curing. The optimum fiber length was 0.197 in (5 mm) for aggregate size up to 0.187 in (4.75 mm).

Kajima-FRC produces carbon fiber reinforced concrete (CFRC) curtain walls. Pitch based carbon fibers with lengths of 0.12 in (3 mm) to 0.394 in (10 mm) are distributed in cement mortar and this mortar is formed to structural elements such as curtain walls. Figure 2.6 shows the test results of the cases with the fiber volume fraction $V_f = 0\%$, 2% and 4%. It is claimed that both tensile and bending strengths are increased by 2–5 times of those of the plain concrete. Both toughness and ductility are increased by 10–20 times, thus making it possible to manufacture larger panels with light weight. The weights of the curtain walls made of CFRC are about 1/2–1/3 of those made of standard

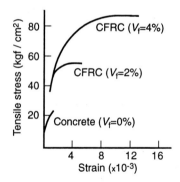

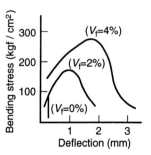

Figure 2.6 The effect of carbon fiber on concrete strength.

concrete with similar properties. Considerably high resistances to freezing and thawing, and degradation are observed. CFRC has good dimensional stability. By using low shrinkage cement or employing controlled cure by autoclave, the shrinkage rate can be controlled within 0.02–0.05%.

2.6.2 Polymer based composites

Short fiber reinforced, polymer based composites can be used for industrial equipment, automotive parts, commercial products, toolings, window frames, building panels and cladding, forms for concreting, and many other applications. Such composite materials are supplied most commonly in the form of compounds such as **sheet molding compound** (SMC) and **bulk molding compound** (BMC).

BMC is thermosetting resin such as unsaturated polyester, vinylester or phenolic, mixed with chopped strand reinforcements, fillers, catalysts, and so on into a viscous compound for compression, transfer or injection molding. SMC is similar to BMC, but compounded and processed into sheet form for

Table 2.8 Typical properties of molded composites (glass fiber, 25–40%)

Composite:	SMC-1	SMC-2	SMC-3	BMC-1	BMC-2	BMC-3
Resin:	Polyester	Polyester	Phenol	Polyester	Polyester	Polyimides
Density	1.82	1.80	1.95	1.91	1.79	1.53
Limiting oxygen (%) Index (IS075)	23	23	100	100	28	
Flexural strength (ksi (MPa))	24.7 (170)	43.5 (300)	24.7 (170)	9.4 (65)	14.5 (100)	16.8 (116)
Flexural modulus (Msi (GPa))	1.67 (11.5)	2.61 (18)	1.74 (12)	1.38 (9.5)	1.45 (10)	
Tensile strength (ksi (MPa))	10.2 (70)	14.5 (100)	8.0 (55)	3.6 (25)	6.5 (45)	12.5 (86)
Impact strength charpy unnotched (kJ/m^2)	65	90	65	13	25	

Table 2.9 Mechanical properties of some of the chemical resistant SMC composites

Resin:	Vinylester		
Fiber:	Glass		
% (wt):	67	65	50
Density	1.84–1.90	1.75	1.8
Flexural strength (ksi (MPa))	70 (483)	58.9 (406)	48.2 (332)
Flexural modulus (Msi (GPa))	2.6 (17.9)	2.19 (15.1)	2.23 (15.3)
Tensile strength (ksi (MPa))	39 (269)	31 (214)	24 (166)
Tensile modulus (Msi (GPa))	2.8 (19.3)	2.29 (15.8)	2.29 (15.8)
Elongation (%)	1.4–1.6	1.4–1.6	1.8

Table 2.10 Mechanical properties of thermoplastic molded composites

Resin:	Nylon 6			Nylon 6.6					Polycarbonate			Polysulphone		Polypropylene
Fiber: % (wt)	Glass 30–50	Glass 30	Carbon (pitch) 20	Carbon 30	Glass 40	(Pan) 40	Kevlar 35	Glass 30	Carbon (Pan) 30	Glass 50	Carbon (Pitch) 20	Glass 30	Carbon 30	Glass 30–50
Density	1.36–1.56	1.37	1.22	1.28	1.45	1.34		1.43	1.33	1.63	1.28	1.46	1.37	1.11–1.33
Tensile strength (ksi (MPa))	22.2–34.5 (152)–(238)	26 (179)	24 (166)	34.9 (241)	32.8 (226)	40.0 (276)	18.3 (126)	22.8 (157)	24.4 (168)	25.1 (173)	19 (131)	18 (124)	22.9 (158)	10–14 (68.9)(96.5)
Modulus (Msi (GPa))	1.3–2.2 (9.0)–(15.2)				1.7 (11.7)		1.2 (8.3)			2.2 (15.1)				0.8–1.4 (5.5)–(9.6)
Elongation (%)	2.1–2.2	3	3	3–4	2.5	3–4	1.9	4–6	2.7	1.5	4	3–4	2–3	2.2–1.3
Flexural strength (ksi (MPa))	35.1–51.2 (242)–(353)	38 (262)	36 (248)	50.9 (351)	49.1 (338)	59.9 (413)	29.4 (203)	27.8 (192)	36 (248)	41.7 (287)	30 (207)	23.9 (165)	31.9 (220)	18.8–22.6 (130)–(156)
Flexural modulus (Msi (GPa))	1.1–1.8 (7.6)–(12.4)	1.3 (9.0)	1.4 (9.6)	2.9 (20)	1.6 (11)	3.4 (23.4)	1.1 (7.6)	1.2 (8.3)	1.9 (13.2)	2.2 (15.2)	1.3 (9.0)	1.2 (8.3)	2.4 (16.5)	0.8–1.3 (5.5)–(9.0)

Resin:	PBT Polyester	PET Polyester	Polyurethane	Polyphenylene sulfide (PPS)				Polyetheretherketone (PEEK)		Polyamide imide (PAI)	
Fiber: % (wt):	Glass 30–50	Glass 30–50	Glass 30–50	Glass 30	Carbon 30	Glass 50	Kevlar 35	Glass 30 (by mass)	Carbon 40 (by mass)	Glass 30	Carbon 30
Density	1.56–1.75	1.61–1.85	1.43–1.63	1.56	1.45	1.72		1.44	1.48		
Tensile strength (ksi (MPa))	20.0–20.8 (138)–(143)	20.2–23.5 (139)–(162)	24.2–34.4 (167)–(237)	20.0 (138)	27.0 (186)	23.2 (160)	12.0 (82.7)	31.2 (215)	32.6 (225)	15.6 (107)	20.6 (142)
Modulus (Msi (GPa))	1.4–2.2 (9.6)–(15.2)	1.7–2.8 (11.7)–(19.3)	1.1–2.1 (7.6)–(14.5)			2.6 (17.9)	1.4 (9.6)		3.7 (25.5)	0.93 (6.41)	2.4 (16.6)
Elongation (%)	1.9–1.1	1.4–0.9	3.0–2.3	3–4	2–3	1.0	1.0				
Flexural strength (ksi (MPa))	32.5–35.5 (224)–(245)	29.3–36.5 (202)–(251)	35.3–52.3 (243)–(360)	29.0 (200)	33.9 (234)	37.3 (257)	21.3 (147)	36 (248)	49 (338)	19 (131)	29.1 (200)
Flexural modulus (Msi (GPa))	1.3–2.2 (9.0)–(15.2)	1.5–2.3 (10.3)–(15.9)	1.1–2.0 (7.6)–(13.8)	1.6 (11)	2.2 (15.2)	2.4 (16.6)	1.2 (8.3)	2.2 (15.5)	2.8 (19.4)	1.1 (7.6)	2.5 (17.2)

easier handling in molding operations, or slugged for injection molding. Mechanical strength is usually higher than that of BMC. When unsaturated polyester resin is used a thermoplastic resin may be added to compensate the shrinkage. The compound formulation varies widely, depending on specific purposes. In order to show the general characteristics, typical properties of glass fiber reinforced molded composites from a few selected companies, made of both SMC and BMC, with different formulations are shown in Table 2.8. Even though generalizing composite properties is not advisable, it is convenient to have the knowledge of the general relationship between the properties and the constituent material designs. One company manufactures vinylester based chemical resistant SMC composites with glass fiber content of 67% by weight, another company with 65% and 50%. The comparison is shown in Table 2.9. It should be noted that the fiber content is higher than those shown in Table 2.8.

A wide variety of thermoplastics are used for molded composites. Properties of thermoplastic composites molded by standard processes are shown in Table 2.10. Some composites for molding are supplied in the form of powder. Such molding compounds are used for transportation industry, pumps, insulators,

Table 2.11 Effect of reinforcement shapes and manufacturing method on mechanical properties of epoxy molding compound

Molding method: Reinforcement form (glass):	Low pressure transfer Granular	Compression high pressure transfer Granular	Compression 1/4″ chopped strand	Compression 1/2″ chopped strand
Density	1.9–2.1	2.0	1.9	1.9
Tensile strength	6.5–12.5	8.5	17	27
(ksi (MPa))	(45)–(86)	(59)	(117)	(186)
Compressive strength	32–24	30	27	42
(ksi (MPa))	(221)–(166)	(207)	(186)	(290)
Flexural strength	13.5–18.5	14.5	39	68
(ksi (MPa))	(93)–(128)	(100)	(269)	(469)
Flexural modulus	1.9–2.1	2.4	3.6	4.1
(Msi (GPa))	(13.1)–(14.4)	(17)	(25)	(28.3)

Table 2.12 Typical mechanical properties of long fiber reinforced composites (epoxy matrix)

Fiber type:[a] V_f (%):	E-Glass 53	S-2 Glass	Aramid 58	Carbon (PAN) 60	Carbon (Pitch) 60
Density	1.9	1.8±	1.45±	1.6±	1.8
Tensile strength	110–150	245	167–200	420–280	200–214
(ksi(MPa))	(760)–(1030)	(1690)	(1150)–(1380)	(2689)–(1930)	(1380)–(1480)
Tensile modulus	6.0	7.6	10.0–15.5	25–19	48–64
(Msi (GPa))	(41)	(52)	(70)–(107)	(172)–(130)	(331)–(440)
Elongation (%)			1.6–1.3		
Flexural strength	210	–	–	231	–
(ksi (MPa))	(1448)			(1593)	
Flexural modulus	6.0	–	–	16	–
(Msi (GPa))	(41)			(110)	
Compression strength (ksi (MPa))	120 (828)	120 (828)	40–38 (276)–(262)	180–215 (1240)–(1482)	74–63 (510)–(434)
Compression modulus	6.0	8.7	9.3–9.6	17	40–57
(Msi (GPa))	(41)	(60)	(64)–(66)	(117)	(276)–(393)

[a]Unidirectonal orientation.

Table 2.13 Mechanical properties of pultruded composite shapes (RYTON-PPS matrix)

Property	Reinforcement	
	Glass	Carbon
Density (g/cm³)	1.92	1.52
Fiber content (vol.%)	55	55
Tensile strength (ksi (MPa))	115(788)	200(1370)
Tensile modulus (Msi (GPa))	6.0(41)	17.0(117)
Elongation (%)	2.0	1.1
Flexural strength (ksi (MPa))	140(959)	170(1165)
Flexural modulus (Msi (GPa))	5.0(35)	14.0(97)
Compression strength (ksi (MPa))	100(685)	130(891)
Compression modulus (Msi (GPa))	6.0(41)	15.0(104)

electric and electronic components, and military purposes. The general tendency is that the shape and quantity of the reinforcement controls the mechanical properties rather than molding methods. This phenomenon is shown in Table 2.11. Table 2.12 shows typical mechanical properties of some long fiber reinforced epoxy matrix composites. Table 2.13 shows the range of the mechanical properties of pultruded sections using thermoplastic PPS. These tables show only the indication of general outlook. A lot of detailed information is available from suppliers (see further reading section). The engineer should scrutinize such details for specific design purposes.

2.7 'Comingle' and 'FIT' – new concepts for thermoplastic composites

Compared with thermoset composites, thermoplastic composites have advantages such as

1. improved environmental resistance;
2. increased interlamina toughness and damage tolerance;
3. short processing time (no chemical reactions – curing);
4. possibility of manufacturing volatile free laminates;
5. possibility of multiple heating cycles which gives advantages in manufacturing complex shapes, joining, reducing scrap rates and repairing.

Regardless of such advantages, the difficulty of impregnating reinforcements has prevented wide use of thermoplastic composites. The fabrication of 3-D composites, which belong to one of the most advanced composite groups, is considerably difficult if the thermoplastic resin is used. In the 'penetration/molten polymer' process, the polymer has to be melted under controlled conditions to avoid oxidation or depolymerization. The high viscosity thermoplastic polymers do not penetrate well into the fibers. In the 'penetration/polymer solution' process, the polymer is supposed to be brought into solution. However, some engineering polymers, especially the crystalline variety, cannot be made soluble. Others such as polyamides dissolve with difficulty. In any case, the solvent has to be removed which means that the fabricator has new problems such as an increased number of process operations, avoiding the formation of voids due to loss of solvent, and some local health regulations on the use of certain solvents.

In the 'powder impregnation' process, the powder has to be melted in order to keep the product in the yarn, and flexibility is lost. The 'fiber/sheet polymer sandwich' process fails to separate the filaments in the tow resulting in prevention of proper penetration by a high viscosity polymer at the processing stage, causing the appearance of voids or a nonuniform composite. All processes explained above yield shapes or semiproducts lacking flexibility, in which the fiber and polymers are already combined to form a composite.

New processes for thermoplastics, free from, or with fewer of, the above problems, have been developed. Some of these are as follows.

2.7.1   Comingled yarn

Comingled yarn is the intermingled yarn of continuous reinforcing fibers and thermoplastic fibers with uniform cross-sectional distribution within the yarn. The composite made by this process has several advantages such as no necessity to cure, easy handling with unlimited shelf life and nontoxicity, and higher toughness, compared with thermosetting matrix composites, and most importantly, drapability and adaptability to textile processing, so that preforming can be realized by using suitable textile processing such as weaving. This process ensures easier impregnation than nonpremixed thermoplastic materials and better penetration than those obtained by 'solution' or 'sandwich' methods. The product is a flexible tow.

Several molding methods can be applicable to comingled yarns. The molding pressure can be lower, and voids are fewer than with other molding processes. It is possible that continuous forming methods such as pultrusion and filament winding will be applied to comingled yarns.

Regardless of the advantages explained above, comingled products have been too expensive for general structural application. However, Toyobo is developing comingled yarn systems with reasonable prices for general use, with success. In addition to the advantages of comingled yarn processes, the test results show mechanical properties comparable to those of conventional composites with thermoset matrices. The reinforcement used is E-glass fiber. The matrices include PET (polyethylene telephthalate), NY (nylon) 6, NY66, PP (polypropylene), PPS (polyphenylene sulfide).

2.7.2   FIT

FIT (Fiber Impregnated with Thermoplastics) is a technology developed by Atochem. In this method, the tow is impregnated with a very finely divided polymer resin powder. The powder particle size is close to the diameter of the individual filaments (10–20 μm). This impregnated tow is then protected by a continuous coating, or jacket, produced without melting the impregnating powder.

In the case of a thermosetting resin, the prepregging operation is carried out after weaving or braiding the tow. The FIT product is a tow which is already impregnated. This can be woven or braided to obtain a fabric preimpregnated with a thermoplastic matrix. Because of the absence of bonding between the fiber, the impregnating powder and the jacket, this prepreg has good flexibility.

Indefinite shelf life is possible as with any thermoplastic. The protective jacket ensures easier weaving and eliminates the need for special precautions in cases of special applications, such as with aramid fiber. From the end-user's viewpoint, this product has all the advantages of using thermoplastics.

Continuous forming systems such as pultrusion and filament winding, in addition to several molding methods including the autoclave process, can be employed to manufacture the end products.

A wide variety of materials can be used for this process. For reinforcement, glass, carbon, aramid and organic fibers, and even metal or metallized fibers and ceramic fibers can be used. Usable polymers include PE, HDPE, PP, PET, PBT, NY6 and NY66 for general purposes, and NY11, NY12 and polyvinylidene fluoride for electrical applications, and PEI, PPS, PEEK and PI for aeronautical and structural applications.

A special size or a polymer additive is used for each fiber–polymer couple for good fiber–matrix bonding. Additives such as UV absorbers, metallics for electrical conduction or electromagnetic properties, and glass beads can be added to produce the properties required.

References

Note: *JISSE 1 = Proceedings 1st Japan International SAMPE Symposium*; SAMPE = Society for the Advancement of Material and Process Engineering; *TCIBC = Textile Composites in Building Construction.*

1. Miller, D. M. (1987) Glass fibers, in *Composites* (eds Dostal, C. A. *et al.*), Engineering Materials Handbook, Vol. 1, ASM International, pp. 45–8.
2. Watson, J. C. and Raghupathi, N. (1987) Glass fibers, in *Composites* (eds Dostal, C. A. *et al.*), Engineering Materials Handbook, Vol. 1, ASM International, pp. 107–11.
3. Lowrie, R. E. (1967) Glass fibers for high-strength composites, in *Modern Composite Materials* (eds Broutman, L. J. and Krock, R. H.), Addison-Wesley, pp. 270–323.
4. Three-dimensional fabrics. Technical bulletin, Shikishima Canvas Co.
5. T-Glass. Technical pamphlet, Nittobo Boseki Co.
6. S2 glass structural fabrics. Technical data, Hexcel.
7. R Glass, high-performance fibers. Technical data, Vetrotex.
8. Texxes Glass. Technical pamphlet, Nittobo Boseki Co.
9. Unifilo (Glass). Technical pamphlet, Vetrotex.
10. Diefendorf, R. J. (1987) Carbon/graphite fibers, in *Composites* (eds Dostal, C. A. *et al.*), Engineering Materials Handbook, Vol. 1, ASM International, pp. 49–53.
11. Dvorak, P. J. (1987) Designing with composites. *Mach. Des.*, 26 Nov.
12. Pitch challenges pan for carbon fibers. *Int. Reinf. Plast.*, May/June 1986, **5**(5), 16.
13. Judd, N. C. W. (1971) The chemical resistance of carbon fibres and a carbon fibre/polyester composite, in *Proceedings of the First International Conference on Carbon Fibres*, Plastics Institute, p. 258.
14. Krueger, W. *et al.* (1988) High performance composites of J-2 thermoplastic matrix reinforced with Kevlar aramid fiber. *Proc. SAMPE*, **33**, 181.
15. Riewald, P. G., Dhingra, A. K. and Chern, T. S. (1987) Recent advances in aramid fiber and composite technology, in *Proceedings ICCM 6*, Vol. 5 (eds Matthews, A. L. *et al.*), London, July, p. 362.
16. Seymour, R. B. (1988) Influence of long-term environmental factors on properties, in *Engineering Plastics*, Engineering Materials Handbook, Vol. 2, ASM International, Metals Park, OH.
17. Ohama, Y., Sekino, K. and Yamamoto, T. (1983) Rapid field polymer impregnation system for concrete and field trial, in *Proceedings of the 26th Japan Congress on*

Materials Research, The Society of Materials Science, Japan, Kyoto, March, pp. 204–8.

18. Fukuchi, T. and Ohama, Y. (1978) Process technology and properties of 2500 kg/cm² strength polymer impregnated concrete, in *Polymers in Concrete, Proceedings of the Second International Congress on Polymers in Concrete* (eds Fowler, D. W. and Paul, D. R.), The University of Texas at Austin, Oct., pp. 45–56.

19. Tsuruda, K. (1990) Development trends of polymer impregnated concrete products, in *Korea–Japan Joint Seminar on 'Development and Application of Concrete Polymer Composites'*, Kangwon University, Aug., pp. 69–79.

20. Kimachi, Y., Suzuki, O., Okayama, Y. and Murakami, K. (1989) Influence of polymer impregnation and addition of steel fibers on behavior of reinforced concrete members. In *JISSE 1* (eds Igata, T. *et al.*), Nov., 1576–81.

21. Ohama, Y. (1990) Application and development direction on concrete–polymer composites, in *Korea–Japan Joint Seminar on Development and Application of Concrete–Polymer Composites* (ed. Yeon, K. S.), Aug., 141–53.

22. American Concrete Institute (1979) Polymers in Concrete, State of the Art Report.

23. Amano, T., Ohama, Y., Takemoto, T. and Takeuchi, Y. (1989) Development of ultrarapid-hardening polymer-modified shotcrete using metal acrylate, in *JISSE 1* (eds Igata, T. *et al.*), Nov., pp. 1551–56.

24. Piasta, Z. and Czarnecki, L. (1989) Analysis of material efficiency of resin concretes, in *Brittle Matrix Composites* 2 (eds Brandt, A. M. and Marshall, I. H.), Elsevier Applied Science, London, pp. 593–602.

25. Ohama, Y., Demura, K. and Shimizu, A. (1987) Process technology and properties of ready mixed polyester concrete, the production performance and potential of polymers in concrete, in *Proceedings of the 5th International Congress on Polymers in Concrete*, Brighton Polytechnic, Brighton, England, pp. 71–4.

26. Fowler, D. W. (1989) Concrete–polymer composites as advanced construction materials, in *JISSE 1* (eds Igata, T. *et al.*), Nov., pp. 1551–6.

27. Three-Dimensional Fibrous Reinforcement Concrete, Forta Corporation.

28. Gale, D. M., Guckert, J. R. and Shelburne, S. S. (1990) Oriented polyethylene pulp for asbestos replacement in cement building products, in *Proceedings TCIBC* (eds Hamelin, P. and Verchery, G.), Vol. 1, Pruralis, pp. 47–56.

29. Chen, P. W. and Chung, D. D. L. (1990) Carbon fiber reinforced concrete, in *Proceedings TCIBC* (eds Hamelin, P. and Verchery, G.), Vol. 1, Pruralis, pp. 77–84.

Further reading

Articles and reports

Akihama, H., Takagi, H. and Tomioka, Y. (1984) Mechanical properties of carbon fiber reinforced cement composite and the application to large domes, Report No. 53, Kajima Institute of Construction Technology, July.

Akihama, S., Suenaga, T. and Banno, T. (1986) Mechanical properties of carbon fiber reinforced cement composite and the application to buildings (Part 2). Report No. 65, Kajima Institute of Construction Technology, Oct.

Anon. (1986) Design and construction guideline for reinforced concrete with epoxy coated steel bars. Concrete Library Vol. 58, Japan Society of Civil Engineers.

Anon. (1987) Aluminium alloys by mechanical alloying. *Aircraft Engng*, Jun., **59**(6), 20.

Anon. (1987) Metal matrix composites advance reported. *Aviat. Wk Space Technol.*, 19 Oct., 52.

Anon. (1990) An electronic material made from whiskey or sake? (Editorial). *SAMPE J.*, **26**(4), Jul./Aug., 2.

Austin, S. A., Robins, P. J. and Beddar, M. (1987) Influence of fiber geometry on the performance of steel fiber refractory concrete, in *Proceedings ICCM 6* (ed. Matthews, F. L.), Vol. 2, July, pp. 80–9.

Bak, D. J. (1986) Vapor deposition improves metal matrix composites. *Des. News*, **42**, 16 Jun., 114–15.

Ballie, C. A. and Bader, M. G. (1991) The influence of fiber surface treatment on the adhesion of carbon fibers in epoxy resin, in *Proceedings ICCM 8* (eds Tsai, S. and Springer, G.), July, p. 11-B.

Banthia, N., Mindess, S. and Bentur, A. (1987) Behaviour of fiber reinforced concrete beams under impact loading, in *Proceedings ICCM 6* (ed. Matthews, F. L.), Vol. 2, July, pp. 70–9.

Barbier, F. and Ambroise, M. H. (1991) Metal/metal composites for automotive components, in *Proceedings ICCM 8* (eds Tsai, S. and Springer, G.), Jul., p. 18-J.

Beever, W. H., Rhodes, V. H. and Wareham, J. R. (1988) Continuous length thermoplastic composite laminates. *SAMPE J.*, Jan/Feb., **24**(1), 8.

Bentur, A. (1988) Interfaces in fibre reinforced cements. *Mater. Res. Soc. Symp. Proc.*, **114**.

Bentur, A. (1989) Silica fume treatments as means for improving durability of glass fiber reinforced cements. *J. Mater. Civ. Engng*, **1**(3), Aug., 1–4.

Berg, M. and Young, J. F. (1989) Introduction to MDF cement composites. *ACBM*, **1**(2), Fall, 167–83.

Brady, D. G. (1986) Aerospace discovers thermoplastic composites. *Mater. Engng*, Sep.

Brandt, A. M. (1985) On the optimal direction of short metal fibres in brittle matrix composites. *J. Mater. Sci.*, **20**(11), 3831–41.

Brandt, A. M. (1987) Influence of the fibre orientation on the mechanical properties of fibre reinforced cement (FRC) specimens, in *Proceedings, 1st International RILEM Congress*, Vol. 2, Versailles, pp. 651–8.

Brandt, A. M. (1990) Cement based composite materials with textile reinforcement, in *Proceedings TCIBC* (eds Hamelin, P. and Verchery, G.), Vol. 1, Pruralis, pp. 39–43.

Bucci, R. J. *et al.* (1988) ARALL laminates: properties and design update, in *Proceedings 33rd International SAMPE Symposium and Exhibition* (eds Carrillo, G. *et al.*), Anaheim, CA., pp. 1239–48.

Bunsell, A. R. (1987) Fiber reinforcements – past, present and future, in *Proceedings ICCM 6* (ed. Matthews, F. L.), Vol. 5, London, p. 1.

Byun, K. J., Choe, H. S. and Lee, S. M. (1990) Development and application of polymer impregnated concrete, in *Korea–Japan Joint Seminar on Development and Application of Concrete Polymer Composites* (ed. Yeon, K. S.), Aug., pp. 13–28.

Carpenter, C. E. and Colton, J. S. (1993) On-line consolidation mechanisms in thermoplastic filament winding (tape laying), in *Proceedings 38th International SAMPE Symposium and Exhibition* (ed. Bailey, V. *et al.*), Anaheim, CA, May, pp. 205–14.

Chamis, C. C. and Shiao, M. C. (1993) Probabilistic assessment of smart composite structures, in *Proceedings 38th International SAMPE Symposium and Exhibition* (eds Bailey, V. *et al.*), Anaheim, CA, May, pp. 1303–21.

Chand, N., Khazanchi, A. C. and Rohatgi, P. K. (1987) Structure of ipomoea carnea and development of polyester and soil cement matrix composite materials with ipomoea, in *Proceedings ICCM 6* (ed. Matthews, F. L.), Vol. 2, July, pp. 90–6.

Chanvillard, G., Banthia, N. and Aitcin, P. C. (1990) Normalized load-deflection curves for fibre reinforced concrete under flexure. *Cem. Concr. Composites*, **12**(1), 41.

Coffenberry, B. S. *et al.* (1993) Low cost alternative: in-situ consolidated thermoplastic composite structures, in *Proceedings 38th International SAMPE Symposium and Exhibition* (eds Bailey, V. *et al.*), Anaheim, CA, May, pp. 391–405.

Cook, J. L. and Mohn, W. R. (1987) Whisker-reinforced MMCs, in *Composites*, Vol. 1, ASM International, Metals Park, OH, pp. 896–902.

Craig, R. J., Mahader, S., Patel, C. C., Viteri, M. and Kertesz, C. Behavior of joints using reinforced fibrous concrete. ACI Publication SP81-6.

Crosby, J. M. (1988) A review of recent advances in discontinuous fiber reinforced thermoplastic composites, in *Proceedings 33rd International SAMPE Symposium and Exhibition* (eds Carrillo, G. *et al.*), Anaheim, CA, pp. 1295–306.

Crouch, K. E. Simulated lightening strike tests on CYCOM MCG fiber panels. Study report for American Cyanamid Company, by Lightening Technologies, Inc.

Currie, B., Neill, M. and McBurney, J. R. (1990) The effect of cyclic exposure to a marine environment and ultra violet light on the flexural behavior of a polypropylene reinforced cement composite, in *Proceedings TCIBC* (eds Hamelin, P. and Verchery, G.), Vol. 1, Pruralis, pp. 181–9.

Debicki, G. and Hamelin, P. (1990) Mechanical aspects of the fiber's role in the mechanisms of crack formation in a fiber reinforced concrete, in *Proceedings TCIBC* (eds Hamelin, P. and Verchery, G.), Vol. 1, Pruralis, pp. 149–56.

Drzal, L. T. and Madhukar, M. (1991) Measurement of fiber-matrix adhesion and its relationship to composite mechanical properties, in *Proceedings ICCM 8* (eds Tsai, S. W. and Springer, G. S.), July, pp. 11-A.

Dudgeon, C. D. (1989) Overview of unsaturated polyester resins-growth opportunities in automotive and marine composites, in *Proceedings 34th International SAMPE Symposium and Exhibition* (eds Zakrewski G. A. *et al.*), Reno, Nevada, May, pp. 2333–43.

Eadara, R., Armbruster, R. and Davies, P. (1989) Epoxies in concrete maintenance, in *Proceedings 34th International SAMPE Symposium and Exhibition* (eds Zakrewski, G. A. *et al.*), Reno, Nevada, May, pp. 1704–14.

English, L. K. (1987) Composite update '87. *Mater. Engng*, Apr.

Fowler, D. W. and Hsu, H. (1984) Static and cyclic behavior of polymer concrete beams, in *Polymers in Concrete–ICPIC '84*, Darmstadt, Germany, pp. 159–64.

Fulmer, R. W. (1980) S-2 glass fiber bridges a gap in the reinforcement spectrum. Paper presented at the National Symposium, SAMPE, May.

Gale, D. M., Guckert, J. R. and Shelburne, S. S. (1990) Oriented polyethylene pulp for asbestos replacement in cement building products, in *International Symposium on Composite Materials with Textile Reinforcement for Use in Building Construction and Related Applications*, Vol. 1, Lyon, France, July, pp. 47–56.

Gardner, C. L. and Street, K. N. (1991) Electromagnetic shielding properties of composite materials, in *Proceedings ICCM 8* (eds Tsai, S. W. and Springer, G. S.), July, p. 16-K.

Goddard, D. M., Burke, P. D., Kizer, D. E., Bacon, R., Harrigan, Jr., W. C. (1987) Continuous graphite fiber MMCs, in *Composites*, Vol. 1 (eds Dostal, C. A. *et al.*), ASM International, Metals Park, OH, pp. 867–73.

Haraki, N., Tsuda, K. and Hojo, H. (1991) Sand erosion behavior of GFRP, in *Proceedings ICCM 8* (eds Tsai, S. and Springer, G.), July, p. 16-O.

Harrigan, W. C. (1987) Discontinuous silicon fiber MMCs, in *Composites*, Vol. 1 (eds Dostal, C. A. *et al.*), ASM International, Metals Park, OH, pp. 889–95.

Hasuo, K. and Okamoto, T. (1989) High-strength concrete using water-absorbent polymer, in *Proceedings JISSE 1* (eds Igata, T. *et al.*), Nov., pp. 1570–5.

Head, A. and Ware, M. (1988) Composite-product design eased by braided fibers. *Mod. Plast. Int.*, **18**(8), 43–5.

Hori, S., Tagaya, K., Taniguchi, Y. and Yoda, K. (1989) Influence of carbon fiber properties on reinforced concrete, in *Proceedings JISSE 1* (eds Igata, T. *et al.*), Nov., pp. 1582–6.

Hughes, B. P. and Guest, J. E. (1975) Polymer modified fiber-reinforced cement composites, in *Proceedings of 1st International Congress on Polymer Concretes*, (eds Cohen, D. H. *et al.*), May, pp. 82–92.

Hughes, D. (1988) Textron unit makes reinforced titanium, aluminum parts. *Aviat. Wk Space Technol.*, **129**(22), 28 Nov., 91–5.

Inman, F. S. (1975) Elevated temperature graphite yarn tensile tests. Company report, Morton, Thiokol, June.

Jelidi, A., Ambroise, J. and Pera, J. (1990) Reinforcement d'une matrice cimentaire par des fibres polyester, in *Proceedings TCIBC* (eds Hamelin, P. and Verchery, G.), Vol. 1, Pruralis, pp. 109–14.

Jindal, R. L. and Hassan, K. A. Behavior of steel fiber reinforced concrete beam–column connections, ACI Publication SP 81-5.

Kenny, J. M. *et al.* (1993) Mathematical modelling of the resin transfer molding of high performance composites, in *Proceedings 38th International SAMPE Symposium and Exhibition* (eds Bailey, V. *et al.*), Anaheim, CA, May, pp. 1263–9.

Kimura, H., Takagi, H. and Tomioka, Y. (1989) CFRP strands for the tension members of structures, in *Proceedings JISSE 1* (eds Igata, T. *et al.*), pp. 1593–8.

Ko, F., Fang, P. and Chu, H. (1988) 3-D braided commingled carbon fiber/PEEK composites, in *Proceedings 33rd International SAMPE Symposium and Exhibition* (eds Carrillo, G. *et al.*), Anaheim, CA, pp. 899–911.

Kurtz, D. *et al.* (1991) Modulus of mesophase pitch carbon fibers, in *Proceedings JISSE 2*, p. 158.

Langston, P. R. Kevlar for aerospace. Aerospace Composites and Materials.

Lei, C. S. C. and Ko, F. K. (1991) Mechanical behavior of 3-D braided hybrid metal matrix composites, in *Proceedings ICCM 8*, July, p. 18-H.

Lin, S. S. (1989) Current status of the carbon fiber industry in Japan. *ONFRE Sci. Info. Bull.*, **14**, Feb.

Maekawa, Z., Yokoyama, A., Hamada, H., Matsuo, T. and Toida, T. (1989) Mechanical properties of commingled yarn composite, in *Proceedings JISSE 1* (eds Igata, T. *et al.*), Nov., pp. 1327–31.

Maikuma, H., Kubomura, K., Ohsone, H., Tsuji, N. and Hirai, T. (1989) A study of processing of thermoplastic powder impregnated carbon fiber prepreg, in *Proceedings JISSE 1* (eds Igata, T. *et al.*), Nov., pp. 1310–14.

Mandil, M. Y., Khalil, H. S., Baluch, M. N. and Azad, A. K. (1990) Performance of epoxy-repaired concrete under thermal cycling. *Cem. Concr. Composites*, **12**(1), 47.

Mapleston, P. (1989) Phenolic composites make a strong bid in high-heat fire-safe applications. *Mod. Plast. Int.*, Sep., 53–7.

Matray, P. and Hamelin, P. (1990) Reinforcement of a magnesia cement mineral matrix by E-glass fabric, in *Proceedings TCIBC* (eds Hamelin, P. and Verchery, G.), Vol. 1, Pruralis, pp. 95–106.

Matsui, J. *et al.* (1989) On the international standardization for carbon fiber materials, in *Proceedings JISSE 1* (eds Igata, T. *et al.*), Tokyo, Nov., p. 935.

Matsuo, T. and Hokudoh, T. (1989) The situation of comingled yarn among several thermoplastic composite technologies and its application, in *Proceedings JISSE 1* (eds Igata, T. *et al.*), Nov., pp. 1321–6.

Matsuo, T. and Hokudoh, T. (1991) New thermoplastic composite material (TCM) system. *Plast. Age*, Jan. Paper presented at 35th FRP Conference, 14 Nov. 1990.

McElman, J. A. (1987) Continuous silicon carbide fiber MMCs, in *Composites*, Engineered Materials Handbook Vol. 1, ASM International, Metals Park, OH, pp. 856–66.

Misra, M. S. and Fishman, S. G. (1991) Metal matrix composites – recent advances, in *Proceedings ICCM 8* (eds Tsai, S. and Springer, G.), July, p. 18-A.

Nakagawa, H., Suenaga, T. and Aikhama, S. (1989) Mechanical properties of three dimensional fabric reinforced concretes and their application to buildings, in *Proceedings JISSE 1* (eds Igata, T. *et al.*), Nov., pp. 1587–92.

Negheimish, A., Wheat, D. and Fowler, D. (1989) Effects of temperature changes on polymer concrete overlays, in *Proceedings JISSE 1* (eds Igata, T. *et al.*), Nov., pp. 1557–63.

Nicholls, R. (1990) Fabric reinforced mortar faces on expanded polystyrene core sandwich construction of quonset buildings, in *Proceedings TCIBC* (eds Hamelin, P. and Verchery, G.), Vol. 1, Pruralis, pp. 115–24.

Nishioko, K., Kakimi, N. and Yamakawa, S. (1975) Effective application of steel fiber reinforced concrete, in *RILEM Symposium 1975, Fiber Reinforced Cement and Concrete*, (ed. Neville, A.), Vol. 1, The Construction Press, pp. 425–34.

Ochi, M. and Abe, T. (1991) Analysis of deformation of rubber composite material, in *Proceedings ICCM 8* (eds Tsai, S. and Springer, G.), July, p. 26-I.

Oh, B. H., Kim, Y. S. and Chung, C. H. (1990) Mechanical properties of fiber reinforced polymer concrete, in *Korea–Japan Joint Seminar on Development and Application of Concrete Polymer Composites* (ed. Yeon, K. S.), Aug., pp. 31–43.

Ohama, Y., Demura, K. and Muranishi, R. (1985) Development of super-high strength concrete made with silica fume addition and polymer impregnation, in *Polymer Concrete, Uses: Materials and Properties*, Publication SP-89, ACI.

Ohno, S., Keer, J. G. and Hannant, D. J. (1990) Configuration of fibrillated polypropylene fiber networks and bonding in polypropylene cement composite, in *Proceedings TCIBC* (eds Hamelin, P. and Verchery, G.), Vol. 1, Pruralis, pp. 137–46.

Park, S. B. (1990) Mechanical properties of carbon fiber reinforced polymer flyashcement composite, in *Korea–Japan Joint Seminar on Development and Application of Concrete Polymer Composites* (ed. Yeon, K. S.), Kangwon University, Aug., pp. 89–104.

Pottick, L. A. (1989) KRATON rubber modified epoxy blends, in *Proceedings 34th International SAMPE Symposium* (ed. Zakrewski, G. A. *et al.*), May, pp. 2243–54.

Prewo, K. M. (1980) Nicalon reinforced glass composites. *J. Mater. Sci.*, **15**, 463–8.

Proctor, B. A. (1990) A review of the theory of GRC. *Cem. Concr. Composites*, **12**(1), 53.

Pruneda, C. O. *et al.* (1985) The impurities in Kevlar 49 fibers, in *Proceedings of the 30th National SAMPE Symposium* (eds Newsoh, N. and Brown, W. D.), Society for the Advancement of Material and Process Engineering, Mar., pp. 1477–85

Reinhardt, H. W., Gerritse, A. and Werner, J. (1990) Arapree, in *11th FIP Congress*, Hamburg, Germany, June.

Romine, J. C. (1987) Continuous aluminum oxide fiber MMCs, in *Composites*, Vol. 1, ASM International, Metals Park, OH, pp. 874–7.

Rouse, N. E. (1988) Optimizing composite design. *Mach. Des.*, **60**(4), 25 Feb., 62–8.

Roy, D. M. (1987) New strong cement materials: chemically bonded ceramics. *Science*, **235**(6), Feb., 4784–90.

So, Y. S., Park, H. S. and Cho, Y. K. (1990) Anti-acid characteristics of super high early strength polymer cement mortar, in *Korea–Japan Joint Seminar on Development and Application of Concrete Polymer Composites* (ed. Yeon, K. S.), Aug., pp. 71–85.

Soroushian, P. and Bayasi, Z. (1986) Prediction of the tensile strength of fiber reinforced concrete: a critique of the composite material concept. *Publication SP-105, Proceedings, ACI Fall Convention*, Baltimore, MD, Nov.

Soroushian, P. and Bayasi, Z. (1987) Mechanical properties of fiber reinforced concrete, in *Proceedings ACI Seminar on Fiber Reinforced Concrete Design Application* (eds Soroushian, P. and Bayasi, Z.) E. Lansing, MI, Feb., p. 28.

Stenzenberger, H. D. *et al.* (1989) Bismaleimide resins, past, present, future, in *Proceedings 34th International SAMPE Symposium* (eds Zakrewski, G. A. *et al.*), May, pp. 1877–88.

Suzuki, Y., Maekawa, Z. and Hamada, W. (1991) Influence of silane coupling agents on interlaminar fracture in glass fiber fabric reinforced unsaturated polyester laminates, in *Proceedings ICCM 8* (eds Tsai, S. and Springer, G.), July, p. 11-K.

Tavakoli, S. M. and Phillips, M. G. (1991) Compatibility and durability in glass/phenolic laminating systems, in *Proceedings ICCM 8* (eds Tsai, S. and Springer, G.), July, p. 16-D.

Thiery, J., Francois-Brazier, J. and Vautrin, A. (1990) Comparative study of the durability of glass fiber reinforced concrete, in *Proceedings TCIBC* (eds Hamelin, P. and Verchery, G.), Vol. 1, Pruralis, pp. 169–80.

Toaz, M. W. (1987) Discontinuous ceramic fiber MMCs, *Composites*, Vol. 1, ASM International, Metals Park, OH, pp. 903–10.

Vasilos, T. (1987) Structural ceramic composites, in *Composites*, Vol. 1, ASM International, Metals Park, OH, pp. 925–32.

Vogelesang, L. B. *et al.* (1989) ARALL laminates developments, in *Proceedings ICCM 7* (ed. Wu, Y.), Vol. 3, pp. 519–24.

Wang, Y., Li, V. C. and Backer, S. (1990) Tensile properties of synthetic fiber reinforced mortar. *Cem. Concr. Composites*, **12**(1), 29.

Wolff, E. G. (1993) Moisture effects on polymer matrix composites. *SAMPE J.*, May/Jun.

Wood, A. S. (1987) Advanced thermoplastic composites get the full automation treatment. *Mod. Plast. Int.*, Apr., **17**(4), 229.

Wood, A. S. (1988) Polyphenylene sulfide: suddenly the options widen. *Mod. Plast. Int.*, **18**(4), 34–7.

Xu, L. Y., Kou, C. H. and Yang, B. X. (1991) An analogous investigation on transverse cracking in composite laminate and concrete pavement, in *Proceedings ICCM 8* (eds Tsai, S. and Springer, G.), July, p. 27-F.

Young, J. F. High performance cement-based materials. C.E. Research Bulletin 347, University of Illinois, at Urbana-Champaign.

Yun, K. S., Kim, K. W. and Hur, N. C. (1990) Influence of coarse aggregate on strength characteristics of polymer concrete, in *Korea–Japan Joint Seminar on Development and Application of Concrete Polymer Composites* (ed. Yeon, K. S.), Aug., pp. 221–8.

Manufacturers' data

Advanced composite reinforcements. Technical data, Hexcel.

Advanced composite thermoplastic materials. Technical data, Polymer Composites Inc.

Advanced fiber braiding and weaving technology. Technical pamphlet, Eurocord B.V.

Advanced materials (P/S 880, PR 969, PR 973, PR 1610, PR 1645, PR 1649, PR 1662, PR 1665, PR 1685, PR 1688, PR 1758, PR 1762, PR 1763, PR 1770, PR 1778, PR 1826, PR 2701, PR 2752, PR 2912). Technical data, PRC.

Advanced tooling systems (Epoxy, Urethane, Polyester, Composites). Product Data Sheets, Advance Polymer Industries, Inc.

Aerospace adhesive products. Bulletin, Hysol Grafil Ltd.

AFRP rod. Technical pamphlet, Teijin.

All-metal-matrix composites. Technical data sheets, DWG Composites Specialties, Inc.

Almax. Alumina fibers, Mitsui Mining Co.

Altair-O, transparent conductive coatings. Product Data, Southwell Technologies.

Alumina fiber composite materials. Technical pamphlet, Avco Specialty Materials.

Aluminized fiberglass. Product guide, Material Industries, Inc.

Ampal, unsaturated polyester moulding compounds. Technical bulletin, Ciba-Geigy.

Apollo, carbon fibers. Technical data, Hysol Grafil Ltd.

Araldite, epoxy resins. Technical data, Ciba-Geigy.

Arapree. Technical pamphlet, AKZO-HBG, May, 1990.

Astroquartz II. Technical data, JPS Glass Fabrics.

Avco continuous silicon carbide fiber metal matrix composite and ceramic composite materials. Technical data, Avco Specialty Materials.

Bekitherm steel fibers. Technical notes, Bakaert Fibre Technologies.

Bennet, polymer modifiers. Product data, High Tech Plastics B.V.

Bonding and tooling materials. Product guide, Airtech International.

Boron composite materials. Technical data, Avco Specialty Materials.

Braided composite structures. Technical notes, Fiber Innovations, Inc.

Braiding. Technical pamphlet, Atkins & Pearce.

Caliber – polycarbonate resins. Technical data, Dow Chemical.

Carbolon. Technical data, Nippon Carbon Co. Ltd.

Carbon epoxy prepreg. Aerospace Technical data, 3M.

Celion carbon fibers. Technical data, BASF.

Ceramic matrix composites. Preliminary engineering data, Du Pont.

CFCC Technical data, Tokyo Rope MFG. Co. Ltd., Oct. 1989.

CFRC curtain-wall. Technical pamphlet, Kajima-FRC.

Comparative properties of fibrils. Technical data, Hyperion Catalysis International.

Compimide, Bismaleimide resins. Technical bulletin, Shell Chemical Company.

Composite mold release application guide, Mc Lube.

Composite structures. Technical pamphlet, Bentley Harris.

Corrosion resistant polyester resins. Technical data, ICI Fiberite.

Crastine – thermoplastic polyester moulding compounds. Ciba-Geigy.

Cycom, nickel coated graphite fiber. American Cyanamid Co.

CYP(G). Technical report, Toyobo.

Data manual for Kevlar 49 aramid. Texile Fiber Department, E.I. Du Pont de Nemours & Company, Inc. 1974, 1986.

Derakane – vinylester resins. Technical note, Dow Chemical Japan.

Designed fabricated components. Technical manual, Tex Tech Industries, Inc.

Dialead. Technical data, Mitsubishi Chemical Industries.

Diaria, shape memory polymer. Product guide data, Mitsubishi Heavy Industries.

DIC-PPS. Technical notes, Danippon Ink and Chemicals, Inc.

Donacarbo-F. Technical pamphlet, Osaka Gas.

Dow epoxy Novolac Resins. Technical data, Dow Chemical.

Dural – silicon carbide aluminum composites. Fact sheets, Dural Aluminum Composites Corporation.

Engineered fiberglass fabrics. Technical pamphlet, Hexcel.

Engineered materials for the transportation industry. Technical bulletin, Fiberite.

Engineering materials profile. Technical data bulletin, G. E. Plastics.

Epon HPT resins and curing agents. Technical bulletins, Shell Chemical Company.

Epon 9000 resin systems. Technical bulletin, Shell Chemical Company.

Epon resins structural reference manual, Shell Chemical Company.

Epoxy and polyurethane systems. Technical data, CONAP, Inc.

Epoxy resin and cyanate ester resin systems. Technical bulletin, Interez Inc.

Epoxy resins for filament winding and resin transfer molding. Technical bulletins, Shell Chemical Company.

Epoxy resins system for pultrusion. Technical bulletin, Shell Chemical Company.

Epoxy-urethane. Technical pamphlet, Imperial Polychemicals, Co.

Fibredux. Advance information sheet, Ciba-Geigy.

Fibrous insulations. Technical data, Manville.

Filament winding with yarns or rovings of Kevlar 49 aramid fiber. Technical bulletin, Du Pont.

Fire PRF_2 resins. Application guide, INDSPEC Chemical Corp.

Fire retardant materials. Product data, Pyroair Technology, Inc.

FIT technology. Technical bulletin, Atochem.

Flame retardants. Technical data, Makhteshim Chemical Works Ltd.

Fortafil. Technical data, Fortafil Fibers Inc.

Fortron PPS fibers. Technical information, Kureha Chemicals.

Fortron, second generation PPS. Technical bulletins, Kureha-Polyplastics.

Frekote, Multiple release systems. Technical bulletin, Hysol Grafil Ltd.

General composite properties. Bulletin ACM-7, Hercules Inc.

Glass epoxy prepreg. Aerospace technical data, 3M.

Glass in building. Data table, Saint-Gobain.

A guide to Dow's polyurethane products and technology. Technical data, Dow Chemical USA.

HDPEX-crosslinked polyethylene. Technical note, Asea Kabel.

Hercules new toughened prepreg system. Technical data package, Hercules.

High performance polymer and products. Product guide, Mitsui Toatsu Chemical Inc.

High purity ceramic materials. Technical information, American Matrix Inc.

High purity SiC ultrafine powder and high purity SiC ceramics. Technical information, Sumitomo Cement Co.

High purity sponge titanium. Technical sheets, Toho Titanium Co. Ltd.

Hipolic, advanced composite materials. Technical data, Mitsui Petrochemical.

Hycomp-PMR-15 polyimide. Technical notes, Hysol Grafil Composite Components Co.

ICI Fiberite materials handbook. ICI Fiberite.

Integrated technology for advanced composites. Technical data, Hysol-Grafil.

Jeffamine – polyoxypropyleneamine curing agents for epoxy resins. Technical databook, Texaco Chemical Company.

Ken-React (Titanate, Zirconate, and Aluminate Coupling Agents). Reference manual, Kenrich Petrochemicals, Inc.

Kevlar, lightweight protective armor. Technical data, Du Pont.

Kevlar. Technical data, Du Pont.

Kevlar 149. Technical data, Du Pont.

Kevlar prepreg. Aerospace technical data, 3M.

Kosca carbon fiber. Technical data, Korea Steel Chemical Co.

Kureha, Carbon fiber. Technical pamphlet, Kureha Chemical Co.

Magnamite, Graphite fibers products data sheets, Hercules.

Melopas, Melamine and melamine-phenolic moulding compounds. Technical bulletin, Ciba-Geigy.

Mechanical property testing of hot molded silicon carbide aluminum. Technical data, Avco Specialty Materials.

Metal metal composites. Technical data, Alcan International Ltd.

Metallized polypropylene film, oriented polypropylene film. Technical information bulletin, Hercules.

Mitsubishi pitch carbon fiber. Technical data, Mitsubishi Kasei.

Mold release. Product guide, Chem-Trend Inc.

Multi-directional woven preforms. Product data, Fiber Materials, Inc.

NEFMAC. Technical pamphlet, NEFCOM Corp.

New materials. Technical data sheets, Mitsui Mining Co.

New Shell resin systems improve the performance of advanced composite parts. Technical bulletin, Shell Chemical Co.

Nicalon, Silicon carbide continuous fibers. Nippon Carbon Co. Ltd.

Novel Z-axis reinforcement for advanced composites. Coats & Clark Inc.

PC thermoplastic composites (hybrid PEEK yarn, hybrid PPS yarn). Technical notes, BASF.

Phenolic sheet moulding compound. Product data, Freeman Chemical Ltd.

Plenco-molding compounds. Technical pamphlet, Plastics Engineering Company.

Polyester moulding compounds. Product data, Permali Premix Ltd.

Polyester moulding compounds (BMC and SMC). Product data, Freeman Chemical Ltd.

Polylac (ABS), Kibisan (AS) Resins. Technical bulletin, Chi-Mei Industrial Co. Ltd.

Powdalloy – aluminium alloy powders. Technical data, Toyo Aluminium K.K.

Prepreg systems for composite tooling. Technical bulletin, Fiberite.

Processing materials (composite, metal bond, toolings). Product guide, Richmond Technology, Inc.

Quartz fiber. Product data, Fiber Materials, Inc.

Quartzel. Technical note, Saint-Gobain.

Reinforced thermoplastic – product information. Technical service bulletin, Hysol Grafil Ltd.

Resin systems for advanced composites (epoxies, BMI, PMR). Data sheets, Hexcel.

RK carbon fibers. Technical data, RK Carbon Fibers, Ltd.

Ryton – polyphenylene sulfide resins, engineering properties. Technical service memorandum, Philips Chemical Company.

Silicon carbide composite materials. Technical pamphlet, Avco Specialty Materials.

Silicon carbide powder reinforced aluminium alloy. Technical report, Furuda R & D Co. Ltd – Tokyo Yogyo Co. Ltd.

Silicone foam tapes. Technical note, Bisco Products, Inc.

Solvay plastics. Technical manual, Solvay & Cie.

SP Cloth. Technical pamphlet, Nittodo Boseki Co.

Specialty aluminum silicates. Technical data, Engelhard Corp.

Specialty engineering plastics. Technical data, ICI Korea.

Specialty fiber products (PEEK, liquid crystal polymer). Technical notes, Celanese.

Spectra, high performance fibers. Technical data, Allied Signal.

Staticure FR. Technical data, Metallized Products, Inc.

Structure in the future. Technical data for carbon, ceramic, Spanply and Kevlar, Hexcel.

S-2 glass fiber high performance/low cost reinforcements. Owens–Corning Fiberglass Corporation, 1981, 1984.

Sumika-Hercules graphite materials briefing. Data sheets, Hercules – Sumitomo.

Surlyn HP – high performance ionomer resins. Technical bulletin, Du Pont.

Tactix performance resins. Technical data, Dow Chemical Co.

Technical bulletins and reports, Forta Corporation.

Technical notes, Coats & Clark Inc.

Technopolymers. Technical data, Azdel Inc.

Technora, high tenacity aramid fiber. Technical information, Teijin Ltd.

Tenax and Twaron (aramid). Technical data, Akzo (Enka).

Textile fiber materials for industry. Publications 1-Gt-1375-A and 5-Tod-8285-C, Owens–Corning Fiber Glass Corporation, 1961, 1980, 1985.

Texxes Hybrid (PEI/T-glass). Technical pamphlet, Nittobo Boseki Co.

Thermal control material and metallized films. Technical data book, Sheldahl.

Thornel. Technical data (pitch and pan), Amoco Performance Products, Inc.

Thornel carbon fiber T-300-3K. Product information sheet 840-1, Hercules, Inc., Aug. 1981.

3D-FRC. Technical pamphlet, Kajima – FRC.

Toho carbon fiber. Technical data, Toho Rayon Co. Ltd.
Toilon – crosslinked P.E. foam. Technical note, Tong Il Ind. Co. Ltd.
Tokawhisker, silicon carbide whiskers. Technical notes, Tokai Carbon Co. Ltd.
Tooling materials. Product guide, Hexcel.
Torayca. Technical pamphlet, Toray Ind. Inc.
TPX-polymethylpentene. Technical data, Mitsui Petrochemical.
Twelve ways to get your aerospace projects off the ground. Technical information, Dow
    Chemical Co.
Weaverite, engineered fabrics. Technical illustration book, Fiberite.
Xycon – hybrid resins. Technical pamphlet, Amoco Chemical Co.
Xydar – high performance thermoplastic engineering resins. Technical notes, Dartco
    Manufacturing Inc.

3 Classical theory of elasticity and mechanics

3.1 Assumptions for classical theory of elasticity

In order to simplify the discussion, we begin from the classical theory of elasticity. Restrictions on the classical theory of elasticity are as follows.

1. The matter of an elastic body is homogeneous.
2. The body is isotropic so that the elastic properties are independent of direction.
3. The material has linear properties.
4. The deformations are small.

3.2 Stress and strain

3.2.1 Components of stress

ΔP is the resultant of all loads acting on ΔA which is a point in a body acted on by the forces $P_1, P_2, P_3, ..., P_i$, under equilibrium (Figure 3.1). We construct the quotient $\Delta P/\Delta A$ and we determine the **stress**, σ, as

$$\sigma = \lim_{\Delta A \to 0} \frac{\Delta P}{\Delta A}$$

being certain that the point always lies within the 'shrinking' ΔA. As we 'shrink' ΔA, there will be a limiting position of the tangent plane, i.e. the tangent to the limit of ΔA.

The projection of σ upon this tangent plane is a **shear stress** and the projection of σ upon the normal to the plane is the **normal stress**. The stress, σ, and its components depend not only upon the location of the chosen point but also upon the 'direction' of the normal to the plane. Because of this dependence upon direction of the (normal to the) reference plane as well as upon location, the stress is mathematically a 'tensor'. Recall that a scalar is a zeroth order tensor, the n-dimension vector space $V_n = (V_1, V_2, ..., V_n)$ is a first order tensor, and a stress or strain is a second order tensor, while an mth order tensor may have n-dimensional vector space in each ith order space.

σ_x, σ_y, σ_z are the (intensities of) normal stresses on planes perpendicular to the axis indicated by the subscript (Figure 3.2). Note that these stresses also act in the direction of the axes indicated by the subscript. A positive (normal) stress is one which occurs when the material on the positive side of a surface (may be an interior surface) exerts a force or traction in the positive direction of the axis to which it is parallel.

The same rule with respect to signs of normal stresses applies to the shear stresses which we denote by τ. The first subscript indicates the plane on which the stress acts (indicating actually, the axis to which this plane is perpendicular)

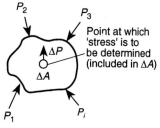

Figure 3.1 A body acted on by forces.

62

Figure 3.2 Sign convention of
stresses. (a) Positive stresses on the
positive faces. (b) Positive stresses
on the negative faces.

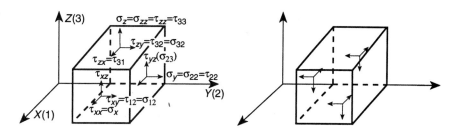

Figure 3.2 Sign convention of stresses. (a) Positive stresses on the positive faces. (b) Positive stresses on the negative faces.

and the second subscript indicates the direction of the stress, i.e. the axis which is parallel to the stress.

The sign convention is as defined in Figure 3.2. In tensorial notation, τ_{ij} or σ_{ij} may be used. Several ways of expressing stresses and coordinate systems are shown in Figure 3.2.

To describe the stresses acting on the six sides of a cubic element, three symbols, σ_x, σ_y and σ_z are necessary for normal stresses and six symbols τ_{xy}, τ_{yx}, τ_{xz}, τ_{zx}, τ_{yz} and τ_{zy} for shearing stresses. By taking the moments of the forces acting on the element, $dxdydz$, at each axis, the following equations are found:

$$\tau_{xy} = \tau_{yx}, \quad \tau_{zx} = \tau_{xz}, \quad \tau_{zy} = \tau_{yz} \tag{3.1}$$

The six quantities are sufficient to describe the stresses acting on the coordinate planes through a point and these are called components of stress at the point.

3.2.2 Components of strain

Let u, v and w be the components of the 'small' displacement of the particle of a deformed body parallel to the coordinate axes, x, y and z, respectively. Consider a small element $dxdydz$ of an elastic body.

After the deformation, the length of dx becomes

$$A'B' = dx + (u + u_x dx) - u = dx + u_x dx$$

The **strain** to the x-axis direction, ε_x, which is the **unit elongation** to the x-axis is

$$\varepsilon_x = \frac{\overline{A'B'} - \overline{AB}}{\overline{AB}} = \frac{dx + u_x\, dx - dx}{dx} = u_x = \frac{\partial u}{\partial x}$$

In the same manner, the strains in the y- and z-directions can be shown by the similar derivatives. In order to get the expressions for the shear strains, we consider the distortion of the angle between AB and AC (Figure 3.3). The displacement of the point B in the y-direction and that of the point C in the x-direction are $v + v_x dx$ and $u + u_y dy$ respectively. Assuming small deformation, the angles α and β in this figure are

$$\alpha = \doteqdot \left(\frac{\partial v}{\partial x}\, dx\right)\bigg/ dx = \frac{\partial v}{\partial x}, \quad \beta \doteqdot \frac{u_y\, dy}{dy} = u_y = \frac{\partial u}{\partial y}$$

Because of deformation, the initially right angle BAC is changed by the angle $\alpha + \beta$. This is the **shearing strain** between planes xz and yz. The shearing strains between the xy and xz planes, and between yx and yz can be obtained in the

Figure 3.3 Deformation of element body.

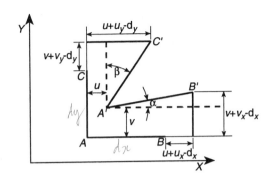

same manner. Using the letter γ for shearing strains, strain components are expressed in terms of deformation components as follows:

$$\varepsilon_x = \frac{\partial u}{\partial x}, \quad \varepsilon_y = \frac{\partial v}{\partial y}, \quad \varepsilon_z = \frac{\partial w}{\partial z}$$

$$\gamma_{xy} = \frac{\partial u}{\partial y} + \frac{\partial v}{\partial x}, \quad \gamma_{xz} = \frac{\partial u}{\partial z} + \frac{\partial w}{\partial x}, \quad \gamma_{yz} = \frac{\partial v}{\partial z} + \frac{\partial w}{\partial y} \tag{3.2}$$

In mathematical theory of elasticity, use of tensorial notation is frequent. The three mutually orthogonal stress components on a face of the element, Figure 3.2, form a vector. This vector is called a **surface traction**.

The strains are of two types. **Dilatational** or **extensional strains** are expressed as ε_{ii}. This is a measure of the change in the dimension of the element body to the subscripted direction due to the normal stress, τ_{ii}, acting on the surfaces of this body. The **shearing strains** are proportional to the change in the angle of the control volume from the original right angle, due to the shear stresses, $\tau_{ij}(i \neq j)$. The shear strain, ε_{ij}, a tensor quantity, is defined as $\varepsilon_{ij} = \gamma_{ij}/2$, and when the Cartesian coordinates are used,

$$\varepsilon_{xy} = \tfrac{1}{2}\gamma_{xy}, \quad \varepsilon_{xz} = \tfrac{1}{2}\gamma_{xz}, \quad \varepsilon_{yz} = \tfrac{1}{2}\gamma_{yz} \tag{3.3}$$

It should be noted that $\gamma_{ij}(i \neq j)$ is the **engineering shear strain** while $\varepsilon_{ij}(i \neq j)$ represents **tensor shear strain**.

For an elastic body, the strain–displacement relations are given by

$$\varepsilon_{ij} = \tfrac{1}{2}(u_{i,j} + u_{j,i}) \tag{3.4}$$

where u_i is the displacement component in the i-direction, and the other subscript after the comma denotes partial differentiation with respect to the variable shown by this subscript. From equation 3.4, the result of equation 3.2 can be obtained with some changes as given by equation 3.3.

3.3   Hooke's law

By Hooke's law, for a one-dimensional body, $\sigma = E\varepsilon$, $\varepsilon = (1/E)\sigma$ where E is the **modulus of elasticity**.

In considering the strength of materials (and perhaps structures), the phenomenon of contraction in the directions normal to the direction of tensile loading should be noted; the ratio of the contraction to the elongation in the loaded direction is called **Poisson's ratio** and frequently denoted by v (Greek nu).

Assuming the validity of superposition (which requires, in this instance, only that the deformations be sufficiently small), the normal strains in the coordinate directions are

$$\varepsilon_x = (1/E)(\sigma_x - v\sigma_y - v\sigma_z) + \alpha(x)\Delta T + \beta(x)\Delta m$$

$$\varepsilon_y = (1/E)(\sigma_y - v\sigma_x - v\sigma_z) + \alpha(y)\Delta T + \beta(y)\Delta m \qquad (3.5)$$

$$\varepsilon_z = (1/E)(\sigma_z - v\sigma_x - v\sigma_y) + \alpha(z)\Delta T + \beta(z)\Delta m$$

The above represents a statement of Hooke's law for an isotropic material in three dimensions. Should one wish to include thermal and hygrothermal effects, the thermal strain $\alpha\Delta T$, and hygrothermal strain, $\beta\Delta m$, may be added as shown, where α and β are the coefficients of thermal expansion and hygrothermal expansion, respectively, and ΔT is the temperature increase and Δm is the increase from zero moisture measured in percentage weight increase. For convenience, we consider the case in which ΔT and Δm are negligible and let us 'order' the previous equations as follows:

$$+\sigma_x - v\sigma_y - v\sigma_z = E\varepsilon_x$$

$$-v\sigma_x + \sigma_y - v\sigma_z = E\varepsilon_y$$

$$-v\sigma_x - v\sigma_y + \sigma_z = E\varepsilon_z$$

Rearranging in a matrix form,

$$\begin{bmatrix} 1 & -v & -v \\ -v & 1 & -v \\ -v & -v & 1 \end{bmatrix} \begin{bmatrix} \sigma_x \\ \sigma_y \\ \sigma_z \end{bmatrix} = E \begin{bmatrix} \varepsilon_x \\ \varepsilon_y \\ \varepsilon_z \end{bmatrix}$$

Then by Cramer's rule,

$$\sigma_x = \frac{E \begin{bmatrix} \varepsilon_x & -v & -v \\ \varepsilon_y & 1 & -v \\ \varepsilon_z & -v & 1 \end{bmatrix}}{\begin{bmatrix} 1 & -v & -v \\ -v & 1 & -v \\ -v & -v & 1 \end{bmatrix}} = \frac{E(1+v)[\varepsilon_x(1-v) + \varepsilon_y v + \varepsilon_z v]}{(1+v)^2(1-2v)}$$

$$= \frac{vE}{(1+v)(1-2v)} e + \frac{E}{1+v}\varepsilon_x$$

where e is the **dilatation** ($e = \varepsilon_x + \varepsilon_y + \varepsilon_z$) and $E/(1+v) = 2G$ where G is the **modulus of rigidity** or **shear modulus**. Also, the term $vE/(1+v)(1-2v)$ is abbreviated, using λ to represent more elaborate expressions. Thus

$$\sigma_x = \lambda e + 2G\varepsilon_x.$$

Sometimes, one further substitution is made, namely $2G \rightarrow \mu$. Then

$$\sigma_x = \lambda e + \mu\varepsilon_x$$

$$\sigma_y = \lambda e + \mu\varepsilon_y \qquad (3.6)$$

$$\sigma_z = \lambda e + \mu\varepsilon_z$$

in which λ and μ are **Lamé's constants**.

The dilatation, e, is also called the **unit volume expansion**. When ΔT and Δm are negligible, from equations 3.5,

$$e = \varepsilon_x + \varepsilon_y + \varepsilon_z = \frac{(1 - 2v)}{E}\,\Theta, \quad \text{where} \quad \Theta = \sigma_x + \sigma_y + \sigma_z.$$

If the point under consideration is under the hydrostatic pressure p, $\sigma_x = \sigma_y = \sigma_z = -p$ and $e = -3(1 - 2v)p/E$. $E/[3/(1 - 2v)]$ is called the **modulus of volume expansion**.

3.4   Two-dimensional (plane) problems

Plane problems are composed of two 'distinct' problems:

1.   plane stress problems;
2.   plane strain problems.

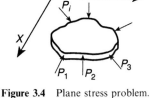

Figure 3.4   Plane stress problem.

Consider an elastic body whose third dimension, say, in the z-axis direction, is small compared to the other two dimensions and which is acted upon by forces at the boundary, parallel to the plane of the body (Figure 3.4). Assume also that the forces are uniformly distributed over the thickness of the body.

If this problem is solved in the x–y plane, assuming that the stress in the z-direction may be neglected, then this is a plane stress solution. Hence, the plane stress solution or approach implies that it has been valid to assume that $\sigma_z = 0$, also $\tau_{xz} = \tau_{yz} = 0$.

Consider a long structure with uniform sections, acted upon by forces uniformly distributed along the long axis, such as a gravity dam, a retaining wall, a culvert or tunnel, tubes or pipes, with internal pressure as well as external forces (Figure 3.5).

If we examine a slice which is remote from the ends of the dam (note that there are loads on the faces of the dam as well as gravity loads, i.e. body forces throughout the slice we are studying) then the assumption that points lying on the two cut faces do not move in the direction of the axis of the dam, i.e. normal to the cut faces, seems tenable.

If we adopt this assumption and proceed to solve the elasticity problem in the plane of the slice, the result is a plane strain solution. In this case, ε_z etc.

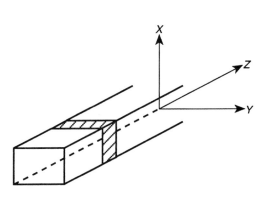

Figure 3.5   An example of plane strain problems.

would be zero. Sometimes, the plane strain problem may be relaxed or generalized by permitting ε_z to be constant over the area in the x–y plane.

From Hooke's law:

$$\varepsilon_x = (1/E)(\sigma_x - v\sigma_y - v\sigma_z)$$

$$\varepsilon_y = (1/E)(\sigma_y - v\sigma_x - v\sigma_z)$$

$$\varepsilon_z = (1/E)(\sigma_z - v\sigma_x - v\sigma_y)$$

In the plane stress problems, we set $\sigma_z = 0$, and

$$\varepsilon_x = (1/E)(\sigma_x - v\sigma_y)$$

$$\varepsilon_y = (1/E)(\sigma_y - v\sigma_x)$$

$$\varepsilon_z = (-v/E)(\sigma_x + \sigma_y)$$

(3.7)

and only the first two of these have major relevance.

In the plane strain problem, with $\varepsilon_z = 0$, the last of equations 3.5 yields

$$\sigma_z = v(\sigma_x + \sigma_y)$$

so that the first two equations become

$$\varepsilon_x = \frac{1+v}{E}\left[(1-v)\sigma_x - v\sigma y\right]$$

$$\varepsilon_y = \frac{1+v}{E}\left[-v\sigma_x + (1-v)\sigma_y\right]$$

(3.8)

or

$$(1-v)\sigma_x - v\sigma_y = \frac{E\varepsilon_x}{1+v}$$

$$-v\sigma_x + (1-v)\sigma_y = \frac{E\varepsilon_y}{1+v}$$

Using Cramer's rule,

$$\sigma_x = \frac{vE}{(1+v)(1-2v)}\left[(1-v)\varepsilon_x + v\varepsilon_y\right] = \frac{vE}{(1+v)(1-2v)}(\varepsilon_x + \varepsilon_y) + \frac{E}{1+v}\varepsilon_x$$

$$\sigma_y = \frac{vE}{(1+v)(1-2v)}(\varepsilon_x + \varepsilon_y) + \frac{E}{1+v}\varepsilon_y$$

(3.9)

which agrees with the form used by Lamé, but with, of course, $\varepsilon_z = 0$.

In the plane stress problem,

$$\sigma_x - v\sigma_y = E\varepsilon_x$$

$$-v\sigma_x + \sigma_y = E\varepsilon_y$$

and again by Cramer's rule,

$$\sigma_x = \frac{E}{(1 - v^2)}(\varepsilon_x + v\varepsilon_y)$$

$$\sigma_y = \frac{E}{(1 - v^2)}(\varepsilon_y + v\varepsilon_x) \tag{3.10}$$

$$\tau_{xy} = \frac{E}{2(1 + v)}\gamma_{xy}$$

3.5 Stress at a point

For all plane problems, the study of stress at a point (i.e. the transformation of stresses from one coordinate system to another) is the same.

To find the stresses on an inclined surface or line (Figure 3.6), we first erect the outer normal, N, and denote the cosines of the angles between N and the coordinate axes by l and m, respectively. Thus

$$\cos \alpha = l$$
$$\cos(90° - \alpha) = m \tag{3.11}$$

The values of l and m are **direction cosines** of N.

Let us indicate by A the area of the face on which we have erected N. Then, the area of the face of the element which is perpendicular to the x-axis is lA and the one perpendicular to the y-axis is mA.

What are the stresses on planes (or lines) having arbitrarily specified directions? In particular, what are the intensities of the stress on a plane or line, the outer normal of which makes an angle α with the x-axis, if the original statement of stresses refers to the x- and y-directions?

It is convenient to calculate the x- and y-components of the resultant stress on the inclined face and, by summing forces in the x-direction, we get X equal

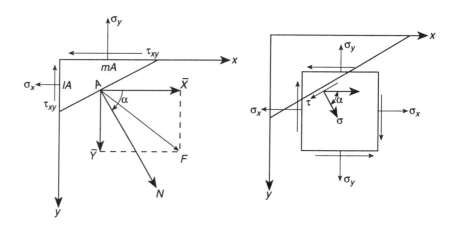

Figure 3.6 Stress at a point.

to the intensity in the x-direction.

$$AX = lA\sigma_x + mA\tau_{xy}$$

similarly

$$AY = mA\sigma_y + lA\tau_{xy}$$

These reduce to

$$X = l\sigma_x + m\tau_{xy}$$
$$Y = m\sigma_y + l\tau_{xy}$$

(3.12)

Having these results, we may calculate the normal stress, σ, and the shear stress, τ, acting on the inclined face, taking the positive direction of τ as shown in Figure 3.6.

$$\sigma = X \cos \alpha + Y \sin \alpha = \sigma_x \cos^2 \alpha + \tau_{xy} \sin \alpha \cos \alpha + \sigma_y \sin^2 \alpha + \tau_{xy} \sin \alpha \cos \alpha$$

$$\tau = Y \cos \alpha - X \sin \alpha = \sigma_y \sin \alpha \cos \alpha + \tau_{xy} \cos^2 \alpha - \sigma_x \sin \alpha \cos \alpha - \tau_{xy} \sin^2 \alpha$$

which reduces to

$$\sigma = \sigma_x \cos^2 \alpha + \sigma_y \sin^2 \alpha + 2\tau_{xy} \sin \alpha \cos \alpha$$
$$\tau = (\sigma_y - \sigma_x)\sin \alpha \cos \alpha + \tau_{xy}(\cos^2 \alpha - \sin^2 \alpha)$$

(3.13)

Recalling that

$$\cos 2\alpha = \cos^2 \alpha - sin^2 \alpha, \quad \cos^2 \alpha = (1/2)(1 + \cos 2\alpha)$$

$$\sin 2\alpha = 2 \sin \alpha \cos \alpha, \quad \sin^2 \alpha = (1/2)(1 - \cos 2\alpha)$$

we obtain

$$\sigma = \tfrac{1}{2}(\sigma_x + \sigma_y) + \tfrac{1}{2}(\sigma_x - \sigma_y)\cos 2\alpha + \tau_{xy} \sin 2\alpha$$
$$\tau = \tfrac{1}{2}(\sigma_y - \sigma_x)\sin 2\alpha + \tau_{xy} \cos 2\alpha$$

(3.14)

The question arises as to the direction of the face (i.e. the magnitude of α) for which the normal stress is a maximum. This we determine by setting the appropriate derivative equal to zero.

$$\frac{d\sigma}{d\alpha} = -2\sigma_x \cos \alpha \sin \alpha + 2\sigma_y \sin \alpha \cos \alpha + 2\tau_{xy} \cos 2\alpha = 0$$

or

$$(\sigma_y - \sigma_x)\sin 2\alpha + 2\tau_{xy} \cos 2\alpha = 0$$

which yields

$$\tan 2\alpha = \frac{\sin 2\alpha}{\cos 2\alpha} = \frac{2\tau_{xy}}{\sigma_x - \sigma_y}$$

(3.15)

The two values of α (representing angles 90° apart) represent the directions for algebraic maximum and algebraic minimum normal stresses.

We may also ask: on what planes does the shear stress vanish? We obtain this information by setting the formula for τ equal to zero:

$$\tau = \tfrac{1}{2}(\sigma_y - \sigma_x)\sin 2\alpha + \tau_{xy} \cos 2\alpha = 0$$

which is precisely the same as the criterion for the directions of the faces on which the normal stress attains its maximum and minimum values. A normal stress which occurs on a 'plane' on which the shear stress is zero is called a **principal stress**, and the direction of this normal stress is called a **principal direction**.

It is frequently convenient to set the x- and y-axes in the direction of the principal stresses. Then, for stresses on planes at arbitrary direction, the transformation formula reduces to

$$\sigma = \sigma_x \cos^2 \alpha + \sigma_y \sin^2 \alpha$$
$$\tau = \tfrac{1}{2}(\sigma_y - \sigma_x)\sin 2\alpha \tag{3.16}$$

which may be written as

$$\sigma = \tfrac{1}{2}(\sigma_x + \sigma_y) + \tfrac{1}{2}(\sigma_x - \sigma_y)\cos 2\alpha$$
$$\tau = \tfrac{1}{2}(\sigma_y - \sigma_x)\sin 2\alpha \tag{3.17}$$

The principal stresses can be written as

$$\sigma_1 = \tfrac{1}{2}(\sigma_x + \sigma_y) + \sqrt{[\tfrac{1}{2}(\sigma_x - \sigma_y)]^2 + \tau_{xy}^2}$$
$$\sigma_2 = \tfrac{1}{2}(\sigma_x + \sigma_y) - \sqrt{[\tfrac{1}{2}(\sigma_x - \sigma_y)]^2 + \tau_{xy}^2} \tag{3.18}$$

The plane on which the maximum (and minimum) shear occurs can be found as

$$\tan 2\alpha_s = \frac{\tfrac{1}{2}(\sigma_x - \sigma_y)}{\tau_{xy}} \tag{3.19}$$

and the maximum and the minimum shear is expressed as

$$\tau_{\substack{\max \\ \min}} = \pm \sqrt{[\tfrac{1}{2}(\sigma_x - \sigma_y)]^2 + \tau_{xy}^2} \tag{3.20}$$

3.6 Mohr's circle

Only for the purpose of drawing a Mohr's circle, we introduce a 'special' sign convention for shear stresses. We define positive shear stress such that it and the shear stress on the opposite face of the rectangular element form a clockwise couple (Figure 3.7). With this convention, the stresses may be represented and transformed graphically.

Any point on (not within) the circle represents the state of stress on a corresponding plane. Coordinates on Mohr's circle are the algebraic values of

Figure 3.7 Sign convention for shear stress.

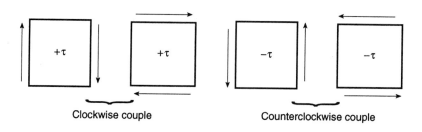

Clockwise couple   Counterclockwise couple

Figure 3.8 Mohr's circle for on-axis stresses.

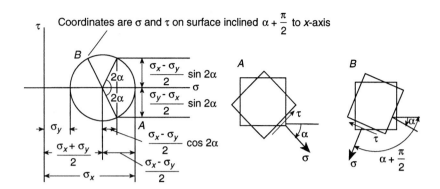

σ and τ on the surface inclined α to the x-axis (A) (Figure 3.8). The special cases such as the principal planes with one of the principal stresses, σ_2, equal to zero, and planes with pure shear, are shown in Figures 3.9 and 3.10. In the general case, there are normal and shear stresses on surfaces parallel to the coordinate axes (Figure 3.11).

For the three-dimensional case (Figure 3.12(a)), given σ_1, σ_2 and σ_3 to be the principal stresses, assuming $\sigma_1 > \sigma_2 > \sigma_3$, all positive stress combinations lie in the shaded area including the boundaries.

The following summary on biaxial stress may help understanding.

1. In the general case of plane stress, the maximum and minimum normal stresses are called principal stresses. The planes on which they occur are called principal planes and are 90° apart.
2. The principal stresses occur on those planes where the shearing stress is equal to zero.
3. The planes of maximum shearing stress make angles of 45° with the principal planes. The normal stresses on the planes of maximum shear are equal.

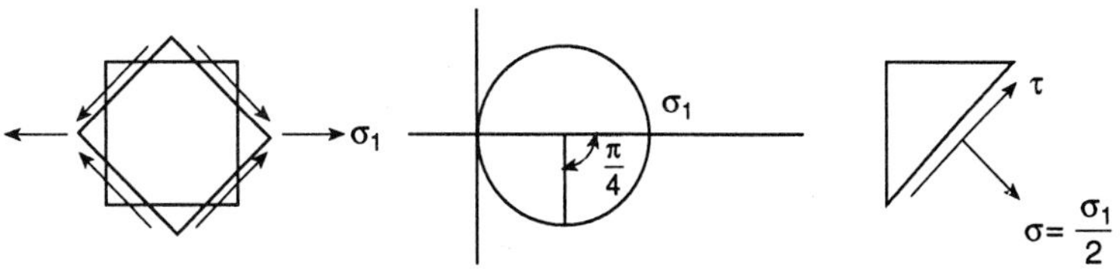

Figure 3.9 Case where $\sigma_1 \neq 0$, $\tau_{12} = 0$, $\sigma_2 = 0$.

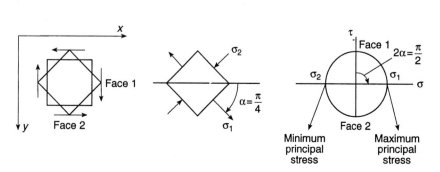

Figure 3.10 Mohr's circle for pure shear.

Figure 3.11 Mohr's circle for off-axis case.

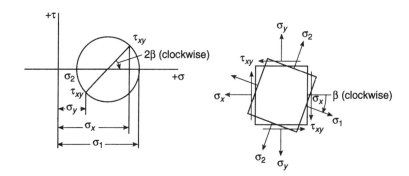

4. The following are useful rules for the directions of the principal stresses and maximum shear.

    (a) In a square element subjected to a two-dimensional stress system, the algebraically maximum principal stress, σ_1, lies in the principal 45° angle which is defined as the 45° angle formed by the shear diagonal and the algebraically greater impressed normal stress, σ_x or σ_y.

    (b) The shear diagonal of the maximum shear stress is parallel to the direction of the σ_1 stress.

    (c) When the normal impressed stresses are equal, the maximum principal stress, σ_1, is parallel to the shear diagonal.

5.
$$\sigma_1 + \sigma_2 = \sigma_x + \sigma_y \qquad (3.21a)$$

$$\tau_{max} = \frac{(\sigma_1 - \sigma_2)}{2} \qquad (3.21b)$$

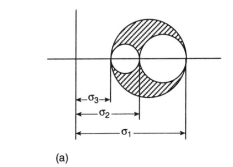

(a)

Figure 3.12 Three-dimensional state of stress.

(b)

(c)

The normal stress on planes of the maximum shear, σ_s, is

$$\sigma_s = \frac{(\sigma_x + \sigma_y)}{2} = \frac{(\sigma_1 + \sigma_2)}{2} \tag{3.21c}$$

3.7 Strain at a point

The strains on a plane whose outer normal makes an angle θ with the reference x-axis, ε_θ and γ_θ, can be obtained by replacing σ by ε_θ, τ by $\gamma_\theta/2$, σ_x by ε_x, σ_y by ε_y, τ_{xy} by $\gamma_{xy}/2$ and α by θ in equations 3.13.
Thus,

$$\varepsilon_\theta = \varepsilon_x \cos^2 \theta + \varepsilon_y \sin^2 \theta + \gamma_{xy} \sin \theta \cos \theta$$
$$\gamma_\theta = \gamma_{xy}(\cos^2 \theta - \sin^2 \theta) + 2(\varepsilon_y - \varepsilon_x)\sin \theta \cos \theta \tag{3.22a}$$

The two planes, 90° apart, on which the shear strain is zero can be found from

$$\tan 2\theta = \frac{\gamma_{xy}}{\varepsilon_x - \varepsilon_y} \tag{3.22b}$$

The normal strains on these planes are called **principal strains**. When Mohr's circle is drawn for strains, the ordinates represent $\gamma_\theta/2$ and the abscissae ε_θ. The principal strains, ε_1, ε_2, are the algebraically greatest and least values of ε_θ as a function of θ. The greatest value of $\gamma_\theta/2$ is represented by the radius of the circle. The maximum shearing strain is shown to be

$$\gamma_{\theta max} = \varepsilon_1 - \varepsilon_2 \tag{3.23}$$

3.8 Pure shear

Suppose that we have a condition of pure shear with shear stress, τ. This deforms a square element into a rhomboid or parallelogram as shown in Figure 3.13.

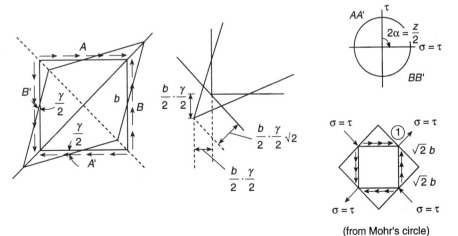

Figure 3.13 An element under pure shear.

(from Mohr's circle)

The strain in the direction of the tensile stress is

$$\varepsilon_1 = \frac{1}{E} \{\sigma - v(-\sigma)\} = \frac{1 + v}{E} \sigma = \frac{1 + v}{E} \tau$$

Consequently, the displacement of point ① relative to the center of the square is

$$e = \varepsilon_1 \left(\frac{1}{2}\sqrt{2}\, b\right) = \frac{1 + v}{E} \tau \frac{1}{2}\sqrt{2}\, b \qquad (i)$$

But from the shear deformation, we had obtained

$$e = \frac{b}{2}\frac{\gamma}{2}\sqrt{2} \qquad \text{and furthermore} \qquad \gamma = \frac{\tau}{G}$$

Hence, from the shear consideration,

$$e = \frac{b}{2}\frac{\tau}{2G}\sqrt{2} \qquad (ii)$$

Equating (i) and (ii),

$$G = \frac{E}{2(1 + v)}$$

This is the shear modulus, and it has been shown that G is not a material but mechanical property.

The dilatation, e, is

$$e = \varepsilon_x + \varepsilon_y + \varepsilon_z = \frac{1}{E}(1 - 2v)(\sigma_x + \sigma_y + \sigma_z)$$

e is zero when $1 - 2v = 0$, which means that when $v = 1/2$ and $G = E/3$, the volume does not change under stress, i.e. the body is incompressible.

3.9   Equilibrium equations and boundary conditions for a two-dimensional problem

Consider a small rectangular block with edge dimensions dx and dy and without loss in generality, with an appropriate unit thickness. X and Y are distributed forces in units of force per unit volume. The stresses acting on the block are as shown in Figure 3.14. Summing the forces in the x-direction,

$$-\sigma_x\, dy + \left(\sigma_x + \frac{\partial \sigma_x}{\partial x}\, dx\right) dy - \tau_{xy}\, dx + \left(\tau_{xy} + \frac{\partial \tau_{xy}}{\partial y}\, dy\right) dx + X\, dx\, dy = 0$$

which reduces to

$$\frac{\partial \sigma_x}{\partial x} + \frac{\partial \tau_{xy}}{\partial y} + X = 0 \qquad (3.24a)$$

Figure 3.14 Two-dimensional
state of stress in Cartesian
coordinates.

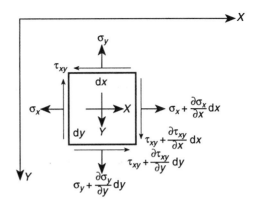

If we sum the forces in the y-direction, we would obtain ultimately,

$$\frac{\partial \sigma_y}{\partial y} + \frac{\partial \tau_{xy}}{\partial x} + Y = 0 \qquad (3.24b)$$

These are the two-dimensional equilibrium equations in Cartesian coordinates.

Equations 3.24 must be satisfied at all points of the volume of the body. At the boundary, the stress components must be in equilibrium with the external forces acting on the boundary of the body. These equilibrium conditions at the boundary can be obtained by equations 3.12. By denoting P_x and P_y the components of the surface forces per unit area of the boundary,

$$\begin{aligned} P_x &= l\sigma_x + m\tau_{xy} \\ P_y &= m\sigma_y + l\tau_{xy} \end{aligned} \qquad (3.25)$$

where l and m are the direction cosines of the normal N to the boundary.

3.10 Compatibility equations

The two-dimensional elasticity problem involves three stress components, σ_x, σ_y, and τ_{xy}, while there are two equilibrium equations. The third equation can be obtained by the condition of compatibility of stress distribution with the existence of continuous functions u, v, defining the deformation. Three strain components in terms of deformation components are

$$\varepsilon_x = \frac{\partial u}{\partial x}, \quad \varepsilon_y = \frac{\partial v}{\partial y}, \quad \gamma_{xy} = \frac{\partial u}{\partial y} + \frac{\partial v}{\partial x}$$

Differentiating γ_{xy} with respect to x and y,

$$\frac{\partial^2 \gamma_{xy}}{\partial x\, \partial y} = \frac{\partial^3 v}{\partial x^2\, \partial y^2} + \frac{\partial^3 u}{\partial x\, \partial y^2}$$

Since

$$\frac{\partial^2 \varepsilon_y}{\partial x^2} = \frac{\partial^3 v}{\partial x^2\, \partial y}, \quad \frac{\partial^2 \varepsilon_x}{\partial y^2} = \frac{\partial^3 u}{\partial x\, \partial y^2}$$

the compatibility equation for two-dimensional elasticity is

$$\frac{\partial^2 \varepsilon_x}{\partial y^2} + \frac{\partial^2 \varepsilon_y}{\partial x^2} = \frac{\partial^2 \gamma_{xy}}{\partial x \, \partial y} \qquad (3.26)$$

For the **plane stress problem**, Hooke's law is

$$\varepsilon_x = \frac{1}{E}(\sigma_x - v\sigma_y)$$

$$\varepsilon_y = \frac{1}{E}(\sigma_y - v\sigma_x)$$

$$\gamma_{xy} = \frac{2(1+v)}{E} \tau_{xy}$$

Substituting these stress–strain relationships into the compatibility equation 3.26,

$$\frac{1}{E}\left(\frac{\partial^2 \sigma_x}{\partial y^2} - \frac{v \, \partial^2 \sigma_y}{\partial y^2}\right) + \frac{1}{E}\left(\frac{\partial^2 \sigma_y}{\partial x^2} - \frac{v \, \partial^2 \sigma_x}{\partial x^2}\right) = \frac{2(1+v)}{E}\frac{\partial^2 \tau_{xy}}{\partial x \, \partial y} \qquad \text{(i)}$$

From the equilibrium equations, we obtain

$$\frac{\partial^2 \sigma_x}{\partial x^2} + \frac{\partial^2 \tau_{xy}}{\partial x \, \partial y} + \frac{\partial X}{\partial x} = 0$$

$$\frac{\partial^2 \sigma_y}{\partial y^2} + \frac{\partial^2 \tau_{xy}}{\partial x \, \partial y} + \frac{\partial Y}{\partial y} = 0$$

Adding the two equations, we obtain

$$2\frac{\partial^2 \tau_{xy}}{\partial x \, \partial y} = -\frac{\partial^2 \sigma_x}{\partial x^2} - \frac{\partial^2 \sigma_y}{\partial y^2} - \frac{\partial X}{\partial x} - \frac{\partial Y}{\partial y} \qquad \text{(ii)}$$

Now, substituting equation (ii) into (i) leads to

$$\frac{\partial^2}{\partial y^2}(\sigma_x - v\sigma_y) + \frac{\partial^2}{\partial x^2}(\sigma_y - v\sigma_x)$$

$$= -(1+v)\left(\frac{\partial^2 \sigma_x}{\partial x^2} + \frac{\partial^2 \sigma_y}{\partial y^2}\right) - (1+v)\left(\frac{\partial X}{\partial x} + \frac{\partial Y}{\partial y}\right)$$

which reduces to

$$\left(\frac{\partial^2}{\partial x^2} + \frac{\partial^2}{\partial y^2}\right)(\sigma_x + \sigma_y) = -(1+v)\left(\frac{\partial X}{\partial x} + \frac{\partial Y}{\partial y}\right) \qquad (3.27)$$

This is the compatibility equation in terms of stresses for the plane stress case.
For **plane strain** problems, we have the requirement that

$$\varepsilon_z = 0 \quad \text{leads to} \quad \sigma_z = v(\sigma_x + \sigma_y).$$

Substituting this value of σ_z into the first two stress–strain relations,

$$\varepsilon_x = \frac{1}{E}[\sigma_x - v\sigma_y - v^2(\sigma_x + \sigma_y)]$$

$$= \frac{1}{E}[(1 - v^2)\sigma_x - v(1 + v)\sigma_y]$$

$$\varepsilon_y = \frac{1}{E}[(1 - v^2)\sigma_y - v(1 - v)\sigma_x]$$

We also have

$$\gamma_{xy} = \frac{2(1 + v)}{E}\tau_{xy} \tag{iii}$$

Substituting (iii) into the compatibility equation in terms of strains (equation 3.26),

$$\frac{(1 - v^2)}{E}\frac{\partial^2\sigma_x}{\partial y^2} - \frac{v(1 + v)}{E}\frac{\partial^2\sigma_y}{\partial y^2} + \frac{(1 - v^2)}{E}\frac{\partial^2\sigma_y}{\partial x^2} - \frac{v(1 + v)}{E}\frac{\partial^2\sigma_x}{\partial x^2} = \frac{2(1 + v)}{E}\frac{\partial^2\tau_{xy}}{\partial x\,\partial y}$$

or

$$(1 - v)\left(\frac{\partial^2\sigma_x}{\partial y^2} + \frac{\partial^2\sigma_y}{\partial x^2}\right) - v\left(\frac{\partial^2\sigma_x}{\partial x^2} + \frac{\partial^2\sigma_y}{\partial y^2}\right) = 2\frac{\partial^2\tau_{xy}}{\partial x\,\partial y}$$

Substituting (ii) into the above, we obtain

$$\left(\frac{\partial^2}{\partial x^2} + \frac{\partial^2}{\partial y^2}\right)(\sigma_x + \sigma_y) = -\frac{1}{1 - v}\left(\frac{\partial X}{\partial x} + \frac{\partial Y}{\partial y}\right) \tag{3.28}$$

This is the compatibility equation in terms of stress for the plane strain case.

In the absence of body force, the compatibility equations in terms of stresses are the same for the plane stress case and the plane strain case. Furthermore, the stress distribution (in the x- and y-direction) will be the same in both cases (for similar bodies, i.e. isotropic), if the boundary tractions are the same.

3.11   Stress function

The equilibrium equations are

$$\frac{\partial\sigma_x}{\partial x} + \frac{\partial\tau_{xy}}{\partial y} + X = 0$$

$$\frac{\partial\sigma_y}{\partial y} + \frac{\partial\tau_{xy}}{\partial x} + Y = 0$$

When the body forces are continuous functions, it is convenient to consider them as being derived from a potential, V, so that

$$X = -\frac{\partial V}{\partial x}, \quad Y = -\frac{\partial V}{\partial y}$$

With this assumption, the equilibrium equations may be written as

$$\frac{\partial}{\partial x}(\sigma_x - V) + \frac{\partial \tau_{xy}}{\partial y} = 0$$

$$\frac{\partial}{\partial y}(\sigma_y - V) + \frac{\partial \tau_{xy}}{\partial x} = 0$$

We make the assumption that the stresses are related to a **stress function**, ϕ (frequently called **Airy's stress function**) by

$$\sigma_x - V = \frac{\partial^2 \phi}{\partial y^2}, \quad \sigma_y - V = \frac{\partial^2 \phi}{\partial x^2}, \quad \tau_{xy} = -\frac{\partial^2 \phi}{\partial x \, \partial y} \tag{3.29}$$

By direct substitution, it can be seen that the stress function so defined satisfies the equilibrium equations. Substituting equation 3.29 into 3.27 for the plane stress case, we obtain

$$\frac{\partial^4 \phi}{\partial x^4} + 2\frac{\partial^4 \phi}{\partial x^2 \, \partial y^2} + \frac{\partial^4 \phi}{\partial y^4} = \nabla^4 \phi = -(1 - v)\left(\frac{\partial^2 V}{\partial x^2} + \frac{\partial^2 V}{\partial y^2}\right) \tag{3.30}$$

where

$$\nabla^4 = \frac{\partial^4}{\partial x^4} + 2\frac{\partial^4}{\partial x^2 \, \partial y^2} + \frac{\partial^4}{\partial y^4}$$

An analogous equation can be obtained for the case of plane strain. When the body forces are absent, we have the following equation for both plane stress and plane strain cases:

$$\nabla^4 \phi = 0 \tag{3.31}$$

3.12 Application of stress function in rectangular coordinate system

In order to demonstrate the method of application, some solutions in polynomial form are examined. If the body forces are zero, we have

$$\nabla^4 \phi = 0$$

Suppose we assume the solution in the form

$$\phi = a_0 + a_1 x + a_2 y$$

By direct substitution, it is easily seen that this will satisfy the biharmonic equation. However, the associated stresses are

$$\sigma_x = \frac{\partial^2 \phi}{\partial y^2} = 0, \quad \sigma_y = \frac{\partial^2 \phi}{\partial x^2} = 0, \quad \tau_{xy} = -\frac{\partial^2 \phi}{\partial x \, \partial y} = 0$$

Therefore, the assumed form of ϕ (linear in x and y) is trivial or irrelevant. We see that in polynomial solutions, one may start with terms not lower than the second power in x and y.

Suppose we assume a solution in the form

$$\phi = \frac{a_2}{2} x^2 + \frac{b_2}{2} xy + \frac{c_2}{2} y^2$$

which satisfies the differential equation. Then

$$\sigma_x = \frac{\partial^2 \phi}{\partial y^2} = c_2, \quad \sigma_y = \frac{\partial^2 \phi}{\partial x^2} = a_2, \quad \tau_{xy} = -\frac{\partial^2 \phi}{\partial x \, \partial y} = -b_2$$

This stress function represents a combination of uniform tension or compression, depending on the sign of coefficients a_2 and c_2 in two perpendicular directions and with a uniform shear.

Assume a solution in the form

$$\phi = \frac{a_3}{6} x^3 + \frac{b_3}{2} x^2 y + \frac{c_3}{2} xy^2 + \frac{d_3}{6} y^3$$

which satisfies the differential equation. Then,

$$\sigma_x = \frac{\partial^2}{\partial y^2} = c_3 x + d_3 y$$

$$\sigma_y = \frac{\partial^2 \phi}{\partial x^2} = a_3 x + b_3 y$$

$$\tau_{xy} = -\frac{\partial^2 \phi}{\partial x \, \partial y} = -b_3 x - c_3 y$$

a_3, b_3, c_3 and d_3 are, at this point, completely arbitrary and independent. Suppose we consider the case $a_3 = b_3 = c_3 = 0$, $d_3 \neq 0$. Then,

$$\sigma_x = d_3 y, \quad \tau_{xy} = \sigma_y = 0$$

If one is dealing with a rectangular plate and places the x-axis at the mid-height, then the foregoing corresponds to pure bending. If the x-axis is not placed at the mid-height, the plate would be subjected to pure bending plus axial load. Many different forms of stress functions are possible, resulting in various combinations of stress states.

3.13   Displacements in a two-dimensional problem

Consider the case

$$u = a + by$$
$$v = c - bx$$

which represent a **linear rigid body displacement** in the x-direction (magnitude a) and in the y-direction (magnitude c) and a small rigid-body rotation about the origin through an angle b. The associated strains are

$$\varepsilon_x = \frac{\partial u}{\partial x} = 0, \quad \varepsilon_y = \frac{\partial v}{\partial y} = 0, \quad \gamma_{xy} = \frac{\partial u}{\partial y} + \frac{\partial v}{\partial x} = (b) + (-b) = 0$$

Since the strains vanish, the stresses associated with this linear motion are identically zero.

Assume that the stress function, $\phi = (d_3/6)y^3$ satisfies the problem under consideration. Then

$$\sigma_x = \frac{\partial^2 \phi}{\partial y^2} = d_3 y, \quad \sigma_y = 0, \quad \tau_{xy} = 0$$

$$\varepsilon_x = \frac{\partial u}{\partial x} = \frac{1}{E}(\sigma_x - v\sigma_y) = \frac{1}{E} d_3 y$$

$$\varepsilon_y = \frac{\partial v}{\partial y} = \frac{1}{E}(\sigma_y - v\sigma_x) = -\frac{v}{E} d_3 y$$

in which d_3 will not be arbitrary but will have a value assigned to satisfy some boundary conditions. Upon integration,

$$u = \frac{d_3}{E} xy + f(y)$$

$$v = -\frac{vd_3}{2E} y^2 + g(x)$$

We also have

$$\gamma_{xy} = \frac{\partial u}{\partial y} + \frac{\partial v}{\partial x} = 0$$

Substituting u and v from above,

$$\frac{d_3 x}{E} + \frac{df(y)}{dy} + \frac{dg(x)}{dx} = 0$$

This implies that

$$\frac{d_3 x}{E} + \frac{dg(x)}{dx} = -\frac{df(y)}{dy} = \text{constant}, \ K$$

or

$$\frac{d_3 x}{E} + \frac{dg(x)}{dx} = K$$

$$\frac{df(y)}{dy} = -K$$

These have the solutions

$$g(x) = -\frac{d_3 x^2}{2E} + Kx + c_1$$

$$f(y) = -Ky + c_2$$

Then

$$u = \frac{d_3}{E} xy - Ky + c_2$$

$$v = \frac{-vd_3}{2E} y^2 - \frac{d_3 x^2}{2E} + Kx + c_1$$

Recalling that the terms $(-Ky + c_2)$ in the first equation and the terms $(Kx + c_1)$ in the second equation represent only rigid body displacements, we may adopt either of the following points of view.

1. These terms may be disregarded, particularly if our only concern is the true stress distribution.
2. We can retain these terms to enable us to locate the body in space on the basis of some specified conditions.

3.14 Application of stress function using Fourier series

The boundary conditions as well as internal stress distribution of an elastic body can be expressed by any combination of analytic functions. The Fourier series is one of the most commonly used functions.

As an example, we consider a rectangular beam with length L, depth $2c$, and uniform unit thickness. We take the Cartesian coordinate system so that the x-axis coincides with the mid-plane of the beam. Suppose the beam is loaded at the upper edge, $y = -c$, by $q_u = A_m \sin(m\pi x/L)$, and at the lower edge, $y = c$, by $q_l = B_m \sin(m\pi x/L)$. Then the boundary conditions may be taken as

- at $y = -c$, $\sigma_y = -A_m \sin\left(\dfrac{m\pi x}{L}\right)$, $\tau_{xy} = 0$

- at $y = c$, $\sigma_y = -B_m \sin\left(\dfrac{m\pi x}{L}\right)$, $\tau_{xy} = 0$

We assume $\phi = f(y)\sin(m\pi x/L)$ which can give a sinusoidal variation of σ_y on the edges since $\sigma_y = (\partial^2\phi/\partial x^2)$ and substitute into the equation $\nabla^4\phi = 0$, assuming no body force. This yields

$$\left[\left(\frac{m\pi}{L}\right)^4 f - 2\frac{d^2 f}{dy^2}\left(\frac{m\pi}{L}\right)^2 + \frac{d^4 f}{dy^4}\right]\sin\left(\frac{m\pi x}{L}\right) = 0$$

which implies

$$\frac{d^4 f}{dy^4} - 2\frac{d^2 f}{dy^2}\left(\frac{m\pi}{L}\right)^2 + \left(\frac{m\pi}{L}\right)^4 f = 0$$

We can rewrite this as

$$\left[D^4 - 2\left(\frac{m\pi}{L}\right)^2 D^2 + \left(\frac{m\pi}{L}\right)^4\right]f = 0 \tag{i}$$

or

$$\left[D^2 - \left(\frac{m\pi}{L}\right)^2\right]^2 = 0, \qquad \text{so that} \qquad D = \pm\frac{m\pi}{L}$$

From elementary mathematics, we have, when

$$\frac{df}{dy} = \beta f, \quad \frac{df}{f} = \beta \, dy$$

and $\log f = \beta y + c$, or $f(y) = e^{\beta y + c} = ce^{\beta y}$. Substituting this into (i),

$$c\left[\beta^4 - 2\left(\frac{m\pi}{L}\right)^2 \beta^2 + \left(\frac{m\pi}{L}\right)^4\right] e^{\beta y} = 0$$

which yields

$$\beta = \pm\left(\frac{m\pi}{L}\right)$$

These correspond to a solution of the form

$$f(y) = c_1 e^{m\pi y/L} + c_2 e^{-m\pi y/L}$$

or in the alternative form (with c_1 and c_2 differing from above),

$$f(y) = c_1 \sinh\left(\frac{m\pi y}{L}\right) + c_2 \cosh\left(\frac{m\pi y}{L}\right)$$

Since the differential equation is fourth order and the roots of the indicial equation are repeated, we obtain two more solutions by multiplying each of the above by the independent variable, thus getting for the solution of the partial differential equation,

$$\phi = \left[c_1 \sinh\left(\frac{m\pi y}{L}\right) + c_2 \cosh\left(\frac{m\pi y}{L}\right) + c_3\left(\frac{m\pi y}{L}\right)\sinh\left(\frac{m\pi y}{L}\right)\right.$$
$$\left. + c_4\left(\frac{m\pi y}{L}\right)\cosh\left(\frac{m\pi y}{L}\right)\right]\sin\left(\frac{m\pi x}{L}\right)$$

which yields, letting $\alpha = m\pi/L$,

$$\sigma_y = \frac{\partial^2 \phi}{\partial x^2} = -\alpha^2[c_1 \sinh \alpha y + c_2 \cosh \alpha y + c_3 \alpha y \sin \alpha y + c_4 \alpha y \cosh \alpha y]\sin \alpha x$$

Hence, the condition on σ_y at the upper $(y = -c)$ and lower $(y = +c)$ boundaries are satisfied if

$$-c_1 \sinh \alpha c + c_2 \cosh \alpha c + c_3 \alpha c \sinh \alpha c - c_4 \alpha c \cosh \alpha c = \frac{A}{\alpha^2}$$

$$c_1 \sinh \alpha c + c_2 \cosh \alpha c + c_3 \alpha c \sinh \alpha c + c_4 \alpha c \cosh \alpha c = \frac{B}{\alpha^2}$$

$\tau_{xy} = -(\partial^2 \phi/\partial x \, \partial y) = 0$, at $y = c$ and $y = -c$, yields

$$c_1 \cosh \alpha c - c_2 \sinh \alpha c + c_3(-\sinh \alpha c - \alpha c \cosh \alpha c)$$
$$+ c_4(\cosh \alpha c + \alpha c \sinh \alpha c) = 0$$
$$c_1 \cosh \alpha c + c_2 \sinh \alpha c + c_3(\sinh \alpha c + \alpha c \cosh \alpha c)$$
$$+ c_4(\cosh \alpha c + \alpha c \sinh \alpha c) = 0$$

By adding, we obtain

$$c_1 = -c_4 \frac{\cosh \alpha c + \alpha c \sinh \alpha c}{\cosh \alpha c} = -c_4(1 + \alpha c \tanh \alpha c)$$

By subtracting, we obtain

$$c_2 = -c_3 \frac{\sinh \alpha c + \alpha c \cosh \alpha c}{\sinh \alpha c} = -c_3(1 + \alpha c \coth \alpha c)$$

By adding the first two boundary equations,

$$c_2 \cosh \alpha c + c_3 \alpha c \sinh \alpha c = \frac{1}{2\alpha^2}(A + B)$$

By subtracting the first from the second,

$$c_1 \sinh \alpha c + c_4 \alpha c \cosh \alpha c = \frac{-1}{2\alpha^2}(A - B)$$

By using the previous equations for c_1,

$$c_4 \left(-\frac{\cosh \alpha c + \alpha c \sinh \alpha c}{\cosh \alpha c} \sinh \alpha c + \alpha c \cosh \alpha c \right) = \frac{-1}{2\alpha^2}(A - B)$$

from which

$$c_4 = \frac{1}{2\alpha^2} \frac{(A - B)\cosh \alpha c}{\cosh \alpha c - \alpha c}$$

$$c_1 = -\frac{1}{2\alpha^2} \frac{(A - B)\cosh \alpha c}{\cosh \alpha c - \alpha c} \frac{(\cosh \alpha c + \alpha c \sinh \alpha c)}{\cosh \alpha c}$$

The stresses are found as

$$\sigma_y = -(A + B) \frac{[(\alpha c \cosh \alpha c + \sinh \alpha c)\cosh \alpha y - \sinh \alpha c \cdot \alpha y \sinh \alpha y]}{\sinh 2\alpha c + 2\alpha c} \sin \alpha x$$

$$+ (A - B) \frac{[(\alpha c \sinh \alpha c + \cosh \alpha c)\sinh \alpha y - \cosh \alpha c \cdot \alpha y \cosh \alpha y]}{\sinh 2\alpha c - 2\alpha c} \sin \alpha x$$

$$\tau_{xy} = -(A + B) \frac{[\alpha c \cosh \alpha c \sinh \alpha y - \sinh \alpha c \cdot \alpha y \cosh \alpha y]}{\sinh 2\alpha c + 2\alpha c} \cos \alpha x$$

$$+ (A - B) \frac{[\alpha c \sinh \alpha c \cosh \alpha y - \cosh \alpha c \cdot \alpha y \sinh \alpha y]}{\sinh 2\alpha c - 2\alpha c} \cos \alpha x$$

In each expression, the first term represents the symmetric component, and the second term, the antisymmetric component.

As another example, consider the concentrated load P at $x = e$. P may be expanded into a Fourier series using either the complete form

$$P \approx a_0 + \sum_{m=2,4,\ldots}^{\infty} a_m \cos\left(\frac{m\pi x}{L}\right) + b_m \sin\left(\frac{m\pi x}{L}\right)$$

or in a half-range form, either of sines or cosines:

$$P \approx \sum_{m=1}^{\infty} b_m \sin\left(\frac{m\pi x}{L}\right)$$

$$P \approx a_0 + \sum_{m=1}^{\infty} a_m \cos\left(\frac{m\pi x}{L}\right)$$

If we choose

$$P \approx \sum_{m=1}^{\infty} b_m \sin\left(\frac{m\pi x}{L}\right)$$

the loading function can be written as

$$q(x) = \begin{cases} 0, & 0 \leq x \leq e - \varepsilon \\ \dfrac{P}{2}\varepsilon, & (e - \varepsilon) \leq x \leq (e + \varepsilon) \\ 0, & (e + \varepsilon) \leq x \leq L \end{cases}$$

in which ε is an arbitrarily chosen small distance from the loading point, $x = e$, in both directions. To evaluate b_m in

$$q(x) = \sum_{m=1}^{\infty} b_m \sin\left(\frac{m\pi x}{L}\right),$$

we multiply both sides by an arbitrary harmonic, say the nth, and integrate to obtain

$$b_n = \frac{2P}{L} \sin\left(\frac{n\pi e}{L}\right)$$

The Fourier half-range series representing P is

$$\sum_{n=1}^{\infty} \frac{2P}{L} \sin\left(\frac{n\pi e}{L}\right)\sin\left(\frac{n\pi x}{L}\right)$$

Hence, for equal loads P applied to the top and bottom edges at the same x-position, $x = e$,

$$A = B = \frac{2P}{L} \sin\left(\frac{n\pi e}{L}\right)$$

and $A - B = 0$.

3.15   Equilibrium equations in terms of displacements

If the displacement components u and v are continuous functions, the equilibrium equations 3.24, together with the strain–displacement relations,

$$\varepsilon_x = \frac{\partial u}{\partial x}, \quad \varepsilon_y = \frac{\partial v}{\partial y}, \quad \gamma_{xy} = \frac{\partial u}{\partial y} + \frac{\partial v}{\partial x}$$

and the boundary conditions, equations 3.25, are sufficient to solve the two-dimensional problems.

For plane stress problems, we have equations 3.10 as

$$\sigma_x = \frac{E}{(1 - v^2)}(\varepsilon_x + v\varepsilon_y)$$

$$\sigma_y = \frac{E}{(1 - v^2)}(\varepsilon_y + v\varepsilon_x)$$

$$\tau_{xy} = \frac{E}{2(1 + v)}\gamma_{xy}$$

Substituting these equations into equations 3.24:

$$\frac{\partial^2 u}{\partial x^2} + \frac{1}{2}(1 - v)\frac{\partial^2 u}{\partial y^2} + \frac{1}{2}(1 + v)\frac{\partial^2 v}{\partial x\,\partial y} = -\frac{(1 - v^2)}{E}X$$

$$\frac{\partial^2 v}{\partial y^2} + \frac{1}{2}(1 - v)\frac{\partial^2 v}{\partial x^2} + \frac{1}{2}(1 + v)\frac{\partial^2 u}{\partial x\,\partial y} = -\frac{(1 - v^2)}{E}Y$$

(3.32)

Since the two displacement components u and v can express the stresses σ_x, σ_y and τ_{xy}, equations 3.32, which are the equilibrium equations in terms of displacements, are sufficient for the analysis.

3.16 Two-dimensional problems in polar coordinates

Many two-dimensional elasticity problems may be solved more easily by using polar coordinates. In polar coordinates, the position of a point is defined by the distance from the origin, r, and by the angle θ between r and a fixed reference axis.

Consider a small element of a plate acted upon by the stresses as shown in Figure 3.15.

K_r and K_θ are body forces into the r- and θ-directions, respectively. Summing the forces in the radial direction,

$$-\sigma_r(r\,d\theta) + \left(\sigma_r + \frac{\partial\sigma_r}{\partial r}\,dr\right)(r + dr)\,d\theta - \tau_{r\theta}\,dr + \left(\tau_{r\theta} + \frac{\partial\tau_{r\theta}}{\partial\theta}\,d\theta\right)dr$$

$$-\sigma_\theta\,dr\,d\theta + K_r r\,dr\,d\theta = 0$$

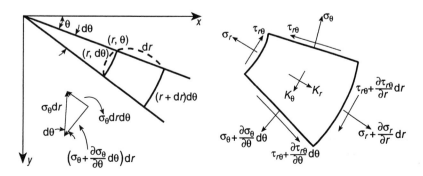

Figure 3.15 Two-dimensional state of stress in polar coordinates.

which reduces to

$$\sigma_r + r\,\frac{\partial \sigma_r}{\partial r} + \frac{\partial \tau_{r\theta}}{\partial \theta} - \sigma_\theta + rK_r = 0,$$

or

$$\frac{\partial \sigma_r}{\partial r} + \frac{1}{r}\frac{\partial \tau_{r\theta}}{\partial \theta} + \frac{\sigma_r - \sigma_\theta}{r} + K_r = 0 \qquad (3.33\text{a})$$

Noting that

$$\sigma_r + r\,\frac{\partial \sigma_r}{\partial r} = \frac{\partial}{\partial r}(r\sigma_r)$$

this equation may be rewritten as

$$\frac{\partial}{\partial r}(r\sigma_r) + \frac{\partial \tau_{r\theta}}{\partial \theta} - \sigma_\theta + rK_r = 0$$

In the tangential direction,

$$-\sigma_\theta\,dr + \left(\sigma_\theta + \frac{\partial \sigma_\theta}{\partial \theta}\,d\theta\right)dr - \tau_{r\theta}r\,d\theta + \left(\tau_{r\theta} + \frac{\partial \tau_{r\theta}}{\partial r}\,dr\right)(r + dr)\,d\theta$$

$$+\,\tau_{r\theta}\,dr\,d\theta + K_\theta r\,dr\,d\theta = 0$$

which reduces to

$$\frac{\partial \sigma_\theta}{\partial \theta} + r\,\frac{\partial \tau_{r\theta}}{\partial r} + 2\tau_{r\theta} + rK_\theta = 0 \qquad (3.33\text{b})$$

For displacements, we define

- $u(r, \theta)$ = displacement in the radial direction ($+u$ is in the direction of increasing r);
- $v(r, \theta)$ = displacement in the tangential direction ($+v$ is in the direction of increasing θ).

Consider the new and the original length of ab (Figure 3.16). The new length is

$$dr + \left(u + \frac{\partial u}{\partial r}\,dr\right) - u = dr + \frac{\partial u}{\partial r}\,dr$$

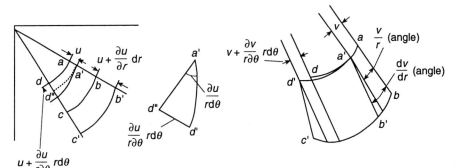

Figure 3.16 Deformed two-dimensional element in polar coordinates.

and its original length is dr. Hence the strain (due to u) in the radial direction is

$$\varepsilon_r = \frac{\left(dr + \dfrac{\partial u}{\partial r} dr\right) - dr}{dr} = \frac{\partial u}{\partial r}$$

The arc $a'd''$ has a length $(r + u)d\theta$. Noting that $d'd'' = (\partial u/\partial\theta)\, d\theta$, the new length of ad, i.e. $a'd'$ is

$$a'd' = \sqrt{(r + u)^2\, d\theta^2 + \left(\frac{\partial u}{\partial\theta}\right)^2 d\theta^2} = d\theta\sqrt{(r^2 + 2ru + u^2) + \left(\frac{\partial u}{\partial\theta}\right)^2}$$

Recalling that

$$(a + b)^n = a^n + (n/1)a^{n-1}b + \cdots + \frac{n(n-1)\cdots(n-r+2)}{(r-1)!}a^{n-r+1}b^{r-1} + \cdots + b^n$$

$$a'd' = r\, d\theta\left[1 + \frac{u}{r} + \frac{u^2}{2r^2} + \frac{1}{2r^2}\left(\frac{\partial u}{\partial\theta}\right)^2 + \cdots\right]$$

Since u and $\partial u/\partial\theta$ are small compared to r, their squares and higher powers may be neglected (for an infinitesimal deformation theory of elasticity – but not for a large deformation theory). Thus, for small deformation theory,

$$a'd' = r\, d\theta\left(1 + \frac{u}{r}\right)$$

and

$$\varepsilon'_\theta = (a'd' - ad)/ad = \frac{r\, d\theta\left(1 + \dfrac{u}{r}\right) - r\, d\theta}{r\, d\theta} = \frac{u}{r}$$

The u-displacement also produces a component of the shear strain, equal to the tangential slope of the circular arc boundary in the deformed state:

$$\gamma'_{r\theta} = \frac{\partial u}{r\,\partial\theta}$$

The v-displacements do not (except for higher-order effects) contribute to ε_r. To calculate the contribution to ε_θ, consider the final and the original length of ad

$$\varepsilon''_\theta = \frac{a'd' - ad}{ad} = \frac{\left[r\, d\theta + \left(v + \dfrac{\partial v}{r\,\partial\theta}r\, d\theta\right) - v\right] - r\, d\theta}{r\, d\theta} = \frac{\partial v}{r\,\partial\theta}$$

The portion v/r (of the angle $\partial v/\partial r$) is associated with rigid-body displacement of the element (i.e. a rotation about the origin) and, therefore, does not produce a shear strain. Therefore

$$\gamma''_{r\theta} = \frac{\partial v}{\partial r} - \frac{v}{r}$$

Adding the appropriate terms, we obtain the following equations for the strains:

$$\varepsilon_r = \frac{\partial u}{\partial r}$$

$$\varepsilon_\theta = \frac{\partial v}{r\, \partial \theta} + \frac{u}{r} \tag{3.34}$$

$$\gamma_{r\theta} = \frac{\partial u}{r\, \partial \theta} + \frac{\partial v}{\partial r} - \frac{v}{r}$$

We also have, for the plane stress case:

$$\varepsilon_r = \frac{1}{E}(\sigma_r - v\sigma_\theta)$$

$$\varepsilon_\theta = \frac{1}{E}(\sigma_\theta - v\sigma_r) \tag{3.35}$$

$$\gamma_{r\theta} = \frac{2(1+v)}{E} \tau_{r\theta}$$

or the equivalents (for plane stress):

$$\sigma_r = \frac{E}{1-v^2}(\varepsilon_r + v\varepsilon_\theta)$$

$$\sigma_\theta = \frac{E}{1-v^2}(\varepsilon_\theta + v\varepsilon_r) \tag{3.36}$$

$$\tau_{r\theta} = \frac{E}{2(1+v)}\gamma_{r\theta}$$

For the plane strain case,

$$\sigma_r = (\lambda + 2G)\varepsilon_r + \lambda\varepsilon_\theta$$
$$\sigma_\theta = (\lambda + 2G)\varepsilon_\theta + \lambda\varepsilon_r \tag{3.37}$$

It can be shown, by direct substitution, that the equilibrium equations 3.33 are satisfied by picking a stress function defined by the following:

$$\sigma_r = \frac{1}{r}\frac{\partial \phi}{\partial r} + \frac{1}{r^2}\frac{\partial^2 \phi}{\partial \theta^2}$$

$$\sigma_\theta = \frac{\partial^2 \phi}{\partial r^2} \tag{3.38}$$

$$\tau_{r\theta} = -\frac{\partial}{\partial r}\left(\frac{\partial \phi}{r\, \partial \theta}\right)$$

Upon eliminating u and v from the expressions of ε_r, ε_θ and $\tau_{r\theta}$, we arrive at the compatibility equation in the form

$$\frac{\partial^2 \varepsilon_\theta}{\partial r^2} + \frac{\partial^2 \varepsilon_r}{r^2\, \partial \theta^2} + 2\frac{\partial \varepsilon_\theta}{r\, \partial r} - \frac{\partial \varepsilon_r}{r\, \partial r} = \frac{\partial^2 \gamma_{r\theta}}{r\, \partial r\, \partial \theta} + \frac{\partial \gamma_{r\theta}}{r^2\, \partial \theta} \tag{3.39}$$

If we now substitute into this form of the compatibility equation the expressions for the strains in terms of the stresses, then substitute into this result the expressions for stresses in terms of the stress function (which satisfy the equilibrium equations), namely equations 3.38, we will arrive at the compatibility equation in the form of $\nabla^2 \nabla^2 \phi = 0$, in the absence of body force, and for the plane stress case, where

$$\nabla^2 = \frac{\partial^2}{\partial r^2} + \frac{1}{r}\frac{\partial}{\partial r} + \frac{1}{r^2}\frac{\partial^2}{\partial \theta^2}$$

We can arrive at the differential equations for the stress function in polar coordinates in another way, i.e. merely by transforming

$$\left(\frac{\partial^2}{\partial x^2} + \frac{\partial^2}{\partial y^2}\right)\left(\frac{\partial^2 \phi}{\partial x^2} + \frac{\partial^2 \phi}{\partial y^2}\right) = 0$$

with transforms

$$r^2 = x^2 + y^2, \quad x = r\cos\theta, \quad y = r\sin\theta, \quad \tan\theta = y/x$$

Suppose we have a function $f(x, y)$, $\partial f/\partial x$, and $\partial f/\partial y$, and we wish to express the derivatives in terms of r and θ:

$$\frac{\partial f}{\partial x} = \frac{\partial f}{\partial r}\frac{\partial r}{\partial x} + \frac{\partial f}{\partial \theta}\frac{\partial \theta}{\partial x}$$

$$\frac{\partial f}{\partial y} = \frac{\partial f}{\partial r}\frac{\partial r}{\partial y} + \frac{\partial f}{\partial \theta}\frac{\partial \theta}{\partial y}$$

From the first transformation equation, we obtain

$$2r\frac{\partial r}{\partial x} = 2x \quad \text{or} \quad \frac{\partial r}{\partial x} = \frac{x}{r} = \frac{r\cos\theta}{r} = \cos\theta$$

$$2r\frac{\partial r}{\partial y} = 2y \quad \text{or} \quad \frac{\partial r}{\partial y} = \frac{y}{r} = \frac{r\sin\theta}{r} = \sin\theta$$

From the second transformation equation,

$$\frac{\partial}{\partial x}(\tan\theta) = \frac{\partial}{\partial x}\frac{y}{x}, \quad \text{or} \quad \sec^2\theta \cdot \frac{\partial \theta}{\partial x} = -\frac{y}{x^2}$$

thus

$$\frac{\partial \theta}{\partial x} = -\frac{y}{x^2}\cos^2\theta = -\frac{r\sin\theta}{r^2\cos^2\theta}\cos^2\theta = -\frac{\sin\theta}{r}$$

$$\frac{\partial}{\partial y}(\tan\theta) = \frac{\partial}{\partial y}\frac{y}{x}$$

$$\sec^2\theta \cdot \frac{\partial \theta}{\partial y} = \frac{1}{x}$$

$$\frac{\partial \theta}{\partial y} = \frac{\cos^2\theta}{x} = \frac{\cos\theta}{r}$$

Hence

$$\frac{\partial f}{\partial x} = \cos\theta\,\frac{\partial f}{\partial r} - \frac{\sin\theta}{r}\frac{\partial f}{\partial\theta}$$

$$\frac{\partial f}{\partial y} = \sin\theta\,\frac{\partial f}{\partial r} + \frac{\cos\theta}{r}\frac{\partial f}{\partial\theta}$$

The governing differential equation in polar coordinates is

$$\left(\frac{\partial^2}{\partial r^2} + \frac{1}{r}\frac{\partial}{\partial r} + \frac{1}{r^2}\frac{\partial^2}{\partial\theta^2}\right)^2 \phi = 0 \tag{3.40}$$

Without considering the more general problem, we can consider the **axisymmetric case**, i.e. the loads and the solution will be independent of θ. In this case, the above reduces to

$$\left(\frac{d^2}{dr^2} + \frac{1}{r}\frac{d}{dr}\right)^2 \phi = 0 \tag{3.41}$$

or

$$\left(\frac{d^2}{dr^2} + \frac{1}{r}\frac{d}{dr}\right)\left(\frac{d^2\phi}{dr^2} + \frac{1}{r}\frac{d\phi}{dr}\right) = 0$$

Note that

$$\frac{d^2 f}{dr^2} + \frac{1}{r}\frac{df}{dr} = \frac{1}{r}\frac{d}{dr}\left(r\frac{df}{dr}\right)$$

so that the equation may be written as

$$\frac{1}{r}\frac{d}{dr}\left\{r\frac{d}{dr}\left[\frac{1}{r}\frac{d}{dr}\left(r\frac{d\phi}{dr}\right)\right]\right\} = 0 \tag{3.42}$$

Assuming that the displacements u and v are continuous, the equilibrium equations 3.33 can be expressed in terms of displacements. Substituting the stress expressions in terms of displacements into the equilibrium equations, the following two equations are obtained for the plane stress case:

$$\frac{\partial^2 u}{\partial r^2} + \frac{\partial u}{r\,\partial r} - \frac{u}{r^2} + \frac{(1-v)}{2r^2}\frac{\partial^2 u}{\partial\theta} - \frac{(3-v)}{2r^2}\frac{\partial v}{\partial\theta} + \frac{(1+v)}{2r}\frac{\partial^2 v}{\partial r\,\partial\theta}$$

$$= -\frac{(1-v^2)}{E}K_r$$

$$\frac{(1-v)}{2}\left(\frac{\partial^2 v}{\partial r^2} + \frac{\partial v}{r\,\partial r} - \frac{v}{r^2}\right) + \frac{\partial^2 v}{r^2\,\partial\theta^2} + \frac{(3-v)}{2r^2}\frac{\partial u}{\partial\theta} + \frac{(1+v)}{2r}\frac{\partial^2 u}{\partial r\,\partial\theta}$$

$$= -\frac{(1-v^2)}{E}K_\theta$$

$$\tag{3.43}$$

We have two equations with two unknown functions u and v; thus, together with given boundary conditions, a complete solution can be obtained. Similar equations can be obtained for the plane strain case.

3.17 Solution of two-dimensional problems in polar coordinates

3.17.1 Axisymmetric case
We can obtain a solution to equation 3.41 in the following way. Let us define
or set

$$\frac{d^2\phi}{dr^2} + \frac{1}{r}\frac{d\phi}{dr} = F(r) \qquad (i)$$

Then equation 3.41 may be written as

$$\frac{d^2 F}{dr^2} + \frac{1}{r}\frac{dF}{dr} = 0 \qquad (ii)$$

or

$$\frac{1}{r}\frac{d}{dr}\left(r\frac{dF}{dr}\right) = 0 \qquad (iii)$$

Multiply both sides of this by r, then

$$\frac{d}{dr}\left(r\frac{dF}{dr}\right) = 0$$

which implies that

$$r\frac{dF}{dr} = C_1, \quad \frac{dF}{dr} = \frac{C_1}{r}, \quad F = C_1 \log r + C_2 \qquad (iv)$$

Substituting (iv) into (i),

$$\frac{1}{r}\frac{d}{dr}\left(r\frac{d\phi}{dr}\right) = C_1 \log r + C_2$$

which implies that

$$\frac{d}{dr}\left(r\frac{d\phi}{dr}\right) = C_1 r \log r + C_2 r$$

Recalling that

$$\int r \log r \, dr = \frac{r^2}{2}\log r - \frac{r^2}{4},$$

after proper integration, we obtain

$$\phi = C_1 r^2 \log r + C_2 \log r + C_3 r^2 + C_4 \qquad (3.44)$$

3.17.2 Non-axisymmetric case
Consider the case in which the loading, or some other phenomenon, varies
according to $\cos \theta$. Obviously, we will be seeking a solution of the same form,
i.e. $\phi = \phi_1 \cos \theta$ (or possibly $\phi = \phi_1 \sin \theta$) in which $\phi_1 = \phi_1(r)$. Substituting this
assumed form of solution into equation 3.40

$$\left(\frac{d^2}{dr^2} + \frac{1}{r}\frac{d}{dr} - \frac{1}{r^2}\right)\left(\frac{d^2\phi_1}{dr^2} + \frac{1}{r}\frac{d\phi_1}{dr} - \frac{\phi_1}{r^2}\right)\cos \theta = 0$$

Noting that

$$\frac{d^2f}{dr^2} + \frac{1}{r}\frac{df}{dr} - \frac{f}{r^2} = \frac{d}{dr}\left[\frac{1}{r}\frac{d}{dr}(rf)\right]$$

let

$$\frac{d^2\phi_1}{dr^2} + \frac{1}{r}\frac{d\phi_1}{dr} - \frac{\phi_1}{r^2} = F_1(r)$$

then the differential equation becomes

$$\frac{d}{dr}\left[\frac{1}{r}\frac{d}{dr}(rF_1)\right] = 0$$

which yields

$$F_1(r) = \frac{C_1 r}{2} + \frac{C_2}{r} \quad \text{or} \quad \frac{d}{dr}\left[\frac{1}{r}\frac{d}{dr}(r\phi_1)\right] = \frac{C_1 r}{2} + \frac{C_2}{r}$$

and upon integrating,

$$\phi_1 = C_1 r^3 + C_2 r \log r + C_3 r + C_4/r$$

or

$$\phi = [C_1 r^3 + C_2 r \log r + C_3 r + C_4/r]\begin{Bmatrix} \sin\theta \\ \cos\theta \end{Bmatrix} \tag{3.45}$$

As a general case, $\phi = \phi_n \cos n\theta$ (or possibly $\phi = \phi_n \sin n\theta$), in which $\phi_n = \phi_n(r)$, may be assumed to be a stress function.

We examine the possibility of a function of θ alone:

$$\frac{1}{r^4}\frac{d^4\phi}{d\theta^4} = 0 \quad \text{or} \quad \frac{d^4\phi}{d\theta^4} = 0$$

which implies

$$\phi = C_1\theta^3 + C_2\theta^2 + C_3\theta + C_4$$

If we take $\phi = C_3\theta$

$$\sigma_\theta = 0, \quad \sigma_r = 0, \quad \tau_{r\theta} = \frac{C_3}{r^2}$$

This is the case of an annulus subjected to tangential shears at the edges.

Recalling that

$$\frac{d^2(f_1 f_2)}{d\theta^2} = f_1\frac{d^2 f_2}{d\theta^2} + 2\frac{df_1}{d\theta}\frac{df_2}{d\theta} + f_2\frac{d^2 f_1}{d\theta^2}$$

$$\frac{d^3(f_1 f_2)}{d\theta^3} = f_1\frac{d^3 f_2}{d\theta^3} + 3\frac{df_1}{d\theta}\frac{d^2 f_2}{d\theta^2} + 3\frac{d^2 f_1}{d\theta^2}\frac{df_2}{d\theta} + f_2\frac{d^3 f_1}{d\theta^3}$$

one may examine the possibility of a function

$$\phi = R(r)\theta \sin \theta \quad \text{or} \quad \phi = R(r)\theta \cos \theta.$$

It can be proved that

$$\phi = C_1 r\theta \sin \theta \quad \text{and} \quad \phi = C_2 r\theta \cos \theta$$

are also valid solutions.

3.17.3 Complete ring

In the case of symmetric problems, we have seen a general solution in the form

$$\phi = A \log r + Br^2 \log r + Cr^2 + D$$

In so far as stresses, and therefore associated strains and relative displacements, are concerned, the constant D is irrelevant and may be dropped. It can be seen that the symmetric problem of this type is really a third-order problem and not a fourth-order problem in as much as we may replace $d\phi/dr$ in the last set of parentheses of equation 3.41 by a new function, say Ψ, so that the differential equation becomes

$$\left(\frac{d^2}{dr^2} + \frac{1}{r}\frac{d}{dr}\right)\left(\frac{d\Psi}{dr} + \frac{\Psi}{r}\right) = 0$$

The stresses are

$$\sigma_\theta = \frac{d^2\phi}{dr^2} = \frac{d\Psi}{dr}, \quad \sigma_r = \frac{1}{r}\frac{d\phi}{dr} = \frac{\Psi}{r}, \quad \tau_{r\theta} = -\frac{\partial}{\partial r}\left(\frac{1}{r}\frac{\partial\phi}{\partial\theta}\right) = 0$$

Since A, B and C are arbitrary, we may inquire into the problems which could be solved if one or two of these constants were assigned zero values. Thus if $B = 0$, $\phi = A \log r + Cr^2$

$$\sigma_\theta = \frac{-A}{r^2} + 2C$$

$$\sigma_r = \frac{A}{r^2} + 2C$$

(3.46)

and $\tau_{r\theta} = 0$, thus confirming the symmetric character of the solution.

Suppose we have a complete ring (Figure 3.17). We can specify two (and only two) conditions; one at the inner boundary, one at the outer boundary, or alternatively, two at either boundary. Thus we specify the boundary condition as

$$\sigma_r = -P_o \quad \text{at} \quad r = a \quad \text{(outer)}$$

$$\sigma_r = -P_i \quad \text{at} \quad r = b \quad \text{(inner)}$$

These will lead to

$$\frac{A}{a^2} + 2C = -P_o$$

$$\frac{A}{b^2} + 2C = -P_i$$

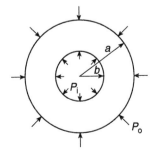

Figure 3.17 Complete ring under symmetric loading.

which yield

$$A = -\frac{(P_i - P_o)}{a^2 - b^2} a^2 b^2, \quad C = \frac{-a^2 P_o + b^2 P_i}{2(a^2 - b^2)}$$

and

$$\sigma_r = \frac{(P_i - P_o)}{(a^2 - b^2)r^2} a^2 b^2 + \frac{b^2 P_i - a^2 P_o}{a^2 - b^2}$$

$$\sigma_\theta = \frac{(P_o - P_i)}{(a^2 - b^2)r^2} a^2 b^2 + \frac{b^2 P_i - a^2 P_o}{a^2 - b^2}$$

(3.47)

Note that $\sigma_r + \sigma_\theta = 4C = $ constant, that is independent of r.

3.17.4   Alternative solution by the use of displacement

An alternative general solution to the symmetric problem of complete rings may be obtained if we formulate the problem in terms of different dependent variables, namely, the radial displacement, $u = u(r)$. In this approach, we utilize the strain–displacement relations,

$$\varepsilon_r = \frac{du}{dr}, \quad \varepsilon_\theta = \frac{u}{r}\left(+\frac{\partial v}{r\,\partial\theta}\right), \quad \text{where} \quad \frac{\partial v}{r\,\partial\theta} = 0$$

From the stress–strain relations, i.e. generalized Hooke's law, we may express the stresses in terms of the strains and therefore in terms of the displacements. We need now only substitute these results into one equilibrium equation to obtain the governing differential equation.

3.17.5   General case

Consider the general case, $\phi = f_n(r)\cos n\theta$. Then

$$\nabla^2 \phi = \left(\frac{d^2 f}{dr^2} + \frac{1}{r}\frac{df}{dr} - \frac{n^2 f}{r^2}\right)\cos n\theta$$

(3.48)

$$\nabla^4 \phi = \left(\frac{d^2}{dr^2} + \frac{1}{r}\frac{d}{dr} - \frac{n^2}{r^2}\right)\left(\frac{d^2 f}{dr^2} + \frac{1}{r}\frac{df}{dr} - \frac{n^2 f}{r^2}\right)\cos n\theta$$

(3.49)

which yields

$$\frac{d^4 f}{dr^4} + \frac{2}{r}\frac{d^3 f}{dr^3} - \frac{(1 + 2n^2)d^2 f}{r^2\,dr^2} + \frac{(1 + 2n^2)}{r^3}\frac{df}{dr} - \frac{n^2(4 - n^2)}{r^4} f = 0$$

(3.50)

Suppose we assume $f = r^\alpha$, then

$$\nabla^2 \nabla^2 \phi = [(\alpha - 2)(\alpha - 3) + (\alpha - 2) - n^2][\alpha(\alpha - 1) + \alpha - n^2]r^{(\alpha - 4)} \cos n\theta = 0$$

which yields

$$\alpha = 2 + n, \quad 2 - n$$

$$\alpha = n, \quad -n$$

so that

$$f = Ar^{n+2} + Br^{-n+2} + Cr^n + \mathrm{D}r^{-n}$$

(3.51)

This is valid for all homogeneous biharmonic equations having a frequency $n(n \neq 1, 0)$, so that

$$\phi = f(r) \begin{Bmatrix} \cos n\theta \\ \sin n\theta \end{Bmatrix} \tag{3.52}$$

This approach can be used for the cases of a partial ring or a sectorial plate by using $n\pi/\beta$ instead of n, where β is the angle between two radial boundaries.

3.17.6   Pure bending

Consider a curved bar with a uniform rectangular cross section with a unit thickness, with its circular axis bent in the plane of curvature by the moment, M, applied at both ends (Figure 3.18). In addition to the differential equation, the problem comprises the following boundary conditions:

$$\sigma_r = 0 \quad \text{on} \quad r = a, b \tag{i}$$

$$\tau_{r\theta} = 0 \quad \text{on} \quad r = a, b \tag{ii}$$

and the ends are to be subjected to a pure couple, or

$$\int_a^b \sigma_\theta r \, dr = -M \tag{iii}$$

The second boundary condition is assured by a symmetric formulation and solution. Timoshenko also indicates that the resultant force on the radial edges, i.e. the ends, must also be zero:

$$\int_a^b \sigma_\theta \, dr = 0$$

and consider this to be a boundary condition. However, this statement or requirement is identically satisfied by the statement of the problem and equilibrium considerations or equilibrium equations.

We have the general solution

$$\phi = A \log r + Br^2 \log r + Cr^2$$

from which we get the expressions for $\sigma_r(b)$ and $\sigma_r(a)$, i.e. (i) and (ii) above, and (iii) in terms of A, B and C. The integration may be performed in a straightforward manner. The result, together with the two relevant boundary conditions ((i) or (ii), above) provide a nonhomogeneous set of three simultaneous algebraic

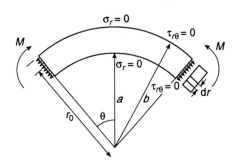

Figure 3.18 Curved bar under pure bending.

equations in A, B and C. Upon solving for A, B and C, and computing stresses, we get

$$\sigma(r) = -\frac{4M}{N}\left[\frac{a^2b^2}{r^2}\log\left(\frac{b}{a}\right) + b^2\log\left(\frac{r}{b}\right) + a^2\log\left(\frac{a}{r}\right)\right]$$

$$\sigma_\theta = -\frac{4M}{N}\left[-\frac{a^2b^2}{r^2}\log\left(\frac{b}{a}\right) + b^2\log\frac{r}{b} + a^2\log\left(\frac{a}{r}\right) + b^2 - a^2\right]$$

(3.53)

where

$$N = (b^2 - a^2)^2 - 4a^2b^2\left[\log\left(\frac{b}{a}\right)\right]^2$$

The type of solution given by equation 3.45 enables us to solve the problem of a (circular) curved beam loaded with a radial load at $\theta = 0$ (Figure 3.19). We must, however, as in the previous cases accept the distribution of stresses on the boundaries which lie on radial lines, but not the resultant of these stresses. We take the solution to be the upper one of the two trigonometric possibilities in equation 3.45. The boundary conditions are

$$\sigma_r(a) = \sigma_r(b) = 0 \quad \text{and} \quad \int_a^b \tau_{r\theta}\,dr = P$$

The solution is found as

$$\phi = (C_1r^3 + C_2r\log r + C_3r + C_4/r)\sin\theta$$

where

$$C_1 = \frac{P}{2(a^2 - b^2) + 2(a^2 + b^2)\log(b/a)}$$

$$C_2 = \frac{-P(a^2 + b^2)}{(a^2 - b^2) + (a^2 + b^2)\log(b/a)}$$

$$C_3 = 0$$

$$C_4 = \frac{-Pa^2b^2}{2(a^2 - b^2) + 2(a^2 + b^2)\log(b/a)}$$

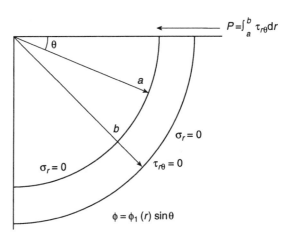

Figure 3.19 Curved beam under radial load.

from which we get

$$\tau_{r\theta(\theta=0)} = -\frac{P\left[r + \dfrac{a^2 b^2}{r^3} - \dfrac{1}{r}(a^2 + b^2)\right]\cos\theta(=1)}{a^2 - b^2 + (a^2 + b^2)\log(b/a)} \tag{3.54}$$

The shear stress distribution depends upon the ratio b/a, approaching the parabolic distribution of elementary beam theory as $b/a \to 1$, i.e. as the radius becomes large compared to the depth of the ring. For larger ratios of b/a, i.e. for rings which are deep in comparison with the radius, the distribution is a sort of 'unsymmetric parabola', having its peak shifted toward the inner edge, and also increased. It also turns out that the normal stress distribution based upon Winkler's theory (plane sections remain plane and the strain distribution considering difference in length of fibers is essentially hyperbolic) gives good agreement with the 'correct' elasticity theory.

3.17.7 Plate with a hole

A solution in the form given by equation 3.52 may be used for the problem of a semi-infinite plate with a hole, whose radius is a, loaded with a stress $+s$ in the $\theta = 0$ direction. Taking $\phi = f(r)\cos 2\theta$, and boundary conditions

$$\sigma_r(a) = 0, \quad \tau_{r\theta}(a) = 0$$

$$\sigma_r(\infty) = (s/2)\cos 2\theta$$

$$\tau_{r\theta}(\infty) = -(s/2)\sin 2\theta$$

and combining the solution for a thick ring, the stresses are found to be

$$\sigma_r = \frac{s}{2}\left(1 - \frac{a^2}{r^2}\right) + \frac{s}{2}\left(1 - 4\frac{a^2}{r^2} + 3\frac{a^4}{r^4}\right)\cos 2\theta$$

$$\sigma_\theta = \frac{s}{2}\left(1 + \frac{a^2}{r^2}\right) - \frac{s}{2}\left(1 + 3\frac{a^4}{r^4}\right)\cos 2\theta \tag{3.55}$$

$$\tau_{r\theta} = -\frac{s}{2}\left(1 + 2\frac{a^2}{r^2} - 3\frac{a^4}{r^4}\right)\sin 2\theta$$

At $\theta = 90°$, and at the edge of the hole,

$$\sigma_\theta = 3S$$

The stress concentration factor due to the hole is 3.

3.18 Problems of concentrated loads

3.18.1 Concentrated load on a semi-infinite elastic body

One of the forms of solution of the biharmonic compatibility equation is

$$\phi = Kr\theta \sin\theta \tag{3.56}$$

where K is an arbitrary constant.

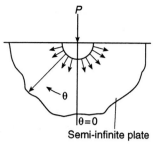

Figure 3.20 Concentrated load on semi-infinite elastic body.

The stresses are

$$\sigma_r = \frac{1}{r}\frac{\partial \phi}{\partial r} + \frac{1}{r^2}\frac{\partial^2 \phi}{\partial \theta^2} = 2K\frac{\cos\theta}{r}$$

$$\sigma_\theta = \frac{\partial^2 \phi}{\partial r^2} = 0$$

$$\tau_{r\theta} = -\frac{\partial}{\partial r}\left(\frac{1}{r}\frac{\partial \phi}{\partial \theta}\right) = 0$$

On a (semi-) circle of radius r, only normal stresses act at the circular boundary (Figure 3.20). Equating the vertical resultant of these stresses to the applied load,

$$\int_{-\pi/2}^{\pi/2} \frac{2K}{r}\cos\theta(r\,d\theta)\cos\theta = -P$$

and we obtain

$$\phi = -\frac{P}{\pi}r\theta\sin\theta$$

$$\sigma_r = -\frac{2P}{\pi}\frac{\cos\theta}{r} \tag{3.57}$$

$$\sigma_\theta = \tau_{r\theta} = 0$$

At $r = 0$, the stresses (according to this theory) are infinitely large. The physical acceptance of this as a plausible conclusion rests upon the fact that the material within a small semicircle directly under the load becomes plastic. We may assume, however, at the circular boundary of this semicircle (or some slightly larger semicircle) a distribution of radial loads is applied which has a resultant equal to the applied load. If the material has a reasonably high yield strength, or if the load P is small, then the radius of this semicircle may be taken as quite small. In any event, by Saint Venant's principle, the solution which has been obtained is valid at distances which are large in comparison with the radius of the deleted semicircle. **Saint Venant's principle** states that if the forces acting on a small portion of the surface of an elastic body are replaced by another statically equivalent system of forces acting on the same portion of the surface, this redistribution of loading produces substantial changes in the stresses locally but has a negligible effect on the stresses at distances which are large in comparison with the linear dimensions of the surface on which the forces are changed.

It is interesting to observe that, for a given σ_r at a point on the vertical line passing through the origin, exactly the same value of σ_r is obtained for all points lying on a circle tangent to the upper boundary at the origin. This results from the fact that $\cos\theta/r$ is a constant for any such circle and is, in fact, the reciprocal of the diameter of the circle.

The calculation of displacements by integration of the appropriate relations between stress, strain and displacement is straightforward. The interesting result is that the displacements in the horizontal direction at the top edge are constant

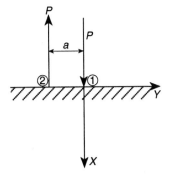

Figure 3.21 Moment on a semi-infinite plate.

in absolute magnitude, being to the left when at a point lying to the right of a compression load, and to the right at a point to the left of the load. If P is a tensile load, the opposite result is obtained.

3.18.2 Moment on a semi-infinite plate

The stresses due to each of a pair of equal but opposite loads at a distance a apart may be obtained by superposition (Figure 3.21). However, if a is quite small, the net condition may be obtained as a differential solution. With the appropriate transformations between the polar and rectangular coordinates, the previously obtained solution may be written as $\phi(x, y)$, $\sigma(x, y)$, etc. and these would apply to load ①. For the second load, we may take the corresponding solution as $-\phi(x, y + a)$, etc. where the minus sign comes from the reversed direction and the a in $y + a$ comes from the fact that all points are at an increased horizontal distance from the load. The resultant or net effect is

$$-\phi(x, y + a) + \phi(x, y)$$

or

$$\frac{-\phi(x, y + a) + \phi(x, y)a}{a}$$

When $a \to 0$, this is precisely $-(\partial\phi/\partial y)\mathrm{d}y$.

3.18.3 Concentrated load parallel to the edge of a semi-infinite plate

We may use the same stress function and stress formulas for a concentrated load parallel to the edge of a semi-infinite plate, if we measure θ from the line of action of the load (Figure 3.22). Since $\sigma_\theta = \tau_{r\theta} = 0$ throughout, the appropriate boundary conditions are satisfied. Moreover, if we consider a (semi)circle of a radius r with center at the load and take the resultants in the parallel and normal directions:

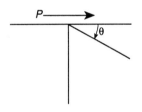

Figure 3.22 Horizontal load on a semi-infinite plate.

- In the direction of the load,

$$\int_0^\pi \frac{2P}{\pi r} \cos\theta(r \sin\theta\ \mathrm{d}\theta) = \frac{2P}{\pi} \int_0^\pi \sin\theta\cos\theta\ \mathrm{d}\theta = 0$$

- Normal to the load,

$$\int_0^\pi \frac{2P}{\pi r} \cos\theta(r \sin\theta\ \mathrm{d}\theta) = \frac{2P}{\pi} \int_0^\pi \sin\theta\cos\theta\ \mathrm{d}\theta = 0$$

Moment equilibrium is automatically satisfied since all stresses, being radial, have zero moment arm about the load point.

3.18.4 Concentrated load in an arbitrary direction

A load in an arbitrary direction may be resolved into normal and parallel components (Figure 3.23):

$$P_1 = P \cos\alpha, \quad P_2 = P \sin\alpha$$

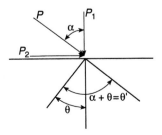

Figure 3.23 Concentrated load in an arbitrary direction.

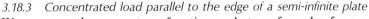

The stresses due to the combination may be obtained by superposition:

- Due to P_1,

$$\sigma_r = \frac{-2P_1}{\pi r} \cos \theta = \frac{-2P \cos \alpha \cos \theta}{\pi r}$$

- Due to P_2,

$$\sigma_r = \frac{-2P_2}{\pi r} \cos\left(\theta + \frac{\pi}{2}\right) = \frac{2P \sin \alpha \sin \theta}{\pi r}$$

Hence, due to the combination or resultant load,

$$\sigma_r = \frac{-2P}{\pi r} (\cos \alpha \cos \theta - \sin \alpha \sin \theta)$$

$$= \frac{-2P}{\pi r} \cos(\theta + \alpha) \tag{3.58}$$

But $\theta + \alpha$ is the angle between the direction of the resultant load and the radial line to the point at which stresses are being computed. Thus

$$\sigma_r = \frac{-2P}{\pi r} \cos \theta'$$

$$\sigma_\theta = \tau_{r\theta} = 0 \tag{3.59}$$

where θ' is measured from the line of action of the oblique load.

3.18.5 Vertical concentrated load at the apex of an infinite wedge

With an appropriately adjusted constant, the 'same' solution may be used to calculate stresses in a wedge which is loaded at the apex (Figure 3.24). We have

$$\sigma_r = \frac{-kP}{\pi r} \cos \theta$$

$$\sigma_\theta = \tau_{r\theta} = 0$$

which satisfies the conditions on the straight boundary. On an arbitrary circular boundary of radius r, the resultant of the stresses in the axial direction is

$$\int_{-\alpha}^{\alpha} \left(-\frac{kP}{\pi r} \cos \theta\right) r \cos \theta \, d\theta = -\frac{kP}{\pi} \int_{-\alpha}^{\alpha} \cos^2 \theta \, d\theta = -\frac{kP}{2\pi}[2\alpha + \sin 2\alpha] = -P$$

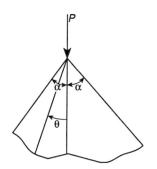

Figure 3.24 Vertical concentrated load at the apex of an infinite wedge.

Hence,

$$k = \frac{2\pi}{2\alpha + \sin 2\alpha}$$

Finally

$$\sigma_r = -\frac{P}{(\alpha + \frac{1}{2} \sin 2\alpha)} \frac{\cos \theta}{r} \tag{3.60}$$

$$\sigma_\theta = \tau_{r\theta} = 0$$

Figure 3.25 Horizontal concentrated load at the apex of an infinite wedge.

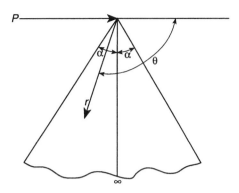

Figure 3.25 Horizontal concentrated load at the apex of an infinite wedge.

3.18.6   Horizontal concentrated load at the apex of an infinite wedge

The fact that the resultant moment of the stresses on a circular boundary vanishes when calculated about the apex is obvious. Furthermore, since the radial stresses are antisymmetric about the bisector of the wedge, the resultant force in this direction vanishes (Figure 3.25).

$$\int_{\pi/2-\alpha}^{\pi/2+\alpha} \frac{kP}{\pi} \cos^2 \theta \; d\theta = P$$

from which we get

$$k = \frac{\pi}{(\alpha - \frac{1}{2}\sin 2\alpha)}$$

$$\sigma_r = \frac{-P}{(\alpha - \frac{1}{2}\sin 2\alpha)} \cdot \frac{\cos \theta}{r} \qquad (3.61)$$

$$\sigma_\theta = \tau_{r\theta} = 0$$

3.18.7   Concentrated load in an arbitrary direction at the apex of an infinite wedge

When the load P is applied in an arbitrary direction as shown in Figure 3.26, equations 3.60 and 3.61 are combined and we obtain

$$\sigma_r = -\frac{P}{r}(k_1 \cos \theta - k_2 \sin \theta) \qquad (3.62)$$

where

$$k_1 = \frac{\cos \beta}{(\alpha + \frac{1}{2}\sin 2\alpha)}$$

$$k_2 = \frac{\sin \beta}{(\alpha - \frac{1}{2}\sin 2\alpha)}$$

The solutions for **concentrated load problems** can be used to handle the problems of column loads, deep beam with concentrated loads, circular disks and others.

Figure 3.26 Concentrated load in an arbitrary direction at the apex of an infinite wedge.

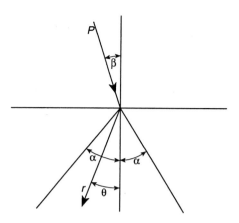

3.18.8 Semi-infinite plate under arbitrary loading

The problem shown in Figure 3.27 can be solved by integrating the solution for concentrated load. Alternatively, the solution to a particular problem may be obtained through the use of a stress function and this solution may be adapted by superposition to the desired case.

Figure 3.27 Semi-infinite plate under arbitrary loading.

3.19 Strain energy and principle of virtual work

3.19.1 Strain energy of a differential element

Consider an elastic differential element $dxdydz$. By Clapeyron's theorem (or by the law of conservation of energy), the strain energy stored in the element as a consequence of the tractions, σ_x, σ_y, σ_z, τ_{xy}, τ_{yz} and τ_{zx} is equal to the work done by these stresses. Thus, the strain energy density is

$$V_0 = \tfrac{1}{2}[\sigma_x\varepsilon_x + \sigma_y\varepsilon_y + \sigma_z\varepsilon_z + \tau_{xy}\gamma_{xy} + \tau_{yz}\gamma_{yz} + \tau_{zx}\gamma_{zx}] \tag{3.63}$$

With the use of Hooke's law

$$V_0 = \frac{1}{2E}(\sigma_x^2 + \sigma_y^2 + \sigma_z^2) - \frac{v}{E}(\sigma_x\sigma_y + \sigma_y\sigma_z + \sigma_z\sigma_x) + \frac{1}{2G}(\tau_{xy}^2 + \tau_{yz}^2 + \tau_{zx}^2) \tag{3.64}$$

Using inversion of Hooke's law (equation 3.6),

$$V_0 = \frac{1}{2}[(\lambda e + 2G\varepsilon_x)\varepsilon_x + (\lambda e + 2G\varepsilon_y)\varepsilon_y + (\lambda e + 2G\varepsilon_z)\varepsilon_z] + \frac{G}{2}(\gamma_{xy}^2 + \gamma_{yz}^2 + \gamma_{zx}^2)$$

or

$$V_0 = \frac{1}{2}\lambda e^2 + G(\varepsilon_x^2 + \varepsilon_y^2 + \varepsilon_z^2) + \frac{G}{2}(\gamma_{xy}^2 + \gamma_{yz}^2 + \gamma_{zx}^2) \tag{3.65}$$

Since all strains appear as squares, it can be seen that the strain energy density is never negative, i.e. a 'positive definite' function. A contrary possibility may exist if $v > 1/2$. From equation 3.64, we may obtain

$$\frac{\partial V_0}{\partial \sigma_x} = \frac{1}{E}(\sigma_x - v\sigma_y - v\sigma_z) = \varepsilon_x$$

so that

$$\frac{\partial V_0}{\partial \sigma_x} = \varepsilon_x, \quad \frac{\partial V_0}{\partial \sigma_y} = \varepsilon_y, \quad \frac{\partial V_0}{\partial \sigma_z} = \varepsilon_z$$

$$\frac{\partial V_0}{\partial \tau_{xy}} = \gamma_{xy}, \quad \frac{\partial V_0}{\partial \tau_{yz}} = \gamma_{yz}, \quad \frac{\partial V_0}{\partial \tau_{zx}} = \gamma_{zx}$$

(3.66)

From equation 3.65, we obtain

$$\frac{\partial V_0}{\partial \varepsilon_x} = \frac{2vEe(\partial e/\partial \varepsilon_x)}{2(1+v)(1-2v)} + \frac{E}{1+v}\varepsilon_x = \lambda e + 2G\varepsilon_x = \sigma_x$$

so that

$$\frac{\partial V_0}{\partial \varepsilon_x} = \sigma_x, \quad \frac{\partial V_0}{\partial \varepsilon_y} = \sigma_y, \quad \frac{\partial V_0}{\partial \varepsilon_z} = \sigma_z$$

$$\frac{\partial V_0}{\partial \gamma_{xy}} = \tau_{xy}, \quad \frac{\partial V_0}{\partial \gamma_{yz}} = \tau_{yz}, \quad \frac{\partial V_0}{\partial \gamma_{zx}} = \tau_{zx}$$

(3.67)

The total strain energy of a body is

$$V = \iiint V_0 \, dx \, dy \, dz$$

(3.68)

3.19.2 Principle of virtual work

Suppose a particle with weight F is moved upward from y to $y + \delta y$ from the datum line. The potential energy, referred to the datum, is increased by the amount $F(\delta y)$. However, the work done by the force F is $-F(\delta y)$.

Consider a particle in equilibrium. Assume that the particle is given a small displacement in an arbitrary direction so that the components of the displacement in the x-, y- and z-directions are δ_u, δ_v and δ_w, respectively. Then the work done by the forces which are acting upon the particle will be

$$\delta W = \left(\sum F_x\right)\delta_u + \left(\sum F_y\right)\delta_v + \left(\sum F_z\right)\delta_w$$

(3.69a)

where $\sum F_x$ is the sum of the forces (or components) which are acting in the x-direction upon the particle and so on. However, since the particle was stated to be in equilibrium,

$$\sum F_x = \sum F_y = \sum F_z = 0$$

and consequently,

$$\delta W = 0$$

(3.69b)

A body is a collection of particles. Since the above equation holds for each particle in the body, it must also hold for the collection of particles, i.e. for the body itself. We therefore have the **principle of virtual work**.

If a body is in equilibrium, the total work done during a virtual displacement (or a virtual change in the configuration of the body) vanishes.

The principle of virtual work may be used in a number of ways, including:

1. to test whether a body or system is in equilibrium;
2. to determine the equilibrium configuration or equilibrium conditions;
3. (as a generalization of 2) to establish the governing differential equations.

3.20   Examples of use of the energy method

3.20.1   Taut string under concentrated load

Consider a taut string loaded by P as shown in Figure 3.28. The length of the deflected string between the left end and the load may be taken as

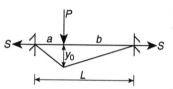

Figure 3.28 Taut string under a concentrated load.

$$l_1 = \sqrt{a^2 + y_0^2} = a\left[1 + \left(\frac{y_0}{a}\right)^2\right]^{1/2} \fallingdotseq a\left[1 + \frac{1}{2}\left(\frac{y_0}{a}\right)^2\right]$$

The increase in length of this segment is

$$l_1 - a = \frac{a}{2}\left(\frac{y_0}{a}\right)^2$$

Similarly the increase in length of the right segment is

$$l_2 - b = \frac{b}{2}\left(\frac{y_0}{b}\right)^2$$

Assuming that the deflections are small, the above relation holds. Furthermore, if we assume that S is sufficiently large, we may assume that S does not change significantly during either the loading or a subsequent virtual displacement of the system. It has been indicated that the increase in length of the deflected string is

$$\lambda = \frac{y_0^2}{2}\left(\frac{1}{a} + \frac{1}{b}\right)$$

Suppose we now assume that a virtual displacement is imposed on the system, and suppose we take this displacement to be a small increase in y_0. Then, the virtual change in length of the initially deflected string will be

$$\delta\lambda = \frac{\partial\lambda}{\partial y_0}\,\delta y_0 = y_0\left(\frac{1}{a} + \frac{1}{b}\right)\delta y_0$$

Consequently, the work done by the internal load, S, during the virtual displacement is

$$-S_0\, y_0\left(\frac{1}{a} + \frac{1}{b}\right)\delta y_0$$

During the virtual displacement, the load P moves through the additional (small) distance δy_0, so that the work done by P during the virtual displacement is $P\delta y_0$.

Now, by the principle of virtual work,

$$P\delta y_0 - S y_0\left(\frac{1}{a} + \frac{1}{b}\right)\delta y_0 = 0, \quad \text{or} \quad \left[P - S y_0\left(\frac{1}{a} + \frac{1}{b}\right)\right]\delta y_0 = 0$$

Since δy_0 is arbitrary,

$$P - S y_0\left(\frac{1}{a} + \frac{1}{b}\right) = 0$$

which yields

$$y_0 = \frac{Pab}{S(a + b)} \tag{3.70}$$

3.20.2 Taut string under general transverse load

As one of the simpler illustrations of the use of the principle of virtual work in the derivation of differential equations, consider, again, a taut string but with a general transverse load, $q = q(x)$. From Figure 3.29,

$$ds = \sqrt{(dx)^2 + \left(\frac{dy}{dx}\right)^2 dx^2} = dx\sqrt{1 + \left(\frac{dy}{dx}\right)^2} \approx dx\left[1 + \frac{1}{2}\left(\frac{dy}{dx}\right)^2\right]$$

For this differential element, therefore, the difference between its actual (inclined) length and its original (projected) length is

$$d\lambda = dx\left[1 + \frac{1}{2}\left(\frac{dy}{dx}\right)^2\right] - dx = \frac{1}{2}\left(\frac{dy}{dx}\right)^2 dx$$

The total change in length for the entire string is

$$\lambda = \int d\lambda = \frac{1}{2}\int_0^L \left(\frac{dy}{dx}\right)^2 dx$$

Assuming that the string is initially stressed to a high value of S so that S may be assumed not to change as a consequence of the small displacements, the work done by S in going to the final configuration may be taken as

$$-S\lambda = -\frac{S}{2}\int_0^L \left(\frac{dy}{dx}\right)^2 dx$$

Let us now superimpose a virtual displacement corresponding to a small change in y continuous over the span and vanishing at the ends. Then virtual

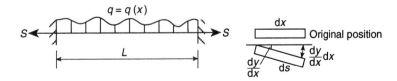

Figure 3.29 Taut string under general transverse load.

work done by S will be

$$-\frac{S}{2}\delta\int_0^L \left(\frac{dy}{dx}\right)^2 dx = -\frac{S}{2}\int_0^L \delta\left(\frac{dy}{dx}\right)^2 dx = -\frac{S}{2}\int_0^L 2\left(\frac{dy}{dx}\right)\delta\left(\frac{dy}{dx}\right) dx$$

$$= -S\int_0^L \left(\frac{dy}{dx}\right)\frac{d(\delta y)}{dx} dx$$

Recall that

$$d(uv) = u\,dv + v\,du \quad \text{and} \quad \int v\,du = uv - \int u\,dv$$

Differentiating by parts, take

$$du = \frac{d(\delta y)}{dx}dx, \quad v = \frac{dy}{dx}, \quad u = \delta y, \quad dv = \frac{d^2 y}{dx^2}dx$$

Thus

$$\int_0^L \left(\frac{dy}{dx}\right)\frac{d(\delta y)}{dx}dx = \left(\frac{dy}{dx}\right)\delta y \bigg|_0^L - \int_0^L \delta y\left(\frac{d^2 y}{dx^2}\right)dx$$

Since the virtual displacement was taken in such a form that δy vanishes at both ends, the first term on the right vanishes, so that

$$\int_0^L \left(\frac{dy}{dx}\right)\frac{d(\delta y)}{dx}dx = -\int_0^L \delta y\left(\frac{d^2 y}{dx^2}\right)dx$$

and the work done by S therefore is

$$S\int_0^L \delta y\left(\frac{d^2 y}{dx^2}\right)dx$$

During the virtual displacement, the distribution of load does work of magnitude

$$\int_0^L q(\delta y)\,dx$$

so that we obtain

$$\int_0^L \left[q + S\left(\frac{d^2 y}{dx^2}\right)\right]\delta y\,dx = 0$$

and

$$q + S\left(\frac{d^2 y}{dx^2}\right) = 0 \tag{3.71}$$

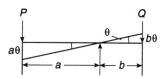

Figure 3.30 The lever.

3.20.3 Lever

To establish the **law of the lever**, we assume that the system (which we postulate to be in equilibrium) is given a virtual displacement corresponding to the small tilt, θ (Figure 3.30). Then the downward displacement of P is $a\theta$ and the upward displacement of Q is $b\theta$ which is opposite to the direction in which Q is acting. The total work done, during this virtual displacement is $P(a\theta) + Q(-b\theta) = (Pa - Qb)\theta$ and by the principle of virtual work, this vanishes. Since θ is arbitrary, the total work must vanish independently of θ, and

$$Pa - Qb = 0$$

or

$$Pa = Qb$$

3.21 Castigliano's theorem

Suppose we have a two-dimensional body (Figure 3.31) which is loaded at the boundary with tractions having x and y components, P_x and P_y (per unit of length).

The direction cosines of the outer normal are denoted by

$$N_x = \cos \alpha = l$$
$$N_y = \sin \alpha = m$$

(3.11)

and

$$dx = ds(\sin \alpha) = m\,ds$$
$$dy = ds(\cos \alpha) = l\,ds$$

Using equation 3.11 for the direction cosines, the boundary conditions are given by equations 3.25 as

$$P_x = l\sigma_x + m\tau_{xy}$$
$$P_y = m\sigma_y + l\tau_{xy}$$

(3.25)

If we change the boundary tractions by the amounts δP_x and δP_y, then the boundary conditions are

$$l(\sigma_x + \delta\sigma_x) + m(\tau_{xy} + \delta\tau_{xy}) = P_x + \delta P_x$$
$$m(\sigma_y + \delta\sigma_x) + l(\tau_{xy} + \delta\tau_{xy}) = P_y + \delta P_y$$

(3.72)

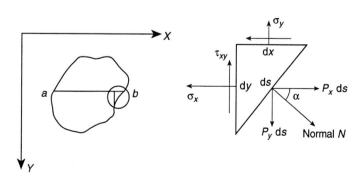

Figure 3.31 Direction of boundary tractions.

since there will be changes in the magnitude of the stresses throughout the body. Subtracting the first set (equations 3.25) from the second (equations 3.72) yields the change in the boundary condition (or conditions) due to a variation in the boundary tractions

$$l\delta\sigma_x + m\delta\tau_{xy} = \delta P_x$$
$$m\delta\sigma_y + l\delta\tau_{xy} = \delta P_y$$

(3.73)

For the original problem, the equilibrium equations (assuming loads applied only at the boundary) are, from equations 3.24 with $X = Y = 0$,

$$\frac{\partial\sigma_x}{\partial x} + \frac{\partial\tau_{xy}}{\partial y} = 0$$
$$\frac{\partial\sigma_y}{\partial y} + \frac{\partial\tau_{xy}}{\partial x} = 0$$

(3.74a)

For the case of an increase in boundary loads, these become

$$\frac{\partial(\sigma_x + \delta\sigma_x)}{\partial x} + \frac{\partial(\tau_{xy} + \delta\tau_{xy})}{\partial y} = 0$$
$$\frac{\partial(\sigma_y + \delta\sigma_y)}{\partial y} + \frac{\partial(\tau_{xy} + \delta\tau_{xy})}{\partial x} = 0$$

(3.74b)

Subtracting the first set (equations 3.74a) from the second (equations 3.74b), the equilibrium equations for the variation of the boundary forces become

$$\frac{\partial\delta\sigma_x}{\partial x} + \frac{\partial\delta\tau_{xy}}{\partial y} = 0$$
$$\frac{\partial\delta\sigma_y}{\partial y} + \frac{\partial\delta\tau_{xy}}{\partial x} = 0$$

(3.75)

We may express the total differential of the strain energy density as

$$\delta V_0 = \frac{\partial V_0}{\partial\sigma_x}\delta\sigma_x + \frac{\partial V_0}{\partial\sigma_y}\delta\sigma_y + \frac{\partial V_0}{\delta\tau_{xy}}\delta\tau_{xy}$$

(3.76)

Furthermore, we have

$$\frac{\partial V_0}{\partial\sigma_x} = \varepsilon_x, \quad \frac{\partial V_0}{\partial\sigma_y} = \varepsilon_y, \quad \frac{\partial V_0}{\delta\tau_{xy}} = \gamma_{xy}$$

Consequently

$$\delta V_0 = \varepsilon_x\delta\sigma_x + \varepsilon_y\delta\sigma_y + \gamma_{xy}\delta\tau_{xy}$$
$$= \frac{\partial u}{\partial x}\delta\sigma_x + \frac{\partial v}{\partial y}\delta\sigma_y + \left(\frac{\partial u}{\partial y} + \frac{\partial v}{\partial x}\right)\delta\tau_{xy}$$

(3.77)

and for the entire body (assuming it to have a unit thickness),

$$\delta V = \iint \delta V_0 \, dx \, dy = \iint \left[\frac{\partial u}{\partial x}\delta\sigma_x + \frac{\partial v}{\partial y}\delta\sigma_y + \left(\frac{\partial u}{\partial y} + \frac{\partial v}{\partial x}\right)\delta\tau_{xy}\right] dx \, dy$$

Consider the first term in the integral,

$$\iint \frac{\partial u}{\partial x} \delta\sigma_x \, dx \, dy = \int dy \int \frac{\partial u}{\partial x} \delta\sigma_x \, dx$$

The second integral may be integrated by parts,

$$\int \frac{\partial u}{\partial x} \delta\sigma_x \, dx = u\delta\sigma_x \Big|_{xa}^{xb} - \int u \frac{\partial \delta\sigma_x}{\partial x} \, dx$$

Consequently,

$$\iint \frac{\partial u}{\partial x} \delta\sigma_x \, dx \, dy = \int [u\delta\sigma_x]_{xa}^{xb} \, dy - \int dy \int u \frac{\partial \delta\sigma_x}{\partial x} \, dx$$

$$= \int [u\delta\sigma_x] l \, ds - \iint u \frac{\partial \delta\sigma_x}{\partial x} \, dx \, dy$$

A similar procedure may be applied to the second and third terms in the integral for δV.

$$\iint \frac{\partial v}{\partial y} \delta\sigma_y \, dx \, dy = \int v\delta\sigma_y m \, ds - \iint \frac{v}{\partial y} \frac{\partial \delta\sigma_y}{\partial y} \, dx \, dy$$

$$\iint \left(\frac{\partial u}{\partial y} + \frac{\partial v}{\partial x}\right) \delta\tau_{xy} \, dx \, dy = \int dy \int \frac{\partial v}{\partial x} \delta\tau_{xy} \, dx + \int dx \int \frac{\partial u}{\partial y} \delta\tau_{xy} \, dy = \int v\delta\tau_{xy} l \, ds$$

$$+ \int u\delta\tau_{xy} m \, ds - \iint v \frac{\partial \delta\tau_{xy}}{\partial x} \, dx \, dy - \iint u \frac{\partial \delta\tau_{xy}}{\partial y} \, dx \, dy$$

Substituting into the expression for the variation of V,

$$\delta V = \int u(l\delta\sigma_x + m\delta\tau_{xy}) \, ds + \int v(m\delta\sigma_y + l\delta\tau_{xy}) \, ds$$

$$- \iint \left[u\left(\frac{\partial \delta\sigma_x}{\partial x} + \frac{\partial \delta\tau_{xy}}{\partial y}\right) + v\left(\frac{\partial \delta\sigma_y}{\partial y} + \frac{\partial \delta\tau_{xy}}{\partial x}\right) \right] dx \, dy$$

By virtue of the equilibrium equations, the integrand in the double integral is zero and therefore the value of the double integral is zero. Furthermore, by virtue of the final form of the boundary conditions equation 3.73, the remaining integral may be written as

$$\delta V = \int (u\delta P_x + v\delta P_y) \, ds \tag{3.78}$$

For discrete loads, $P_1, P_2, ..., P_n$, etc., on the boundary having components P_x^i, P_y^i in the x- and y-directions,

$$\delta V = \sum [u_i \delta P_x^i + v_i \delta P_y^i] \tag{3.79}$$

By setting all but one (or one term) equal to zero, we can solve for the desired displacement (at the load),

$$\frac{\delta V}{\delta P_x^j} = u_j$$

or

$$\frac{\partial V}{\partial P_x^j} = u_j \tag{3.80}$$

This is called **Castigliano's theorem**.

3.22 Equilibrium equations in three dimensions

Figure 3.32 shows stresses in the positive directions. Taking X, Y and Z to be the volumetric intensity of body forces, and summing all forces in the three coordinate directions, we obtain

$$\frac{\partial \sigma_x}{\partial x} + \frac{\partial \tau_{yx}}{\partial y} + \frac{\partial \tau_{zx}}{\partial z} + X = 0$$

$$\frac{\partial \sigma_y}{\partial y} + \frac{\partial \tau_{zy}}{\partial z} + \frac{\partial \tau_{xy}}{\partial x} + Y = 0 \tag{3.81}$$

$$\frac{\partial \sigma_z}{\partial z} + \frac{\partial \tau_{xz}}{\partial x} + \frac{\partial \tau_{yz}}{\partial y} + Z = 0$$

By taking moments about lines parallel to each of the coordinate axes, we would merely verify that the shear stresses on two faces normal to each other are equal, i.e. $\tau_{xy} = \tau_{yx}$, etc. (equation 3.1). The stress–strain relations are given by the generalized Hooke's law (equations 3.5). If the temperature and

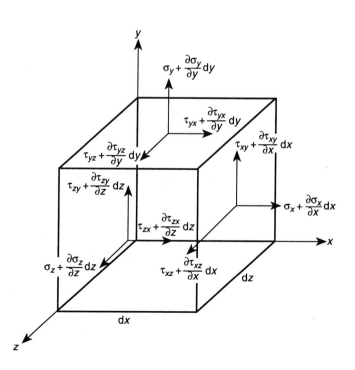

Figure 3.32 Positive direction of stresses.

hygrothermal effects are neglected, we have the expressions of stresses in terms of strains as shown by equations 3.6:

$$\sigma_x = \lambda e + 2G\varepsilon_x$$
$$\sigma_y = \lambda e + 2G\varepsilon_y \quad (3.6)$$
$$\sigma_z = \lambda e + 2G\varepsilon_z$$

where $G = \frac{1}{2}\mu$.

In addition to the stress–strain relations for the normal stresses and strains, we have the shear stress–strain relations,

$$\tau_{xy} = G\gamma_{xy}, \quad \tau_{yz} = G\gamma_{yz}, \quad \tau_{zx} = G\gamma_{zx}$$
$$\gamma_{xy} = \frac{1}{G}\tau_{xy}, \quad \gamma_{yz} = \frac{1}{G}\tau_{yz}, \quad \gamma_{zx} = \frac{1}{G}\tau_{zx} \quad (3.82)$$

The relations between displacements and strains are given by equations 3.2,

$$\varepsilon_x = \frac{\partial u}{\partial x}, \quad \varepsilon_y = \frac{\partial v}{\partial y}, \quad \varepsilon_z = \frac{\partial w}{\partial z}$$

$$\gamma_{xy} = \frac{\partial u}{\partial y} + \frac{\partial v}{\partial x}, \quad \gamma_{yz} = \frac{\partial v}{\partial z} + \frac{\partial w}{\partial y}, \quad \gamma_{zx} = \frac{\partial w}{\partial x} + \frac{\partial u}{\partial z} \quad (3.2)$$

Substituting these relations into the first of equations 3.81, we get

$$G\left(\frac{\partial^2 u}{\partial x^2} + \frac{\partial^2 u}{\partial y^2} + \frac{\partial^2 u}{\partial z^2}\right) + \lambda\frac{\partial e}{\partial x} + G\left(\frac{\partial^2 u}{\partial x^2} + \frac{\partial^2 v}{\partial x\,\partial y} + \frac{\partial^2 w}{\partial x\,\partial z}\right) + X = 0$$

Noting that

$$\frac{\partial^2 u}{\partial x^2} + \frac{\partial^2 v}{\partial x\,\partial y} + \frac{\partial^2 w}{\partial x\,\partial z} = \frac{\partial}{\partial x}\left(\frac{\partial u}{\partial x} + \frac{\partial v}{\partial y} + \frac{\partial w}{\partial z}\right) = \frac{\partial e}{\partial x}$$

and using the abbreviation

$$\nabla^2 = \frac{\partial^2}{\partial x^2} + \frac{\partial^2}{\partial y^2} + \frac{\partial^2}{\partial z^2}$$

the foregoing may be written as

$$G\nabla^2 u + (\lambda + G)\frac{\partial e}{\partial x} + X = 0$$

In a similar manner, or by permutation of the coordinates, we may obtain

$$G\nabla^2 v + (\lambda + G)\frac{\partial e}{\partial y} + Y = 0$$
$$\quad (3.83)$$
$$G\nabla^2 w + (\lambda + G)\frac{\partial e}{\partial z} + Z = 0$$

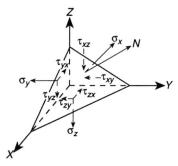

Figure 3.33 Boundary tractions.

3.23 Boundary conditions in three-dimensional problems

In addition to the **equilibrium equations** (and the **compatibility equations** if the problem is to be formulated in terms of stresses) the **boundary conditions** must be stated. These can be of two general forms:

- displacements specified on the boundary;
- tractions (i.e. stresses) specified on the boundary (Figure 3.33).

There is also the possibility of '**mixed**' boundary conditions, tractions specified on a portion of the boundary, and displacements specified on the remainder of the boundary, or the conditions such that a part of condition specified by displacements.

If the boundary values of the tractions are given, we will have at each point on the boundary a distributed loading having a resultant **intensity** of the same prescribed magnitude and direction. This intensity may be resolved into **components** (of intensity) having magnitudes X, Y, Z in the directions of x-, y- and z-axes. Note that this can be done even though the element of the surface at the point in question is not normal to any of the coordinate directions.

Let N be the outer normal to the (boundary) surface, and let the direction cosines of N (with respect to the x-, y- and z-axes, respectively) be l, m and n. Then, if dA is the area of the element of the surface on the boundary, the areas of the three faces normal to the coordinate planes are $l\,dA$, $m\,dA$ and $n\,dA$, where $l\,dA$ is the area of the face which is normal to the x-axis, $m\,dA$ is the area of the face normal to the y-axis, and $n\,dA$ is the area of the face normal to the z-axis.

If we sum forces in the x-direction,

$$X\,dA - \sigma_x l\,dA - \tau_{yx} m\,dA - \tau_{zx} n\,dA = 0$$

or

$$l\sigma_x + m\tau_{yx} + n\tau_{zx} = X$$

and similarly,

$$l\tau_{xy} + m\sigma_y + n\tau_{zy} = Y$$
$$l\tau_{xz} + m\tau_{yz} + n\sigma_z = Z$$

(3.84)

These are the boundary conditions in terms of tractions.

3.24 Compatibility equations in three-dimensional problems

Compatibility equations can be deduced by integrating around an arbitrary closed path. They can also be deduced by operating upon the strains (in terms of displacements) as was done in two-dimensional elasticity. We have the strain–displacement relationship (equation 3.2).

The first shear strain formula implies

$$\frac{\partial^2 \gamma_{xy}}{\partial x\,\partial y} = \frac{\partial^3 u}{\partial x\,\partial y^2} + \frac{\partial^3 v}{\partial x^2\,\partial y} = \frac{\partial^2 \varepsilon_x}{\partial y^2} + \frac{\partial^2 \varepsilon_y}{\partial x^2}$$

Hence, by permutation of subscripts, we get three compatibility equations:

$$\frac{\partial^2 \gamma_{xy}}{\partial x \, \partial y} = \frac{\partial^2 \varepsilon_x}{\partial y^2} + \frac{\partial^2 \varepsilon_y}{\partial x^2}$$

$$\frac{\partial^2 \gamma_{yz}}{\partial y \, \partial z} = \frac{\partial^2 \varepsilon_y}{\partial z^2} + \frac{\partial^2 \varepsilon_z}{\partial y^2} \qquad (3.85\text{a})$$

$$\frac{\partial^2 \gamma_{zx}}{\partial x \, \partial z} = \frac{\partial^2 \varepsilon_z}{\partial x^2} + \frac{\partial^2 \varepsilon_x}{\partial z^2}$$

To obtain three additional equations of compatibility, we proceed in the following manner:

$$\frac{\partial^2 \varepsilon_x}{\partial y \, \partial z} = \frac{\partial^3 u}{\partial x \, \partial y \, \partial z}$$

$$\frac{\partial \gamma_{yz}}{\partial x} = \frac{\partial}{\partial x}\left(\frac{\partial v}{\partial z} + \frac{\partial w}{\partial y} \right) = \frac{\partial^2 v}{\partial x \, \partial z} + \frac{\partial^2 w}{\partial x \, \partial y}$$

$$\frac{\partial \gamma_{xz}}{\partial y} = \frac{\partial^2 w}{\partial x \, \partial y} + \frac{\partial^2 u}{\partial y \, \partial z}$$

$$\frac{\partial \gamma_{xy}}{\partial z} = \frac{\partial^2 u}{\partial y \, \partial z} + \frac{\partial^2 v}{\partial x \, \partial z}$$

Note that

$$\frac{\partial \gamma_{xy}}{\partial z} - \frac{\partial \gamma_{yz}}{\partial x} = \frac{\partial^2 u}{\partial y \, \partial z} + \frac{\partial^2 v}{\partial x \, \partial z} - \frac{\partial^2 v}{\partial x \, \partial z} - \frac{\partial^2 w}{\partial x \, \partial y} = \frac{\partial^2 u}{\partial y \, \partial z} - \frac{\partial^2 w}{\partial x \, \partial y},$$

and

$$\frac{\partial \gamma_{xy}}{\partial z} - \frac{\partial \gamma_{yz}}{\partial x} + \frac{\partial \gamma_{xz}}{\partial y} = \frac{\partial^2 u}{\partial y \, \partial z} - \frac{\partial^2 w}{\partial x \, \partial y} + \frac{\partial^2 w}{\partial x \, \partial y} + \frac{\partial^2 u}{\partial y \, \partial z} = 2 \frac{\partial^2 u}{\partial y \, \partial z}$$

Hence

$$2 \frac{\partial^2 \varepsilon_x}{\partial y \, \partial z} = 2 \frac{\partial^3 u}{\partial x \, \partial y \, \partial z} = \frac{\partial}{\partial x}\left(-\frac{\partial \gamma_{yz}}{\partial x} + \frac{\partial \gamma_{zx}}{\partial y} + \frac{\partial \gamma_{xy}}{\partial z} \right)$$

Again, permutation gives two additional equations, so that

$$2 \frac{\partial^2 \varepsilon_x}{\partial y \, \partial z} = \frac{\partial}{\partial x}\left(-\frac{\partial \gamma_{yz}}{\partial x} + \frac{\partial \gamma_{zx}}{\partial y} + \frac{\partial \gamma_{xy}}{\partial z} \right)$$

$$2 \frac{\partial^2 \varepsilon_y}{\partial x \, \partial z} = \frac{\partial}{\partial y}\left(\frac{\partial \gamma_{yz}}{\partial x} - \frac{\partial \gamma_{zx}}{\partial y} + \frac{\partial \gamma_{xy}}{\partial z} \right) \qquad (3.85\text{b})$$

$$2 \frac{\partial^2 \varepsilon_z}{\partial x \, \partial y} = \frac{\partial}{\partial z}\left(\frac{\partial \gamma_{yz}}{\partial x} + \frac{\partial \gamma_{zx}}{\partial y} - \frac{\partial \gamma_{xy}}{\partial z} \right)$$

The compatibility equations may be written in terms of stresses. By Hooke's law,

$$\varepsilon_y = \frac{1}{E}\left[(1 + v)\sigma_y - v\Theta\right]$$

in which $\Theta = \sigma_x + \sigma_y + \sigma_z$. Similarly

$$\varepsilon_z = \frac{1}{E}\left[(1 + v)\sigma_z - v\Theta\right]$$

We also have

$$\gamma_{yz} = \frac{2(1 + v)}{E}\tau_{yz}$$

Now, consider the first of the compatibility equations,

$$\frac{\partial^2 \varepsilon_y}{\partial z^2} + \frac{\partial^2 \varepsilon_z}{\partial y^2} = \frac{\partial^2 \gamma_{yz}}{\partial y\, \partial z}$$

and substitute the above stress–strain relations. Then

$$(1 + v)\left(\frac{\partial^2 \sigma_y}{\partial z^2} + \frac{\partial^2 \sigma_z}{\partial y^2}\right) - v\left(\frac{\partial^2 \Theta}{\partial z^2} + \frac{\partial^2 \Theta}{\partial y^2}\right) = 2(1 + v)\frac{\partial^2 \tau_{yz}}{\partial y\, \partial z} \qquad \text{(i)}$$

Recalling the second and third of the equilibrium equations 3.81, we have

$$\frac{\partial \tau_{zy}}{\partial z} = -\frac{\partial \sigma_y}{\partial y} - \frac{\partial \tau_{xy}}{\partial x} - Y$$

$$\frac{\partial \tau_{yz}}{\partial y} = -\frac{\partial \sigma_z}{\partial z} - \frac{\partial \tau_{xz}}{\partial x} - Z$$

We differentiate each of the above and add to get

$$2\frac{\partial^2 \tau_{yz}}{\partial y\, \partial z} = -\frac{\partial^2 \sigma_y}{\partial y^2} - \frac{\partial^2 \sigma_z}{\partial z^2} - \frac{\partial^2 \tau_{xy}}{\partial x\, \partial y} - \frac{\partial^2 \tau_{xz}}{\partial x\, \partial z} - \frac{\partial Y}{\partial y} - \frac{\partial Z}{\partial z}$$

or

$$2\frac{\partial^2 \tau_{yz}}{\partial y\, \partial z} = -\frac{\partial^2 \sigma_y}{\partial y^2} - \frac{\partial^2 \sigma_z}{\partial z^2} - \frac{\partial}{\partial x}\left(\frac{\partial \tau_{xy}}{\partial y} + \frac{\partial \tau_{xz}}{\partial z}\right) - \frac{\partial Y}{\partial y} - \frac{\partial Z}{\partial z}$$

Now, the equilibrium equation in the x-direction may be written as

$$\frac{\partial \tau_{yx}}{\partial y} + \frac{\partial \tau_{zx}}{\partial z} = -X - \frac{\partial \sigma_x}{\partial x}$$

so that, upon substituting into the previous equation, we get

$$2\frac{\partial^2 \tau_{yz}}{\partial y\, \partial z} = \frac{\partial^2 \sigma_x}{\partial x^2} - \frac{\partial^2 \sigma_y}{\partial y^2} - \frac{\partial^2 \sigma_z}{\partial z^2} + \frac{\partial X}{\partial x} - \frac{\partial Y}{\partial y} - \frac{\partial Z}{\partial z}$$

Substituting this result into the right side of equation (i), yields

$$(1 + v)\left(\frac{\partial^2\sigma_y}{\partial z^2} + \frac{\partial^2\sigma_z}{\partial y^2}\right) - v\left(\frac{\partial^2\Theta}{\partial y^2} + \frac{\partial^2\Theta}{\partial z^2}\right)$$

$$= (1 + v)\left(\frac{\partial^2\sigma_x}{\partial x^2} - \frac{\partial^2\sigma_y}{\partial y^2} - \frac{\partial^2\sigma_z}{\partial z^2} + \frac{\partial X}{\partial x} - \frac{\partial Y}{\partial y} - \frac{\partial Z}{\partial z}\right) \quad \text{(ii)}$$

In a similar manner, we may obtain

$$(1 + v)\left(\frac{\partial^2\sigma_z}{\partial x^2} + \frac{\partial^2\sigma_x}{\partial z^2}\right) - v\left(\frac{\partial^2\Theta}{\partial x^2} + \frac{\partial^2\Theta}{\partial z^2}\right)$$

$$= (1 + v)\left(\frac{\partial^2\sigma_y}{\partial y^2} - \frac{\partial^2\sigma_x}{\partial x^2} - \frac{\partial^2\sigma_z}{\partial z^2} + \frac{\partial Y}{\partial y} - \frac{\partial X}{\partial x} - \frac{\partial Z}{\partial z}\right)$$

and

$$(1 + v)\left(\frac{\partial^2\sigma_x}{\partial y^2} + \frac{\partial^2\sigma_y}{\partial x^2}\right) - v\left(\frac{\partial^2\Theta}{\partial x^2} + \frac{\partial^2\Theta}{\partial y^2}\right)$$

$$= (1 + v)\left(\frac{\partial^2\sigma_z}{\partial z^2} - \frac{\partial^2\sigma_x}{\partial x^2} - \frac{\partial^2\sigma_y}{\partial y^2} + \frac{\partial Z}{\partial z} - \frac{\partial X}{\partial x} - \frac{\partial Y}{\partial y}\right)$$

Adding the three equations,

$$(1 + v)\left|\begin{matrix}\frac{\partial^2\sigma_x}{\partial x^2} + \frac{\partial^2\sigma_y}{\partial x^2} + \frac{\partial^2\sigma_z}{\partial x^2} \\[2mm] + \frac{\partial^2\sigma_x}{\partial y^2} + \frac{\partial^2\sigma_y}{\partial y^2} + \frac{\partial^2\sigma_z}{\partial y^2} \\[2mm] + \frac{\partial^2\sigma_x}{\partial z^2} + \frac{\partial^2\sigma_y}{\partial z^2} + \frac{\partial^2\sigma_z}{\partial z^2}\end{matrix}\right| - 2v\left|\begin{matrix}\frac{\partial^2\Theta}{\partial x^2} \\[2mm] + \frac{\partial^2\Theta}{\partial y^2} \\[2mm] + \frac{\partial^2\Theta}{\partial z^2}\end{matrix}\right| = -(1 + v)\left(\frac{\partial X}{\partial x} + \frac{\partial Y}{\partial y} + \frac{\partial Z}{\partial z}\right)$$

or using the abbreviation

$$\nabla^2 = \frac{\partial^2}{\partial x^2} + \frac{\partial^2}{\partial y^2} + \frac{\partial^2}{\partial z^2}$$

$$(1 + v)\nabla^2\Theta - 2v\nabla^2\Theta = -(1 + v)\left(\frac{\partial X}{\partial x} + \frac{\partial Y}{\partial y} + \frac{\partial Z}{\partial z}\right)$$

or

$$(1 - v)\nabla^2\Theta = -(1 + v)\left(\frac{\partial X}{\partial x} + \frac{\partial Y}{\partial y} + \frac{\partial Z}{\partial z}\right)$$

Rewriting,

$$\nabla^2\Theta = -\frac{(1 + v)}{(1 - v)}\left(\frac{\partial X}{\partial x} + \frac{\partial Y}{\partial y} + \frac{\partial Z}{\partial z}\right) \quad \text{(iii)}$$

Writing equation (ii) in the form

$$(1 + v)\left[-\frac{\partial^2 \sigma_x}{\partial x^2} + \left(\frac{\partial^2 \sigma_y}{\partial y^2} + \frac{\partial^2 \sigma_y}{\partial z^2} \right) + \left(\frac{\partial^2 \sigma_z}{\partial y^2} + \frac{\partial^2 \sigma_z}{\partial z^2} \right) - v\nabla^2\Theta + v\frac{\partial^2\Theta}{\partial x^2} \right]$$

$$= (1 + v)\left(\frac{\partial X}{\partial x} - \frac{\partial Y}{\partial y} - \frac{\partial Z}{\partial z} \right)$$

we have

$$\nabla^2\Theta - (1 + v)\nabla^2\sigma_x - \frac{\partial^2\Theta}{\partial x^2} = (1 + v)\left(\frac{\partial X}{\partial x} - \frac{\partial Y}{\partial y} - \frac{\partial Z}{\partial z} \right)$$

Substituting $\nabla^2\Theta$ of equation (iii) into the above,

$$\nabla^2\sigma_x + \frac{1}{1 + v}\frac{\partial^2\Theta}{\partial x^2} = -\frac{v}{1 - v}\left(\frac{\partial X}{\partial x} + \frac{\partial Y}{\partial y} + \frac{\partial Z}{\partial z} \right) - 2\frac{\partial X}{\partial x} \qquad (3.86)$$

Two similar equations are obtained by interchanging subscripts and co-ordinates. Three equations are also obtained by starting from the remaining compatibility equations. They will have the form

$$\nabla^2\tau_{yz} + \frac{1}{1 + v}\frac{\partial^2\Theta}{\partial y\,\partial z} = -\left(\frac{\partial Z}{\partial y} + \frac{\partial Y}{\partial z} \right) \qquad (3.87)$$

or in general,

$$\nabla^2\tau_{ij} + \frac{1}{1 + v}\frac{\partial^2\Theta}{\partial x_i\,\partial x_j} = \cdots \qquad (3.86, 3.87)$$

When the body forces vanish, the set of compatibility equations becomes

$$(1 + v)\nabla^2\sigma_x + \frac{\partial^2\Theta}{\partial x^2} = 0, \quad (1 + v)\nabla^2\tau_{xy} + \frac{\partial^2\Theta}{\partial x\,\partial y} = 0,$$

$$(1 + v)\nabla^2\sigma_y + \frac{\partial^2\Theta}{\partial y^2} = 0, \quad (1 + v)\nabla^2\tau_{yz} + \frac{\partial^2\Theta}{\partial y\,\partial z} = 0, \qquad (3.88)$$

$$(1 + v)\nabla^2\sigma_z + \frac{\partial^2\Theta}{\partial z^2} = 0, \quad (1 + v)\nabla^2\tau_{zx} + \frac{\partial^2\Theta}{\partial x\,\partial z} = 0$$

or

$$(1 + v)\nabla^2\tau_{ij} + \frac{\partial^2\Theta}{\partial x_i\,\partial x_j} = 0$$

If a solution to a (three-dimensional) elasticity problem is obtained in terms of stresses, it may be desirable (or, in the case of mixed boundary conditions, necessary) to compute displacements. To do so, one first relates strains to stresses by Hooke's law in its generalized form (equations 3.5 and 3.82). The displacement–strain relations (equations 3.2) yield a set of differential equations in terms of the strains.

3.25   Saint Venant's solution of the problem of torsion of a prismatic bar

The stresses on the ends of the bar (Figure 3.34) must be distributed over these areas precisely as the stresses are distributed over an interior section normal to the z-axis.

Saint Venant's assumptions were equivalent to

$$\sigma_z = 0$$

$$\sigma_x = 0$$

$$\sigma_y = 0$$

$$\tau_{xy} = 0$$

From equilibrium equations,

$$\frac{\partial \sigma_x}{\partial x} + \frac{\partial \tau_{yx}}{\partial y} + \frac{\partial \tau_{zx}}{\partial z} = 0$$

$$\frac{\partial \sigma_y}{\partial y} + \frac{\partial \tau_{xy}}{\partial x} + \frac{\partial \tau_{zy}}{\partial z} = 0$$

$$\frac{\partial \sigma_z}{\partial z} + \frac{\partial \tau_{xz}}{\partial x} + \frac{\partial \tau_{yz}}{\partial y} = 0$$

Since $\sigma_x = \tau_{xy} = \sigma_y = 0$,

$$\frac{\partial \tau_{zx}}{\partial z} = \frac{\partial \tau_{zy}}{\partial z} = 0$$

which means that τ_{zx} and τ_{zy} are independent of z. The last of the above equations yields

$$\frac{\partial \tau_{xz}}{\partial x} + \frac{\partial \tau_{yz}}{\partial y} = 0$$

Assume a stress function, ϕ, such that $\tau_{xz} = \partial\phi/\partial y$, $\tau_{yz} = -\partial\phi/\partial x$. We see that ϕ satisfies the differential equation.

Since there is no stress (in any direction) on the lateral boundary, equilibrium requires, by summing forces in the z-direction,

$$\tau_{yz}dx + \tau_{xz}(-dy) = 0$$

Note that $\sigma_z = 0$ was implicitly taken into consideration. Note the equilibrium in the x- and y-directions for the boundary element are satisfied by the assumption previously made,

$$\sigma_x = \sigma_y = \tau_{xy} = 0$$

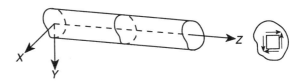

Figure 3.34 Bar with noncircular section.

In terms of the stress function, the boundary condition becomes

$$\frac{\partial \phi}{\partial x} dx + \frac{\partial \phi}{\partial y} dy = 0 \quad \text{or} \quad \frac{\partial \phi}{\partial x} \frac{dx}{ds} + \frac{\partial \phi}{\partial y} \frac{dy}{ds} = = 0$$

$$\tau_{yz} \frac{dx}{ds} - \tau_{xz} \frac{dy}{ds} = 0 \quad \text{or} \quad \frac{\partial \phi}{\partial s} = 0$$

which states that ϕ is constant on the boundary of the cross-section.

In terms of stresses, the six compatibility equations for no body force are

$$(1 + v)\nabla^2 \tau_{ij} + \frac{\partial^2 \Theta}{\partial x_i \, \partial x_j} = 0 \tag{3.88}$$

Of these, the first three and the last are satisfied by the assumption $\sigma_x = \sigma_y = \sigma_z = \tau_{xy} = 0$. Hence, the relevant compatibility equation reduces to

$$\frac{\partial^2 \tau_{yz}}{\partial x^2} + \frac{\partial^2 \tau_{yz}}{\partial y^2} = 0 \rightarrow \frac{-\partial}{\partial x}\left(\frac{\partial^2 \phi}{\partial x^2} + \frac{\partial^2 \phi}{\partial y^2}\right) = 0$$

$$\frac{\partial^2 \tau_{xz}}{\partial x^2} + \frac{\partial^2 \tau_{xz}}{\partial y^2} = 0 \rightarrow \frac{\partial}{\partial y}\left(\frac{\partial^2 \phi}{\partial x^2} + \frac{\partial^2 \phi}{\partial y^2}\right) = 0 \tag{3.89}$$

which indicates that

$$\frac{\partial^2 \phi}{\partial x^2} + \frac{\partial^2 \phi}{\partial y^2} = \text{Const.} = F = -2G\theta \tag{3.90}$$

where θ is the angle of twist.

M_t, in Figure 3.35 is the resultant applied torque. Considering the upper boundary surface,

$$\iint [-y(\tau_{zx}) + x(\tau_{zy})] \, dx \, dy = M_t$$

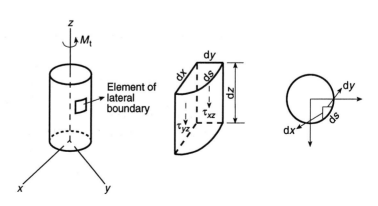

Figure 3.35 An element of the lateral boundary.

or in terms of the stress functions, $\tau_{zx} = \partial\phi/\partial y$ and $\tau_{zy} = -\partial\phi/\partial x$

$$-\iint\left(\frac{\partial\phi}{\partial y}y + \frac{\partial\phi}{\partial x}x\right)dx\,dy = M_t$$

$$\int_{y_1}^{y_2}\frac{\partial\phi}{\partial y}y\,dy = \phi y|_{y_1}^{y_2} - \int\phi\,dy$$

therefore

$$\iint\frac{\partial\phi}{\partial y}y\,dx\,dy = \int\left(\phi y|_{y_1}^{y_2} - \int\phi\,dy\right)dx$$

The first term, namely $\phi y|_{y_1}^{y_2}$ involves only the values of ϕ at the boundary point (having y-coordinates y_1 and y_2). However, from the previous analysis of the boundary conditions on the lateral boundary, it was established that ϕ has a constant magnitude on the boundary, and it may be shown that any constant value on the boundary may be assigned. If we assign zero, then

$$\iint\frac{\partial\phi}{\partial y}y\,dx\,dy = -\iint\phi\,dx\,dy$$

$$\iint\frac{\partial\phi}{\partial x}x\,dx\,dy = -\iint\phi\,dx\,dy$$

Therefore

$$-\iint\left(\frac{\partial\phi}{\partial y}y + \frac{\partial\phi}{\partial x}x\right)dx\,dy = 2\iint\phi\,dx\,dy$$

or

$$M_t = 2\iint\phi\,dx\,dy \tag{3.91}$$

In addition to equations 3.90 and 3.91 we have

$$\phi = 0 \tag{3.92}$$

on the boundary.

3.26 Membrane analogy for torsional problems

Exact analysis of torsion in bars having other than circular cross-section is very difficult. L. Prandtl, in 1903, introduced the membrane analogy which is very useful in approximate mathematical solutions of torsional problems. This analogy is based on the similarity of the differential equation of torsion to that of a membrane subjected to uniform lateral pressure, P. The differential equation of the stretched membrane is

$$\frac{\partial^2 w}{\partial x^2} + \frac{\partial^2 w}{\partial y^2} = \nabla^2 w = -\frac{P}{S} \tag{3.93}$$

where w and S are deflections and membrane tensions, respectively. If we let

$$2G\theta = kP/S$$

where k is a constant of proportionality, the stress function, $\phi = kw$. Then

$$\tau_{zx} = \frac{\partial \phi}{\partial y} = k\frac{\partial w}{\partial y}, \quad \text{and} \quad \tau_{zy} = -k\frac{\partial w}{\partial x} \tag{3.94}$$

and the torque

$$M_t = 2 \times (\text{volume under deflected membrane}) \tag{3.95}$$

In the case of thin-walled hollow members, we assume $2G\theta = P/S$. Since the walls of the bar are thin, the slope of the membrane in the direction of the thickness of the wall of the cross-section is nearly constant and may be approximated at any point by q/t, where q is the height of the membrane plateau, the magnitude of which is commonly called **shear flow**, and t is the thickness of the tube at that point. This is equivalent to assuming the shearing stress is uniform throughout the wall (uniform thickness) of the thin walled bar. The volume under the membrane may be approximated by qA_c, where A_c is the area closed by the center line of the tube wall. Then

$$M_t = 2 \times (\text{volume under membrane}) = 2qA_c \tag{3.96}$$

Compatibility conditions around the closed path along the center line of the tube wall yields

$$\oint \tau_{zs}ds = \oint (q/t)ds = 2G\theta A_c \tag{3.97}$$

For the given torque, M_t, equations 3.96 and 3.97 enable us to obtain the torsional shearing stress at any point in the tube and the angle of twist of the tube per unit length.

3.27 Asymmetric bending of prismatic beams

With the following assumptions:

1. the material is homogeneous,
2. the material obeys Hooke's law,
3. transverse stresses may be neglected,
4. plane sections (normal to the longitudinal or z-axis) before bending remain plane during bending (Bernoulli–Euler assumptions),
    we can say

$$\sigma = ax + by + c \tag{3.98}$$

where σ is the longitudinal stress and positive values of σ represent tensile stress.
From Figure 3.36, let

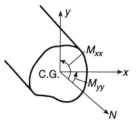

Figure 3.36 Direction of positive moment.

* N be the resultant of the longitudinal force taken to be applied at the center of gravity ($+N$ represents tensile force).
* M_{xx}, M_{yy} be the bending moments about the x- and y-axis, respectively ($+M$ tends to produce compressive stress at locations corresponding to positive values of the other coordinate).

For equilibrium, we must have

$$\int_A \sigma \, dA = N$$

$$\int_A \sigma x \, dA = -M_{yy} \qquad (3.99)$$

$$\int_A \sigma y \, dA = -M_{xx}$$

where A is the cross-sectional area. Substituting the general expression for σ:

$$a \int x \, dA + b \int y \, dA + c \int dA = N$$

$$a \int x^2 \, dA + b \int xy \, dA + c \int x \, dA = -M_{yy} \qquad (3.100)$$

$$a \int xy \, dA + b \int y^2 \, dA + c \int y \, dA = -M_{xx}$$

$\int x \, dA$ and $\int y \, dA$ represent the static moments of the cross-sectional area about the y- and x-axis, respectively, and since these axes pass through the centroid, these integrals vanish. We also observe that

$$\int_A dA = A$$

$$\int_A x^2 \, dA = I_{yy} = \text{moment of inertia about } y\text{-axis}$$

$$\int_A y^2 \, dA = I_{xx} = \text{moment of inertia about } x\text{-axis}$$

$$\int_A xy \, dA = I_{xy} = \text{product of inertia}$$

The equilibrium equations thus become

$$cA = N$$
$$aI_{yy} + bI_{xy} = -M_{yy}$$
$$aI_{xy} + bI_{xx} = -M_{xx}$$

Solving for a, b and c,

$$c = N/A$$

$$a = -\frac{M_{yy}I_{xx} - M_{xx}I_{xy}}{I_{xx}I_{yy} - I_{xy}^2}$$

$$b = -\frac{M_{xx}I_{yy} - M_{yy}I_{xy}}{I_{xx}I_{yy} - I_{yy}^2}$$

Finally, therefore,

$$\sigma = -\frac{M_{yy}I_{xx} - M_{xx}I_{xy}}{I_{xx}I_{yy} - I_{xy}^2} x - \frac{M_{xx}I_{yy} - M_{yy}I_{xy}}{I_{xx}I_{yy} - I_{xy}^2} y + \frac{N}{A} \qquad (3.101)$$

This equation was obtained by Swain, in 1870.

3.28 Asymmetric bending of a beam with longitudinal stringers

Consider a thin-walled bar reinforced by the stringer longerons as shown in Figure 3.37. The longerons are assumed to carry axial and bending loads and the thin walls are assumed to carry shear stress only. Let F_i be the axial force due to axial load and bending of the entire section on the ith member, so that

$$F_i = \sigma_i a_i \qquad (3.102)$$

where σ_i is the stress at the center of gravity of the ith longeron, and a_i is the cross-sectional area of the ith longeron. $q_{i,i+1}$ is defined as the 'shear flow' in the web between the ith and $(i+1)$th longeron, and is equal to the average shear stress times the thickness. Summing forces in the longitudinal direction,

$$\left(\frac{dF_i}{dz} dz + F_i\right) - F_i + (q_{i,i+1}) dz - (q_{i-1,i}) dz = 0$$

or

$$\Delta q_i = (q_{i,i+1}) - (q_{i-1,i}) = -\frac{dF_i}{dz} \qquad (3.103)$$

where Δq_i is the 'jump' or increase in shear flow produced at or by the ith longeron. Recalling the formula for stress due to bending moment (equation 3.101),

$$F_i = -a_i\left(\frac{M_{yy}I_{xx} - M_{xx}I_{xy}}{I_{xx}I_{yy} - I_{xy}^2} x + \frac{M_{xx}I_{yy} - M_{yy}I_{xy}}{I_{xx}I_{yy} - I_{xy}^2} y\right)$$

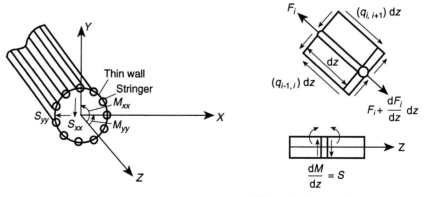

Figure 3.37 Beams with stringers.

Differential element of a general longeron
and adjoining web or skin

it follows that

$$\frac{dF_i}{dz} = -a_i\left(\frac{\frac{dM_{yy}}{dz}I_{xx} - \frac{dM_{xx}}{dz}I_{xy}}{I_{xx}I_{yy} - I_{xy}^2}x + \frac{\frac{dM_{xx}}{dz}I_{yy} - \frac{dM_{yy}}{dz}I_{xy}}{I_{xx}I_{yy} - I_{xy}^2}y\right)$$

We have

$$\frac{dM_{xx}}{dz} = S_{xx}, \quad \frac{dM_{yy}}{dz} = S_{yy}$$

so that

$$\Delta q_i = a_i\left(\frac{S_{yy}I_{xx} - S_{xx}I_{xy}}{I_{xx}I_{yy} - I_{xy}^2}x + \frac{S_{xx}I_{yy} - S_{yy}I_{xy}}{I_{xx}I_{yy} - I_{xy}^2}y\right) \tag{3.104}$$

3.29   Torsion of thin walled structures with longitudinal stringers

The magnitude, direction and sense of the resultant of a constant shear flow acting on an arbitrary curve are given by the closing line of the curve. To determine the position of the resultant, we may take a moment about a suitably chosen point (Figure 3.38). It has been shown that the moment due to a constant shear flow is

$$M_t = 2Aq \tag{3.96}$$

where A is the area enclosed by the curve and the two radii from the reference or moment point to the extremities of the curve. If we take the reference point at one end of the curve, and denote the length of the closing line by b, and let e be the normal distance from the closing line to the resultant, R, then

$$e = \frac{M_t}{R} = \frac{2Aq}{bq} = \frac{2A}{b} \tag{3.105}$$

which is twice the average ordinate of the curvature, using the closing line as a reference axis.

As the radius vector moves so that its outer end is in contact with successive points on the upper part of the curve from a to b (Figure 3.39(a)), it sweeps out continuously increasing central angles (and areas) about the reference point, Q. Thus, for these curve points, we may take dA as positive. As we move along the lower curve from b to a, the radius vector now sweeps out negative angles

Figure 3.38   (a) Locating the shear center. (b) Shear flow and shear center.

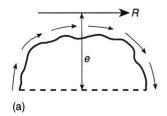
(a)

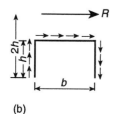
(b)

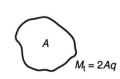

Figure 3.39 A scheme to obtain
$T = 2qA$.

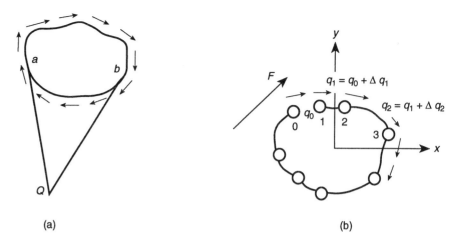

(a)   (b)

(in general) so that here the dAs are (in general) negative. Thus in taking the moment or torque about the exterior point, Q, the result is

$$T = 2q \int dA = 2qA$$

where the final A is just the area enclosed by the curve.

The force F can be resolved into components S_{xx} and S_{yy} perpendicular to x- and y-axes respectively. The 'jump' or increment in shear flow at each longeron may now be calculated by means of the formula given by equation 3.104. While this formula gives us the increments, it does not determine the absolute value of the shear flows. However, this is precisely the generality required to satisfy all equilibrium conditions. q_0 is taken as an unknown quantity and, therefore at this stage, the absolute shear flows are undetermined. However, the shear flow in a segment between the nth and $(n + 1)$th longerons may be written as

$$q_n = q_0 + \sum_{i=1}^{n} \Delta q_i \qquad (3.106)$$

Now, it will be found that the Δqs will satisfy the conditions $\sum F_x = 0$, $\sum F_y = 0$ and, furthermore, the added constant shear flow q_0 will not upset these two conditions. We are thus free to select q_0 so that (in combination with the Δqs) the normal equilibrium condition is satisfied.

This is done in the following way. An arbitrary (moment) reference point is chosen. (This may, for convenience, be the centroid or it may be a point on the line of action of the applied load, F.) Suppose we take the centroid (or some other interior point). We draw the radii to the longerons and measure or compute each A_i. We also measure e, the moment arm of F about the reference point. For moment equilibrium, we must now have

$$2\sum q_n A_n = Fe$$

But

$$q_n = q_0 + \sum_{i=1}^{n} \Delta q_i$$

Hence

$$2\sum\left(q_0 + \sum_{i=1}^{n} \Delta q_i\right)A_n = Fe, \quad \text{or} \quad 2\sum q_0 A_n + 2\sum A_n\left(\sum_{i=1}^{n} \Delta q_i\right) = Fe$$

Here, $\sum A_n$ = total enclosed area A, hence

$$q_0 = \frac{1}{2A}\left[Fe - 2\sum A_n\left(\sum_{i=1}^{n} \Delta q_i\right)\right]$$

where the $\sum$ without limit means the entire set.

3.30 Determination of rate of twist of a cell

It is assumed that the shear flows due to the applied loading are known. If we let U be the strain energy (for the 1 cm length of the shell in Figure 3.40) due to the actual shear flows plus the shear flow due to a virtual torque, T, and calculate $\partial U/\partial T$, the result, when properly evaluated, will be the rate of twist. We have the shear strain, $\gamma = \tau/G$ where $\tau = q/t$ is the shear stress and t is the thickness of the wall. Hence $\gamma = q/Gt$.

The strain energy density or strain energy per square centimeter is $\frac{1}{2}q\gamma = q^2/2Gt$. The strain energy (for a 1 cm length of beam) around the entire cell is

$$U = \frac{1}{2}\oint \frac{q^2}{Gt}\, ds \tag{3.107}$$

where the line integral goes around the entire cell.

Let

$$q = q_a + q_T$$

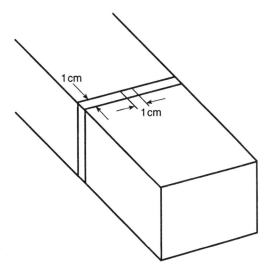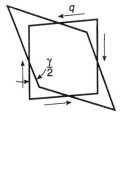

Figure 3.40   Angle of twist of a box cell.

where q_a is the actual shear flow and q_T the shear flow due to a superimposed (and generally nonexistent or virtual) torque $T = T/2A$. Hence

$$U = \frac{1}{2} \oint \frac{(q_a + T/2A)^2}{Gt} \, ds$$

Now, by Castigliano's theorem,

$$\theta = \text{rate of twist} = \frac{\partial U}{\partial T}$$

$$= \frac{1}{2} \oint \frac{2(q_a + T/2A)}{Gt} \frac{1}{2A} \, ds$$

Since the real shear flows are represented by q_a, we may drop the subscript. Furthermore, T is a virtual torque and hence has zero value. Thus

$$\theta = \frac{1}{2A} \oint \frac{q \, ds}{Gt} \tag{3.108}$$

3.31   Shear center of beams with longerons

If the load, S (Figure 3.41), is applied in a direction normal to the axis of symmetry, we can determine a set of shear flows for which we put the system in vertical and horizontal equilibrium by using the Δq formula. The resultant of these shear flows will then be equivalent to the applied loads S (in fact, S

Figure 3.41 Locating the shear center of a beam with stringers.

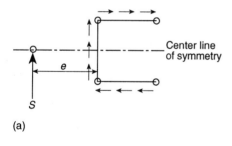

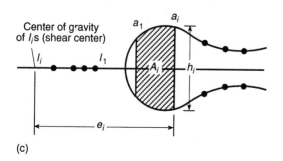

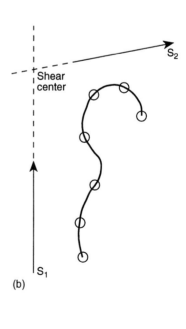

need not be applied normal to the axis of symmetry in order to satisfy this part of the equilibrium condition). In general, the computed shear flows will produce a certain couple about an arbitrary point, and by dividing this couple by the value of S, the position of the applied load will thus be found for which no 'torsional' stress distribution is required to be superimposed.

The intersection of the required line of action of S and the axis of symmetry is called (for this symmetric section) the 'shear center' (Figure 3.41(b)). To find the shear center for an asymmetric section, we first apply a load S_1, in an arbitrarily chosen direction (it may be convenient to take the direction of the x- or y-axis), then using the Δq formula and calculating the moment or the shear flows about some reference point, determine the line of action of S_1. Repeat for a second load S_2 not parallel to S_1. The intersection of the lines of action is the shear center.

Associated with the symmetric pair of stringers having areas a_i, there is an enclosed area A_i, a height h_i and a moment of inertia $I_i = a_i h_i^2 / 2$. There is also an 'eccentricity' $e_i = 2A_i / h_i$. If we place a quantity I_i at the position of the axis of symmetry (Figure 3.41(c)), a distance e_i from the line joining the ith pair of stringers, and do this for all pairs of stringers, then the center of gravity of this set of I_is is the shear center of the section.

3.32 Shear flow distribution in multicell beams with longerons

The problem is handled in the following way (Figure 3.42). A 'cut' is made in each cell so that the shear flows due to the applied resultant shear may be calculated by means of a formula of the type equation 3.104, i.e.

$$\Delta q_i = \frac{S_{yy}I_{xx} - S_{xx}I_{xy}}{I_{xx}I_{yy} - I_{xy}^2}\, a_i x_i + \frac{S_{xx}I_{yy} - S_{yy}I_{xy}}{I_{xx}I_{yy} - I_{xy}^2}\, a_i y_i$$

Suppose we designate the shear flow due to Δq_i as q_0^*. We now super-impose two constant shear flows for a two-cell beam around closed paths, q_1^* and q_2^*. Note that, since these are constant and form complete closed paths, they do not modify vertical or horizontal equilibrium. If we now pick a reference point, we may write a torsional equilibrium equation

$$2\sum q_{0i}^* A_i + 2(A_1^* + A_2^*)q_1^* + 2A_2^* q_2^* = T \tag{3.109}$$

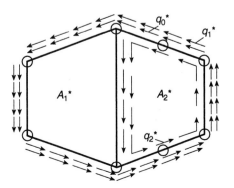

Figure 3.42 Shear flow in multi-cell beam with stringers.

where T is the applied torque about the reference point. This provides one equation in the two unknown shear flows, q_1^* and q_2^*. A second equation, which may be termed a compatibility equation, states that the twist (or the rate of twist) of the two cells is equal,

$$\frac{1}{2A_1^*}\sum\frac{q_i\Delta S_i}{G_it_i}=\frac{1}{A_2^*}\sum\frac{q_i\Delta S_i}{G_it_i}\qquad(3.110)$$

where q_i is the total shear flow, i.e. in the left-hand cell (except for the dividing wall),

$$q_i = q_{0i}^* + q_1^*$$

and in the right-hand cell (except for the dividing wall),

$$q_i = q_{0i}^* + q_1^* + q_2^*$$

and in the dividing wall, the following two values are used:

$$q_i = -(q_0^* + q_2^*)$$

when considered a part of the left-hand cell, and

$$q_i = (q_0^* + q_2^*)$$

when considered a part of the right-hand cell.

Solving equations 3.109 and 3.110 simultaneously yields the value of the superimposed shear flows, q_1^* and q_2^*.

3.33 Out-of-plane bending of curved girders

If the cross-section (Figure 3.43) is compact, or is a closed tube, warping effects probably will not be significant except, perhaps, at points of attachment to the wall or at locations of concentrated load, and the problem may be solved by appealing to the principles of elementary bending and twisting theory. Note that, if the principal axis of the cross-sections does not lie in the plane of the ring, there will be a coupling between the effects of load normal to the plane of the ring and loads in the plane of the ring. If, however, the principal axis lies in the plane of the ring, the loads in that plane may be handled independently of the loads normal to the plane by an arch analysis.

For out-of-plane bending, we may make a cut at an appropriate or convenient location such as at one end, which reduces the structure to a statically determinate state with arbitrary redundant forces and couples applied at the cut faces. The magnitudes of these redundants will be adjusted to maintain continuity at the cut under the applied load.

If the ring is not circular in plane, it will probably become necessary to evaluate the integrals, which are involved, by numerical methods, such as Simpson's rule or others. For the purposes of clarity, we will assume that the ring under consideration is circular in plane.

Let θ be the angular coordinate of a generic cross-section. Let M_s and T_s be the bending moment and torsion, respectively, due to the applied loads and

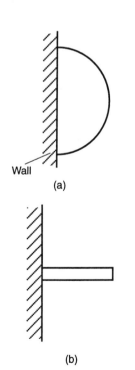

Wall

(a)

(b)

Figure 3.43 Curved girder. (a) Plan view. (b) Side view.

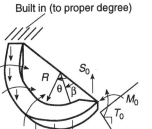

Figure 3.44 Applied loads and couples.

couples (Figure 3.44), if any, acting on the cross-section. Then the total bending moment and torsion are

$$M = M_s + M_0 \cos \theta + S_0 R \sin \theta + T_0 \sin \theta$$
$$T = T_s + T_0 \cos \theta - M_0 \sin \theta - S_0 R(1 - \cos \theta)$$

(3.111)

If shear deformation is significant, a similar general expression for shear should also be written. For the purposes of the present discussion, we will assume that shear deformation is negligible and we need therefore consider only flexural and twisting deformation. The total strain energy is then

$$U = \frac{1}{2} \int \frac{M^2}{EI} R \, d\theta + \frac{1}{2} \int \frac{T^2}{GJ'} R \, d\theta$$

(3.112)

The conditions to be satisfied are, at the cut face,

1. the vertical displacement must be zero;
2. the circumferential slope must be zero;
3. the angular (twisting) displacement must be zero.

In terms of the strain energy, by use of Castigliano's theorem, these become

$$\frac{\partial U}{\partial T_0} = \int \frac{M}{EI} \frac{\partial M}{\partial T_0} R \, d\theta + \int \frac{T}{GJ'} \frac{\partial T}{\partial T_0} R \, d\theta = 0$$

$$\frac{\partial U}{\partial M_0} = \int \frac{M}{EI} \frac{\partial M}{\partial M_0} R \, d\theta + \int \frac{T}{GJ'} \frac{\partial T}{\partial M_0} R \, d\theta = 0$$

(3.113)

$$\frac{\partial U}{\partial S_0} = \int \frac{M}{EI} \frac{\partial M}{\partial S_0} R \, d\theta + \int \frac{T}{GJ'} \frac{\partial T}{\partial S_0} R \, d\theta = 0$$

If the radial dimension of the girder is not small compared to the radius, corrections to the above expressions might be required, these being analogous to the corrections which are involved in the bending of curved beams in their own plane. It might also be mentioned that, if the depth of the section (normal to the plane of the ring) is not small compared to the radius and to the developed length, shear deformation might be significant compared to the other two deformations, and it would be appropriate to add to U

$$\frac{1}{2} \int \frac{S^2}{A'G} R \, d\theta$$

where A' is an 'adjusted' cross-sectional area. Equation 3.112 should be rewritten as

$$U = \frac{1}{2} \int \frac{M^2}{EI} R \, d\theta + \frac{1}{2} \int \frac{T^2}{GJ'} R \, d\theta + \frac{1}{2} \int \frac{S^2}{A'G} R \, d\theta$$

(3.114)

For solid sections,

- rectangular bar: $A' = \dfrac{A}{1.2}$, $J' = \dfrac{bh^3}{3}\left(1 - 0.63\dfrac{h}{b}\right)$ or $J' = \dfrac{hb^3}{3}\left(1 - 0.63\dfrac{b}{h}\right)$

- circular bar: $A' = \dfrac{A}{1.11}$, $J = J'$

The three continuity equations (for the case of a circularly curved girder) become

$$R \int \frac{(M_s + M_0 \cos\theta + T_0 \sin\theta + S_0 R \sin\theta)\sin\theta \, d\theta}{EI}$$

$$+ R \int \frac{[T_s - M_0 \sin\theta + T_0 \cos\theta - S_0 R(1 - \cos\theta)]\cos\theta \, d\theta}{GJ'} = 0$$

$$R \int \frac{(M_s + M_0 \cos\theta + T_0 \sin\theta + S_0 R \sin\theta)\cos\theta \, d\theta}{EI}$$

$$- R \int \frac{[T_s - M_0 \sin\theta + T_0 \cos\theta - S_0 R(1 - \cos\theta)]\sin\theta \, d\theta}{GJ'} = 0$$

$$R^2 \int \frac{(M_s + M_0 \cos\theta + T_0 \sin\theta + S_0 R \sin\theta)\sin\theta \, d\theta}{EI}$$

$$+ R^2 \int \frac{[T_s - M_0 \sin\theta + T_0 \cos\theta - S_0 R(1 - \cos\theta)](-1 + \cos\theta) \, d\theta}{GJ'}$$

$$+ R \int \frac{S_s + S_0}{A'G} \, d\theta = 0$$

Subtracting the first equation from the last simplifies the latter by dropping out the entire first integral as well as the cosine term in the second integral, and similarly, we get the other two continuity equations. We thus obtain three simultaneous equations in the three redundants. The integrals may be evaluated formally in certain 'regular' cases or numerically.

For a downward-acting element of load $qR \, d\beta$ at β, the bending moment at θ is

$$dM_s = -qR \, d\beta[R\sin(\theta - \beta)]$$

and the torsional moment at θ is

$$dT_s = qR \, d\beta\{R[1 - \cos(\theta - \beta)]\}$$

Therefore,

$$M_s(\theta) = -\int_0^\theta qR^2 \sin(\theta - \beta) \, d\beta$$

$$T_s(\theta) = \int_0^\theta qR^2 [1 - \cos(\theta - \beta)] \, d\beta \tag{3.115}$$

in which $q = q(\beta)$.

If q is constant, and the ring is circular,

$$M_s = -qR^2 \int_0^\theta \sin(\theta - \beta) \, d\beta = -[qR^2 \cos(\theta - \beta)]_0^\theta$$

$$= +qR^2[\cos\theta - \cos 0] = -qR^2(1 - \cos\theta)$$

$$T_s = -qR^2 \int_0^\theta d\beta - qR^2 \int_0^\theta \cos(\theta - \beta) \, d\beta = qR^2(\theta - \sin\theta)$$

and

$$M = M_0 \cos\theta + T_0 \sin\theta + S_0 R \sin\theta - qR^2(1 - \cos\theta)$$

$$T = -M_0 \sin\theta + T_0 \cos\theta - S_0 R(1 - \cos\theta) + qR^2(\theta - \sin\theta)$$

(3.116)

We have three unknowns, M_0, T_0, S_0, with three equations in the form of equations 3.113, and using Castigliano's theorem this problem is solved.

3.34 Alternative formulation for out-of-plane bending of curved beams with constant radius (may be extended to non-circular girders)

In this method, differential equations of equilibrium are written for a differential element of the beam. We will consider the case in which a principal axis of each cross-section lies in the plane of the ring (Figure 3.45). The method may, however, be extended to the case in which the principal axes are inclined to the plane of the ring. The cross-section need not be rectangular, or circular, or indeed regular.

Summing forces in the direction normal to the plane of the ring, taking moments about a radial axis and tangential axis, and dropping the terms of the second order in the differential,

$$-Q + \left(Q + \frac{\mathrm{d}Q}{\mathrm{d}\theta}\,\mathrm{d}\theta\right) + qR\,\mathrm{d}\theta = 0$$

$$-M + \left(M + \frac{\mathrm{d}M}{\mathrm{d}\theta}\,\mathrm{d}\theta\right) - QR\,\mathrm{d}\theta + T\,\mathrm{d}\theta = 0$$

$$-T + \left(T + \frac{\mathrm{d}T}{\mathrm{d}\theta}\,\mathrm{d}\theta\right) + mR\,\mathrm{d}\theta - M\,\mathrm{d}\theta = 0$$

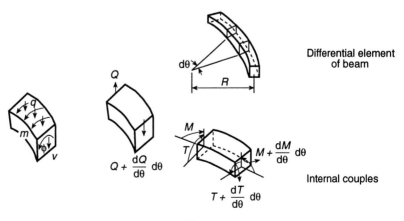

Figure 3.45 Out-of plane bending of curved beams.

Applied loads and positive directions of displacements

Internal forces

Upon simplifying

$$\frac{dQ}{R\,d\theta} = -q \tag{3.117a}$$

$$\frac{dM}{R\,d\theta} + \frac{T}{R} - Q = 0 \tag{3.117b}$$

$$\frac{dT}{R\,d\theta} - \frac{M}{R} = -m \tag{3.117c}$$

Eliminating Q between the first and the second equations leaves the system in the following form:

$$\frac{d^2M}{R^2\,d\theta^2} + \frac{1}{R^2}\frac{dT}{d\theta} = -q \tag{3.118a}$$

$$\frac{M}{R^2} - \frac{1}{R^2}\frac{dT}{d\theta} = \frac{m}{R} \tag{3.118b}$$

Eliminating T between the above equations,

$$\frac{d^2M}{R^2\,d\theta^2} + \frac{M}{R^2} = \frac{m}{R} - q \tag{3.119}$$

which is readily solved for M when m and q are defined as functions of θ. Assuming that M thus becomes known, we can obtain T by integration of equation 3.118b. Combining equation 3.117c and equation 3.119 to eliminate M,

$$\frac{d^3T}{d\theta^3} + \frac{dT}{d\theta} = -R^2\frac{d^2m}{d\theta^2} - qR^2 \tag{3.120}$$

The horizontal component of the displacement of the fiber (for small values of ϕ) is $z\phi$ and therefore the radius of the fiber is reduced by the amount $z\phi$ (Figure 3.46). The circumferential strain is

$$\varepsilon = \frac{[2\pi(r - z\phi) - 2\pi r]}{2\pi r} = \frac{-z\phi}{r}$$

The associated circumferential stress is

$$E\varepsilon = -E\frac{\phi z}{r}$$

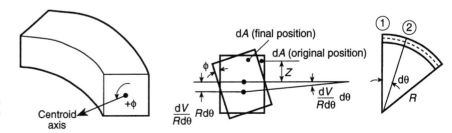

Figure 3.46 Displacements of a curved beam under out-of-plane loading.

The stresses, which are dependent upon z, vary from compression in the upper half to tension in the lower half. Since the reference axis passes through the centroid of the section, the resultant force vanishes:

$$F = \int \left(-E \frac{\phi z}{r} \right) \mathrm{d}A = -\frac{E\phi}{r} \int z \, \mathrm{d}A = 0$$

where r has been taken from under the integral because we will assume that the width of the ring is small compared to its radius. The moment of these stresses about the radial axis is

$$M = -\int \sigma z \, \mathrm{d}A = -\int \left(\frac{-E\phi}{r} z \right) z \, \mathrm{d}A = \frac{E\phi}{r} \int z^2 \, \mathrm{d}A = \frac{E\phi}{R} I$$

where I is the moment of inertia of the cross-section about the radial axis. We also obtain a contribution to the vertical bending moment as a result of vertical displacement,

$$M = -EI \frac{\mathrm{d}^2 v}{R^2 \, \mathrm{d}\theta^2}$$

Consequently, the total bending moment (in the vertical direction) is

$$M = -EI \left(\frac{\mathrm{d}^2 v}{R^2 \, \mathrm{d}\theta^2} - \frac{\phi}{R} \right) \tag{3.121}$$

Suppose we turn the entire 'piece' of a ring a rotation about axis ① by $\mathrm{d}v/R\,\mathrm{d}\theta$. The resulting vertical displacement at section ② (on the mean circle) is $(\mathrm{d}v/R \, \mathrm{d}\theta)R \, \mathrm{d}\theta$. The slope of the radial line at section ② is

$$\left[\left(\frac{\mathrm{d}v}{R \, \mathrm{d}\theta} \right) R \, d\theta \right] \frac{1}{R} = \frac{\mathrm{d}v}{R \, \mathrm{d}\theta} \, d\theta$$

Hence, a rotation of the cross-section in the 'positive' direction would have to be imposed if ϕ is to be zero at each section. Since the foregoing rotation has taken place over a circumferential length $R\,\mathrm{d}\theta$, the rate of twist is

$$\frac{1}{R \, \mathrm{d}\theta} \left(\frac{\mathrm{d}v}{R \, \mathrm{d}\theta} \, d\theta \right) = \frac{\mathrm{d}v}{R^2 \, \mathrm{d}\theta}$$

In order to bring the cross-sections back to zero rotation, an effective rate of twist $\mathrm{d}v/R^2\mathrm{d}\theta$ must be imposed. Hence, due to twist and vertical displacement,

$$T = C \left(\frac{\mathrm{d}\phi}{R \, \mathrm{d}\theta} + \frac{\mathrm{d}v}{R^2 \, \mathrm{d}\theta} \right) \tag{3.122}$$

Consider the case

$$q = q_n \sin n\theta$$
$$m = m_n \sin n\theta \tag{3.123}$$

Then, we may assume

$$Q = Q_n \cos n\theta, \quad v = v_n \sin n\theta$$
$$M = M_n \sin n\theta, \quad \phi = \phi_n \sin n\theta \tag{3.124}$$
$$T = T_n \cos n\theta$$

Substituting into equation 3.117a,

$$Q_n = \frac{q_n R}{n}$$

Substituting into equation 3.117b, and substituting the appropriate expression into equation 3.117c,

$$nM_n \cos n\theta + T_n \cos n\theta = \frac{q_n R^2}{n} \cos n\theta$$

$$-M_n \sin n\theta - nT_n \sin n\theta = -m_n R \sin n\theta$$

or

$$nM_n + T_n = \frac{q_n R^2}{n}$$

$$M_n + nT_n = m_n R$$

which yields

$$M_n = \frac{1}{n^2 - 1}(q_n R^2 - m_n R) \tag{3.125a}$$

$$T_n = \frac{1}{n^2 - 1}\left(nm_n R - \frac{q_n R^2}{n}\right) \tag{3.125b}$$

These results could have been obtained from equations 3.119 and 3.120. Now, substituting M_n and T_n into equations 3.121 and 3.122,

$$-\frac{n^2}{R^2}v_n \sin n\theta - \frac{\phi_n \sin n\theta}{R} = \frac{-1}{EI(n^2 - 1)}(q_n R^2 - m_n R)\sin n\theta$$

$$\frac{n}{R^2}v_n \cos n\theta + \frac{n\phi_n}{R}\cos n\theta = \frac{1}{C(n^2 - 1)}\left(nm_n R - \frac{q_n R^2}{n}\right)\cos n\theta$$

The solution of these algebraic equations is

$$v_n = \frac{R^2}{EI(n^2 - 1)^2}(q_n R^2 - m_n R) - \frac{R^2}{Cn(n^2 - 1)^2}\left(nm_n R - \frac{q_n R^2}{n}\right) \tag{3.126a}$$

$$\phi_n = \frac{nR}{C(n^2 - 1)^2}\left(nm_n R - \frac{q_n R^2}{n}\right) - \frac{R}{EI(n^2 - 1)^2}(q_n R^2 - m_n R) \tag{3.126b}$$

Identical results would be obtained for loadings defined by

$$q = q_n \cos n\theta$$
$$m = m_n \cos n\theta \tag{3.127}$$

with corresponding changes such as

$$m = m_n \cos\theta \text{ etc.}$$

If we have a general loading (for both q and m), it may be represented by a Fourier series, and similar steps can be taken term by term, as we did with either or both of equations 3.123 and 3.127.

3.35 Bending in the plane of the ring

Let w be the inward displacement and v the tangential displacement, positive in the direction of increasing θ (Figure 3.47). If the differential element of the ring is given a displacement w while $v = 0$, the length of the element is reduced by the amount $w\,d\theta$. Hence, the circumferential strain is

$$\varepsilon_\theta = \frac{-w\,d\theta}{R\,d\theta} = \frac{-w}{R}$$

Restricting the problem to small strains, superposition applies, so that we may add the strain due to v-displacements. If the tangential displacement at θ is v, then the tangential displacement at the other end of the element is $v + (dv/d\theta)d\theta$. The 'new' length of the element is

$$R\,d\theta + \left(v + \frac{dv}{d\theta}\,d\theta\right) - v$$

and the change in length is $(dv/d\theta)d\theta$. Hence the strain due to v is

$$\varepsilon_\theta = \left(\frac{dv}{d\theta}\,d\theta\right)\bigg/ R\,d\theta = \frac{dv}{R\,d\theta}$$

Then, the total circumferential strain is

$$\varepsilon_\theta = \frac{dv}{R\,d\theta} - \frac{w}{R} \tag{3.128}$$

The circumferential force is

$$N_\theta = EA\varepsilon_\theta = EA\left(\frac{dv}{R\,d\theta} - \frac{w}{R}\right) \tag{3.129}$$

If we say that the circumferential strains are **negligibly** small, i.e. identically zero, we get immediately a relation between tangential and radial displacements,

$$\frac{dv}{d\theta} = w \tag{3.130}$$

Due to the second derivative, we get (obviously)

$$M = -EI\,\frac{d^2w}{R^2\,d\theta^2}$$

that is, the change in curvature due to the second derivative is $d^2w/R^2d\theta^2$.

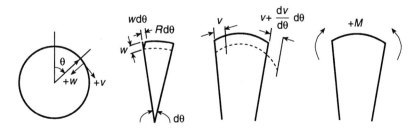

Figure 3.47 In-plane bending of a ring.

Now, we consider the change in curvature due to a pure w-displacement:

- The original curvature is $1/R$.
- The new curvature is $1/(R - w)$.

The change in curvature is

$$\frac{1}{(R - w)} - \frac{1}{R} = \frac{w}{R^2 - Rw}$$

and, if w is very small, as we are assuming, compared to R, the change in curvature is w/R^2. Hence, we have a contribution to the bending moment equal to $M = -EI(w/R^2)$. The total moment is

$$M = -EI\left(\frac{d^2w}{R^2\,d\theta^2} + \frac{w}{R^2}\right) \tag{3.131}$$

Consider the ring loaded by two diametric loads as shown in Figure 3.48. We can calculate M_0, the redundant moment, by any of the usual methods. Having M_0, we can express M as a function of θ. We then substitute M into equation 3.131, and solve for w. Next, if we wish, we substitute w into equation 3.130 and solve for v.

We may solve the above problem in a different way. We can assume the deflection to be represented, in the present symmetric problem, by the half range series,

$$w \approx \sum_{0}^{\infty} w_n \cos n\theta$$

or

$$w \approx w_0 + \sum_{1}^{\infty} w_n \cos n\theta$$

Our problem is to determine the w_ns, on the basis of the assumption that the ring deforms only as a result of flexture. If, as simplified by the above assumption, axial, or hoop deformation is negligible, then w_0 must be zero. One

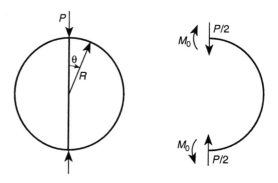

Figure 3.48 Ring under symmetric concentrated loads.

may evaluate the remaining w_ns by an energy method. Noting that

$$M = -\frac{EI}{R^2}\left(\frac{d^2 w}{d\theta^2} + w\right)$$

$$= \frac{EI}{R^2}\sum_{n=1}^{\infty}(n^2 - 1)w_n \cos n\theta$$

we observe that the strain energy is

$$U = \int_0^{2\pi}\frac{M^2}{2EI}R\,d\theta = \frac{EI\pi}{2R^3}\sum_{n=1}^{\infty}(n^2 - 1)^2 w_n^2$$

The work done by the loads, P, is

$$W = \frac{P}{2}\left[\sum_{n=1}^{\infty}w_n\cos n(0) + \sum_{n=1}^{\infty}w_n\cos n\pi\right] = P\sum_{2,4,6,\ldots}^{\infty}w_n$$

By the principle of virtual work, since the system is presumed to be in equilibrium in the loaded and deformed state,

$$\delta(W - U) = 0,$$

or

$$P\delta w_j = \frac{EI\pi}{2R^3}(j^2 - 1)^2 2w_j\delta w_j,$$

where j is any even integer. This yields

$$w_j = \frac{PR^3}{\pi EI(j^2 - 1)^2} \tag{3.132}$$

Note that the P-term vanishes entirely for odd j. Hence this equation needs to be considered only for even j, and therefore

$$w \approx \frac{PR^3}{\pi EI}\sum_{2,4,6,\ldots}^{\infty}\frac{\cos n\theta}{(n^2 - 1)^2} \tag{3.133}$$

Suppose the loading is $P = P_n\cos n\theta$. The work done during a virtual displacement is

$$\delta W = P_n\delta w_n\int_0^{2\pi}\cos^2 n\theta R\,d\theta = P_n\delta w_n R\pi$$

We have

$$\delta U = \frac{\partial U}{\partial w_n}\delta w_n = \frac{EI\pi}{R^3}(n^2 - 1)^2 w_n\delta w_n$$

Since $\delta W - \delta U = 0$, we obtain

$$w_n = \frac{P_n R^4}{EI(n^2 - 1)^2} \tag{3.134}$$

Therefore, if we have a general loading, it may be represented by a Fourier series and, if symmetric, this will have the form

$$P = P_0 + \sum_{2}^{\infty} P_n \cos n\theta$$

Then, the deflections may be represented by a similar series

$$w = \sum_{2}^{\infty} w_n \cos n\theta \tag{3.135}$$

where w_n is given by equation 3.134.

3.36 Elementary theory of thin shell

If this problem were done by the 'usual' method of analysis, the shear flow distribution in the shell would be

$$q = PQ/I$$

where Q and I refer to the shell, and

$$J = \int_0^{2\pi} (Rt\ d\theta)R^2 = 2\pi R^3 t$$

$$I = J/2 = \pi R^3 t$$

$$Q = \int_0^\theta (tR\ d\phi)R \cos \phi = tR^2 \int_0^\theta \cos \phi\ d\phi = tR^2 \sin \theta$$

so that

$$q = \frac{P(tR^2 \sin \theta)}{\pi R^3 t} = \frac{P \sin \theta}{\pi R} = q_1 \sin \theta \tag{3.136}$$

On the basis of elementary beam theory as applied to the shell, the shear flow distribution is found to be as equation 3.136 and the equilibrium picture of the ring would be as shown by Figure 3.49b without cut.

The shear flow on a differential element is $[(P/\pi R)\sin\phi]R\,d\phi$. Its moment arm about the point defined by the angle θ is $R[1 - \cos(\theta - \phi)]$. Hence, the

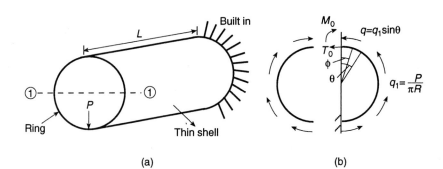

Figure 3.49   Cylindrical shell.   (a)   (b)

bending moment due to q up to any point θ is

$$Mq = \frac{P}{\pi R} \int_0^\theta R \sin \phi R[1 - \cos(\theta - \phi)] \, d\phi = q_1 R^2 \left(\frac{\theta}{2} \sin \theta + \frac{1}{4} \cos \theta - \frac{1}{2} \right)$$

$$(3.137)$$

We assume a more general form of shear flow in the shell,

$$q = \frac{P}{\pi R} \sin \theta + \sum_{n=2}^{\infty} q_n \sin n\theta \qquad (3.138)$$

We can calculate the shear strain energy

$$U_s = \frac{L}{2G} \int_0^{2\pi} \frac{q^2 R \, d\theta}{t} \qquad (3.139)$$

since

$$\text{shear stress} = q/t$$

$$\text{shear strain} = q/Gt$$

$$\text{strain energy density} = \tfrac{1}{2} q \times \text{strain} = \frac{q^2}{2Gt}$$

Then

$$U_s = \frac{\pi RL}{2Gt} \sum_{n=1}^{\infty} q_n^2 \qquad (3.140)$$

in which $q_1 = P/\pi R$.

Consider a longitudinal strip of the shell, as shown in Figure 3.50b. At any pair of opposite points, the difference in the shear flow is

$$dq = \frac{dq}{d\theta} \, d\theta = \left(\frac{P}{\pi R} \cos \theta + \sum_{n=2}^{\infty} n q_n \cos n\theta \right) d\theta$$

At any section a distance x from the ring end, noting that the longitudinal stress in the strip should be zero at the ring end, the longitudinal stress is

$$\sigma_x = x \left(\frac{P}{\pi R} \cos \theta + \sum_{n=2}^{\infty} n q_n \cos n\theta \right) \frac{d\theta}{(R \, d\theta)t}$$

$$= \frac{x}{Rt} \left(\frac{P}{\pi R} \cos \theta + \sum_{n=2}^{\infty} n q_n \cos n\theta \right) \qquad (3.141)$$

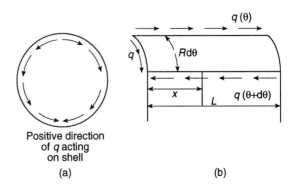

Positive direction
of q acting
on shell

(a)   (b)

Figure 3.50 Segment of the shell.

The strain energy in the shell due to the longitudinal stress is

$$U_1 = \frac{1}{2} \int_0^L \int_0^{2\pi} t\sigma_x \varepsilon_x R \; d\theta \; dx = \frac{Rt}{2} \int_0^L \int_0^{2\pi} \frac{\sigma_x^2}{E} \; d\theta \; dx$$

$$= \frac{\pi R t L^3}{6E(Rt)^2} \sum n^2 a_n^2$$

where $a_1 = P/\pi R$, $a_{2,3,\dots} = q_{2,3,\dots}$, so that

$$U_1 = \frac{\pi L^3}{6ERt} \left(\frac{P^2}{\pi^2 R^2} + \sum_{n=2}^{\infty} n^2 q_n^2 \right) \tag{3.142}$$

The total strain energy of the shell is

$$U = U_s + U_1 = \frac{P^2}{\pi^2 R^2} \left(\frac{\pi R L}{2Gt} + \frac{\pi L^3}{6ERt} \right) + \sum_{n=2}^{\infty} q_n^2 \left(\frac{\pi R L}{2Gt} + \frac{n^2 \pi L^3}{6ERt} \right) \tag{3.143}$$

The 'vertical' resultant force due to a sinusoidally varying shear flow is

$$V = \int_0^{2\pi} (q_n \sin n\theta)(R \; d\theta) \sin \theta$$

$$= \begin{cases} 0 & \text{if } n \neq 1 \\ \pi R q_n & \text{if } n = 1 \end{cases} \tag{3.144}$$

Hence, a load on the ring is not necessary for a shear flow corresponding to $n > 1$ (a set of loads which have a zero resultant is called a 'self-equilibrating system'). However, if $n = 1$, the ring must have a 'vertical' load.

The redundant moment and hoop force at the top (M_0 and T_0) may be found by any of the standard techniques of indeterminate structural analysis. V_0 may be calculated by summing forces along the radial line which bisects the angle α.

$$V_0 \cos \frac{\alpha}{2} = R \int_0^{\alpha/2} q_n \sin n\phi \sin \phi \; d\phi \tag{3.145}$$

If we state that all deformation is due to bending, it then follows that

$$\varepsilon_\theta = \frac{dv}{R \; d\theta} - \frac{w}{R} = 0 \quad \text{or} \quad \frac{dv}{d\theta} = w \tag{3.146}$$

so that we may assume,

$$w = \sum_{n=2}^{\infty} w_n \cos n\theta, \quad \frac{dv}{d\theta} = \sum_{n=2}^{\infty} w_n \cos n\theta$$

$$v = \sum_{n=2}^{\infty} \frac{w_n}{n} \sin n\theta + C \tag{3.147}$$

We take as boundary condition, $v = 0$ at $\theta = 0$, hence $C = 0$.

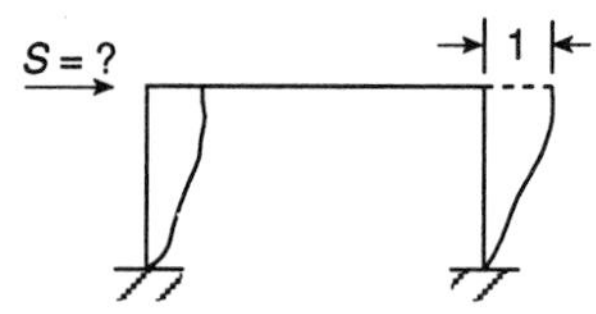

3.37 Space frame

If the structure comprises a collection of vertical frames, each properly connected to a rigid slab or diaphragm (Figure 3.51), one first calculates the stiffness

Figure 3.51 Frame stiffness.

of each frame with respect to a shear applied at the top in its own plane, i.e. the shear required to produce a unit displacement of the frame.

Let $S_i = K_i$ = stiffness of the frame. One now resolves these K_is into x and y components in the plane of the frame, finds the centers of gravity of these, and also computes

$$J = \sum r_i^2 k_i$$

where r_i is the normal distance from the frame to the center of gravity. Any system of horizontal forces applied at the slab level may be reduced to a concentration at the center of gravity, having components X and Y, and a couple T, about the center of gravity. The component X is distributed to each frame in proportion to the x-component of its stiffness K_i. The component Y is distributed to each frame in proportion to the y-component of its stiffness, K_i. The couple T is distributed to each frame as shear in the following way,

$$S_i = T \frac{r_i k_i}{J} \tag{3.148}$$

At an arbitrary joint in a space structure, assuming that the joint is a rigid joint, six components of displacements are possible:

1. three linear displacements, u, v and w, in the direction of the x, y and z axes;
2. three rotations, one each about the x, y and z axes.

For any member, there will be six components of generalized forces at each end, i.e. three forces and three couples.

3.38 Space trusses

Let

- F_{ij} be the axial load in the member connecting joints i and j;
- L_{ij} the length of the member;
- h_{ij} the projected horizontal length of the member;
- v_{ij} the projected vertical length of the member;
- T_{ij} the tension coefficient for the member, F_{ij}/L_{ij}.

Then the horizontal component of the force in the member is

$$H_{ij} = T_{ij} h_{ij} \tag{3.149a}$$

and similarly, the vertical component is

$$V_{ij} = T_{ij} v_{ij} \tag{3.149b}$$

At every joint, the sum of the vertical components of the load, including members and applied forces, must vanish, and similarly for the horizontal forces

$$\sum_j T_{ij} h_{ij} + \sum \bar{H}_i = 0$$

$$\sum_J T_{ij} v_{ij} + \sum \bar{V}_i = 0 \tag{3.150}$$

where $\bar{H}_i$ and $\bar{V}_i$ are applied horizontal and vertical forces, respectively, acting at the ith joint. Writing these equations for all joints, solving 'simultaneously', then substituting into

$$F_{ij} = T_{ij} L_{ij} \tag{3.151}$$

one obtains the forces in all members.

In plane trusses, the basic (stable and statically determinate) unit is a triangle. Thus, to construct a plane truss for simple truss action, one starts with a triangle (three joints, three bars, $j = 3$, $n = 3$). To extend this by one joint, two additional bars are required. This also applies to all successive joints which are to be included. Hence a necessary (but not sufficient) relation between the number of bars n and number of joints j is

$$2(j - 3) = n - 3$$

In the case of a space truss, the basic structural unit is the tetrahedron ($j = 4$, $n = 6$). An additional joint requires three additional bars (members). Hence, a necessary (but not sufficient) condition for a space truss is

$$3(j - 4) = n - 6$$

Suppose we have a space truss meeting the above requirement, also having these members properly distributed for stability and determinancy, and attached to a reactive system in a stable and determinate way. Let P_x, P_y, P_z be the components of external load at an arbitrary joint. Then the following system of equations must be satisfied at all joints

$$P_{xi} + \sum_j T_i X_{ij} = 0$$

$$P_{yi} + \sum_j T_i Y_{ij} = 0 \tag{3.152}$$

$$P_{zi} + \sum_j T_i Z_{ij} = 0$$

in which X_{ij} is the x-component of the length of the member extending from joint i to joint j with proper sign. Essentially, X_{ij} etc. are differences in coordinates of the two joints, such that $X_{ik} = X_k - X_i$ etc. if forces (internal and external) are taken positive when acting in the positive coordinate directions.

When approaching a joint at which there are more than three unknown Ts, the 'extra' Ts may be represented by their symbols and the analysis proceeds beyond this joint retaining these symbols. Ultimately, if the system is determinate, an equation will be found among the usual joint equations, for each 'extra' T.

3.39 Cable

The cable has a uniform weight and may also be subjected to a uniformly distributed load (Figure 3.52). The tension T is assumed to be sufficiently great so that the sag f is not large. Since the curve is reasonably flat, the weight of

Figure 3.52 Cable under uniform load.

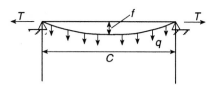

the cable may be replaced, with no practical error, by a uniform loading q. Then

$$f = \frac{qC^2}{8T}$$
(3.153)

Since the curve is flat, the 'true' length of the cable is

$$L = C + \lambda$$

where

$$\lambda = \frac{1}{2} \int_0^L (y')^2 \, dx, \quad \text{in which} \quad y = \frac{4fx(C - x)}{C^2}$$
(3.154)

From the foregoing formulas, one obtains

$$\lambda = \frac{8}{3} \frac{f^2}{C}$$

The **unstressed** true length would be

$$L' = C + \lambda - \frac{TC}{AE}$$
(3.155)

In the field, T may be determined by any of several methods, including, where appropriate:

- tension meter
- strain gage
- measuring f
- measuring frequency

Using subscript 'i' for the initial state,

$$L' = C_i + \frac{8f_i^2}{3C_i} - \frac{T_i C_i}{AE} = C_i + \frac{q^2 C_i^3}{24T_i^2} - \frac{T_i C_i}{AE}$$
(3.156a)

For the final state,

$$L' = C_f + \frac{q^2 C_f^3}{24T_f^2} - \frac{T_f C_f}{AE}$$
(3.156b)

Since the last two terms in the foregoing expressions are small compared to the first terms, i.e. C_i and C_f, we may replace either by the other, i.e. replace C_f by C_i. Then, equating L' expressions of equations 3.156,

$$C_f + \frac{q^2 C_i^3}{24T_f^2} - \frac{T_f C_i}{AE} = C_i + \frac{q^2 C_i^3}{24T_i^2} - \frac{T_i C_i}{AE}$$

or
(3.157)

$$C_f - C_i = \frac{q^2 C_i^3}{24} \left(\frac{1}{T_i^2} - \frac{1}{T_f^2} \right) + \frac{C_i}{AE} (T_f - T_i)$$

3.40   Guyed tower

We may assume that the inclined guy (Figure 3.53) behaves as it would if it were horizontal and subjected to the appropriate components of load, and that we may take the effective tension as the average of the two end tensions, T_1 and T_2. The component of sag normal to the chord is $f^* \cos \alpha$ (Figure 3.54(a)), and if the loads are gravity loads, their component of the distributed weight of the cable is $q \cos \alpha$ where q is the weight per unit length.

Assuming a parabolic shape and no bending rigidity of the cable,

$$Tf^* \cos \alpha = \tfrac{1}{8}(q \cos \alpha)C^2$$

or   (3.158)

$$Tf = \tfrac{1}{8}(q \cos \alpha)C^2 = \tfrac{1}{8}qCb$$

If T is changed from T_i to T_f,

$$e = \frac{(T_f - T_i)C}{AE} + \frac{q^2C^3}{24}\left(\frac{1}{T_i^2} - \frac{1}{T_f^2}\right)$$ (3.159)

If we assume no vertical displacement of the attachment point,

$$\delta = e/\cos \alpha$$ (3.160)

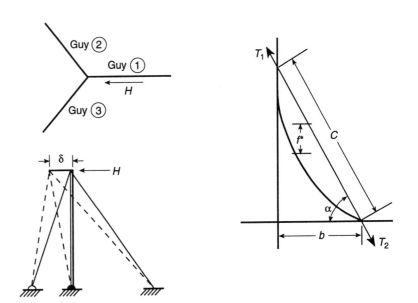

Figure 3.53   Guyed tower.

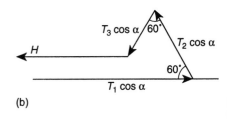

Figure 3.54.   (a) Projection of a guy normal to its axis. (b) Vector diagram of forces in horizontal plane.

(a)   (b)

valid for δ and e in the same plane. For our problem, this is valid for the single guy on the 'windward' side.

For the pair of guys on the leeward side, we must establish the relation between δ and the horizontal component of the quantity e appropriate to each guy. Letting h be this horizontal component of e, and noting it to be the same, by symmetry, for each of the two leeward guys, $h_2 = h_3 = \delta/2$. Consequently, e for each in the leeward direction is equal to $\frac{1}{2}\delta \cos \alpha$. The vector diagram of forces in the horizontal plane is as shown in Figure 3.54(b), and

$$(T_1 - T_2)\cos \alpha = H \tag{3.161}$$

If the guys were under the same initial tension, T_i, then

$$T_1 = T_i + \Delta T_1$$
$$T_2 = T_i + \Delta T_2 \tag{3.162}$$

and the foregoing becomes

$$(\Delta T_1 - \Delta T_2)\cos \alpha = H \tag{3.163}$$

3.41 Review of beam theory

The Bernouilli–Euler theory of beams assumed that plane sections before bending remain plane and remain normal to the axis of the beam, after bending, the latter assumption being equivalent to the assumption that shear strains can be neglected. These assumptions lead to the following differential equation governing the deflection of beams so long as the deflections are small, or, more correctly, so long as the curvatures are small:

$$EI \frac{\mathrm{d}^2 y}{\mathrm{d}x^2} = -M \tag{3.164}$$

where $M = M(x)$ is the bending moment and, for the negative sign as shown, the positive directions of y and the bending moment are as indicated in Figure 3.55.

The positive direction of y is immaterial. Since in civil engineering problems, the most important class of loads are gravity loads and in most cases the resulting deflections of interest are downward, it is convenient to take the positive direction of y to be downward as shown in Figure 3.55(a).

Figure 3.55 Sign convention of moment. (a) For y positive downward, $+M$ produces compression at the top of the beam. (b) For y positive upward, $+M$ produces tension at the top of the beam.

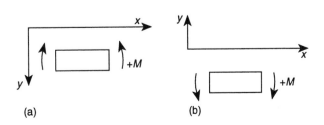

(a)   (b)

Figure 3.56 A differential element of a beam.

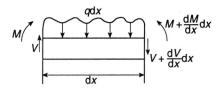

If we consider a differential element of a beam under loading acting in the conventional direction (Figure 3.56), and take the moment and shear to be acting in the conventionally positive directions, we may write the following equilibrium equations by summing vertical forces and by taking moments, respectively, dropping higher order terms in the process.

$$\left(V + \frac{dV}{dx}\,dx\right) - V + q\,dx = 0$$

in which q is the average intensity of the distributed load and in the limit, of course, becomes the actual intensity, and

$$M - \left(M + \frac{dM}{dx}\,dx\right) + V\,dx = 0$$

These reduce to

$$\frac{dV}{dx} = -q, \quad \frac{dM}{dx} = V \tag{3.165}$$

Substitution of the second equation into the first gives

$$\frac{d^2M}{dx^2} = -q \tag{3.166}$$

If we now substitute the expression for the bending moment, in terms of the approximation to the curvature, into the latter equation, we obtain

$$\frac{d^2}{dx^2}\left(EI\,\frac{d^2y}{dx^2}\right) = q \tag{3.167}$$

and if we substitute the same basic expression into the formula relating M and V, we obtain

$$\frac{d}{dx}\left(EI\,\frac{d^2y}{dx^2}\right) = -V \tag{3.168}$$

Note that, only when EI is independent of x, at least over an interval, can we write

$$EI\,\frac{d^4y}{dx^4} = q$$

$$EI\,\frac{d^3y}{dx^3} = -V \tag{3.169}$$

Consider a beam with arbitrary end conditions loaded by axial force P at both ends in addition to the transverse load $q(x)$ (Figure 3.57). If we take a

Figure 3.57 Deformed shape of a differential element of a beam.

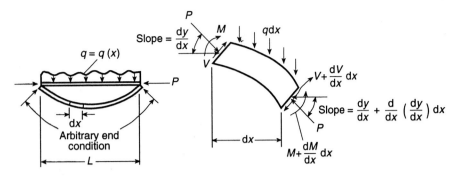

differential element of the beam and set up equilibrium conditions, we obtain

$$\frac{dV}{dx} - P\frac{d^2y}{dx^2} = -q$$

$$\frac{dM}{dx} = V \tag{3.170}$$

from which we get

$$\frac{d^2M}{dx^2} - P\frac{d^2y}{dx^2} = -q \tag{3.171}$$

Substituting equation 3.164 into equation 3.171,

$$\frac{d^2M}{dx^2} + \frac{P}{EI}M = -q \tag{3.172}$$

or alternatively,

$$\frac{d^2}{dx^2}\left(EI\frac{d^2y}{dx^2}\right) + P\frac{d^2y}{dx^2} = q \tag{3.173}$$

If EI is constant,

$$EI\frac{d^4y}{dx^4} + P\frac{d^2y}{dx^2} = q \tag{3.174}$$

We have equations applicable to practical problems with axial loadings.

3.42 Plates with irregular shapes and arbitrary boundary conditions

3.42.1 General remarks

Unless the boundary conditions are ideal, obtaining an analytical solution for the plates is very difficult. Any analytical function on a boundary can be expressed by a linear combination of other functions or numbers. If a plate of any shape, for which an analytical solution is possible, is under self-equilibrium, we can obtain moments, stresses and displacements at any internal point. If a closed cut is made inside the plate, displacements and tractions on this line can be considered as the boundary conditions while those inside of this line can represent the displacements and tractions corresponding to given 'boundary conditions'. If we change the self-equilibrating loadings, we get different 'boundary conditions' and a resulting 'internal' solution. After changing the loadings several times, we can express the 'real' boundary conditions by a linear combination of real numbers, with the degree of accuracy we want. The solution of this problem is, then, the linear combination of the internal solution corresponding to each of the self-equilibrating loadings, with the coefficients of combination the same as those on the boundary [1].

As an illustration, a disk of isotropic material is considered. Assuming thin plate theory is applicable, we use polar coordinates and consider the in-plane problem separately from the normal to the plane (plate) problem.

3.42.2 In-plane problem
The strains in terms of displacements are

$$\varepsilon_r = \frac{\partial u}{\partial r}$$

$$\varepsilon_\theta = \frac{\partial v}{r\,\partial \theta} + \frac{u}{r} \tag{3.34}$$

$$\gamma_{r\theta} = \frac{\partial u}{r\,\partial \theta} + \frac{\partial v}{\partial r} - \frac{v}{r}$$

The stresses in terms of displacements are

$$\sigma_r = \frac{E}{1-v^2}\left[\frac{\partial u}{\partial r} + v\left(\frac{\partial v}{r\,\partial \theta} + \frac{u}{r}\right)\right]$$

$$\sigma_\theta = \frac{E}{1-v^2}\left(v\frac{\partial u}{\partial r} + \frac{\partial v}{r\,\partial \theta} + \frac{u}{r}\right) \tag{3.36}$$

$$\tau_{r\theta} = \frac{E}{2(1+v)}\left(\frac{\partial v}{\partial r} + \frac{\partial u}{r\,\partial \theta} - \frac{v}{r}\right)$$

The equilibrium equations in terms of stresses are

$$\frac{\partial \sigma_r}{\partial r} + \frac{1}{r}\frac{\partial \tau_{r\theta}}{\partial \theta} + \frac{\sigma_r - \sigma_\theta}{r} + k_r = 0$$

$$\frac{\partial \sigma_\theta}{\partial \theta} + r\frac{\partial \tau_{r\theta}}{\partial r} + 2\tau_{r\theta} + rk_\theta = 0 \tag{3.33}$$

Substituting equations 3.36 into equations 3.33, the equilibrium equations in terms of the displacements are obtained as

$$\frac{\partial^2 u}{\partial r^2} + \frac{\partial u}{r\,\partial r} - \frac{u}{r^2} + \frac{(1-v)}{2}\frac{\partial^2 u}{r^2\,\partial \theta^2} - \frac{(3-v)}{2}\frac{\partial v}{r^2\,\partial \theta} + \frac{(1+v)}{2}\frac{\partial^2 v}{r\,\partial r\,\partial \theta}$$

$$= -\frac{(1-v^2)}{E}k_r$$

$$\frac{(1-v)}{2}\left(\frac{\partial^2 v}{\partial r^2} + \frac{\partial v}{r\,\partial r} - \frac{v}{r^2}\right) + \frac{\partial^2 v}{r^2\,d\theta^2} + \frac{(3-v)}{2}\frac{\partial u}{r^2\,\partial \theta} + \frac{(1+v)}{2}\frac{\partial^2 u}{r\,\partial r\,\partial \theta}$$

$$= -\frac{(1-v^2)}{E}k_\theta \tag{3.43}$$

Because of difficulties with boundary conditions, the sector problem with finite radius under arbitrary boundary conditions has not been solved by completely analytical means. The general solution of the biharmonic equation of compatibility was obtained by Michell [2]. This solution is rigorous for a wedge with infinite length. The case of rotational symmetry in cylindrical coordinates can be solved by using the Hankel transform [3, 4]. Tranter [5, 6] used the Mellin transform to find the stress distribution in an infinite wedge. Complex variables were used as stress functions by Brahtz [7, 8] to solve for the stresses at the base of a gravity dam. Silverman [9] used the variational method to obtain an expression for an approximate stress function for determining the effect of a third boundary to an infinite triangular wedge. A similar method was used to solve the circular sector problem by Horvay and Hansen [10]. The finite Fourier–Hankel transform was used by Nomachi to solve the stress problems in cylindrical coordinates [11, 12]. This transformation scheme could be used for the circular sector with arbitrary boundary conditions at circular boundaries and with arbitrary shear stresses and tangential displacements at the radial boundaries.

The solution to equations 3.43 can be separated into the homogeneous parts, u_h and v_h, and the particular solutions, u_p and v_p. The homogeneous part of displacement functions may be assumed to be

$$u_h = \sum_{n=0}^{\infty} U(\theta) r^{kn} \tag{3.175}$$

$$v_h = \sum_{n=0}^{\infty} V(\theta) r^{kn} \tag{3.176}$$

in which U and V are functions of θ only, k is any real number and n is an integer. Let

$$\alpha = kn \tag{3.177}$$

and substitute equations 3.175 and 3.176 into equations 3.43 with the right-hand side equal to zero, to obtain

$$u_h = a_{10} \sin\theta + a_{20} \cos\theta - a_{30} \sin\theta + a_{40} \cos\theta + r^k[a_{11}\sin(k+1)\theta$$
$$+ a_{21}\cos(k+1)\theta + a_{31}\sin(k-1)\theta + a_{41}\cos(k-1)\theta]$$
$$+ \sum_{n=2}^{\infty} r^\alpha[a_{1n}\sin(\alpha+1)\theta + a_{2n}\cos(\alpha+1)\theta \tag{3.178}$$
$$+ a_{3n}\sin(\alpha-1)\theta + a_{4n}\cos(\alpha-1)\theta]$$

$$v_h = b_{10} \sin\theta + b_{20} \cos\theta - b_{30} \sin\theta + b_{40} \cos\theta + r^k[b_{11}\sin(k+1)\theta$$
$$+ b_{21}\cos(k+1)\theta + b_{31}\sin(k-1)\theta + b_{41}\cos(k-1)\theta]$$
$$+ \sum_{n=2}^{\infty} r^\alpha[b_{1n}\sin(\alpha+1)\theta + b_{2n}\cos(\alpha+1)\theta \tag{3.179}$$
$$+ b_{3n}\sin(\alpha-1)\theta + b_{4n}\cos(\alpha-1)\theta]$$

Since the nature of the two sets of second-order partial differential equations is similar to one fourth-order partial differential equation, and one orthogonal function is assumed to be known, the number of arbitrary coefficients is four. A careful study of equations 3.43 indicates that there are certain relations between the cosine function part of u and the sine function part of v and *vice versa*. Furthermore, the values of u and v, for each n, must satisfy these equilibrium equations. Hence the relation between a_{in} and b_{jn} is found by substituting a sine term of u and a corresponding cosine term of v into the two equilibrium equations. In this way, the following results are obtained:

$$a_{1n} = b_{2n}, \quad a_{2n} = -b_{1n}$$

$$b_{3n} = \frac{(3 - v) + \alpha(1 + v)}{(3 - v) - \alpha(1 + v)} a_{4n}$$

$$b_{4n} = -\frac{(3 - v) + \alpha(1 + v)}{(3 - v) - \alpha(1 + v)} \alpha_{3n}$$

(3.180)

Letting

$$\beta_{12,\alpha} = -\frac{(3 - v) + \alpha(1 + v)}{(3 - v) - \alpha(1 + v)}$$

(3.181)

the displacements can be written as

$$u_h = \sum_{n=0}^{\infty} r^\alpha [a_{1n} \sin(\alpha + 1)\theta + a_{2n} \cos(\alpha + 1)\theta$$

$$+ a_{3n} \sin(\alpha - 1)\theta + a_{4n} \cos(\alpha - 1)\theta]$$

(3.182)

$$v_h = \sum_{n=0}^{\infty} r^\alpha [-a_{2n} \sin(\alpha + 1)\theta + a_{1n} \cos(\alpha + 1)\theta$$

$$- \beta_{12,\alpha} a_{4n} \sin(\alpha - 1)\theta + \beta_{12,\alpha} a_{3n} \cos(\alpha - 1)\theta]$$

(3.183)

These solutions include only four arbitrary constants while there are, in general, eight boundary conditions. If any four boundary conditions are satisfied by these solutions, at either $\theta = \phi_i$, $\theta = \phi_f$, or $r = r_i$, $r = r_f$, where ϕ_i, ϕ_f, r_i and r_f are the first and the second radial boundaries, and the internal and the external circular boundaries, respectively, the other four boundary conditions should be satisfied by other types of solutions and both types of the solution functions should be combined.

A circular disk, loaded as shown in Figure 3.58 is considered. Two radial lines, $0a$ and $0b$ with a given opening angle, ϕ, are cut such that no external force is acting between them and solutions to the disk problems are evaluated on these lines. The sector which is made by these two cut lines is in equilibrium and the results of the solution to the disk problem evaluated on these lines can represent arbitrary boundary conditions of this sector. When the loadings are changed, the new calculated values represent other conditions. By combining several different conditions on these lines from several different loadings, it is possible to represent specified radial boundary conditions. If the internal values as well as the values on $0a$ and $0b$ are multiplied by the coefficients of the combinations and added, the combination represents the solution of the finite length sectorial plate problem under arbitrary boundary conditions.

Figure 3.58 Symmetrically loaded disk.

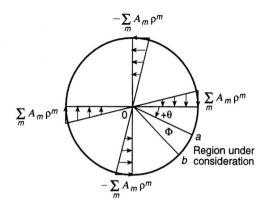

Region under consideration

Symmetric case

The disk with radius r_f is acted upon by partially distributed body forces which are in equilibrium (Figure 3.58). The body force may act in either the r- or θ-direction, or in an arbitrary direction. For the present illustration, we let the body forces in the θ-direction acting on the volume $\frac{1}{2}r_f^2\,\delta\theta\cdot h$ be applied as shown in this figure. h is the thickness of the disk. If the disk is cut by two radial lines, the results of the solution to the disk problem evaluated on these two cut lines may represent symmetric or arbitrary conditions depending on where the cuts are made. The loadings – i.e. the body forces, can be expressed by suitable functions, such as ordinary polynomials, Fourier–Bessel series, Tchebycheff polynomials, hypergeometric functions and so on.

Let an 'in plane force' be expressed by polynomials in r:

$$F(\theta) = \sum_m A_m r^m \tag{3.184}$$

This force may be expanded over the whole plane in terms of a double series in r and θ;

$$F(\theta) = \sum_m \sum_n A_m' r^m G(n, \theta) \tag{3.185}$$

Let r be fixed and consider the elements of a body $r\delta\theta\cdot dr\cdot h$. The body force k_θ at r and any particular line, say $\theta = 0$, is

$$k_{\theta 0} = \frac{\sum A_m r^m}{r\delta\theta\cdot dr\cdot h} \tag{3.186}$$

Consider a particular mth term, and for convenience let

$$A_m = 1$$
$$\delta\theta = \varepsilon \tag{3.187}$$
$$dr = 1$$

The expression of the body force as shown in Figure 3.58 is then

$$k_{\theta, m} = \begin{cases} \rho^{m-1}/(h\varepsilon) & -\varepsilon/2 \le \theta \le \varepsilon/2 \\ -\rho^{m-1}/(h\varepsilon) & (\pi/2) - (\varepsilon/2) \le \theta \le (\pi/2) + (\varepsilon/2) \\ \rho^{m-1}/(h\varepsilon) & \pi - (\varepsilon/2) \le \theta \le \pi + (\varepsilon/2) \\ -\rho^{m-1}/(h\varepsilon) & (3\pi/2) - (\varepsilon/2) \le \theta \le (3\pi/2) + (\varepsilon/2) \end{cases} \tag{3.188}$$

where ρ is the nondimensional radial distance. $k_{\theta,m}$ can be expanded into a Fourier series in θ

$$k_{\theta,m} = \frac{8\rho^{m-1}}{\pi h \varepsilon} \sum_{n=2,6,10,\dots}^{\infty} \frac{1}{n} \sin \frac{n\varepsilon}{2} \cos n\theta \qquad (3.189)$$

If $\varepsilon \to 0$, $\sin n\varepsilon/2 \to n\varepsilon/2$, and

$$k_{\theta,m} = \frac{4\rho^{m-1}}{\pi h} \sum_{n=2,6,10,\dots}^{\infty} \cos n\theta \qquad (3.190)$$

Equations 3.43 indicate that a particular solution can be assumed as

$$u_p = \sum_{n=2,6,10,\dots}^{\infty} u_n \rho^{m+1} \sin n\theta \qquad (3.191)$$

$$v_p = \sum_{n=2,6,10,\dots}^{\infty} v_n \rho^{m+1} \cos n\theta \qquad (3.192)$$

in which u_n and v_n are the coefficients to be found. Substituting equations 3.190, 3.191, and 3.192 into equations 3.43 leads to

$$\beta_{19,n} u_n - \beta_{20,n} v_n = 0 \qquad (3.193)$$

$$\beta_{21,n} u_n + \beta_{22,n} v_n = D_{mn} r_f \qquad (3.194)$$

where

$$\beta_{19,n} = m(m+2) - \tfrac{1}{2} n^2 (1-v)$$

$$\beta_{20,n} = n(1-v) - \tfrac{1}{2} mn(1+v)$$

$$\beta_{21,n} = n[4 + m(1+v)]$$

$$\beta_{22,n} = m(m+2)(1-v) - 2n^2$$

$$D_{mn} = -8(1-v^2)/(E\pi h)$$

From equations 3.193 and 3.194,

$$u_n = \frac{\beta_{20,n} D_{mn} r_f}{\beta_{19,n} \beta_{22,n} - \beta_{20,n} \beta_{21,n}}$$

$$v_n = \frac{\beta_{19,n} D_{mn} r_f}{\beta_{19,n} \beta_{22,n} - \beta_{20,n} \beta_{21,n}}$$

Introducing the abbreviation

$$\theta_{E,m} = \frac{D_{mn} r_f}{\beta_{19,n} \beta_{22,n} - \beta_{20,n} \beta_{21,n}}$$

Equations 3.191 and 3.192 may be written as

$$u_p = -\rho^{m+1} \sum_{n=2,6,10,\dots}^{\infty} \beta_{20,n} \theta_{E,m} \sin n\theta \qquad (3.195)$$

$$v_p = \rho^{m+1} \sum_{n=2,6,10,\dots}^{\infty} \beta_{19,n} \theta_{E,m} \cos n\theta \qquad (3.196)$$

A singularity condition may occur when $m = n$. To avoid this possibility, a fractional value can be taken for m.

For the homogeneous solution, equations 3.178 and 3.179 can be used. Letting $n = \alpha_1 + 1$,

$$u_1 = \mu_{1n} r^{\alpha_1} \sin(\alpha_1 + 1)\theta$$
$$v_1 = v_{1n} r^{\alpha_1} \cos(\alpha_1 + 1)\theta \tag{3.197}$$

Letting $n = \alpha_2 - 1$,

$$u_2 = n_{2n} r^{\alpha_2} \sin(\alpha_2 - 1)\theta$$
$$v_2 = v_{2n} r^{\alpha_2} \cos(\alpha_2 - 1)\theta \tag{3.198}$$

From the results of equations 3.180,

$$v_{1n} = u_{in}$$
$$v_{2n} = \beta_{12,n} u_{2n} \tag{3.199}$$

where $\beta_{12,n}$ is found by substituting $\alpha = n + 1$ into equation 3.181 as

$$\beta_{12,n} = \frac{4 + n(1 + v)}{n(1 + v) - 2(1 - v)}$$

The complete solution is

$$u = u_{\mathrm{h}} + u_{\mathrm{p}} = \sum_{n=2,6,10,\ldots}^{\infty} (u_{1n}\rho^{n-1} + u_{2n}\rho^{n+1} - \beta_{20,n}\theta_{\mathrm{E},m}\rho^{m+1})\sin n\theta \tag{3.200}$$

$$v = v_{\mathrm{h}} + v_{\mathrm{p}} = \sum_{n=2,6,10,\ldots}^{\infty} (u_{1n}\rho^{n-1} + \beta_{12,n}u_{2n}\rho^{n+1} + \beta_{19,n}\theta_{\mathrm{E},m}\rho^{m+1})\cos n\theta$$
$$\tag{3.201}$$

where u_{1n} and u_{2n} are coefficients to be determined by the boundary conditions at $\rho = 1$ or $r = r_{\mathrm{f}}$.

As an example, consider the case where the circular boundary at $\rho = 1$ is free so that $\sigma_r = 0$, $\tau_{r\theta} = 0$. Letting

$$\beta_{18,n} = \frac{1}{\beta_{13,n}\beta_{17,n} - \beta_{14,n}\beta_{16,n}},$$
$$\beta_{1,n} = -\beta_{18,n}(\beta_{23,n}\beta_{17,n} - \beta_{24,n}\beta_{14,n}) \tag{3.202}$$
$$\beta_{2,n} = -\beta_{18,n}(\beta_{24,n}\beta_{13,n} - \beta_{23,n}\beta_{16,n})$$

where

$$\beta_{13,n} = (n - 1)(1 - v)$$

$$\beta_{14,n} = \frac{(n - 2)(n + 1)(1 - v^2)}{n(1 + v) - 2(1 - v)}$$

$$\beta_{16,n} = 2(n - 1)$$

$$\beta_{17,n} = \frac{2n(n + 1)(1 + v)}{n(1 + v) - 2(1 - v)}$$

$$\beta_{23,n} = -[\beta_{20,n}(m + 1) + v(n\beta_{19,n} + \beta_{20,n})]$$

$$\beta_{24,n} = -n\beta_{20,n} + m\beta_{19,n}$$

and substituting equations 3.200 and 3.201 into the boundary condition expressions

$$\sigma_r = 0$$

$$\tau_{r\theta} = 0 \quad \text{at} \quad \rho = 1$$

we obtain

$$u_{1n} = \beta_{1,n}\theta_{E,m}$$
$$u_{2n} = \beta_{2,n}\theta_{E,m}$$

(3.203)

Since the circular disk problem under a specific polynomial type force distribution is solved, the next step is to determine the coefficients A_m so that $\sum_m A_m\rho^m$ of equation 3.184 can express the specified radial boundary conditions.

Antisymmetric case

The disk as shown in Figure 3.59 is in equilibrium and, if the first quarter-plane is considered, an antisymmetric case is encountered.

Proceeding as in the symmetric case,

$$K_{\theta,m} = \frac{\rho^{m-1}}{\pi h}\left(8\sum_{n=4,12,20,\ldots}^{\infty}\cos n\theta - 4\sum_{n=2,6,10,\ldots}^{\infty}\sin\frac{3n\pi}{4}\sin n\theta\right)$$

(3.204)

and the solution can be obtained as

$$u = \sum_{n=4,12,20,\ldots}^{\infty}(u_{1n}\rho^{n-1} + u_{2n}\rho^{n+1} - \beta_{20,n}\theta_{EC,m}\rho^{m+1})\sin n\theta$$

$$+ \sum_{n=2,6,10,\ldots}^{\infty}[u_{3n}\rho^{n-1} + u_{4n}\rho^{n+1} - \beta_{20,n}\theta_{ES,m}\rho^{m+1}]\cos n\theta$$

(3.205)

and

$$v = \sum_{n=4,12,20,\ldots}^{\infty}[u_{1n}\rho^{n-1} + \beta_{12,n}u_{2n}\rho^{n+1} + \beta_{19,n}\theta_{EC,m}\rho^{m+1}]\cos n\theta$$

$$+ \sum_{n=2,6,10,\ldots}^{\infty}[-u_{3n}\rho^{n-1} - \beta_{12,n}u_{4n}\rho^{n+1} - \beta_{19,n}\theta_{ES,m}\rho^{m+1}]\sin n\theta$$

(3.206)

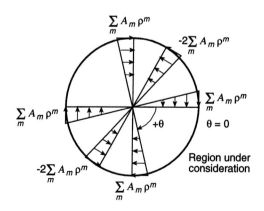

Figure 3.59 Antisymmetrically loaded disk.

where

$$\theta_{EC,m} = \frac{2D_{mn}r_f}{\beta_{19,n}\beta_{22,n} - \beta_{20,n}\beta_{21,n}} = 2\theta_{E,m}$$

and

$$\theta_{ES,m} = \frac{\sin\dfrac{3n\pi}{4} D_{mn}r_f}{\beta_{19,n}\beta_{22,n} - \beta_{20,n}\beta_{21,n}} = \sin\frac{3n\pi}{4}\theta_{E,m}$$

If the outer circular boundary is free,

$$u_{1n} = \beta_{1,n}\theta_{EC,m}$$
$$u_{2n} = \beta_{2,n}\theta_{EC,m}$$
$$u_{3n} = \beta_{1,n}\theta_{ES,m}$$
$$u_{4n} = \beta_{2,n}\theta_{ES,m}$$

(3.207)

obtained by the boundary conditions, $\sigma_r = \tau_{r\theta} = 0$.

Mathematically, any radial boundary condition can now be expressed by combining the general results of the symmetric and the antisymmetric cases.

3.42.3 Normal to the plane problem

The normal to the plane problem is to find solutions to the classical plate equation

$$\nabla^4 w = \frac{q_p}{D} \tag{3.208}$$

where

- w is the deflection;
- q_p is the normal load per unit surface area of the plate;
- D is the flexural rigidity of the plate expressed as $Eh^3/[12(1 - v^2)]$.

In a polar coordinate system, the plate equation is

$$\left(\frac{\partial^2}{\partial r^2} + \frac{1}{r}\frac{\partial}{\partial r} + \frac{1}{r^2}\frac{\partial^2}{\partial\theta^2}\right)\left(\frac{\partial^2 w}{\partial r^2} + \frac{\partial w}{r\,\partial r} + \frac{\partial^2 w}{r^2\,\partial\theta^2}\right) = q_p/D \tag{3.209}$$

The slopes, moments, shear forces, and Kirchhoff forces are expressed as

$$S_r = \frac{\partial w}{\partial r}$$

$$S_t = \frac{\partial w}{r\,\partial\theta}$$

$$M_r = -D\left(\frac{\partial^2 w}{\partial r^2} + v\left(\frac{\partial w}{r\,\partial r} + \frac{\partial^2 w}{r^2\,\partial\theta^2}\right)\right)$$

$$M_t = -D\left(v\frac{\partial^2 w}{\partial r^2} + \frac{\partial w}{r\,\partial r} + \frac{\partial^2 w}{r^2\,\partial\theta^2}\right)$$

$$M_{rt} = (1 - v)D\left(\frac{\partial^2 w}{r\,\partial r\,\partial\theta} - \frac{\partial w}{r^2\,\partial\theta}\right) = -M_{tr}$$

$$Q_r = -D\frac{\partial}{\partial r}(\nabla^2 w) = -D\frac{\partial}{\partial r}\left(\frac{\partial^2 w}{\partial r^2} + \frac{\partial w}{r\,\partial r} + \frac{\partial^2 w}{r^2\,\partial\theta^2}\right)$$

$$Q_t = -D\frac{\partial}{r\,\partial\theta}(\nabla^2 w) = -D\frac{\partial}{r\,\partial\theta}\left(\frac{\partial^2 w}{\partial r^2} + \frac{\partial w}{r\,\partial r} + \frac{\partial^2 w}{r^2\,\partial\theta^2}\right)$$

$$V_r = Q_r - \frac{\partial M_{rt}}{r\,\partial\theta} = -D\left(\frac{\partial}{\partial r}(\nabla^2 w) + (1 - v)\left(\frac{1}{r^2}\frac{\partial^3 w}{\partial r\,\partial\theta^2} - \frac{1}{r^3}\frac{\partial^2 w}{\partial\theta^2}\right)\right)$$

$$V_t = Q_t + \frac{\partial M_{tr}}{\partial r} = -D\left(\frac{\partial}{r\,\partial\theta}(\nabla^2 w) + (1 - v)\left(\frac{\partial^3 w}{r\,\partial r^2\,\partial\theta} - \frac{2}{r^2}\frac{\partial^2 w}{\partial r\,\partial\theta} + \frac{2}{r^3}\frac{\partial w}{\partial\theta}\right)\right)$$

$$(3.210)$$

According to Timoshenko [2], the general solution to the biharmonic equation, $\nabla^4 w = 0$, for a circular plate, in the form of a Fourier series in θ, was given by Clebsch as

$$w = R_0 = \sum_{m=1}^{\infty} (A_m r^m + B_m r^{-m} + C_m r^{m+2} + D_m r^{-m+2})\cos m\theta$$

$$+ \sum_{m=1}^{\infty} (A'_m r^m + B'_m r^{-m} + C'_m r^{m+2} + D'_m r^{-m+2})\sin m\theta \qquad (3.211)$$

By means of inversion, Michell [13] solved the problem of a partially loaded circular plate. He also solved for the deflections of a circular plate which was clamped at the radial edges and subjected to a concentrated load, by using a Green's function for $\nabla^4 w = 0$, combined with the method of inversion [14]. Nadai solved the simply supported sectorial plate under uniform load, using the Clebsch solution and a method similar to inversion [15]. Carrier and Shaw [16] reported a simple, approximate method for a sectorial plate under various loadings and boundary conditions. A method credited to Weinstein was used by Hasse [17] to solve the semicircular plate under uniform load. Uflyand is known to have solved similar problems by using integral transforms [18]. The plate with two radial boundaries and two circular arcs was treated by Dixon [19] and Carrier [20]. Considering a sector of infinite radius and using integral transforms, Woinowsky-Krieger [18, 21] solved the problem of bending of a sectorial plate with arbitrary conditions along the radial edges for the case of a single load. This assumption of an infinite radius was advantageous as long as the question of stress distribution near the corner point of the plate was the only concern and it was found that no singularity condition existed when the opening angle was less than 90°. The method of eigenfunctions was used by Williams [22] to find the singularity condition at the angular corner of the sector. He found that the most pronounced singularity condition appeared in the case of a simply supported sectorial plate with an opening angle greater than 90° and that no singularity occurred when either the edges were clamped or the opening angle was less than 90°. Kawai [23] treated the sectorial plate with fixed circumferential and simply supported radial boundaries using Four-ier–Bessel double series, single series, Green's functions and conformal mapping.

By Green's functions it was found that the moments and the corner reaction did not appear at the apex when the opening angle of the pertinent sector was less than 90°. This fact confirms the results obtained by Williams. Reisman [24], Godfrey [25], and Scherer [1, 26] also worked on various sectorial plate problems.

Because of the nature of the biharmonic equation, it is possible to assume another form of the solution such as

$$w = \sum_{n=0}^{\infty} w_n(\theta) r^n \tag{3.212}$$

Substituting this expression for w into equation 3.209 with $q_p = 0$, we obtain

$$w = \sum_{n=0}^{\infty} r^n [w_{n1} \sin n\theta + w_{n2} \cos n\theta + w_{n3} \sin(n-2)\theta + w_{n4} \cos(n-2)\theta] \tag{3.213}$$

While the Clebsch solution is well suited to the circular boundary problems, this solution is better suited to radial boundary problems.

Because of the difficulties with boundary conditions, none of the previously mentioned methods is satisfactory for a sectorial plate with elastic supports. For the case of the arbitrary support, the method used for in-plane problems can be employed.

Antisymmetric case

Consider the plate as shown in Figure 3.60. If two radial lines are properly chosen in the first quarter of the plate, the sector bounded by the radial lines represents the plate under antisymmetric conditions. Proceeding as the in-plane problem,

$$q_p(r, \theta) = \frac{4\rho^{m-1}}{\pi} \sum_{n=2,6,10,\ldots}^{\infty} \cos n\theta \tag{3.214}$$

and

$$w_p = \rho^{m+3} \sum_{n=2,6,10,\ldots}^{\infty} a_n \cos n\theta \tag{3.215}$$

where

$$a_n = \frac{4 r_f^4}{\pi D [n^2 - (m+3)^2][n^2 - (m+1)^2]}$$

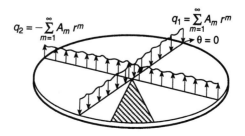

$$q_2 = -\sum_{m=1}^{\infty} A_m r^m \qquad\qquad q_1 = \sum_{m=1}^{\infty} A_m r^m$$
$$\theta = 0$$

Figure 3.60 Antisymmetric plate.

Since the plate under consideration is circular, the Clebsch solution can be used for the homogeneous solution. Since the deflection at the origin is considered to be finite, the terms which may give singular conditions at this point are discarded. Then

$$w = \sum_{n=2,6,10,...}^{\infty} [a_{1n} \cdot \rho^n + a_{2n} \cdot \rho^{n+2} + a_n \cdot \rho^{m+3}] \cos n\theta \qquad (3.216)$$

If the circular boundary at $\rho = 1$ is assumed as free, as an example,

$$M_r(\rho = 1) = \frac{-D}{r_f^2} \sum_{n=2,6,10...}^{\infty} [\beta_{5,n} a_{1n} \rho^{n-2} + \beta_{9,n} a_{2n} \rho^n$$

$$+ \beta_{28,n} Q_{p,m} \rho^{m+1}] \cos n\theta = 0$$

$$V_r = \frac{D}{r_f^3} \sum_{n=2,6,10,...}^{\infty} [n\beta_{5,n} a_{1n} \rho^{n-3} + \beta_{11,n} a_{2n} \rho^{n-1} - \beta_{29,n} Q_{p,m} \rho^m] \cos \theta$$

$$= 0$$

$$(3.217)$$

in which

$$\beta_{5,n} = n(n-1)(1-v)$$
$$\beta_{9,n} = (n+1)[(n+1) - v(n-2)]$$
$$\beta_{11,n} = n(n+1)[n(1-v) - 4]$$
$$\beta_{28,n} = (m+3)(m+2) + v[(m+3) - n^2]$$
$$\beta_{29,n} = (m+3)^2(m+1) - (m+3)n^2(1-v) - n^2 v - mn^2$$
$$Q_{p,m} = a_n$$

From equation 3.217, we can obtain a_{1n} and a_{2n} for specific n:

$$a_{1n} = \beta_{3,n} Q_{p,m}$$
$$a_{2n} = \beta_{4,n} Q_{p,m}$$

$$(3.218)$$

where

$$\beta_{3,n} = -\beta_{27,n}[\beta_{11,n}\beta_{28,n} + \beta_{9,n}\beta_{29,n}]$$
$$\beta_{4,n} = \beta_{27,n}\beta_{5,n}[\beta_{29,n} + n\beta_{28,n}]$$
$$\beta_{27,n} = 1/[\beta_{5,n}(\beta_{11,n} - n\beta_{9,n})]$$

Symmetric case

If we draw two radial lines properly in the first quarter of the disk shown in Figure 3.61, the resulting sectorial plate is under symmetric conditions.

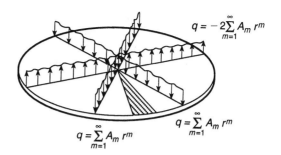

Figure 3.61 Symmetric plate.

By using procedures similar to those employed in the previous case, the following result can be found:

$$w = \sum_{n=4,12,20,\ldots}^{\infty} [as_{1n}\rho^n + as_{2n}\rho^{n+2} + Q_{\mathrm{pc},m}\rho^{m+3}]\cos n\theta$$

$$+ \sum_{n=2,6,10,\ldots}^{\infty} [as_{3n}\rho^n + as_{4n}\rho^{n+2} + Q_{\mathrm{ps},m}\rho^{m+2}]\sin n\theta \qquad (3.219)$$

where

$$Q_{\mathrm{pc},m} = \frac{8r_{\mathrm{f}}^4}{\pi D[n^2 - (m+3)^2][n^2 - (m+1)^2]}$$

$$Q_{\mathrm{ps},m} = \frac{-4\sin(3n\pi/4)r_{\mathrm{f}}^4}{\pi D[n^2 - (m+3)^2][n^2 - (m+1)^2]}$$

$$as_{1n} = \beta_{3,n}\theta_{\mathrm{pc},m}$$

$$as_{2n} = \beta_{4,n}\theta_{\mathrm{pc},m}$$

$$as_{3n} = \beta_{3,n}\theta_{\mathrm{ps},m}$$

$$as_{4n} = \beta_{4,n}\theta_{\mathrm{ps},m}$$

3.42.4   On convergence of solutions

Since the lines on which the loads are applied can be avoided when the equations are evaluated, all terms of the equations converge. However, it is found that the series for some of the quantities converge slowly. Certain methods can be employed to accelerate the convergence. One such method is to separate the slow converging part from those terms with n^2 or higher powers. For example, consider the term $n\beta_{5,n}a_{1n}\rho^{n-3}\cos n\theta$ in equation 3.217. We may write

$$n\beta_{5,n}a_{1n} = -\frac{2v(1-v)}{\pi(3+v)} + \frac{8v}{n\pi(3+v)} - \frac{2(1-v)[(m+3)(1-v)+v+m]}{\pi(3+v)n}$$

$$+ f(n^2 \text{ or higher})$$

It can be shown that [27]

$$\sum_{n=1,2,3,\ldots}^{\infty} \rho_1^n \cos n\theta_1 = \frac{\rho_1(\cos\theta_1 - \rho_1)}{1 - 2\rho_1\cos\theta_1 + \rho_1^2} \tag{3.220}$$

Letting $\rho^2 = \rho_1$ and $2\theta = \theta_1$, this becomes

$$\sum_{n=2,4,6,\ldots}^{\infty} \rho^n \cos n\theta = \frac{\rho^2(\cos 2\theta - \rho^2)}{1 - 2\rho^2\cos 2\theta + \rho^4} \tag{3.221}$$

Subtracting equation 3.221 from 3.220,

$$\sum_{n=1,3,5,\ldots}^{\infty} \rho^n \cos n\theta = \frac{\rho(\cos\theta - \rho)}{1 - 2\rho\cos\theta + \rho^2} - \frac{\rho^2(\cos 2\theta - \rho^2)}{1 - 2\rho^2\cos 2\theta + \rho^4} \tag{3.222}$$

Letting $\rho^2 = \rho$, $2\theta = \theta$ again,

$$\sum_{n=2,6,10,\ldots}^{\infty} \rho^n \cos n\theta = \frac{\rho^2(\cos 2\theta - \rho^2)}{1 - 2\rho^2\cos 2\theta + \rho^4} - \frac{\rho^4(\cos 4\theta - \rho^4)}{1 - 2\rho^4\cos 4\theta + \rho^8} \tag{3.223}$$

Such a scheme will make the series solution converge faster.

Another practical scheme is to use the loadings distributed over certain areas, instead of line loadings.

References

1. Goldberg, J. E. and Kim, D. H. (1967) The effect of neglecting the radial moment terms in analyzing a sectorial plate by means of finite differences, in *Proceedings 7th International Symposium on Space Technology and Science* (eds Kuroda, Y. et al.), Tokyo, pp. 267–78.
2. Michell, J. H. (1899) On the direct determination of stress in an elastic solid with application to the theory of plates. *London Math. Soc. Proc.*, **31**, 100.
3. Sneddon, I. N. (1951) *Fourier Transforms*, 2nd edn, McGraw-Hill, New York.
4. Watson, G. N. (1922) *A Treatise on the Theory of Bessel Functions*, The University Press, Cambridge.
5. Tranter, C. J. (1948) The use of Mellin transform in infinite wedge. *Q. J. Mech. Appl. Maths*, **1**, 125.
6. Tranter, C. J. (1951) *Integral Transformations in Mathematical Physics*, John Wiley, New York, p. 53.
7. Brahtz, J. H. A. (1933) Stress distribution in a reentrant corner. *Trans. Am. Soc. Mech. Engrs*, **55**, 31–7.
8. Brahtz, J. H. A. (1936) The stress function and photoelasticity applied to dams. *Trans. Am. Soc. Civ. Engrs*, **101**, 1240.
9. Silverman, I. K. (1955) Approximate stress functions for triangular wedge. *Trans. Am. Soc. Civ. Engrs*, **77**.
10. Horvay, G. and Hansen, K. L. (1957) The sector problem. *Trans. Am. Soc. Mech. Engrs, J. Appl. Mech.*, Dec.
11. Nomachi, S. G. (1960) On one method of solving stress problems in cylindrical coordinates by means of finite Fourier Hankel transforms Part I. *Mem. Muroran Inst. Technol.*, **3**(3), June.
12. Nomachi, S. G. (1960) Method of solving stress problems in cylindrical coordinates by means of finite Fourier Hankel transforms, Part II. *Mem. Muroran Inst. Technol.*, **3**(4), June.
13. Michell, J. H. (1901) The inversion of plane stress. *Proc. London Math. Soc.*, **34**, Oct., 134.

14. Michell, J. H. (1902) The flexure of a circular plate. *Proc. London Math. Soc.*, **34**, Apr., 223.
15. Timoshenko, S. and Woinowsky-Krieger, S. (1959) *Theory of Plates and Shells*, McGraw-Hill, New York.
16. Carrier, G. F. and Shaw, F. S. (1950) Some problems in bending of thin plates. *Proc. Symp. Appl. Math.*, **3**, 125.
17. Hasse, H. R. (1950) The bending of uniformly loaded clamped plate in the form of circular sector. *Q. J. Mechs Appl. Maths*, **3**, 271.
18. Woinowsky-Krieger, S. (1953) The bending of a wedge shaped plate. *J. Appl. Mechs, Trans. Am. Soc. Mech. Engrs*, **78**, 77.
19. Dixon, A. C. (1920) Theory of thin elastic plate bounded by two circular arcs and clamped straight edges. *Proc. Math. Soc. London*, **19**, 373.
20. Carrier, G. F. (1944) Bending of clamped sectorial plate. *J. Appl. Mechs, Am. Soc. Mech. Engrs*, **66**, A134.
21. Woinowsky-Krieger, S. (1952) Uber die Anwendung der Mellin Transformation zur Lösung einer Aufgabe der Plattenbiegung. *Ing. Archiv.*, **20**, 391.
22. Williams, M. L. Surface stress singularities resulting from various boundary conditions in angular corner of plates under bending. *Proceedings First US National Congress of Applied Mechanics*, p. 325.
23. Kawai, T. (1958) On the bending of a sectional plate. *Int. Assoc. Bridge Struct. Engrs*, **18**, 63.
24. Reisman, M. Bending of clamped wedge plates. *J. Appl. Mechs*, **75**, 141.
25. Godfrey, D. E. R. (1955) Normal loading of a wedge shaped plate. *Aeronaut. Q.*, **6**(6), 196–205.
26. Scherer, H. (1957) Einflussflächen einer Dreiecks Platte mit Aufpunkt am Freien Rand. *Ing. Archiv.*, **26**(4), 255.
27. Tolke, F. (1950) *Praktische Funktionenlehre*, Vol. 1, 2nd edn, Springer, Berlin.

Further reading

Lass, H. (1950) *Vector and Tensor Analysis*, McGraw-Hill, New York.

Love, A. E. H. (1944) *A Treatise on Mathematical Theory of Elasticity*, 4th edn, Dover Publications, New York.

Shu, S. S. (1987) *Boundary Value Problems of Linear Partial Differential Equations for Engineers and Scientists*, World Scientific, Singapore.

Sokolnikoff, I. S. (1951) *Tensor Analysis, Theory and Applications*, John Wiley, New York.

Sokolnikoff, I. S. (1956) *Mathematical Theory of Elasticity*, 2nd edn, McGraw-Hill, New York.

Struik, D. J. (1950) *Lectures on Classical Differential Geometry*, Addison–Wesley, Cambridge, MA.

Timoshenko, S. (1958) *Strength of Materials*, Parts I and II, Van Nostrand, New York.

Timoshenko, S. and Goodier, J. N. (1951) *Theory of Elasticity*, 2nd edn, McGraw-Hill, New York.

4 Eigenvalue problems of beams and frames of isotropic materials

In Chapter 3, the methods of analysis for the maximum deflections and stresses in several isotropic structural elements are illustrated to be used for determining critical stress and strain criteria needed in design procedure. The methods used for isotropic elements are useful for composite material elements, in principle, and can be applied directly for composite material members for preliminary design phases.

In addition to deflections and stresses, there are two other situations in which the structure may fail; one is the response to dynamic loads that cause deflections and, consequently, stresses that are too large, and the other is the elastic instability or buckling.

4.1 Stability of beams and frames

4.1.1 Critical load of a column

Consider a beam as shown in Figure 4.1. Assuming the deflection is small,

$$\frac{d^2y}{dx^2} + k^2 y = \frac{-QC}{EIL} x, \quad x < L - C$$

$$\frac{d^2y}{dx^2} + k^2 y = \frac{-Q(L-C)(L-x)}{EIL}, \quad x \geq L - C$$

where $k = \sqrt{P/EI}$. Solutions to these equilibrium equations may be written as

$$y = A \sin kx + B \cos kx - \frac{QC}{PL} x$$

$$y = a \sin kx + b \cos kx - \frac{Q(L-C)(L-x)}{PL}$$

Applying the boundary conditions, $y = 0$, at $x = 0$ and $x = L$, $y^{(-)} = y^{(+)}$, at $x = L - C$, and

$$\frac{dy^{(-)}}{dx} = \frac{dy^{(+)}}{dx}, \quad \text{at } x = L - C$$

$$y = \frac{Q \sin kC}{Pk \sin kL} \sin kx - \frac{QCx}{PL} \quad \text{for } x < L - C \qquad (4.1)$$

$$y = \frac{Q}{Pk} \frac{\sin k(L-C)}{\sin kL} \sin(L - x) - \frac{Q(L-C)(L-x)}{PL} \quad \text{for } x \geq L - C$$

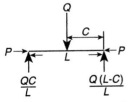

Figure 4.1 Beam under single transverse load and an axial load.

Alternatively, the second equation can be obtained by replacing x by $(L - x)$ in the first equation. As a general case, if the beam as shown in Figure 4.1 is under n transverse loadings $Q_1, Q_2, ..., Q_n$, the deflection at a point which lies between Q_m and Q_{m+1} is

$$y = \sum_{i=1}^{m} \left[\frac{Q_i}{Pk} \frac{\sin k(L - C_i)}{\sin kL} \sin k(L - x) - \frac{Q_i(L - C_i)(L - x)}{PL} \right]$$

$$+ \sum_{i=m+1}^{n} \left[\frac{Q_i}{Pk} \frac{\sin kC_i}{\sin kL} \sin kx - \frac{Q_i C_i x}{PL} \right] \tag{4.2}$$

If we examine both equations 4.1 and 4.2, the deflection becomes unbound if $\sin KL \to 0$, or $KL \to n\pi$. This indicates that if the axial load P approaches $(n^2\pi^2 EI)/L^2$, the deflection increases unbound without an increase in P.

By definition, the **critical load** is that load which is just able to hold the column in equilibrium in the deflected state and is expressed as

$$P_{cr} = \frac{n^2\pi^2 EI}{L^2} \tag{4.3}$$

The minimum of P_{cr} is the **Euler load** for buckling given as

$$P_e = P_{cr(n=1)} = \frac{\pi^2 EI}{L^2} \tag{4.4}$$

This Euler's formula for a uniform, pin-ended straight member loaded by only an axial load P, can be obtained directly from equation 3.164, which gives us

$$M = Py = -EI \frac{d^2y}{dx^2} \quad \text{or} \quad \frac{d^2y}{dx^2} + \frac{P}{EI} y = 0$$

The solution can be written as

$$y = A \sin\left(\sqrt{\frac{P}{EI}} x\right) + B \cos\left(\sqrt{\frac{P}{EI}} x\right)$$

Applying the boundary conditions,

$$y = 0 \quad \text{at } x = 0 \to B = 0$$

$$y = 0 \quad \text{at } x = L \to A \sin\left(\sqrt{\frac{P}{EI}} L\right) = 0$$

If A is zero, the solution becomes trivial and we have to have

$$\sin\left(\sqrt{\frac{P}{EI}} L\right) = 0 \quad \text{or} \quad \sqrt{\frac{P}{EI}} L = n\pi$$

so that

$$P_{cr} = \frac{n^2\pi^2 EI}{L^2}$$

$$P_e = P_{cr(n=1)} = \frac{\pi^2 EI}{L^2}$$

Figure 4.2 Critical load of columns with different end conditions.

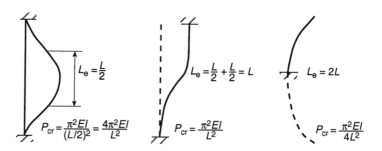

$$L_e = \frac{L}{2} \qquad L_e = \frac{L}{2} + \frac{L}{2} = L \qquad L_e = 2L$$

$$P_{cr} = \frac{\pi^2 EI}{(L/2)^2} = \frac{4\pi^2 EI}{L^2} \qquad P_{cr} = \frac{\pi^2 EI}{L^2} \qquad P_{cr} = \frac{\pi^2 EI}{4L^2}$$

If a beam has other than pin-ended supports, one may locate the inflection points of the deflected shape and obtain the 'equivalent' length of the beam, and obtain P_{cr} from equation 4.4. Some examples are shown in Figure 4.2.

4.1.2 Critical column load above elastic limit

The Euler formula has been developed on the assumption that the material is perfectly elastic. If the stress exceeds the proportional limit, a formula similar to the Euler formula may be used to determine the critical load.

If the load **continues to increase** during buckling, the appropriate modulus is the tangential modulus, E_t, which is the local slope of the stress–strain curve (Figure 4.3). If the load is not permitted to increase during the buckling operation, then the correct modulus, generally referred to as the reduced modulus, E_r, lies between the elastic modulus and the tangential modulus and is hence greater than the tangential modulus. However, whereas the tangential modulus is independent of the shape of the cross-section, the reduced modulus depends on the shape of the cross-section. Thus, the modified critical load is either

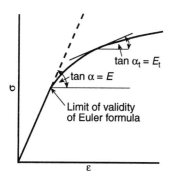

Figure 4.3 Stress–strain curve.

$$P_{cr} = \frac{\pi^2 E_r I}{L^2} \quad \text{or} \quad P_{cr} = \frac{\pi^2 E_t I}{L^2} \tag{4.5}$$

4.1.3 Energy method for stability problems

There are several methods to solve stability problems. The energy method is one of the often used methods. If we let λ be the distance through which the axial load P travels during the buckling process,

$$\lambda = \frac{1}{2} \int_0^L \left(\frac{dy}{dx}\right)^2 dx \tag{4.6}$$

L is the original length of the straight column, but it also designates the arc length of the buckled column. The work done by the load P through λ is

$$W = P\lambda$$

The strain energy is

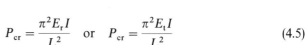

$$U = \int_0^L \frac{M^2\,dx}{2EI} = \int_0^L \frac{(Py)^2\,dx}{2EI} \quad \text{or if the column is uniform,} \quad \frac{EI}{2}\int_0^L \left(\frac{d^2y}{dx^2}\right)^2 dx \tag{4.7}$$

P_{cr} can be obtained by setting $U = W$.

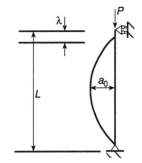

Figure 4.4 Simple column.

If the assumed shape of y is not correct, it is more accurate to use $M = Py$, instead of

$$M = -EI \frac{d^2 y}{dx^2}$$

In order to illustrate this fact as well as the application of the energy method, we consider the column as shown in Figure 4.4.

We assume the deflected shape as

$$y = \frac{4a_0}{L^2} x(L - x), \text{ then}$$

$$\lambda = \frac{1}{2} \int_0^L \left(\frac{dy}{dx}\right)^2 dx = \left(\frac{8a_0^2}{3}\right)/L$$

$$W = P\lambda = \frac{8}{3}\frac{a_0^2}{L} P$$

1. When we use $M = -EI(d^2 y/dx^2)$

$$U = \frac{EI}{2} \int_0^L \left(\frac{d^2 y}{dx^2}\right) dx = \frac{32EIa_0^2}{L^3}$$

Letting $W = U$,

$$P_{cr} = \frac{12EI}{L^2}$$

2. When $M = Py$ is used,

$$U = \int_0^L \frac{(Py)^2 \, dx}{2EI} = \frac{4}{15}\frac{P^2 a_0^2 L}{EI}$$

From $W = U$, we get

$$P_{cr} = \frac{10EI}{L^2}$$

Note that the true $P_{cr} = \pi^2 EI/L^2 = 9.876EI/L^2$. Note that P_{cr} obtained by the energy method is **generally higher** than the true one, thus it is on the unsafe side.

When the energy equation in the form $U - P\lambda = 0$ does not yield enough information, we may use a more fundamental and general approach. Since the buckled system is in equilibrium, the **principle of virtual work** applies. We may arbitrarily take the virtual displacement to be the small change in the configuration of the buckled system corresponding to a small change (or variation, δa) in the magnitude of a, when the deflected mode shape is assumed in the form of $a_n f_n(x)$.

Using the principle of virtual work to analyze the stability problem:

1. Assume a deflection curve,

$$y = a_1 f_1(x) + \cdots + a_n f_n(x)$$

The *as* are assumed to be those which give the best possible approximation to the actual deflection curve with chosen *fs*.

2. Let the amplitudes of the functions used in defining the deflection curve undergo a 'variation',

$$\delta H_i = \frac{\partial H}{\partial a_i}\,\delta a_i \quad \text{etc.}$$

3. Solve the resulting equations for P_{cr}.

For the **Rayleigh–Ritz procedure**:

1. Assume the deflection curve

$$y = a_1 f_1(x) + a_2 f_2(x) + \cdots + a_n f_n(x)$$

where $f_1, ..., f_n$ are different specified functions each of which satisfies the boundary conditions and $a_1, a_2, ..., a_n$ are undetermined coefficients.

2. Adjust the coefficients $a_1, a_2, ..., a_n$ to obtain the best possible representation of the deflection curve with the specified functions,

$$f_1, f_2, ..., f_n \quad \text{by} \quad \frac{\partial(U - W)}{\partial a} = 0$$

One may compute the strain energy of the buckled column by means of

$$U = \frac{1}{2}\int_0^L EI\left(\frac{d^2 y}{dx^2}\right)^2 dx$$

U will turn out to be a quadratic expression in the *as*; $a_1^2, a_2^2, ..., a_1 a_2, ...,$ each of which will be multiplied by a reference value of EI/L. Also λ and hence $P\lambda$ will turn out to be a quadratic expression in the *as*. Write the energy equation, $U - W = 0$. Since the structure is in equilibrium both before and after buckling occurs, taking the derivative of this equation with respect to the *as*, we obtain a set of linear homogeneous algebraic equations in the *as*:

$$A_1 a_1 + A_2 a_2 + \cdots + A_n a_n = 0$$
$$\vdots \qquad\qquad \vdots \qquad\qquad\qquad (4.8)$$
$$N_1 a_1 + N_2 a_2 + \cdots \quad N_n a_n = 0$$

In order for these to be a nontrivial solution for the *as*, the determinant of this set of equations must be equal to zero.

From equations 4.6 and 4.7.

$$U = \frac{1}{2}\int_0^L EI\left(\frac{d^2 y}{dx^2}\right)^2 dx$$

$$W = P\lambda = \frac{P}{2}\int_0^L \left(\frac{dy}{dx}\right)^2 dx$$

Substituting into the energy equation, $U - W = 0$, and solving for P_{cr}, we obtain

$$P_{cr} = \frac{\displaystyle\int_0^L EI\left(\frac{d^2 y}{dx^2}\right)^2 dx}{\displaystyle\int_0^L \left(\frac{dy}{dx}\right)^2 dx} \qquad (4.9)$$

Since this is an energy method, we know that any value of P_{cr} which is calculated by this means is never too low. We therefore seek, among all possibilities, which remain available to us following our choice of $f_1, ..., f_n$, the proper combinations of as which will give a minimum value of P_{cr}. Thus

$$\frac{\partial P_{cr}}{\partial a_i} = 0$$

Substituting the expression of P_{cr} given in equation 4.9, and again substituting

$$P_{cr} \int_0^L \left(\frac{dy}{dx}\right)^2 dx \quad \text{for} \quad \int_0^L EI\left(\frac{d^2y}{dx^2}\right)^2 dx$$

term during the integral operation, we get

$$\int_0^L \left(\frac{dy}{dx}\right)^2 dx \frac{\partial}{\partial a_i}\left[\int_0^L EI\left(\frac{d^2y}{dx^2}\right)^2 dx - P_{cr} \int_0^L \left(\frac{dy}{dx}\right)^2 dx\right] = 0 \quad (4.10)$$

4.1.4 Converging approximations

According to Timoshenko [1], Friedrich Engesser offered a method for calculating critical buckling loads by successive approximations. To get an approximate solution, he suggested the adoption of some shapes for the deflection curve which satisfies the end conditions. This curve gives a bending-moment diagram, and the corresponding deflections can be calculated by using the area-moment method. Comparing this calculated deflection curve with the assumed one, an equation for determining the critical value of the load can be found. To get a better approximation, the calculated curve is taken as a new approximation for that of the buckled bar and repeats the above calculation. L. Vianello offered a graphical method in which the initially assumed curve could be graphically expressed, instead of taking an analytical expression, and the successive approximation could be made in a graphical way (section 4.1.5).

Consider the column as shown in Figure 4.5. The M/EI diagram is also shown in this figure. By the moment area theorem, the new deflection, δ_{02}, is

$$\delta_{02} = \frac{5}{48} \frac{PL^2 \delta_0}{EI}$$

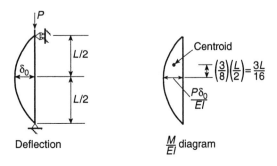

Figure 4.5 Simply supported column.

Deflection $\frac{M}{EI}$ diagram

If we set $\delta_{02} = \delta_0$, we obtain

$$P_{cr} = \frac{9.6EI}{L^2}$$

If we use δ_{02} as the new deflection, instead of δ_0, and proceed as above, we can improve the accuracy.

We may proceed by using a continuous function. Let us assume the deflected shape in the form of a parabola given as

$$y_1 = \frac{4\delta_0}{L^2} x(L - x)$$

Then

$$M = Py_1 = \frac{4P\delta_0}{L^2} x(L - x) = -EI \frac{d^2y}{dx^2}$$

Integrating

$$\frac{dy}{dx} = -\frac{4P\delta_0}{EIL^2}\left(\frac{Lx^2}{2} - \frac{x^3}{3}\right) + C_1$$

$$y = -\frac{4P\delta_0}{EIL^2}\left(\frac{Lx^3}{6} - \frac{x^4}{12}\right) + C_1 x + C_2$$

Applying boundary conditions $y = 0$ at $x = 0$ gives us $C_2 = 0$, and $y = 0$ at $x = L$ gives us

$$C_1 = \frac{P\delta_0 L}{3EI}$$

so that

$$y_2 = -\frac{4P\delta_0}{EIL^2}\left(\frac{Lx^3}{6} - \frac{x^4}{12}\right) + \frac{P\delta_0 L}{3EI} x$$

$$y_{2, x=L/2} = \frac{5}{48}\frac{P\delta_0 L^2}{EI}$$

$$y_{1, x=L/2} = \delta_0$$

Letting $y_{2, x = L/2} = y_{1, x = L/2}$, we get

$$P_{cr} = \frac{9.6EI}{L^2}$$

If we take y_2 to be the 'new' deflection,

$$M = Py_2 = \frac{P^2\delta_0}{3EI}\left[-\frac{4}{L^2}\left(\frac{Lx^3}{2} - \frac{x^4}{4}\right) + Lx\right] = -EI \frac{d^2y_3}{dx^2}$$

Integrating,

$$\frac{dy_3}{dx} = -\frac{P^2\delta_0}{3(EI)^2}\left[-\frac{4}{L^2}\left(\frac{Lx^4}{8} - \frac{x^5}{20}\right) + \frac{Lx^2}{2}\right] + C_1$$

$$y_3 = \frac{P^2\delta_0}{3(EI)^2}\left[-\frac{4}{L^2}\left(\frac{Lx^5}{40} - \frac{x^6}{120}\right) + \frac{Lx^3}{6}\right] + C_1 x + C_2$$

Applying the boundary conditions, we get

$$y_3 = \frac{P^2 \delta_0}{3(EI)^2} \left[-\frac{4}{L^2} \left(\frac{Lx^5}{40} - \frac{x^6}{120} \right) + \frac{Lx^3}{6} + \frac{L^3 x}{10} \right]$$

Now, letting $y_{3,x=L/2} = y_{2,x=L/2}$, we get

$$P_{cr} = 9.82 \frac{EI}{L^2}$$

If we need more accuracy, we may proceed by further steps. If we choose points other than $x = L/2$, the result is as follows. For example,

- at $x = L/8$, $y_1 = y_2$ yields $P_{cr} = 10.8 EI/L^2$;
- at $x = L/4$, $y_1 = y_2$ yields $P_{cr} = 10.1 EI/L^2$;
- at $x = 3L/8$, $y_1 = y_2$ yields $P_{cr} = 9.72 EI/L^2$.

4.1.5 Numerical procedure for columns with nonuniform cross-section

If the cross-section and the end conditions as well as loadings are not simple, the concept explained above combined with the numerical procedure may be very efficient, especially when the use of a computer is possible.

The method consists essentially of determining the mode shape by successive approximations of iteration at a certain number of points on the column [2]. If the mode shape, as determined, is sufficiently accurate, then the relative deflections at these points will remain unchanged under the critical load. In other words, the critical load is that load which is just sufficient to maintain the relative deflections of the mode shape. Generally it suffices to use, for this computation, an approximate mode shape. The resulting value will be somewhat approximate, but with care can be taken to be within the limits of engineering accuracy.

The procedure is as follows. We divide the column by a certain number of segments. We select a trial mode shape given at each joining point of two neighboring segments, which may be based upon experience or upon previous cycles of the iterative procedure. The deflection curve corresponding to this trial mode shape becomes the moment diagram since the moment is merely the axial load times the deflection in the case of a pin-ended column. On the basis of the resulting M/EI diagram or values at each point, the deflections are calculated in terms of the axial load P. When the column is not uniform, the actual values of $M/[E(i)I(i)]$ at each segment are used. Equating the newly calculated deflection at a point with the previous value at the same point furnishes a criterion for determining the magnitude of the axial load, i.e. the value of the critical load at which this equality is possible. As an example, consider a column as shown in Figure 4.6.

We may use any method to obtain a 'new' deflection. It may be helpful to recall the analogy between

- loads and angle changes
- shears and slopes
- moments and deflections

when the moment-area method is used.

Figure 4.6 Numerical procedure to obtain the buckling load.

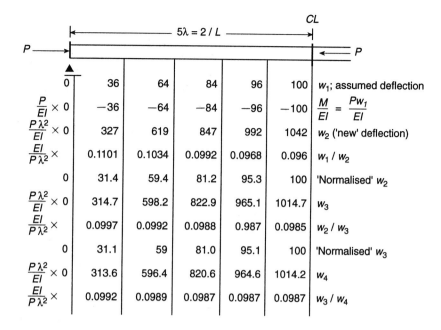

Since the values of the ratio w_1/w_2 are not constant, it is evident that the assumed deflections did not represent the correct mode shape, and that additional cycles of iteration will be required to obtain convergence. However, if the range of the ratios is not great, these may be used to obtain approximate values or estimates of the critical load.

On the basis of the results just obtained, we can calculate an approximate value using the results at the **center line**,

$$w_1/w_2 = 0.0960 EI/P\lambda^2, \quad \lambda = L/10$$

and setting this ratio of deflection equal to unity yields $P_{cr} = 9.60 EI/L^2$. As an alternative calculation, we may use the **average of the ratios**, thus obtaining

$$P_{cr} = 10.17 EI/L^2$$

or, as a second calculation, we may use the **ratio of the sums of the deflection** and we would obtain

$$P_{cr} = 9.98 EI/L^2.$$

If we are willing to disregard the fact that we have no determinate information in the neighborhood of the end, we can state that for our mathematical model, by Schwarz's inequality, the correct value of the critical load lies between the highest and lowest values associated with the highest and lowest values of the ratios w_1/w_2. Thus

$$0.0960 \frac{EI}{\lambda^2} \le P_{cr} \le 0.1101 \frac{EI}{\lambda^2}$$

If, barring numerical errors, the difference between the upper and lower bounds is sufficiently small so that the indicated maximum error lies within the desired limits of engineering accuracy, we may terminate the calculation at this point. If the range is too great, for example, in the above calculation it is about 14%, at least one more cycle should be computed. This would be done by using the last computed values of w_2s at each point (or numbers proportional to these) as starting values for the next cycle.

As Figure 4.6 shows, by the use of w_2/w_3,

$$0.0997 \frac{EI}{\lambda^2} \leq P_{cr} \leq 0.0985 \frac{EI}{\lambda^2}$$

and by w_3/w_4,

$$0.0992 \frac{EI}{\lambda^2} \leq P_{cr} \leq 0.0987 \frac{EI}{\lambda^2}$$

It is noted that at five points near the center of the whole beam,

$$P_{cr} = \frac{9.87EI}{L^2}$$

4.1.6 Application of slope–deflection method for buckling analysis of frames

This method had been taught by John E. Goldberg at his classes since 1953, and has been quite effective for this problem [3]. Consider a beam–column as shown in Figure 4.7. The moment at point x is

$$M = M_1 - \frac{M_1 + M_2}{L} x + Py = -EI \frac{d^2y}{dx^2}$$

or

$$\frac{d^2y}{dx^2} + \frac{P}{EI} y = \frac{-M_1}{EI} + \frac{M_1 + M_2}{EIL} x$$

The solution to this differential equation may be written as

$$y = a \sin kx + b \cos kx - \frac{M_1}{P} + \frac{M_1 + M_2}{PL} x \qquad (4.11)$$

where

$$k = \sqrt{\frac{P}{EI}}$$

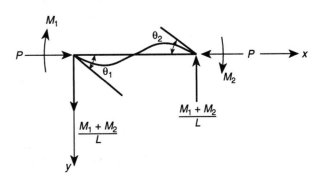

Figure 4.7 Sign convention for slope–deflection equation.

From boundary conditions $y_{x=0} = 0$ and $y_{x=L} = 0$,

$$b = \frac{M_1}{P}$$

$$a = -\frac{M_2 + M_1 \cos kL}{P \sin kL}$$

so that

$$y = -\frac{M_2 + M_1 \cos kL}{P \sin kL} \sin kx + \frac{M_1}{P} \cos kx - \frac{M_1}{P} + \frac{M_1 + M_2}{PL} x \quad (4.12)$$

Since

$$\frac{dy}{dx} = ak \cos kx - bk \sin kx + \frac{M_1 + M_2}{PL} \quad (4.13)$$

$$\theta_1 = \frac{dy}{dx}\bigg|_{x=0} = -\frac{(M_2 + M_1 \cos kL)}{P \sin kL} k + \frac{M_1 + M_2}{PL}$$

$$\theta_2 = \frac{dy}{dx}\bigg|_{x=L} = -\frac{(M_2 + M_1 \cos kL)}{P \sin kL} k \cos kL - \frac{M_1}{P} k \sin kL + \frac{M_1 + M_2}{PL}$$

$$(4.14)$$

Solving for M_1, we obtain

$$M_1 = \frac{EI}{L}\left(\frac{\sin kL - kL\cos kL}{(2/PL)(1 - \cos kL) - \sin kL}\theta_1 - \frac{\sin kL - kL}{(2/PL)(1 - \cos kL) - \sin kL}\theta_2\right)$$

which may be written in the form

$$M_1 = -\frac{EI}{L}(A\theta_1 + B\theta_2) \quad (4.15)$$

where A and B may be expressed as

$$A, B = f(kL) = f\left(\sqrt{\frac{PL^2}{EI}}\right) = f_1\left(\frac{P_{cr}}{P_e}\right) = f_1(\rho) \quad (4.16)$$

in which

$$\rho = \frac{P_{cr}}{P_e} = \frac{PL^2}{\pi^2 EI}$$

Equation 4.15 may be written in general form so that for uniform members with axial loads, we have

$$M = -\frac{EI}{L}\left[A\theta_1 + B\theta_2 - \frac{\Delta}{L}(A + B)\right] \quad (4.17)$$

If the member is not uniform, we have the following type of equation by John M. Hayes [4].

$$M_{CD} = -\frac{\theta_{dd}}{\theta_{cc}\theta_{dd} - \theta_{cd}^2}\theta_C - \frac{\theta_{cd}}{\theta_{cc}\theta_{dd} - \theta_{cd}^2}\theta_D + \frac{\theta_{dd} + \theta_{cd}}{\theta_{cc}\theta_{dd} - \theta_{cd}^2}\frac{\Delta}{L} \quad (4.18)$$

Table 4.1 Slope deflection coefficient for uniform members under axial load

ρ	A	B	$A + B$	$A' = A - \dfrac{B^2}{A}$
3.9	−78.34	78.58	0.24	0.48
3.8	−39.05	39.54	0.49	0.99
3.7	−24.69	25.39	0.70	1.42
3.6	−17.87	18.79	0.92	1.89
3.5	−13.73	14.86	1.13	2.35
3.4	−10.91	12.24	1.33	2.82
3.3	−8.86	10.40	1.54	3.35
3.2	−7.30	9.02	1.72	3.85
3.1	−6.05	7.96	1.91	4.42
3.0	−5.03	7.12	2.09	5.05
2.8	−3.449	5.884	2.435	6.589
2.6	−2.252	5.019	2.767	8.934
2.5	−1.749	4.678	2.929	10.763
2.4	−1.300	4.383	3.083	13.477
2.2	−0.519	3.901	3.382	28.802
2.0	0.143	3.521	3.664	−86.552
1.8	0.717	3.224	3.941	−13.780
1.6	1.224	2.980	4.204	−6.031
1.5	1.457	2.873	4.330	−4.208
1.4	1.673	2.778	4.451	−2.940
1.2	2.090	2.610	4.700	−1.169
1.0	2.468	2.468	4.936	0
0.9	2.645	2.404	5.049	0.460
0.8	2.816	2.346	5.162	0.862
0.7	2.981	2.291	5.272	1.220
0.6	3.140	2.241	5.381	1.541
0.5	3.295	2.194	5.489	1.834
0.4	3.444	2.150	5.594	2.102
0.3	3.589	2.109	5.698	2.350
0.2	3.730	2.070	5.800	2.581
0.1	3.865	2.033	5.898	2.796
0	4.000	2.000	6.000	3.000
−0.1	4.131	1.968	6.099	3.193
−0.2	4.255	1.938	6.193	3.372
−0.3	4.384	1.910	6.294	3.552
−0.4	4.502	1.883	6.385	3.714

in which θ_{cc} and θ_{cd} are the rotations at the ends C and D of the member CD, respectively, caused by a unit moment applied at C and so on. These can be calculated by several methods such as conjugate beam analogy.

Goldberg made tables of A, B, and ρ to aid use of equation 4.17 as given in Table 4.1.

Example 1
Consider the frame as shown in Figure 4.8(a). Writing slope deflection equations,

$$M_{21} = -\frac{EI_2}{h} A\theta_2$$

$$M_{23} = -\frac{EI_1}{b} 4\theta_2$$

Figure 4.8 Column as a part of a frame.

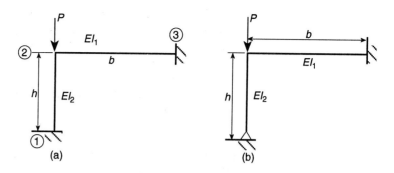

(a)   (b)

The joint equation is $M_{21} + M_{23} = 0$, from which we get

$$A = \frac{-4EI_1 h}{EI_2 b}$$

If $EI_2 = \frac{1}{2}EI_1$ and $h = b$, $A = -8$. From Table 4.1, $\rho = 3.25$ so that $P_{cr} = 3.25 P_e = 3.25\pi^2 EI_2/h^2$.

Example 2
For the frame of Figure 4.8(b),

$$M_{12} = -\frac{EI_2}{h}(A\theta_1 + B\theta_2) = 0, \quad \theta_1 = -\frac{B}{A}\theta_2$$

$$M_{21} = -\frac{EI_2}{h}(A\theta_2 + B\theta_1) = -\frac{EI_2}{h}\left(A - \frac{B^2}{A}\right)\theta_2$$

$$M_{23} = -\frac{EI_1}{b}4\theta_2$$

The joint equation $M_{21} + M_{23} = 0$ yields

$$A - \frac{B^2}{A} = -\frac{4EI_1}{EI_2}\frac{h}{b}$$

As an example, if $EI_2 = \frac{1}{2}EI_1$ and $h = b$,

$$A - \frac{B^2}{A} = -8$$

From Table 4.1, $\rho = 1.65$, so that

$$P_{cr} \approx \frac{16EI_2}{h^2}$$

The steps in using this method are illustrated for Figure 4.9.

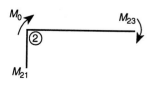

Figure 4.9 Direction of moments.

1. We assume a value for P_{cr}, i.e. the value of ρ which then defines the appropriate A and B values for the respective members.
2. Now, assume an arbitrary rotation at joint ②, say, a unit rotation.
3. Compute M_{21} and M_{23}, and solve for the magnitude of M_0 which is required to maintain equilibrium in the deflected, i.e. buckled state from the equilibrium equation, $M_0 + M_{21} + M_{23} = 0$.

4. If M_0 turns out to be zero, then the assumed value of P actually is the critical load. If not, select a new value of P and repeat.
5. Note that if M_0 is positive (for positive θ_2 in the same direction, i.e. if both M_0 and θ_2 are in the same direction), $P < P_{cr}$, and should be increased for the next trial.
6. Conversely, if M_0 and θ_2 are of opposite sign, $P > P_{cr}$, and should be reduced for the next trial.

Example 3
Given that $\alpha = \tan^{-1}2 = 63.45°$, (Figure 4.10), and all members are equal, the vertical and horizontal components of P are
We have

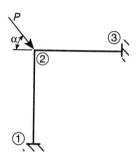

Figure 4.10 Column as a part of a frame (example 3).

$$P_V = P \sin \alpha = 0.895P$$

$$P_H = P \cos \alpha = 0.446P$$

$$M_{21} = -\frac{EI}{L}(A_1\theta_2) \quad M_{23} = -\frac{EI}{L}(A_2\theta_2)$$

$$\sum M_2 = M_0 + M_{21} + M_{23} = 0 \text{ yields}$$

$$\frac{M_0}{(EI/L)\theta_2} = A_1 + A_2.$$

1. We start from $P = 3P_e$:

$$P_V = 2.69P_e, \quad \rho_1 = 2.69, \quad A_1 = -2.850$$

$$P_H = 1.34P_e, \quad \rho_2 = 1.34, \quad A_2 = 1.77$$

$$\frac{M_0}{(EI/L)\theta_2} = -2.850 + 1.777 \rightarrow \text{negative, therefore } P > P_{cr}$$

2. Let $P = 2P_e$:

$$P_V = 1.79P_e, \quad \rho_1 = 1.79, \quad A_1 = 0.717$$

$$P_H = 0.892P_e, \quad \rho_2 = 0.892, \quad A_2 = 2.645$$

$$\frac{M_0}{(EI/L)\theta_2} = 0.717 + 2.645 \rightarrow \text{positive, therefore } P < P_{cr}$$

3. Let $P = 2.83P_e$:

$$P_V = 2.53P_e, \quad \rho_1 = 2.53, \quad A_1 = -1.8999 \Big\}$$
$$P_H = 1.26P_e, \quad \rho_2 = 1.26, \quad A_2 = 1.965 \quad \Big\} + 0.0651$$

$$\frac{M_0}{(EI/L)\theta_2} = -2.850 + 1.777 \rightarrow \text{negative, therefore } P > P_{cr}$$

4. Finally $P = 2.84P_e$, $M_0 \approx 0$, so that

$$P_{cr} = \frac{2.84\pi^2 EI}{L^2(\text{or } h^2)} = \frac{28EI}{h^2}$$

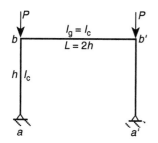

Figure 4.11 Columns of a frame (hinged).

Example 4

Find P_{cr} for asymmetric buckling (Figure 4.11):

$$\theta_b = \theta_{b'}, \quad M_{ab} = M_{a'b'} = 0$$

Moment equations are

$$M_{ab} = -\frac{EI_c}{h}\left[A\theta_a + B\theta_b - (A + B)\frac{\delta}{h}\right] = 0$$

$$M_{ba} = -\frac{EI_c}{h}\left[A\theta_b + B\theta_a - (A + B)\frac{\delta}{h}\right] \qquad \text{(i)}$$

$$M_{bb'} = -\frac{EI_c}{2h}[4\theta_b + 2\theta_{b'}] = -\frac{3EI_c}{h}\theta_b$$

From (i), $\theta_a = -\dfrac{B}{A}\theta_b + \left(1 + \dfrac{B}{A}\right)\dfrac{\delta}{h}$

$\sum M_b = 0$ yields

$$-\frac{EI_c}{h}\left[\left(A - \frac{B^2}{A} + 3\right)\theta_b + \left(\frac{B^2}{A} - A\right)\frac{\delta}{h}\right] = 0. \qquad \text{(a)}$$

From $M_{ba} - Sh - P\delta = 0$,

$$S = -\frac{EI_c}{h^2}\left[\left(A - \frac{B^2}{A}\right)\theta_b + \left(\frac{B^2}{A} - A\right)\frac{\delta}{h} + \rho\pi^2\frac{\delta}{h}\right] = 0 \qquad \text{(b)}$$

The nontrivial solution exists when the determinant of equations (a) and (b) vanishes. Thus

$$3\rho\pi^2 - (3 - \rho\pi^2)A' = 0 \qquad \text{(c)}$$

where

$$A' = A - \frac{B^2}{A}$$

1. Assume

$$\rho = 0.25, \quad A' = 2.465, \quad 7.4 - (1.36) \neq 0$$

2. Assume

$$\rho = 0.15, \quad A' = 2.698, \quad 4.44 - 4.1 \neq 0$$

3. Finally

$$\rho = 0.14 \text{ makes (c)} \approx 0,$$

so that

$$P_{cr} = \frac{0.14\pi^2 EI_c}{h^2} = \frac{1.38EI_c}{h^2}$$

Example 5

We consider the asymmetric mode of the frame in Figure 4.12. The moment equations are

$$M_{ab} = -\frac{EI_h}{h}\left[B\theta_b - (A + B)\frac{\Delta}{h}\right]$$

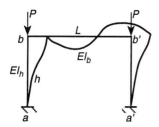

Figure 4.12 Columns of a frame (fixed).

$$M_{ba} = -\frac{EI_h}{h}\left[A\theta_b - (A+B)\frac{\Delta}{h}\right]$$

$$M_{bb'} = -\frac{EI_b}{L}(4\theta_b + 2\theta_{b'}) = -\frac{EI_b}{L}6\theta_b$$

Note that $\theta_b = \theta_{b'}$. Writing the equilibrium equation $M_{ba} + M_{bb'} = 0$ yields

$$\left(\frac{EI_h}{h}A + 6\frac{EI_b}{L}\right)\theta_b - \frac{EI_h}{h}(A+B)\frac{\Delta}{h} = 0 \qquad (i)$$

From $M_{ba} + M_{ab} - Sh - P\Delta = 0$,

$$S = -(A+B)\theta_b + 2\left(A + B - \frac{\pi^2\rho}{2}\right)\frac{\Delta}{h} = 0 \qquad (ii)$$

For the nontrivial solution to exist, the determinant of both equations (i) and (ii) is set to zero, so that

$$\frac{EI_h}{h}(A+B)^2 - 2\left(\frac{EI_h}{h}A + \frac{6EI_b}{L}\right)\left(A + B - \frac{\pi^2\rho}{2}\right) = 0$$

If we choose ρ so as to cause the left-hand side of this equation to go to zero, we can get P_{cr}. Alternatively, solving for θ_b in (i), and substituting into (ii), with S assumed nonzero,

$$\frac{h}{\Delta}S = -\left(\frac{EI_h}{h^2}\right)^2\frac{(A+B)^2}{(EI_h/h)A + (6EI_b/L)} + 2\left(A + B - \frac{\pi^2\rho}{2}\right)\frac{EI_h}{h^2}$$

By choosing ρ and the corresponding A and B from Table 4.1, which will make this equation go to zero, we can obtain $P_{cr} \approx 0.6P_e$.

Example 6

The frame in Figure 4.13 is restrained by springs at supports, a and a'. We will find the As, Bs and K, and the critical load for symmetric buckling. Since $\theta_b = -\theta_{b'}$, $\delta = 0$, the moment equations are

$$M_{ab} = -\frac{EI_{ab}}{h}[A\theta_a + B\theta_b] = k\theta_a$$

$$M_{ba} = -\frac{EI_{ab}}{h}[A\theta_b + B\theta_a] \qquad (a)$$

$$M_{bb'} = -\frac{EI_{ab}}{2h}[4\theta_b + 2\theta_{b'}] = -\frac{EI_{ab}}{h}\theta_b$$

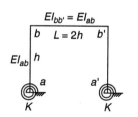

$EI_{bb'} = EI_{ab}$

$b \quad L = 2h \quad b'$

$EI_{ab} \quad h$

$a \qquad a'$

$K \qquad K$

Figure 4.13 Columns of a frame (restrained by springs).

From equation (a),

$$\theta_a = -\frac{(EI_{ab}B/h)\theta_b}{k + (EI_{ab}A/h)}$$

From $\sum M_b = 0$,

$$\frac{M_0}{\theta_b} = A - \frac{(EI_{ab}B^2/h)}{k + (EI_{ab}A/h)} + 1$$

$$k = \frac{1}{2}\frac{EI_{ab}}{h}$$

Taking the values of A and B which make M_0 zero, we can get ρ, i.e. P_{cr}. When $\rho = 1.3$, $A = 1.90$, $B = 2.65$, $M_0/\theta_b \approx 0$. Thus

$$P_{cr} = \frac{1.3\pi^2 EI}{h^2}$$

4.1.7 Summary of stability analysis of beams and frames

As a summary of the methods of stability analysis, we consider the structure in Figure 4.14. We have

$$EI_1\frac{d^2y}{dx^2} + Py = 0 \quad (0 < x < L_1)$$

$$EI_2\frac{d^2y}{dx^2} + Py = 0 \quad (L_1 < x < L_1 + L_2)$$

The boundary conditions are

$$y = 0 \quad \text{at} \quad x = 0$$
$$y_{L_1}^- = y_{L_1}^+$$
$$y_{L_1}'^- = y_{L_1}'^+$$
$$y' = 0 \quad \text{at} \quad X = L/2$$

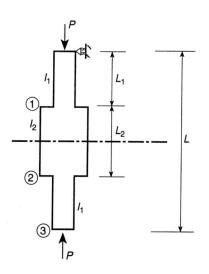

Figure 4.14 Column with variable cross-section.

1. Substitution of general solutions of the above equations into the given boundary conditions yields a stability criterion which may be expressed in several forms; for example, as a fourth-order determinant whose value vanishes when $P = P_{cr}$.
2. An alternative 'crude' procedure assumes $y = a \sin(\pi x/L)$ and uses an energy approach.
3. A better approximation is as follows.

    (a) Assume $y = a_1 \sin(\pi x/L) + a_3 \sin(3\pi x/L)$.
    (b) Construct the energy equation $P\lambda - U = 0$. This equation is homogeneous quadratic in the a's.
    (c) Construct linear homogeneous equations in the a's by

    $$\frac{\partial}{\partial a_1}(P\lambda - U) = 0$$

    $$\frac{\partial}{\partial a_3}(P\lambda - U) = 0$$

    (d) The determinant made up of the coefficients of the as must vanish. This will give a transcendental equation for P_{cr}.
4. Successive approximation assumes y and let

    $$Py = -EI_i \frac{d^2 y}{dx^2}$$

    This equation is integrated twice, and new and old deflected curves are equated.
5. Use of successive approximation combined with numerical procedure can be used for columns with complex shapes and boundary conditions. The present problem is in fact best suited for this method.
6. Slope deflection method (modified) deals with this problem as a structure with three members. The joint equilibrium equations are written in terms of θ_1 and θ_2 to solve this problem.

4.2 Structural vibration of beams and frames

4.2.1 Basic terms

The basic terms frequently used in vibration studies are explained using Figure 4.15.

-   **Period**. The time for the plumb bob to move from position 1 to 2 then back to 3($= 1$); also the time it takes to move from a to b to c to $d(= a)$.
-   **Frequency**. The reciprocal of the period; the frequency represents the number of times the configuration is repeated in a unit of time.
-   **Circular frequency**. Equal to 2π times the frequency, and a somewhat more useful and convenient quantity than the latter.
-   **Classification of vibration**. (a) Free vibration; (b) forced vibration – continual (but not necessarily continuous) introduction of force or forces which cause the system to move.

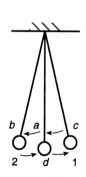

Figure 4.15 Pendulum.

- **Transient vibration**. First phase of motion which occurs upon applying or abruptly changing the forces.
- **Steady state (steady vibration)**. That motion which persists after transient or uncharacteristic motion has been damped out.
- **Node**. A point which does not move during vibration.
- **Antinode**. The point at which the maximum displacement (or maximum amplitude) occurs.

4.2.2   Vibration of beams

Consider a simply supported beam as shown in Figure 4.16. For purposes of approximation, using Rayleigh's principle, we assume the mode shape to be a parabola, given as

$$y = \frac{4a}{L^2} (Lx - x^2), \quad \frac{d^2y}{dx^2} = -\frac{8a}{L^2} \tag{4.19}$$

Note that this violates the condition that the boundary moment $(-EId^2y/dx^2)$ should vanish at the ends. When the beam is deflected into the assumed mode so that each point is at its maximum displacement y, then each point on the beam is instantaneously at rest, and the energy of the system is entirely of the potential type. That is, the energy of the beam consists entirely of bending strain energy, hence,

$$U_{\text{max}} = \frac{1}{2} \int_0^L \frac{M^2}{EI} \, dx = \frac{EI}{2} \int_0^L \left(\frac{d^2y}{dx^2}\right)^2 \, dx$$

In this particular case, assuming the mode shape as above,

$$U_{\text{max}} = \frac{32EIa^2}{L^3}$$

If a particle of mass m oscillates with single harmonic motion at a circular frequency ω and an amplitude y, its maximum velocity is ωy and its maximum kinetic energy (which occurs when the displacement, y, is zero) is

$$K_{\text{max}} = \frac{m}{2} (\omega y)^2 \tag{4.20}$$

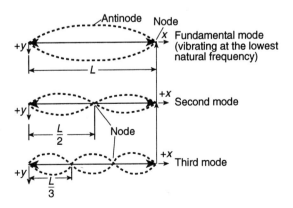

Figure 4.16   Vibration modes.

In the case of a beam, we replace m by μdx where μ is the mass per unit length, and the total kinetic energy or T_{max} is

$$T_{max} = \frac{1}{2} \int_0^L \mu(\omega y)^2 \, dx \tag{4.21a}$$

In the case of a uniform beam

$$T_{max} = \frac{\mu\omega^2}{2} \int_2^L y^0 \, dx \tag{4.21b}$$

For the mode shape we assumed, this becomes

$$T_{max} = \frac{8}{30} \mu L a^2 \omega^2$$

Now, T_{max} occurs when U as taken above is zero. The law of conservation of energy states that the sum of the potential energy and kinetic energy is constant, i.e. independent of time, so that

$$T + U = \text{const.}$$

and

$$\frac{d}{dt}(T + U) = 0 \tag{4.22}$$

Therefore, by this law, $T_{max} = U_{max}$ from which we obtain

$$\omega^2 = \frac{120EI}{\mu L^4}$$

As another example, we consider the simply supported uniform beam, assuming the mode shape to be

$$y = a \sin\left(\frac{\pi x}{L}\right) \quad \text{and} \quad \frac{d^2 y}{dx^2} = -a\left(\frac{\pi}{L}\right)^2 \sin\left(\frac{\pi x}{L}\right)$$

$$U_{max} = \frac{EI}{4} a^2 \left(\frac{\pi}{L}\right)^4 L \quad \text{and} \quad T_{max} = \frac{\mu}{4}\omega^2 a^2 L \tag{4.23}$$

Equating $T_{max} = U_{max}$,

$$\omega^2 = \frac{97.4EI}{\mu L^4}$$

which happens to be correct.

As a further example, we take the dead load deflection curve as an assumed mode shape. For a simply supported beam with uniform load q, the deflection curve is

$$y = \frac{q}{24EI}(L^3 x - 2Lx^3 + x^4) \tag{4.24}$$

Let $\mu = q/g$ be the mass per unit length. q may include both uniform line load and dead load. Then

$$\frac{dy}{dx} = \frac{\mu g}{24EI}(L^3 - 6Lx^2 + 4x^3), \quad \frac{d^2y}{dx^2} = \frac{\mu g}{24EI}(-12Lx + 12x^2)$$

$$U_{max} = \frac{144\mu^2 g^2 L^5}{60(24)^2 EI}$$

and

$$T_{max} = \frac{1}{2}mv_{max}^2 = \frac{1}{2}\int_0^L \mu(\omega y)^2 \, dx = \frac{\omega^2 \mu \mu^2 g^2 L^9}{2(24)^2(EI)^2}\left(\frac{31}{630}\right)$$

Letting $T_{max} = U_{max}$, we get

$$\omega^2 = \frac{97.6EI}{\mu L^4} = (2\pi f)^2$$

Compare this value with the previous ones, i.e. 97.4 vs 97.6 and 97.4 vs 120. The frequency is

$$f = \frac{9.88}{2\pi L^2}\sqrt{\frac{EI}{\mu}} = \frac{4.94}{\pi L^2}\sqrt{\frac{EI}{\mu}}$$

At the maximum displacement $(+y)$ of each particle, the particle has its maximum acceleration, which is numerically equal to $\omega^2 y$ and, furthermore, is **upward**. The upward force upon the particle, which is being exerted by the beam is $m\omega^2 y$. Since the force exerted by the beam upon the particle is upward, then it follows that the force which the particle exerts upon the beam is **downward** and, of course, numerically equal.

Consequently, when the vibrating beam has reached the maximum displacement at all points then the appropriate differential equation for the displacement is

$$EI\frac{d^4y}{dx^4} = m\omega^2 y \quad \text{or} \quad \frac{d^4y}{dx^4} - \frac{m\omega^2}{EI}y = 0 \tag{4.25}$$

Consider the simply supported beam. The boundary conditions are

at $x = 0$, $\qquad\qquad\qquad y = 0$ (i), $\qquad\qquad \frac{d^2y}{dx^2} = 0$ (ii)

and

at $x = L$, $\qquad\qquad\qquad y = 0$ (iii), $\qquad\qquad \frac{d^2y}{dx^2} = 0$ (iv)

Letting $k = (m\omega^2/EI)^{1/4}$, we can write the solution to equation 4.25 as

$$y = A_1 \sin kx + A_2 \cos kx + A_3 \sinh kx + A_4 \cosh kx$$

Applying the boundary conditions, $A_2 = A_3 = A_4 = 0$ and $A_1 \sin kL = 0$. If $A_1 = 0$, $y = 0$, i.e. there is no vibration so that we must have $\sin kL = 0$, which

Figure 4.17 Taut string with two masses.

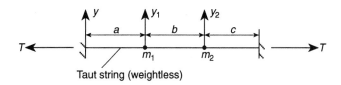

implies that $kL = n\pi$. This will yield

$$\omega^2 = \frac{n^4\pi^4EI}{mL^4} \tag{4.26}$$

n shows the mode shapes. Note that m is the mass per unit length μ in the case of a beam. Figure 4.16 shows the first three modes ($n = 1, 2, 3$).

4.2.3 Systems with two degrees of freedom

Consider a weightless taut string as shown in Figure 4.17. If the shape of the displaced system at the extreme displacements or excursions of the masses is repeated in each cycle, then the system is undergoing free vibration in a 'normal mode'. T is the tension in the string. The displacement will be assumed to be small so that the tension does not exceed T by any significant amount during the motion. If we write the equation of motion in the y-direction for each particle, we may start by considering that particle as a free body (Figure 4.18):

$$m_1\frac{\partial^2 y_1}{\partial t^2} = -T\frac{y_1}{a} - T\frac{y_1 - y_2}{b} + \{F_1(t)\}$$

$$m_2\frac{\partial^2 y_2}{\partial t^2} = -T\frac{y_2 - y_1}{b} - T\frac{y_2}{c} + \{F_2(t)\} \tag{4.27}$$

For free vibration in a normal (or natural) mode, we may write

$$y_1 = Y_1 \sin \omega t$$

$$y_2 = Y_2 \sin \omega t$$

without loss in generality.

Substituting these expressions into the equations of motion (equations 4.27),

$$-m_1\omega^2 Y_1 \sin \omega t + \frac{T}{a}Y_1 \sin \omega t + \frac{T}{b}(Y_1 - Y_2)\sin \omega t = 0$$

$$-m_2\omega^2 Y_2 \sin \omega t + \frac{T}{b}(Y_2 - Y_1)\sin \omega t + \frac{T}{c}Y_2 \sin \omega t = 0$$

Since $\sin \omega t$ is not, in general, zero, this factor may be cancelled in each equation.

Figure 4.18 Forces acting on the masses.

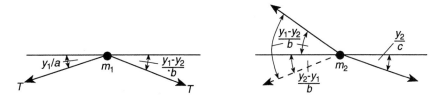

Grouping remaining terms, we obtain

$$\left(\frac{T}{a} + \frac{T}{b} - m_1\omega^2\right)Y_1 - \frac{T}{b}Y_2 = 0$$

$$-\frac{T}{b}Y_1 + \left(\frac{T}{b} + \frac{T}{c} - m_2\omega^2\right)Y_2 = 0$$

(4.28)

We obtain two expressions of Y_1/Y_2 from both equations, and equating both, we get

$$m_1 m_2 \omega^4 - \left[m_1\left(\frac{T}{b} + \frac{T}{c}\right) + m_2\left(\frac{T}{a} + \frac{T}{b}\right)\right]\omega^2$$

$$+ \left(\frac{T}{a} + \frac{T}{b}\right)\left(\frac{T}{b} + \frac{T}{c}\right) - \left(\frac{T}{b}\right)^2 = 0 \quad (4.29)$$

which is a quadratic equation in ω^2 and may be solved by the quadratic formula to obtain ω_1^2 and ω_2^2. The lower of the two natural frequencies corresponds to the case when the masses move in phase, and the higher one corresponds to out-of-phase movement.

An alternative viewpoint is to inquire as to the conditions under which the pair of simultaneous equations has a nontrivial solution. In our case, the right-hand side of the set of the equations 4.28 is zero, and therefore the numerator determinants are zero for Y_1 and Y_2, when Cramer's rule is used to solve for Y_1 and Y_2. A nontrivial solution exists if and only if the determinant formed by the coefficients of the Ys in the equation of motion vanishes:

$$\begin{vmatrix} \left(\frac{T}{a} + \frac{T}{b} - m_1\omega^2\right) & -\frac{T}{b} \\ -\frac{T}{b} & \left(\frac{T}{b} + \frac{T}{C} - m_2\omega^2\right) \end{vmatrix} = 0$$

The expansion of this determinant yields the same quadratic equation in ω^2, which is called the **frequency equation**, as was obtained previously (equation 4.29).

4.2.4 Influence coefficient method

At the maximum displacement, or excursion, the force exerted upon the string by the first mass is $m_1\omega^2 Y_1$, and by the second mass, $m_2\omega^2 Y_2$. The maximum displacements may be written as

$$Y_1 = a_{11}m_1\omega^2 Y_1 + a_{12}m_2\omega^2 Y_2$$

$$Y_2 = a_{21}m_1\omega^2 Y_1 + a_{22}m_2\omega^2 Y_2$$

(4.30)

where a_{ij}s are the deflection influence coefficients, meaning the deflection at i, due to a unit load at j.

Consider the taut string in Figure 4.17. Let $m_1 = m_2$, $a = c = l = \frac{1}{2}b$. To obtain the influence coefficients a_{11} and a_{21}, we apply a unit load at ① (Figure 4.19) in the direction of the normal displacement and solve the resulting

Figure 4.19   Taut string.

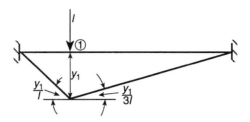

problem. In this particular example, only statics is involved. However, in the more general case, for example in beam problems, elastic behavior will also be involved. By summing vertical forces at ①,

$$T\left(\frac{y_1}{l} + \frac{y_1}{3l}\right) = 1, \quad \text{or} \quad T\left(\frac{4y_1}{3l}\right) = 1$$

which yields

$$y_1 = \frac{3l}{4T}$$

It follows that

$$y_2 = \frac{l}{4T} = \frac{y_1}{3}$$

Hence

$$a_{11} = \frac{3l}{4T}, \quad a_{21} = \frac{l}{4T}$$

In this particular case, we may take advantage of the symmetry of the elastic system to obtain, without further calculation,

$$a_{22} = \frac{3l}{4T}$$

Note that $a_{12} = a_{21}$.

When the system is vibrating freely and all masses are at their extreme positions, X_1 and X_2, then the masses have instantaneous zero velocities and they exert upon the string loads equal to

$$m_i \frac{\partial^2 X_i}{\partial t^2}$$

This holds for all positions of X, but, for our purpose, we will evaluate these inertia forces at the maximum values of the respective Xs.

Since the motion is simple harmonic, we write

$$X_1 = X_1 \sin \omega t$$
$$X_2 = X_2 \sin \omega t$$

(4.31)

so that the loads imposed upon the string by the beads are

$$F_1 = m_1\omega^2 X_1$$
$$F_2 = m_2\omega^2 X_2 \tag{4.32}$$

when the displacements are the maximums.

We can also write, using influence coefficients,

$$X_1 = a_{11}F_1 + a_{12}F_2$$
$$X_2 = a_{21}F_1 + a_{22}F_2 \tag{4.33}$$

Substituting the expressions for F_1 and F_2, equations 4.32, and the values of the influence coefficients into equations 4.33, we obtain

$$\frac{3l}{4T}m_1\omega^2 X_1 + \frac{l}{4T}m_2\omega^2 X_2 = X_1$$

$$\frac{l}{4T}m_1\omega^2 X_1 + \frac{3l}{4T}m_2\omega^2 X_2 = X_2 \tag{4.34}$$

A nontrivial solution can exist if and only if

$$\left(3m_1 - \frac{4T}{l\omega^2}\right)\left(3m_2 - \frac{4T}{l\omega^2}\right) - m_1 m_2 = 0$$

which reduces to

$$\left(\frac{4T}{l\omega^2}\right)^2 - 3(m_1 + m_2)\frac{4T}{l\omega^2} + 8m_1 m_2 = 0 \tag{4.35}$$

which can be solved for $4T/l\omega^2$ by the quadratic formula.

When $m_1 = m_2 = m$, we get

$$\omega^2 = \begin{cases} \dfrac{T}{lm} \\[2mm] \dfrac{2T}{lm} \end{cases}$$

The two modes are

- $X_1/X_2 = 1$ for the lower frequency ($\omega^2 = T/lm$);
- $X_1/X_2 = -1$ for the higher frequency ($\omega^2 = 2T/lm$).

4.2.5 Orthogonality theorem

A theorem in the theory of small (i.e. linear) oscillations states that the mode shapes are orthogonal with a weighting function equal to the mass. In the case of two masses, this takes the form

$$m_1 X_1^{(1)}X_1^{(2)} + m_2 X_2^{(1)}X_2^{(2)} = 0 \tag{4.36}$$

where $X_1^{(1)}$ is the amplitude of the first mass point (subscript 1) in the first mode (superscript 1) etc. In the present case $X_1^{(1)} = -X_1^{(2)}$, $X_2^{(2)} = -X_2^{(1)}$ and $m_1 = m_2 = m$. We confirm the orthogonality theorem

$$mX_2^{(1)}(-X_2^{(2)}) + mX_2^{(1)}X_2^{(2)} = 0$$

We consider the same problem with $m_1 = m$, $m_2 = 2m$. From equation 4.34, we obtain

$$3X_1 + 2X_2 = \frac{4T}{lm\omega^2} X_1$$

$$X_1 + 6X_2 = \frac{4T}{lm\omega^2} X_2$$

Suppose we assume a mode shape, $X_1 = 1$, $X_2 = 1$ and substitute into the left-hand side of the frequency equations.

$$3(1) + 2(1) = 5$$
$$(1) + 6(1) = 7$$

then

$$X_2 = \tfrac{7}{5}X_1 = 1.4X_1$$

Evidently, the computed quantities 5 and 7 are roughly proportional to X_1 and X_2 since they are presumably 'equivalent' to the original right-hand sides. Let us treat these (or numbers proportional to these) as new ordinates and again substitute into the left-hand sides:

$$3(1) + 2(1.4) = 5.8$$
$$(1) + 6(1.4) = 9.4$$

then

$$\frac{X_1}{X_2} = \frac{5.8}{9.4} = \frac{1}{1.65}$$

We continue using at least three decimal places to convergence and substitute into an 'actual' equation involving $4T/lm\omega^2$ and solve for ω_1^2.

$$3(1) + 2(1.65) = 6.3 = X_1$$
$$(1) + 6(1.65) = 10.9 = X_2, \rightarrow X_2 = 1.73X_1$$

After two more steps, we get $X_2 = 1.780X_1$,

$$3(1) + 2(1.780) = \frac{4T}{lm\omega^2}(1) = 6.56$$

or

$$(1) + 6(1.780) = \frac{4T}{lm\omega^2}(1.78) = 11.68 \rightarrow \frac{4T}{lm\omega^2} = 6.56$$

From this we obtain

$$\omega_1^2 = \frac{4T}{(6.56)lm} = \frac{T}{1.64lm} = 0.61\frac{T}{lm}.$$

Then, we use orthogonality conditions to obtain the second mode. From the orthogonality theorem,

$$m_1 X_1^{(1)} X_1^{(2)} + m_2 X_2^{(1)} X_2^{(2)} = 0$$

or

$$m(1)x_1^{(2)} + 2m(1.780)X_2 = 0$$

so that

$$\frac{X_1^{(2)}}{X_2^{(2)}} = -3.56$$

Putting into the original equation,

$$3(X_1) + 2X_2 = 3(3.56) + 2(-1) = \frac{4T(3.56)}{lm\omega_2^2}$$

from which

$$\omega_2^2 = \frac{4T}{2.43lm} = 1.646 \frac{T}{lm}$$

This is summarized in Figure 4.20. The orthogonality theorem for a high degree of freedom can be written as

$$\sum_{k=1}^{n} m_k X_k^{(i)} X_k^{(j)} = m_1 X_1^{(i)} X_1^{(j)} + m_2 X_2^{(i)} X_2^{(j)} + \cdots = 0 \qquad (4.37)$$

4.2.6 Proof of orthogonality theorem for normal modes

The orthogonality theorem may be proved for linear systems in general, but it is more convenient to prove the theorem for a specific case. Suppose we have an elastic beam having a mass distribution $\mu = \mu(x)$ where x is the longitudinal coordinate of the beam ($0 \le x \le L$) and suppose the beam is vibrating in the mth normal mode at the corresponding frequency, ω_m. At the instant of the maximum displacements, the displacement at any point is

$$y = y_m = y_m(x)$$

and these define the mth normal mode. The inertia loads on the beam at the instant of the maximum displacements are

$$P = P(x) = \mu \omega_n^2 y_n$$

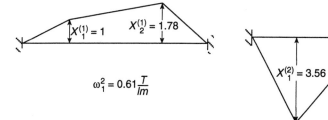

Figure 4.20 Mode shapes.

and the beam is instantaneously in equilibrium under this set of loads, thus the law of virtual work applies. For the displacements, we may, without loss in validity, pick the mode shape and displacements of the nth normal mode, $y = y_n = y_n(x)$. The external work done during the virtual displacement is

$$\Delta W = \int_0^L P y_n \, dx = \omega_m^2 \int_0^L \mu y_m y_n \, dx$$

The bending moment at any point in the beam at the instant when the beam has reached its maximum displacements may be written as

$$EI \frac{d^2 y_m}{dx^2}$$

The change in strain energy (= potential energy in bending) during the same virtual displacement is

$$\Delta U = \int_0^L \left(EI \frac{d^2 y_m}{dx^2} \right) \frac{d^2 y_n}{dx^2} \, dx.$$

By the law of virtual work, $\Delta \bar{W} = \Delta \bar{U}$, or

$$\omega_m^2 \int_0^L \mu y_m y_n \, dx = \int_0^L EI \frac{d^2 y_m}{dx^2} \frac{d^2 y_n}{dx^2} \, dx$$

Suppose we now start with the nth mode and use the mth mode for generating the virtual displacements. The application of the law of virtual work, in a manner exactly analogous to the foregoing, now yields

$$\omega_n^2 \int_0^L \mu y_n y_m \, dx = \int_0^L EI \frac{d^2 y_n}{dx^2} \frac{d^2 y_m}{dx^2} \, dx$$

The integrals are seen to have the same value as the corresponding integrals of the first case. Subtracting the second equation from the first, therefore, yields

$$-(\omega_n^2 - \omega_m^2) \int_0^L \mu y_m y_n \, dx = 0$$

Since $\omega_m \neq \omega_n$, it follows that

$$\int_0^L \mu y_m y_n \, dx = 0 \qquad\qquad (4.38)$$

4.2.7 Proof of the Rayleigh minimum frequency principle

If a mode shape is chosen arbitrarily as an approximation to the fundamental mode, we may expand the assumed shape as an expansion in terms of the normal modes. Thus, if $y = y(x)$ is the assumed shape, we may write

$$y(x) = y_1(x) + a_2 y_2(x) + a_3 y_3(x) + \cdots \qquad (4.39)$$

where $y_1(x)$, $y_2(x)$, etc. are the first, second, third, etc. modes, and a_2, a_3, etc. are appropriate but undetermined constants or multipliers.

If the deflection at some point x in the ith mode is $y_i(x)$, then the load on the body or system is $\mu\omega_i^2 y_i$ where μ is the mass per unit of length. That is, a loading of $\mu\omega_i^2 y_i$ produces maximum deflections y_i. For the assumed mode shape, the corresponding loading at any point is

$$P(x) = \mu\omega_1^2 y_1(x) + \mu a_2 \omega_2^2 y_2(x) + \mu a_3 \omega_3^2 y_3(x) + \cdots \qquad (4.40)$$

The work done by the external load, if one views this in the context of beams, is

$$W = \frac{1}{2} \int_0^L P(x)y(x)\, dx \qquad (4.41)$$

and this work is equal to the potential energy of the system:

$$U_{max} = W$$

$$= \frac{1}{2} \int_0^L (\mu\omega_1^2 y_1 + \mu a_2 \omega_2^2 y_2 + \mu a_3 \omega_3^2 y_3 + \cdots)(y_1 + a_2 y_2 + a_3 y_3 + \cdots)\, dx$$

$$(4.42)$$

In view of the orthogonality of the modes, which we have proved previously, the cross product terms vanish upon integration and

$$U_{max} = \frac{\omega_1^2}{2} \int_0^L \mu y_1^2\, dx + \frac{a_2 \omega_2^2}{2} \int_0^L \mu y_2^2\, dx + \frac{a_3 \omega_3^2}{2} \int_0^L \mu y_3^2\, dx + \cdots \quad (4.43)$$

The kinetic energy of the system when oscillating with circular frequency ω is

$$T_{max} = \frac{1}{2} mv^2 = \frac{1}{2} \int_0^L \mu\omega^2 y^2\, dx \qquad (4.44)$$

Substituting the normal mode expansion for y (equation 4.39) into this, we obtain

$$T_{max} = \frac{\omega^2}{2} \int_0^L \mu(y_1 + a_2 y_2 + a_3 y_3 + \cdots)^2\, dx \qquad (4.45)$$

Again making use of the orthogonality condition, this becomes

$$T_{max} = \frac{\omega^2}{2} \left(\int_0^L \mu y_1^2\, dx + a_2^2 \int_0^L \mu y_2^2\, dx + a_3^2 \int_0^L \mu y_3^2\, dx + \cdots \right) \quad (4.46)$$

To apply Rayleigh's method, we equate T_{max} and U_{max}, and solve the resulting equation for ω^2.

$$\omega^2 = \omega_1^2 \frac{\int_0^L \mu y_1^2\, dx + a_2^2 \left(\frac{\omega_2}{\omega_1}\right)^2 \int_0^L \mu y_2^2\, dx + a_3^2 \left(\frac{\omega_3}{\omega_1}\right)^2 \int_0^L \mu y_3^2\, dx + \cdots}{\int_0^L \mu y_1^2\, dx + a_2^2 \int_0^L \mu y_2^2\, dx + a_3^2 \int_0^L \mu y_3^2\, dx + \cdots} \qquad (4.47)$$

Since $\omega_2 > \omega_1$, $\omega_3 > \omega_1$, etc., we observe that each term in the numerator is either equal to or greater than the corresponding term in the denominator, and, therefore, the quotient is either equal to or greater than unity, i.e. cannot be less than unity. Therefore, since

$$\omega^2 = \omega_1^2(\text{quotient} \geq 1)$$

ω^2 cannot be less than ω_1^2. A quotient of this type is called **Rayleigh's quotient**.

4.2.8 Frequency of vibration of taut string by Rayleigh's method

In the deflected configuration (Figure 4.21), the string has obviously stretched by an amount related to the magnitude of the displacement, so that

$$dx = ds \cos\alpha$$

where $\alpha = \tan^{-1}(dy/dx) \approx dy/dx$ for small slopes. The difference in developed length between ds and dx, i.e. the stretch of the element, is

$$d\lambda = ds - dx = ds\,(1 - \cos\alpha).$$

The Taylor series for cosine of α is

$$\cos\alpha = 1 - \frac{\alpha^2}{2!} + \frac{\alpha^4}{4!} - \frac{\alpha^6}{6!} + \cdots$$

or, for sufficiently small angles,

$$\cos\alpha \approx 1 - \frac{\alpha^2}{2} \quad \text{and} \quad 1 - \cos\alpha \approx \frac{\alpha^2}{2}$$

so that

$$d\lambda \approx ds\left(\frac{\alpha^2}{2}\right)$$

Since there is very little difference between ds and dx, we may write, with sufficient accuracy for our purposes if the angle α is always sufficiently small,

$$d\lambda = \frac{\alpha^2}{2}\,dx$$

Taking $\alpha = dy/dx$, this becomes

$$d\lambda = \frac{1}{2}\left(\frac{dy}{dx}\right)^2 dx$$

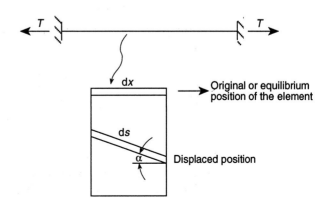

Figure 4.21 Taut string.

For a rather shallow or flat curve defined by $y = y(x)$ over the interval 0–L, the difference between the developed length and its projected length on the x-axis is

$$\lambda = \frac{1}{2}\int_0^L \left(\frac{dy}{dx}\right)^2 dx \tag{4.48}$$

Consequently, the change in potential energy of the deflected string, if T is sufficiently large and dy/dx is sufficiently small, is

$$\delta U = T\lambda = \frac{T}{2}\int_0^L \left(\frac{dy}{dx}\right)^2 dx$$

If we take the potential energy (PE) of the straight string as a datum, then the above expression may be taken as the PE. Furthermore, if y is taken as the mode,

$$\text{Max PE} = \frac{T}{2}\int_0^L \left(\frac{dy}{dx}\right)^2 dx \tag{4.49}$$

If the mass per unit length of the string is μ, then the maximum kinetic energy (KE), which occurs when the system is passing through its equilibrium configuration (when the potential energy above the datum is zero), is

$$\text{Max KE} = \frac{\omega^2}{2}\int_0^L \mu y^2 \, dx \tag{4.44}$$

Since the maximum PE occurs when the KE is zero, then, by the law of conservation of energy, we obtain

$$\omega^2 = \frac{T\int_0^L \left(\frac{dy}{dx}\right)^2 dx}{\int_0^L \mu y^2 \, dx} \tag{4.50}$$

Note that the right-hand side without T is a Rayleigh quotient.

We may, for the case of a uniform string, take the mode shape to be $y = a_{11}\sin n\pi x/L$ which happens to be mathematically correct. Substituting into the Rayleigh quotient,

$$\omega^2 = \frac{T\left(\frac{n\pi}{L}\right)^2 \int_0^L \cos^2\left(\frac{n\pi x}{L}\right) dx}{\mu \int_0^L \sin^2\left(\frac{n\pi x}{L}\right) dx}$$

Note that a_n^2 cancels since the frequency is independent of the amplitude if the system remains in the linear range. It can be ascertained that

$$\int_0^L \sin^2\left(\frac{n\pi x}{L}\right) dx = \int_0^L \cos^2\left(\frac{n\pi x}{L}\right) dx = \frac{L}{2}$$

Therefore

$$\omega^2 = \frac{n^2\pi^2}{\mu L^2} T \qquad (4.51)$$

where n is the 'order' or number of the mode.

4.2.9 Vibration of elastic body

Vibrating string
For the taut string in Figure 4.22 the displacement is a function of x and time t, i.e. $y = y(x, t)$. Summing forces in the $+y$ direction,

$$S\left(\frac{\partial y}{\partial x} + \frac{\partial^2 y}{\partial x^2}\,\mathrm{d}x\right) - S\frac{\partial y}{\partial x} = \text{mass} \times \text{acceleration in } +y\text{-direction}$$

$$= \mu\,\mathrm{d}x\,\frac{\partial^2 y}{\partial t^2}$$

which reduces to

$$\frac{\partial^2 y}{\partial x^2} = \frac{\mu}{S}\frac{\partial^2 y}{\partial t^2} \qquad (4.52)$$

where μ is the mass per unit length.

Assuming the solution to be $y = X(x)T(t)$, and substituting into equation 4.52,

$$\frac{1}{X}\frac{\mathrm{d}^2 X}{\mathrm{d}x^2} = \frac{\mu}{S}\frac{1}{T}\frac{\mathrm{d}^2 T}{\mathrm{d}t^2} = \text{const.} = a^2$$

or

$$\frac{\mathrm{d}^2 X}{\mathrm{d}x^2} - a^2 X = 0, \quad \frac{\mathrm{d}^2 T}{\mathrm{d}t^2} - \frac{a^2 S}{\mu} T = 0 \qquad (4.53)$$

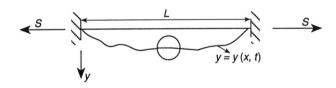

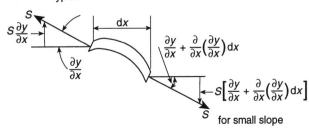

Figure 4.22 Taut string under distributed load.

We may get the solution to these equations, satisfying the boundary conditions, in the form

$$y = C_n \sin\left(\frac{n\pi x}{L}\right)\sin\left(\frac{n\pi}{L}\right)\sqrt{\frac{S}{\mu}}\,t \qquad (4.54)$$

In this equation

$$\omega = \frac{n\pi}{L}\sqrt{\frac{S}{\mu}}$$

is the circular frequency.

We may write a solution in the form

$$y = F\left(x - \sqrt{\frac{S}{\mu}}\,t\right) \qquad (4.55)$$

where F is an arbitrary function of the 'argument'

$$x - \sqrt{\frac{S}{\mu}}\,t = X \qquad (4.56)$$

Equation 4.52 can be written as

$$\frac{\mathrm{d}^2 F}{\mathrm{d}X^2}\left(\frac{\partial X}{\partial x}\right)^2 - \frac{\mu}{S}\frac{\mathrm{d}^2 F}{\mathrm{d}X^2}\left(\frac{\partial X}{\partial t}\right)^2 = 0$$

but

$$\frac{\partial X}{\partial t} = -\sqrt{\frac{S}{\mu}}$$

so that

$$\left(\frac{\partial X}{\partial t}\right)^2 = \frac{S}{\mu} \quad \text{and} \quad \frac{\partial X}{\partial x} = 1$$

thus we have

$$\frac{\mathrm{d}^2 F}{\partial X^2} - \frac{\mu}{S}\frac{\mathrm{d}^2 F}{\mathrm{d}X^2}\frac{S}{\mu} = 0$$

This verifies that the assumed form of solution satisfies the differential equation. Suppose X is constant, i.e. prescribed. Then, x and t are related as equation 4.56. With X prescribed, for any specified value of t, there is a corresponding value of x. Since x is the distance and t is the time, $\sqrt{S/\mu}$ must be a velocity. Usually this velocity is denoted by C and is defined by

$$C^2 = \frac{S}{\mu} \qquad (4.57)$$

The differential equation now may be written as

$$\frac{\partial^2 y}{\partial x^2} = \frac{1}{C^2}\frac{\partial^2 y}{\partial t^2} \qquad (4.58)$$

and, in this form, is known as the '**wave equation**' and its solution may be written in the form

$$y = F(x - ct)$$

or (4.59)

$$y = F(x + ct)$$

Longitudinal oscillation of bars

$u = u(x, t)$ is the displacement in the longitudinal direction (Figure 4.23). The longitudinal strain is related to the displacement by $\varepsilon = \partial u/\partial x$, and we have $\sigma = E\varepsilon = E\partial u/\partial x$. The total force acting on the section at X is, for a bar of small cross-section,

$$F = A\sigma = EA \frac{\partial u}{\partial x}$$

If the bar is uniform, the sum of the forces acting in the positive x-direction upon the element is

$$EA \frac{\partial^2 u}{\partial x^2} \, dx$$

Hence the equation of motion of the element is

$$\rho A \frac{\partial^2 u}{\partial t^2} = EA \frac{\partial^2 u}{\partial x^2}$$

where ρ is the mass density. We may write this as

$$\frac{\partial^2 u}{\partial x^2} = \frac{\rho}{E} \frac{\partial^2 u}{\partial t^2} \tag{4.60}$$

which is the standard form of the 'wave equation' and the velocity of propagation is

$$c = \sqrt{\frac{E}{\rho}} \tag{4.61}$$

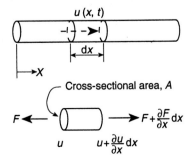

Figure 4.23 Bar under axial force.

The solution of the above differential equation may be written in several ways. Under certain boundary conditions, the solution may be assumed to be of the product form

$$u(x, t) = X(x)T(t) \qquad (4.62)$$

Under other conditions, the wave form may be more convenient,

$$u(x, t) = U_1(x - ct) + U_2(x + ct) \qquad (4.63)$$

where U_1 and U_2 are arbitrary functions of the indicated arguments, $(x - ct)$ and $(x + ct)$, and c, as before, is the wave velocity.

Beam with arbitrary geometrical boundary conditions
The equilibrium equation and equations of motion for the beam of Figure 4.24 are, neglecting the rotatory inertia effects,

$$(\mu \, dx) \frac{\partial^2 y}{\partial t^2} = \frac{\partial V}{\partial x} \, dx, \quad V \, dx = \frac{\partial M}{\partial x} \, dx, \quad \text{or} \quad V = \frac{\partial M}{\partial x}$$

We also have

$$M = -EI \frac{d^2 y}{dx^2}$$

If $EI = $ constant, we get

$$\frac{\partial^4 y}{\partial x^4} = -\frac{\mu}{EI} \frac{\partial^2 y}{\partial t^2} \qquad (4.64)$$

For many problems, a product form of solution is satisfactory, i.e.

$$y = X(x)T(t).$$

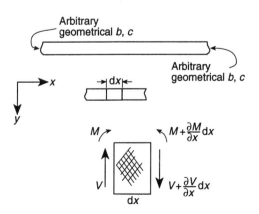

Figure 4.24 Beam with arbitrary boundary conditions.

Waves in isotropic elastic media

For an infinitesimal elastic body, under inertial forces, and without the body forces, the equations of equilibrium in terms of displacements are

$$(\lambda + G)\frac{\partial e}{\partial x} + G\nabla^2 u - \rho\frac{\partial^2 u}{\partial t^2} = 0$$

$$(\lambda + G)\frac{\partial e}{\partial y} + G\nabla^2 v - \rho\frac{\partial^2 v}{\partial t^2} = 0 \qquad (4.65)$$

$$(\lambda + G)\frac{\partial e}{\partial z} + G\nabla^2 w - \rho\frac{\partial^2 w}{\partial t^2} = 0$$

If the volume expansion, e, is zero and the deformation consists of shear distortion and rotation only, the above equations become

$$G\nabla^2 u - \rho\frac{\partial^2 u}{\partial t^2} = 0$$

$$G\nabla^2 v - \rho\frac{\partial^2 v}{\partial t^2} = 0 \qquad (4.66)$$

$$G\nabla^2 w - \rho\frac{\partial^2 w}{\partial t^2} = 0$$

These are equations for **waves of distortion**.

If the deformation is irrotational, equations 4.65 become

$$(\lambda + 2G)\nabla^2 u - \rho\frac{\partial^2 u}{\partial t^2} = 0$$

$$(\lambda + 2G)\nabla^2 v - \rho\frac{\partial^2 v}{\partial t^2} = 0 \qquad (4.67)$$

$$(\lambda + 2G)\nabla^2 w - \rho\frac{\partial^2 w}{\partial t^2} = 0$$

These are equations for **irrotational waves** or **waves of dilatation**.

At a great distance from the center of the disturbance, we consider such waves as plane waves. When all particles are moving parallel to the direction of wave propagation, these longitudinal waves are called waves of dilatation. If all particles move perpendicular to the direction of wave propagation, these transverse waves are called waves of distortion.

For the first case, since $v = w = 0$, we have

$$\frac{\partial^2 u}{\partial x^2} = C_1^2\frac{\partial u}{\partial t^2} \qquad (4.68)$$

which is the same as equation 4.60, except that $C_1 = \sqrt{E/\rho}$ (equation 4.61), is replaced by

$$C_1 = \sqrt{\frac{\lambda + 2G}{\rho}} = \sqrt{\frac{E(1 - v)}{(1 + v)(1 - 2v)\rho}} \qquad (4.69)$$

For the second case,

$$C_2 = \sqrt{\frac{G}{\rho}} = C_1 \sqrt{\frac{1-2v}{2(1-v)}} \tag{4.70}$$

4.2.10 Vibration of framed structures

As an illustration, we consider the three-story one-bay frame as shown in Figure 4.25. For this typical building, let 0 be the foundation. We define

- S_i story shear at ith story;
- M_i story moment at ith story;
- $M_{g,i}$ moment of girder at ith story;
- $M_{c,i,i+1}$ moment of column between ith and $(i+1)$th stories;
- Δ_i horizontal deflection at ith story;
- h_i story height at ith story;
- $R_i = \Delta_i/h_i$;
- C_i relative stiffness of column;
- G_i relative stiffness of girder;
- 'True' $\Delta_i = R_i h_i \dfrac{C_i}{(2EI/h)_i}$;
- θ_i angle of rotation at both ends of the ith story girder. θ_i is assumed to be equal at both ends of the girder in the case of simple lateral vibration.

If we write the slope–deflection equations,

$$
\begin{aligned}
M_{g,i} &= -G(3\theta_i) \\
M_{c,i,i+1} &= C_{i+1}(-2\theta_i - \theta_{i+1} + 3R_{i+1}) \\
M_{c,i,i-1} &= C_i(-2\theta_i - \theta_{i-1} + 3R_i) \\
\frac{S_i h_i}{2} &= \frac{M_i}{2} = M_{c,i,i-1} + M_{c,i-1,i}
\end{aligned}
\tag{4.71}
$$

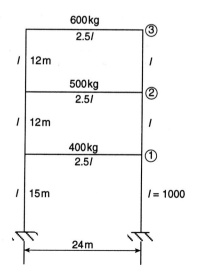

Figure 4.25 One bay frame.

Figure 4.26 Notations for one bay frame.

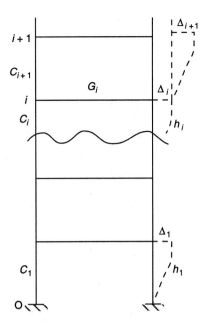

Note that we have assumed that the columns and girders are uniform for this particular problem. However, if these members are not uniform, we can handle such a problem by the use of equation 4.18 by Hayes. From the last three of equations 4.71 we obtain

$$R_i = \frac{\Delta_i}{h_i} = \frac{M_i}{12C_i} + \frac{1}{2}\theta_i + \frac{1}{2}\theta_{i-1} \tag{4.72}$$

Setting the sum of the moments at joint i to zero, we get

$$-C_{i+1}\theta_{i+1} + A_i\theta_i - C_i\theta_{i-1} = \frac{M_i + M_{i+1}}{2} \tag{4.73}$$

in which $A_i = 6G_i + C_{i+1} + C_i$.

For a uniform section, the bending stiffness is $2EI/L$, and C and G are proportional to I/L. Therefore, we may assign arbitrary relative values for Cs and Gs as follows:

$$G_1 = G_2 = G_3 = 5$$

$$C_1 = 5 \times \frac{1}{2.5} \times \frac{24}{15} = 3.2$$

$$C_2 = C_3 = 5 \times \frac{1}{2.5} \times \frac{24}{12} = 4$$

from which

$$A_1 = 6G_1 + C_1 + C_2 = 37.2$$

$$A_2 = 6G_2 + C_2 + C_3 = 38$$

$$A_3 = 6G_3 + C_3 = 34$$

By the use of equation 4.73

$$37.2\theta_1 - 4\theta_2 = \frac{M_1 + M_2}{2}$$

$$-4\theta_1 + 38\theta_2 - 4\theta_3 = \frac{M_2 + M_3}{2}$$

$$-4\theta_2 + 34\theta_3 = \frac{M_3}{2}$$

Letting $L = 12$ m, we apply unit horizontal force at each joint alternatively. The story shear and moment for each case are as shown in Figure 4.27. With the above equations and the result as shown in Figure 4.27, we have

$$\begin{bmatrix} 37.2 & -4 & 0 \\ -4 & 38 & -4 \\ 0 & -4 & 34 \end{bmatrix} \begin{bmatrix} \theta_1 \\ \theta_2 \\ \theta_3 \end{bmatrix} = \begin{bmatrix} 9/8 & 9/8 & 5/8 \\ 1 & 1/2 & 0 \\ 1/2 & 0 & 0 \end{bmatrix} L$$

From this matrix equation, the θ_is for each case are obtained as

$$\begin{bmatrix} \theta_1 \\ \theta_2 \\ \theta_3 \end{bmatrix} = 10^{-4}L \begin{bmatrix} 336 & 320 & 169 \\ 317 & 167 & 18.7 \\ 184 & 19.7 & 2.1 \end{bmatrix}$$

Using the expression of R_i given by equation 4.72, the 'true displacements' are

$$\Delta_i = \frac{k}{h_i}\left[\frac{M_i}{4} + 1.5C_i(\theta_i + \theta_{i-1})\right] \qquad (4.74)$$

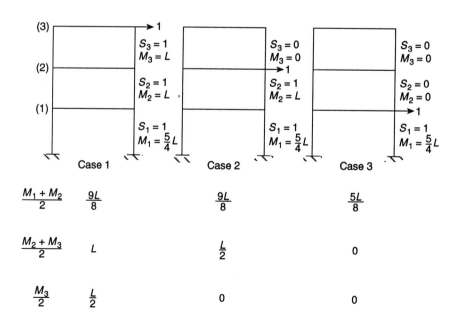

Figure 4.27 The story shear and moment of the frame.

where $k = h^3/6EI$ and $h = L = 12$ m. Thus, we obtain the influence coefficient, $\Delta_{i,j}$, where the subscript j denotes the case number, as follows:

$$\Delta_{i,j} = a_{i,j} = \begin{bmatrix} 0.616 & 0.729 & 0.741 \\ 0.729 & 1.268 & 1.380 \\ 0.741 & 1.380 & 1.932 \end{bmatrix} k$$

We assume, $y_i = y_i \sin \omega t$. Then

$$F_i = m_i \frac{\partial^2 y}{\partial t^2} = -m_i \omega^2 y_i \sin \omega t$$

The maximum force acting on the beam is $+m_i\omega^2 y_i$, of which the negative sign on the particle means the positive sign on the beam.

In general, we have

$$y_i = a_{i,j} F_j$$

as the deflection caused by the force system, F_j so that

$$\begin{bmatrix} y_1 \\ \vdots \\ y_n \end{bmatrix} = \omega^2 \begin{bmatrix} a_{11} & \cdots & a_{1n} \\ \vdots & & \vdots \\ a_{n1} & \cdots & a_{nn} \end{bmatrix} \begin{bmatrix} m_1 & & \cdots & & 0 \\ 0 & m_2 & \cdots & & \vdots \\ \vdots & & & & \\ 0 & & & \cdots & m_n \end{bmatrix} \begin{bmatrix} y_1 \\ \vdots \\ y_n \end{bmatrix} \qquad (4.75)$$

This is obviously an eigenvalue problem in ω^2, and making the determinant of matrix coefficients equal to zero, we can get ω_1^2, ω_2^2 and ω_3^2 for our illustrative problem.

Assuming $m_1 = 400$ kg, $m_2 = 500$ kg, and $m_3 = 600$ kg, we have

$$\begin{bmatrix} y_1 \\ y_2 \\ y_3 \end{bmatrix} = 100k\omega^2 \begin{bmatrix} 0.616 & 0.729 & 0.741 \\ 0.729 & 1.268 & 1.380 \\ 0.741 & 1.380 & 1.932 \end{bmatrix} \begin{bmatrix} 4 & 0 & 0 \\ 0 & 5 & 0 \\ 0 & 0 & 6 \end{bmatrix} \begin{bmatrix} y_1 \\ y_2 \\ y_3 \end{bmatrix}$$

If the iteration method and the orthogonality theorem are used, the procedure is as follows.

1. Iterate the above equation.
2. From the converged value, get the first mode shape.
3. By the orthogonality theorem,

$$\sum m_i y_i^{(1)} y_i^{(2)} = m_1 y_1^{(1)} y_1^{(2)} + m_2 y_2^{(1)} y_2^{(2)} + m_3 y_3^{(1)} y_3^{(2)} = 0$$

   Express $y_2^{(2)}$ in terms of $y_1^{(1)}$, $y_1^{(2)}$ and $y_3^{(2)}$.
4. Substitute $y_2^{(2)}$ into the first and third equations of the original system.
5. This gives us two equations in $y_3^{(2)}$ and $y_1^{(2)}$.
6. Compute $\omega_n^{(2)}$ and the second mode shape.
7. $\sum m_i y_i^{(2)} y_i^{(3)} = 0$, thus $y_3^{(3)} = f[y_1^{(3)}, y_2^{(3)}, y_i^{(2)}]$. Call this equation (a).
8. $\sum m_i y_i^{(1)} y_i^{(3)} = 0$, so that $y_1^{(3)} = f[y_2^{(3)}, y_3^{(3)}, y_i^{(1)}]$. Substitute this into the second equation of the first mode to get equation (b).
9. Take the first equation from the second mode, and let this be equation (c).
10. Substitute equation (a) into (b) and (c).
11. Iterate these.
12. Find $\omega_n^{(3)}$ and the third mode shape.

The final result for this problem is as follows.

- First mode

$$\omega_1^2 = \frac{1}{1840k}, \quad k = \frac{h^3}{6EI}$$

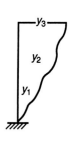

Mode shape

$$y_3^{(1)} = 2.175$$
$$y_2^{(1)} = 1.732$$
$$y_1^{(1)} = 1$$

- Second mode

$$\omega_2^2 = \frac{1}{151.4k}$$

Mode shape

$$y_3^{(2)} = -0.695$$
$$y_2^{(2)} = 0.588$$
$$y_1^{(2)} = 1$$

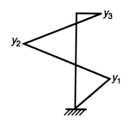

- Third mode

$$\omega_3^2 = \frac{1}{6k}$$

Mode shape

$$y_3^{(3)} = 0.68$$
$$y_2^{(3)} = -1.485$$
$$y_1^{(3)} = 1$$

As another illustration, we consider the same frame as the previous one except that the stiffnesses of the girders are very large so that these are assumed to be infinity (Figure 4.28). If the rigid slabs are well connected to the girders, the

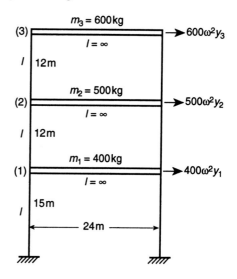

Figure 4.28 Frame with stiff girders.

stiffnesses of the horizontal members can be considered very large. We assume, as before, $y_i = y_i \sin \omega t$ so that

$$F_i = m_i \frac{\partial^2 y_i}{\partial t^2} = -m_i \omega^2 y_i \sin \omega t$$

The maximum force is $F_i = -m_i \omega^2 y_i$ on the particle, $F_i = +m_i \omega^2 y_i$ on the frame. We also have for a given Δ,

$$M = \frac{6EI\Delta}{L^2}$$

$$V = \frac{2M}{L} = \frac{12EI\Delta}{L^3} = k\Delta$$

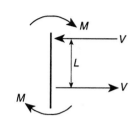

Figure 4.29 Column moments.

where $k = 12EI/L^3$, for this problem (Figure 4.29).
    For story 1,

$$k_1 = \frac{12EI}{(\frac{15}{12}L)^3} = 0.512k$$

The shear forces acting on each girder are as shown in Figure 4.30. Taking equilibrium at each floor,

$$2k(y_3 - y_2) = 600\omega^2 y_3$$

$$2k(-y_1 + 2y_2 - y_3) = 500\omega^2 y_2$$

$$3.024ky_1 - 2ky_2 = 400\omega^2 y_1$$

Letting $\alpha = 100\omega^2/k$, we have

$$\begin{bmatrix} (1.512 - 2\alpha) & -1 & 0 \\ -1 & (2 - 2.5\alpha) & -1 \\ 0 & -1 & (1 - 3\alpha) \end{bmatrix} \begin{bmatrix} y_1 \\ y_2 \\ y_3 \end{bmatrix} = \begin{bmatrix} 0 \\ 0 \\ 0 \end{bmatrix}$$

    We have a nontrivial solution when the determinants of the coefficients are equal to zero, so that

$$\alpha^3 - 1.89\alpha^2 + 0.79\alpha - 0.0342 = 0$$

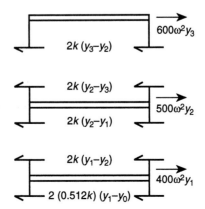

Figure 4.30 Shear forces acting on each girder.

Solving for αs, we have the following result.

- First mode

$$\alpha_1 = 0.049 = \frac{100\omega_1^2}{k} \quad \text{or} \quad \omega_1^2 = 0.049\,\frac{k}{100}$$

and letting $y_1 = 1.0$,

$$y_2 = 1.414$$
$$y_3 = 1.66$$

- Second mode

$$\omega_2^2 = 0.533k/100$$
$$y_1 = 1$$
$$y_2 = 0.446$$
$$y_3 = -0.743$$

- Third mode

$$\omega_3^2 = 1.308k/100$$
$$y_1 = 1$$
$$y_2 = -1.104$$
$$y_3 = 0.378$$

If we check by the orthogonality theorem,

$$\sum m_i y_i^{(1)} y_i^{(2)} = 100a^2[4(1.0)(1.0) + 5(1.414)(0.446) + 6(1.66)(-0.745)] \approx 0.$$

4.2.11   Vibration analysis of irregularly shaped structural elements

Introduction
Design and analysis of any structural member involves calculating influence coefficients, from which displacements, moments and shears can be obtained, under all kinds of anticipated loadings. After materials and cross-sections of each member are designed according to the results of such calculations, vibration and stability problems are analyzed. If the result is unsatisfactory, the engineer has to start from the beginning. In the case of a laminated composite plate, with boundary conditions other than Navier or Levy solution types, or irregular cross-section, an analytical solution is very difficult to obtain. Numerical methods for eigenvalue problems are also very much involved in seeking such a solution.

Structural engineers are quite often faced with the problems of designing high rise towers, such as elevated water tanks, transmission as well as observation towers, landmarks, sea berths, offshore oil drilling platforms, and many other miscellaneous types of structures. Unfortunately, they have no access to rigorous and practical methods to handle the vibration problems of such structures.

The author encountered solving vibration and buckling problems when he designed, in 1970, the 270 m high Seoul Tower for which no existing method of solution was available. The method of analysis of buckling of columns due to Vianello was useful to study the stability problem. Since both buckling of columns and vibration of beams are, mathematically, eigenvalue problems, similar concepts could be applied for both cases. In buckling problems, the deflection is caused by the axial load. In the case of vibration, the deflection is caused by the inertia force. A method of calculating the natural frequency and mode shape corresponding to the first mode of vibration of structural members with variable cross-section was developed for this tower [5]. This method was examined extensively for several types of beams with various conditions and found to be very simple in calculation, and was proved to be very accurate and efficient. This method was extended to two-dimensional problems including composite laminates and proved to be as good as for the one-dimensional case [6–11].

Method of analysis

A natural frequency of a structure is the frequency under which the deflected mode shape corresponding to this frequency begins to diverge under the resonance condition. From the deflection caused by the free vibration, the force required to make this deflection can be found, and from this force, the resulting deflection can be obtained. Starting from an arbitrarily assumed mode shape, the force required for this mode shape can be found, and the 'new' mode shape corresponding to this force system is obtained. If the mode shape as determined by the series of this process is sufficiently accurate, then the relative deflections (maximum) of both the converged and the previous shape should remain unchanged under the inertial force related to this natural frequency.

Vibration of a structure is a harmonic motion and the amplitude may contain a part expressed by a trigonometric function. Considering only the first mode as a start, the deflected shape of a structural member can be expressed as

$$w = W(x, y)F(t) = W(x, y)\sin \omega t \qquad (4.76)$$

where W is the maximum amplitude, ω is the circular frequency of vibration and t is time.

By Newton's law, the dynamic force of the vibrating mass m is

$$F = m \frac{\partial^2 w}{\partial t^2}$$

Substituting equation 4.76 into this, the force acting on the particle is,

$$F = -m(\omega)^2 W(x, y)\sin \omega t = \begin{cases} -m(\omega)^2 w \text{ (particle)} \\ +m(\omega)^2 w \text{ (beam)} \end{cases}$$

In this expression, ω and W are unknowns.

In order to obtain the natural circular frequency ω, the following process is taken. The magnitude of the maximum deflection at a certain number of points is arbitrarily given as

$$w(i, j)^{(1)} = W(i, j)^{(1)} \qquad (4.77)$$

where (i, j) denotes the point under consideration. This is absolutely arbitrary but educated guessing is good for accelerating convergence. The dynamic force acting on the structure corresponding to this (maximum) amplitude is

$$F(i, j)^{(1)} = +m(i, j)[\omega(i, j)^{(1)}]^2 w(i, j)^{(1)} \tag{4.78}$$

The 'new' deflection caused by this force is a function of F and can be expressed as

$$w(i, j)^{(2)} = f\{m(i, j)[\omega(i, j)^{(1)}]^2 w(i, j)^{(1)}\}$$
$$= \sum^{k, l} \Delta(i, j, k, l)\{+m(i, j)[\omega(i, j)^{(1)}]^2 w(k, l)^{(1)}\} \tag{4.79}$$

where Δ is the deflection influence surface. The relative (maximum) deflections at each point under consideration of a structural member under the resonance condition, $w(i, j)^{(1)}$ and $w(i, j)^{(2)}$, have to remain unchanged and the following condition must hold:

$$w(i, j)^{(1)}/w(i, j)^{(2)} = 1 \tag{4.80}$$

From this equation, $\omega(i, j)^{(1)}$ at each point of (i, j) can be obtained, but they are not equal in most cases. Since the natural frequency of a structural member has to be equal at all points of the member, i.e. $\omega(i, j)$ should be equal for all (i, j), this step is repeated until sufficient equal magnitude of $\omega(i, j)$ is obtained at all (i, j) points. However, in most cases, the difference between the maximum and the minimum values of $\omega(i, j)$ is sufficiently negligible for engineering purposes. The accuracy can be improved by simply taking the average of the maximum and the minimum, or by taking the value of $\omega(i, j)$ where the deflection is the maximum. For the second cycle, if necessary, the absolute numerics of $w(i, j)^{(2)}$ can be used for $w(i, j)^{(2)}$ in

$$w(i, j)^{(3)} = f\{m(i, j)[\omega(i, j)^{(2)}]^2 w(i, j)^{(2)}\} \tag{4.81}$$

for convenience.

In the case of a structural member with irregular section, including a composite one, and nonuniformly distributed mass, regardless of the boundary conditions, it is convenient to consider the member as divided by a finite number of elements. The accuracy of the result is proportional to the accuracy of the deflection calculation.

Numerical illustration

As an illustration of this method, a simply supported beam with uniform flexural rigidity, EI, is considered (Figure 4.31). The length of the beam is 10 m. The weight of the beam is assumed as 500 kg/m. The weight acts as the mass when

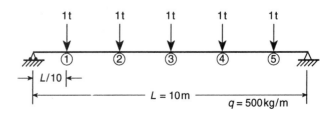

Figure 4.31 Simply supported beam.

Table 4.2 Influence coefficients, $EI\Delta_{i,j}$, for simple beam

i	j				
	1	2	3	4	5
1	2.7	5.8	6.2	4.5	1.6
2	5.8	14.7	16.5	12.3	4.5
3	6.2	16.5	20.8	16.5	6.2
4	4.5	12.3	16.5	14.7	5.8
5	1.6	4.5	6.2	5.8	2.7

the beam vibrates and is treated as concentrated loads at five equally spaced points. Since a beam is one-dimensional, one subscript, i, is used.

The set of influence coefficients, $\Delta_{i,j}$, where i is the point under consideration and j is the loading point (unit load), is given in Table 4.2. The initially guessed maximum amplitude, $W(i)^{(1)}$, can be arbitrary and the following values are given:

$$W(1)^{(1)} = W(5)^{(1)} = 40$$

$$W(2)^{(1)} = W(4)^{(1)} = 80$$

$$W(3)^{(1)} = 100$$

$$m(i) = 1 \text{ Mg per gram at all of } i$$

These values are substituted into equations 4.77 and 4.78, and from equation 4.79, the following result is obtained:

$$w(1)^{(2)} = 1616m(1)[\omega(1)^{(1)}]^2/EI$$

$$w(2)^{(2)} = 4222m(2)[\omega(2)^{(1)}]^2/EI$$

$$w(3)^{(2)} = 5216m(3)[\omega(3)^{(1)}]^2/EI$$

Letting $w(i)^{(1)}/w(i)^{(2)} = 1$, we get

$$\omega(1)^{(1)} = 0.157\ 3A(1)$$

$$\omega(2)^{(1)} = 0.137\ 6A(2)$$

$$\omega(3)^{(1)} = 0.138\ 5A(3)$$

where

$$A(i) = \sqrt{\frac{EI}{m(i)}}$$

Since all $\omega(i)^{(1)}$s should be equal at all i points, this process has to be repeated. For the second cycle, only the relative magnitude of the amplitude is necessary, and $W(i)^{(2)}$s are assigned as follows:

$$W(1)^{(2)} = W(5)^{(2)} = 16.2$$

$$W(2)^{(2)} = W(4)^{(2)} = 42.2$$

$$W(3)^{(2)} = 52.2$$

The same influence coefficient for the first cycle is repeatedly used, and the 'new' amplitude, $w(i)^{(3)}$, is obtained as

$$w(1)^{(3)} = 827.76[\omega(1)^{(2)}]^2/[A(1)]^2$$

$$w(2)^{(3)} = 2167.08[\omega(2)^{(2)}]^2/[A(2)]^2$$

$$w(3)^{(3)} = 2678.64[\omega(3)^{(2)}]^2/[A(3)]^2$$

From $w(i)^{(2)}/w(i)^{(3)} = 1$,

$$\omega(1)^{(3)} = 0.139\ 575A(1)$$

$$\omega(2)^{(3)} = 0.139\ 575A(2)$$

$$\omega(3)^{(3)} = 0.139\ 575A(3)$$

One more process is executed in order to obtain the better result as follows.

$$\omega(1)^{(3)} = 0.139\ 575A(1)$$

$$\omega(2)^{(3)} = 0149\ 575A(2)$$

$$\omega(3)^{(3)} = 0.139\ 575A(3)$$

Note that all As are the same, and

$$\omega = 0.139\ 575A$$

The result obtained by the 'exact' theory is $\omega = 0.139\ 575A$.

It is noted that the result of the first cycle is good enough for engineering purposes. If ω at the point of the maximum deflection, $\omega(3)^{(1)}$, is considered, it is only 0.77% away from the 'exact' result.

In the case of a variable cross-section, including materials, $A(i) = E(i)(I_i)/m(i)$ should be used. Influence coefficients can be found with relative ease in any case.

As the second illustration, we consider the cantilever beam as shown in Figure 4.32. The uniform load of 500 kg/m is treated as five concentrated loads as shown. The influence coefficients are shown in Table 4.3. If necessary, the number of points under consideration may be changed and the two adjacent points may be made closer, i.e. nonuniform spacing, when the geometry of mode shape requires us to do so. The initially guessed maximum amplitude is

$$W(1)^{(1)} = 5$$

$$W(2)^{(1)} = 15$$

$$W(3)^{(1)} = 50$$

$$W(4)^{(1)} = 80$$

$$W(5)^{(1)} = 100$$

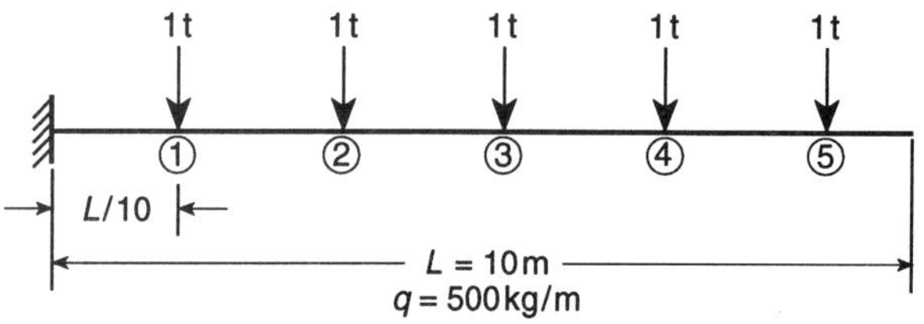

Figure 4.32   Cantilever beam.

Table 4.3 Influence coefficients, $EI\Delta_{i,j}$, for cantilever beam

			j		
i	1	2	3	4	5
1	0.33	1.33	2.33	3.33	4.33
2	1.33	9.00	18.00	27.00	36.00
3	2.33	18.00	41.33	66.67	91.67
4	3.33	27.00	66.67	114.33	163.33
5	4.33	36.00	91.67	163.33	243.00

Proceeding as the previous example, we obtain

$$w(1)^{(2)} = 829m(1)[\omega(1)^{(1)}]^2/EI$$

$$w(2)^{(2)} = 6809m(2)[\omega(2)^{(1)}]^2/EI$$

$$w(3)^{(2)} = 16\ 865m(3)[\omega(3)^{(1)}]^2/EI$$

$$w(4)^{(2)} = 29\ 235m(4)[\omega(4)^{(1)}]^2/EI$$

$$w(5)^{(2)} = 45\ 845m(5)[\omega(5)^{(1)}]^2/EI$$

Letting $w(i)^{(1)}/w(i)^{(2)} = 1$, we obtain

$$\omega(1)^{(1)} = 0.077A(1)$$

$$\omega(2)^{(1)} = 0.0469A(2)$$

$$\omega(3)^{(1)} = 0.0544A(3)$$

$$\omega(4)^{(1)} = 0.052A(4)$$

$$\omega(5)^{(1)} = 0.0468A(5)$$

After the second cycle of calculation,

$$\omega(1)^{(2)} = 0.0494A(1)$$

$$\omega(2)^{(2)} = 0.049A(2)$$

$$\omega(3)^{(2)} = 0.0494A(3)$$

$$\omega(4)^{(2)} = 0.049A(4)$$

$$\omega(5)^{(2)} = 0.0507A(5)$$

The result by 'exact' solution is

$$\omega = 0.496A$$

Remarks
This method can be applied to any structural element with variable stiffnesses and loadings, and with any boundary conditions, including deep beams and thick plates for which an analytical solution is difficult to obtain [10, 11]. The accuracy of the result is proportional to that of deflection calculation. Calculation of the deflection influence surface is the fundamental first step in any structural analysis and design. Attention should be given to the fact that this

Figure 4.33 Simply supported uniform beam with different mass distribution (1).

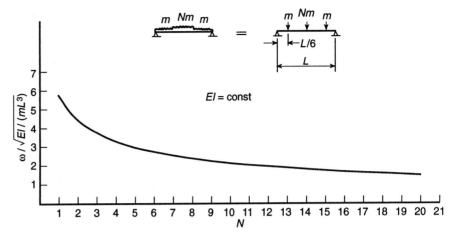

method utilizes the deflection influence surfaces which are used at the beginning of the analysis and design.

This method was used to study the effects of mass distribution and moment inertia on natural frequencies of simply supported beams, fixed beams and cantilever beams, and tower type structures. The results are shown in Figures 4.33–4.41.

In the case of cantilevered beams, increase of mass near the support does not significantly affect the vibration characteristics (Figure 4.39). On the other hand, increase of moment of inertia near the end has a similar effect (Figure 4.40). Increase of moment of inertia near the support influences the natural frequencies quite profoundly (Figures 4.40 and 4.41).

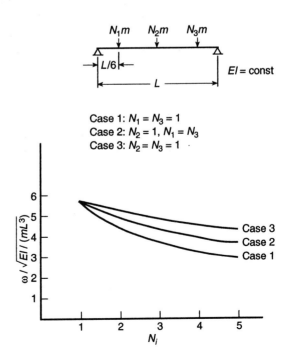

Figure 4.34 Simply supported uniform beam with different mass distribution (2).

Table 4.4 Critical circular frequency of cantilevered uniform beam with different mass distribution

N_i	Case 1	Case 2	Case 3	Case 4
1	2.08333	2.08333	2.08333	2.08333
2	1.53795	1.92913	2.08028	1.92140
3	1.27469	1.80224	2.07723	1.78464
4	1.11236	1.69606	2.07417	1.66763
5	0.99952	1.60584	2.07200	1.56629

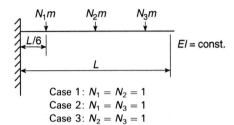

Case 1: $N_1 = N_2 = 1$
Case 2: $N_1 = N_3 = 1$
Case 3: $N_2 = N_3 = 1$

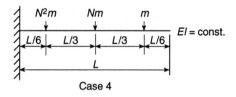

Case 4

For practical design purposes, it is desirable to simplify the vibration analysis procedure. One of the methods is to neglect the weight of the beam. The effect of neglecting the weight (thus mass) of the beam is studied as follows.

If a weightless beam is acted upon by a load P, the critical circular frequency of this beam is

$$\omega = \sqrt{g/\delta_{st}} \tag{4.82}$$

where δ_{st} is the static deflection.

For a massless simply supported beam with uniform EI throughout the length L, acted upon by a load P, at the center

$$\omega = \sqrt{\frac{48EIg}{PL^3}} \tag{4.83}$$

For a fixed beam with similar condition,

$$\omega = \sqrt{\frac{192EIg}{PL^3}} \tag{4.84}$$

In the case of a cantilever beam loaded at the free end,

$$\omega = \sqrt{\frac{3EIg}{PL^3}} \tag{4.85}$$

Figure 4.35 Uniformly loaded simple beams with different cross-section.

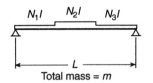

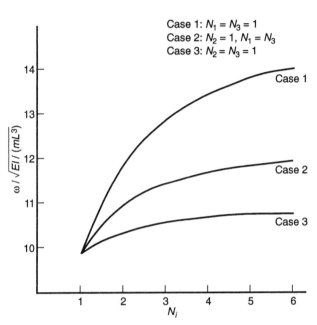

Figure 4.36 Fixed uniform beam with different mass distribution (1).

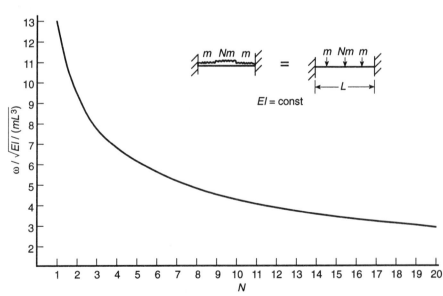

Figure 4.37 Fixed uniform beam with different mass distribution (2).

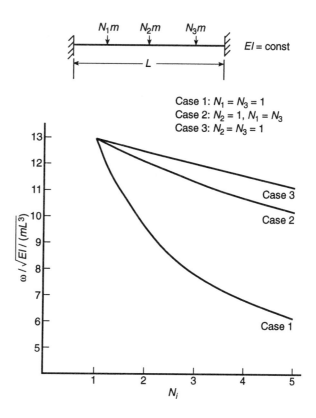

Case 1: $N_1 = N_3 = 1$
Case 2: $N_2 = 1$, $N_1 = N_3$
Case 3: $N_2 = N_3 = 1$

Table 4.5 Effect of neglecting weight of beams

				$\omega/\sqrt{EI/(mL^3)}$					
N	A1	A2	Difference (%)	B1	B2	Difference (%)	C1	C2	Difference (%)
1	5.6919	6.9282	21.72	12.9454	13.8564	7.04	2.0833	2.2769	9.29
2	4.4045	4.8990	11.23	9.4714	9.7980	3.45	1.5380	1.6099	4.68
5	2.9627	3.0984	4.58	6.1137	6.1968	1.36	0.9995	1.0182	1.87
10	2.1416	2.1909	2.30	4.3523	4.3818	0.68	0.7134	0.7199	0.92
15	1.7618	1.7889	1.54	3.5617	3.5777	0.45	0.5843	0.5879	0.61
20	1.5315	1.5492	1.16	3.0880	3.0984	0.34	0.5068	0.5091	0.46

Replacing P by Nm, N is gradually increased and equations 4.83, 4.84 and 4.85 are calculated for each of N. The results are compared with those of a previous study (Figures 4.33, 4.36 and 4.37), and some of the values are given in Table 4.5. It is noted that N does not directly indicate the ratio of the weight of the concentrated load to the total weight of a uniform load. For example, $N = 10$ indicates that the ratio is $(10 - 1)/3 = 3$, i.e. the weight of P is three times the total weight of the beam. Thus, in the case of a uniform simple beam with a concentrated load at the center of the span, the weight of which is three times that of the beam, the critical frequency difference between the correct value, obtained by considering the weight of the beam, and the approximate one, obtained by neglecting this weight, is 2.30%. In the case of a fixed beam

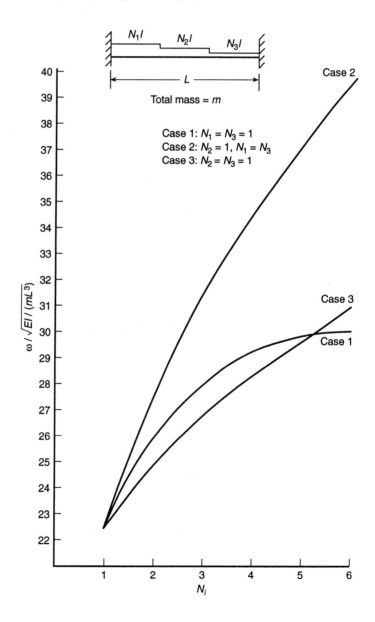

Figure 4.38 Uniformly loaded fixed beams with different cross-section.

Figure 4.39 Cantilevered uniform beam with different mass distribution.

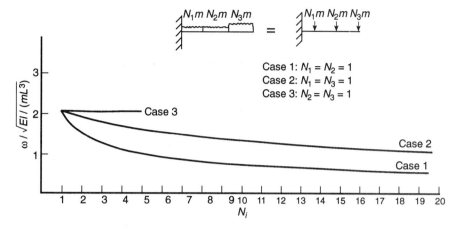

with similar condition, the difference is 0.68%. Study of a similar cantilever beam loaded at the end (actually $5L/6$ point from the end) shows that the difference is 0.92%.

In this method, approximation of mass distribution and the number of points under consideration does 'not' affect much the vibration characteristics of beams. It is the influence coefficients which generate most differences. When 'correct' influence coefficients are used, the results of six segmented (five pointed) beams and that of four segmented (three pointed) beams have exactly the same values to the accuracy of the computer truncating errors.

To illustrate the significant role of the influence coefficients, consider a cantilevered conical bar. The 'exact' formula of natural period of vibration by Kirchhoff is available as

$$\omega_k = 0.719 L^2 \sqrt{\frac{mA_0}{EI_0 g}} \tag{4.86}$$

where m is the unit weight, and A_0 and I_0 are the cross-sectional area and the moment of inertia at the supported base of the cone, respectively.

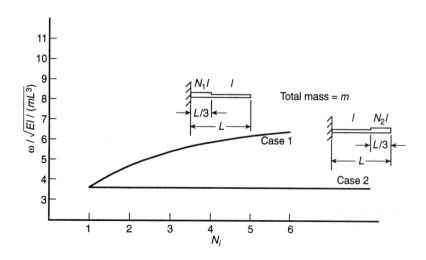

Figure 4.40 Uniformly loaded cantilever beams with different cross-section (1).

Figure 4.41 Uniformly loaded cantilever beams with different cross-section (2).

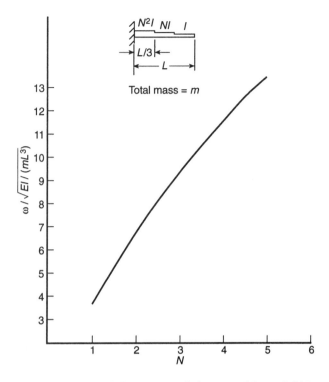

The diameter of the base of the cone, D, is increased from $0.02L$ to $0.1L$ with the increment of $0.02L$ to see the generality of the phenomenon. Each cone is segmented first by five points and by ten points for the next. Each case is then analysed twice by considering the diameter at each point and by considering that at the base of each trapezoidal segment. The result is compared with the 'exact' solution of each case, ω_k as shown in Figure 4.42. Note that, if D_B

Table 4.6 Some of the influence coefficients of a conical bar

	Case		
	1	2	3
1, 5	0.8565×10^4	0.6811×10^4	0.6810×10^4
2, 5	0.9737×10^5	0.6867×10^5	0.7351×10^5
3, 5	0.3669×10^6	0.2323×10^6	0.2692×10^6
4, 5	0.1158×10^7	0.6247×10^6	0.7801×10^6
5, 5	0.4892×10^7	0.1668×10^7	0.2342×10^7

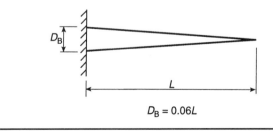

$D_B = 0.06L$

Figure 4.42 Analysis of canti-levered conical bars. Case 1: five segments, D at each point. Case 2: five segments, D at base of each segment. Case 3: 10 segments, D at base of each segment.

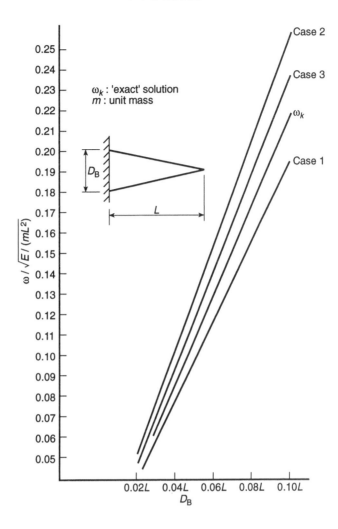

increases, the structure no longer behaves as a beam, and the accuracy of $\Delta_{i,j}$ decreases. It is interesting to note that all cases have almost straight lines. Furthermore, the ratio of the natural frequency of case i, ω_i, where the subscript i indicates the case number, to the 'exact' value ω_k, is the same for all values of the diameter of the base of the cone, D_B:

$$\omega_1/\omega_k = 0.89$$

$$\omega_2/\omega_k = 1.182$$

$$\omega_3/\omega_k = 1.085$$

This characteristic is very useful for practical applications where the 'exact' solution is not available. The ratio ω_i/ω_k can be anticipated (with rough approximation) if the influence coefficients of each case are studied.

As an example, consider the conical bar with $D_B = 0.06L$. Some of the influence coefficients, $\delta^l_{i,j}$, where the superscript, l, indicates the case number in Figure 4.42, are shown in Table 4.6. The ω_1s are inversely proportional to the square root of the ratio of $\delta^l_{i,j}$s:

$$\frac{\omega_2}{\omega_1} = \frac{0.1549}{0.1168} \doteqdot 1.32$$

Note that

$$\sqrt{\frac{\delta^1_{3.5}}{\delta^2_{3.5}}} = \sqrt{\frac{3669}{2323}} \doteqdot 1.26$$

$$\sqrt{\frac{\delta^1_{4.5}}{\delta^2_{4.5}}} = \sqrt{\frac{11\,580}{6247}} \doteqdot 1.36$$

Because of complexity of the geometry or loading condition, no 'exact' theory is available for vibration analysis of most of the practical design problems. Such problems can be solved with relative ease by this method. If the accuracy of the influence coefficients is in question because of the complexity of the geometry and of the mass distribution, the number of finite elements to obtain these coefficients can be changed and by Schwarz's inequality the range in which the exact solution lies can be found.

Examples of two-dimensional problems including composite laminates are handled in section 7.8.3 dealing with composite vibration.

This method may be extended to stability analysis of complex structural elements.

References

Note: *ICCM = International Conference on Composite Materials*; SAMPE = Society for the Advancement of Material and Process Engineering.

1. Timoshenko, S. P. (1953) *History of Strength of Materials*, McGraw-Hill, New York.
2. Lecture notes by Professor John E. Goldberg and several other professors, Purdue University, 1960–1965.
3. Goldberg, J. E. (1953) Slope–deflection coefficients for uniform member under axial load, Purdue University.
4. Hayes, J. M. (1960) Slope–deflection procedures for members, with varying moment of inertia, Purdue University.
5. Kim, D. H. (1974) A method of vibration analysis of irregularly shaped structural elements, in *International Symposium on Engineering Problems in Creating Coastal Industrial Sites* (ed. ISWACO), Seoul, Korea, pp. 39–63.
6. Kim, D. H. *et al.* (1989) A simple method of vibration analysis of irregularly shaped composite structural elements, in *1st Japan International SAMPE Symposium* (eds Igata, T. *et al.*), 28 Nov.–1 Dec., pp. 863–8.
7. Kim, D. H. *et al.* (1990) Vibration analysis of irregularly shaped composite structural members – for higher modes, in *Eighth Structural Congress* (eds Corotis, R. B. and Ellingwood, B.), American Society of Civil Engineers, Baltimore, MD, USA, 30 Apr.–3 May, pp. 63–4.
8. Kim, D. H. (1990) A new simple method of vibration analysis of laminated composite materials. Korea Society of Composite Materials, Seoul, Korea.
9. Kim, D. H. (1991) Vibration analysis of laminated thick composite plates, in *3rd EASEC*, Vol. 2 (eds Tian, L. C. *et al.*), Shanghai, China, Apr., pp. 1249–54.

10. Kim, D. H. (1991) Vibration analysis of irregularly shaped laminated thick composite plates, in *ICCM 8* (eds Tsai, S. W. and Springer, G. T. S.), Honolulu, July, p. 30-J.
11. Kim, D. H., Park, J. S. and Kim, K. J. (1991) Vibration analysis of irregularly shaped laminated thick composite plates II, in *2nd Japan International SAMPE Symposium and Exhibition* (eds Kimpara, T. *et al.*), Dec., pp. 1310–17.

Further reading

Shu, S. S. (1987) *Boundary Value Problems of Linear Partial Differential Equations for Engineers and Scientists*, World Scientific, Singapore.
Timoshenko, S. P. and Gere, J. M. (1963) *Theory of Elastic Stability*, McGraw-Hill, New York.

5 Anisotropic elasticity

5.1 Introduction

As shown in Chapter 2, composites with short fibers or particulates have general isotropic properties, and for such composites, conventional theories combined with some knowledge on material characteristics are acceptable for analysis and design. Such composites are mainly used for low cost, high production systems. However, the maximum specific strength and/or stiffness can be obtained when long straight fibers are used. Most of the composite material structural elements are in the form of laminated beam, plate and shell types.

In reality, structural design is commenced mostly by conventional theories as discussed in previous chapters, based on the assumption of isotropy and homogeneity of materials to be used. Even if composite materials are used partially or for all of the system, a preliminary study may be carried out by the use of classical theories to select, initially, the materials and dimensions of the structure. This concept is indirectly supported by the recent papers of Verchery [1–3]. Then, rigorous analysis is essential, taking into account the anisotropic nature of the materials, i.e. the materials' directional characteristics. Even with three-dimensional structures, one may reduce such structures to either two- or one-dimensional problems for simple analysis and calculation. For example, a building is a three-dimensional structure. Structural engineers try to design such a structure by reducing this to one-dimensional beams and columns, or two-dimensional frames, and slabs and bearing walls. In doing so, the engineer must have clear knowledge on the three-dimensional behavior of the whole structure as well as the material to be used.

When composite materials are used, one may start to 'compose' the structure using, to the maximum, the benefits of the materials to be used. Simple classical theory may yield good results for selecting 'initial' sections for preliminary design, if one 'modifies' initially selected stiffnesses of each member, taking into account that selection of materials and reinforcement types/directions will be done for optimum design. By doing so, rigorous analysis, which will follow such steps, may be concluded at the early stage of repeated calculation cycles.

The equilibrium equations, strain–displacement relations and compatibility equations used for classical theories for isotropic materials, with the assumption of the small deformation and linear elastic nature of such structures, are the same for the anisotropic composite materials. It is the stress–strain relationships, also called constitutive relations, which are very much different from one to the other. It is stressed that an engineer must have thorough knowledge on, and ability to apply, at will, conventional or classical theories of elasticity and mechanics, but, at the same time, must be flexible enough to employ the diverse characteristics of composite materials to the maximum advantage. A clear picture of the constitutive relations as well as properties of each of the major materials is very important.

5.2 Stress–strain relations of anisotropic materials

Consider the control element with dimension dx, dy and dz as shown in Figure 3.2. The positive directions and the notations for three mutually orthogonal components of a surface traction are as shown in this figure. It was explained in section 3.1 that the number of independent stresses acting on the surfaces of a control element is six. The strain–displacement relations are given in equation 3.2. The relation between the engineering shear strain $\gamma_{ij}(i \neq j)$, and the tensor shear strain, $\varepsilon_{ij}(i \neq j)$, is given as $\varepsilon_{ij} = \frac{1}{2}\gamma_{ij}$ in equation 3.3. 'Generalized' Hooke's law for an isotropic material is given in equation 3.5.

For composite materials, we need generalized Hooke's law relating stresses to strains considering the nature of anisotropy. In the mathematical theory of elasticity, such a relationship is expressed as

$$\sigma_{ij} = C_{ijkl}\varepsilon_{kl} \tag{5.1}$$

As explained in Chapter 3, both stresses and strains are second-order tensor quantities and, in three-dimensional space, they have nine components. But it is shown that the number of independent components of stress as well as strain is six because of symmetry in both stress and strain, so that the elasticity tensor, C_{ijkl}, has 36 components instead of 81. To avoid confusion, we define notations as follows:

$$
\begin{array}{ll}
\sigma_{11} = \sigma_1 & \varepsilon_{11} = \varepsilon_1 \\
\sigma_{22} = \sigma_2 & \varepsilon_{22} = \varepsilon_2 \\
\sigma_{33} = \sigma_3 & \varepsilon_{33} = \varepsilon_3 \\
\sigma_{23} = \tau_{23} = \sigma_4 & 2\varepsilon_{23} = \gamma_{23} = \varepsilon_4 \\
\sigma_{31} = \tau_{31} = \sigma_5 & 2\varepsilon_{31} = \gamma_{31} = \varepsilon_5 \\
\sigma_{12} = \tau_{12} = \sigma_6 & 2\varepsilon_{12} = \gamma_{12} = \varepsilon_6
\end{array}
\tag{5.2}
$$

By considering the derivatives of the strain energy [4, 5], it can be shown that the stiffness matrix, $[C_{ij}]$ (not tensor quantities), and the compliance matrix, $[S_{ij}]$ (not tensor quantities) are symmetric, so that

$$
\begin{aligned}
[C_{ij}] &= [C_{ji}] \\
[S_{ij}] &= [S_{ji}]
\end{aligned}
\tag{5.3}
$$

the number of independent elasticity constants is 21. We can always suppose that the quadratic form, $W = \frac{1}{2}C_{ij}\varepsilon_i\varepsilon_j$ with the property, $\partial W/\partial \varepsilon_i = \sigma_i$ is symmetric, and $\sigma_i = C_{ij}\varepsilon_j$, where $C_{ij} = C_{ji}$.

The general expression for stress–strain relationships for a linear elastic system can be written as

$$
\begin{bmatrix} \sigma_1 \\ \sigma_2 \\ \sigma_3 \\ \tau_{23} \\ \tau_{31} \\ \tau_{12} \end{bmatrix}
=
\begin{bmatrix}
C_{11} & C_{12} & C_{13} & C_{14} & C_{15} & C_{16} \\
C_{12} & C_{22} & C_{23} & C_{24} & C_{25} & C_{26} \\
C_{13} & C_{23} & C_{33} & C_{34} & C_{35} & C_{36} \\
C_{14} & C_{24} & C_{34} & C_{44} & C_{45} & C_{46} \\
C_{15} & C_{25} & C_{35} & C_{45} & C_{55} & C_{56} \\
C_{16} & C_{26} & C_{36} & C_{46} & C_{56} & C_{66}
\end{bmatrix}
\begin{bmatrix} \varepsilon_1 \\ \varepsilon_2 \\ \varepsilon_3 \\ \gamma_{23} \\ \gamma_{31} \\ \gamma_{12} \end{bmatrix}
\tag{5.4}
$$

or $\tau_i = C_{ij}\varepsilon_i$. A similar expression for strain–stress relations is

$$\begin{bmatrix} \varepsilon_1 \\ \varepsilon_2 \\ \varepsilon_3 \\ \gamma_{23} \\ \gamma_{31} \\ \gamma_{12} \end{bmatrix} = \begin{bmatrix} S_{11} & S_{12} & S_{13} & S_{14} & S_{15} & S_{16} \\ S_{12} & S_{22} & S_{23} & S_{24} & S_{25} & S_{26} \\ S_{13} & S_{23} & S_{33} & S_{34} & S_{35} & S_{36} \\ S_{14} & S_{24} & S_{34} & S_{44} & S_{45} & S_{46} \\ S_{15} & S_{25} & S_{35} & S_{45} & S_{55} & S_{56} \\ S_{16} & S_{26} & S_{36} & S_{46} & S_{56} & S_{66} \end{bmatrix} \begin{bmatrix} \sigma_1 \\ \sigma_2 \\ \sigma_3 \\ \tau_{23} \\ \tau_{31} \\ \tau_{12} \end{bmatrix} \tag{5.5}$$

or $\varepsilon_i = S_{ij}\tau_j$. These two equations characterize anisotropic materials and such a material is alternatively called a triclinic material.

The 'real' materials have three mutually orthogonal planes and are called orthogonally anisotropic, or, briefly, orthotropic materials. Such materials have only nine independent components in the stiffness matrix as shown

$$\begin{bmatrix} \sigma_1 \\ \sigma_2 \\ \sigma_3 \\ \tau_{23} \\ \tau_{31} \\ \tau_{12} \end{bmatrix} = \begin{bmatrix} C_{11} & C_{12} & C_{13} & 0 & 0 & 0 \\ C_{12} & C_{22} & C_{23} & 0 & 0 & 0 \\ C_{13} & C_{23} & C_{33} & 0 & 0 & 0 \\ 0 & 0 & 0 & C_{44} & 0 & 0 \\ 0 & 0 & 0 & 0 & C_{55} & 0 \\ 0 & 0 & 0 & 0 & 0 & C_{66} \end{bmatrix} \begin{bmatrix} \varepsilon_1 \\ \varepsilon_2 \\ \varepsilon_3 \\ \gamma_{23} \\ \gamma_{31} \\ \gamma_{12} \end{bmatrix} \tag{5.6}$$

It is noted that there is no 'coupling' between the normal stresses and the shear strains, as well as between the shear stress and the normal strains. The strain–stress relations are

$$\begin{bmatrix} \varepsilon_1 \\ \varepsilon_2 \\ \varepsilon_3 \\ \gamma_{23} \\ \gamma_{31} \\ \gamma_{12} \end{bmatrix} = \begin{bmatrix} S_{11} & S_{12} & S_{13} & 0 & 0 & 0 \\ S_{12} & S_{22} & S_{23} & 0 & 0 & 0 \\ S_{13} & S_{23} & S_{33} & 0 & 0 & 0 \\ 0 & 0 & 0 & S_{44} & 0 & 0 \\ 0 & 0 & 0 & 0 & S_{55} & 0 \\ 0 & 0 & 0 & 0 & 0 & S_{66} \end{bmatrix} \begin{bmatrix} \sigma_1 \\ \sigma_2 \\ \sigma_3 \\ \tau_{23} \\ \tau_{31} \\ \tau_{12} \end{bmatrix} \tag{5.7}$$

If the material is transversely isotropic, i.e. there is one isotropic plane at every point of the material, the number of independent constants is five, and we have the following equations. For this example, if the plane of isotropy is the 1–2 plane:

$$\begin{bmatrix} \sigma_1 \\ \sigma_2 \\ \sigma_3 \\ \tau_{23} \\ \tau_{31} \\ \tau_{12} \end{bmatrix} = \begin{bmatrix} C_{11} & C_{12} & C_{13} & 0 & 0 & 0 \\ C_{12} & C_{11} & C_{33} & 0 & 0 & 0 \\ C_{13} & C_{13} & C_{33} & 0 & 0 & 0 \\ 0 & 0 & 0 & C_{44} & 0 & 0 \\ 0 & 0 & 0 & 0 & C_{44} & 0 \\ 0 & 0 & 0 & 0 & 0 & (C_{11}-C_{12})/2 \end{bmatrix} \begin{bmatrix} \varepsilon_1 \\ \varepsilon_2 \\ \varepsilon_3 \\ \gamma_{23} \\ \gamma_{31} \\ \gamma_{12} \end{bmatrix} \tag{5.8}$$

$$\begin{bmatrix} \varepsilon_1 \\ \varepsilon_2 \\ \varepsilon_3 \\ \gamma_{23} \\ \gamma_{31} \\ \gamma_{12} \end{bmatrix} = \begin{bmatrix} S_{11} & S_{12} & S_{13} & 0 & 0 & 0 \\ S_{12} & S_{11} & S_{13} & 0 & 0 & 0 \\ S_{13} & S_{13} & S_{33} & 0 & 0 & 0 \\ 0 & 0 & 0 & S_{44} & 0 & 0 \\ 0 & 0 & 0 & 0 & S_{44} & 0 \\ 0 & 0 & 0 & 0 & 0 & 2(S_{11}-S_{12}) \end{bmatrix} \begin{bmatrix} \sigma_1 \\ \sigma_2 \\ \sigma_3 \\ \tau_{23} \\ \tau_{31} \\ \tau_{12} \end{bmatrix} \tag{5.9}$$

If the material has an infinite number of planes of symmetry, it is the isotropic material, and there are only two independent constants in the stiffness matrix, as follows:

$$
\begin{bmatrix} \sigma_1 \\ \sigma_2 \\ \sigma_3 \\ \tau_{23} \\ \tau_{31} \\ \tau_{12} \end{bmatrix} =
\begin{bmatrix}
C_{11} & C_{12} & C_{12} & 0 & 0 & 0 \\
C_{12} & C_{11} & C_{12} & 0 & 0 & 0 \\
C_{12} & C_{12} & C_{11} & 0 & 0 & 0 \\
0 & 0 & 0 & (C_{11}-C_{12})/2 & 0 & 0 \\
0 & 0 & 0 & 0 & (C_{11}-C_{12})/2 & 0 \\
0 & 0 & 0 & 0 & 0 & (C_{11}-C_{12})/2
\end{bmatrix}
\times
\begin{bmatrix} \varepsilon_1 \\ \varepsilon_2 \\ \varepsilon_3 \\ \gamma_{23} \\ \gamma_{31} \\ \gamma_{12} \end{bmatrix}
\tag{5.10}
$$

$$
\begin{bmatrix} \varepsilon_1 \\ \varepsilon_2 \\ \varepsilon_3 \\ \gamma_{23} \\ \gamma_{31} \\ \gamma_{12} \end{bmatrix} =
\begin{bmatrix}
S_{11} & S_{12} & S_{12} & 0 & 0 & 0 \\
S_{12} & S_{11} & S_{12} & 0 & 0 & 0 \\
S_{12} & S_{12} & S_{11} & 0 & 0 & 0 \\
0 & 0 & 0 & 2(S_{11}-S_{12}) & 0 & 0 \\
0 & 0 & 0 & 0 & 2(S_{11}-S_{12}) & 0 \\
0 & 0 & 0 & 0 & 0 & 2(S_{11}-S_{12})
\end{bmatrix}
\times
\begin{bmatrix} \sigma_1 \\ \sigma_2 \\ \sigma_3 \\ \tau_{23} \\ \tau_{31} \\ \tau_{12} \end{bmatrix}
\tag{5.11}
$$

The compliance matrix, $[S_{ij}]$, is the adjoint matrix, $[C_{ij}^+]$, which is the transpose of the cofactor matrix of $[C_{ij}]$, the stiffness matrix, divided by the determinant of the matrix $[C_{ij}]$. The compliance matrix can be expressed as

$$
[S_{ij}] = \frac{[C_{ij}^+]}{|C_{ij}|}
\tag{5.12}
$$

This follows directly by considering equations 5.4 and 5.5.

5.3 Engineering constants for orthotropic materials

The components of both the stiffness matrix, $[C_{ij}]$, and the compliance matrix, $[S_{ij}]$, are simply the mathematical symbols relating stresses and strains. Clear understanding of each constant is necessary for practicing engineers.

We consider the standard tensile test in the X_1 direction. The stress and strain tensors are

$$\sigma_{ij} = \begin{pmatrix} \sigma_{11} & 0 & 0 \\ 0 & 0 & 0 \\ 0 & 0 & 0 \end{pmatrix}$$

$$\varepsilon_{ij} = \begin{pmatrix} \varepsilon_{11} & 0 & 0 \\ 0 & -v_{12}\varepsilon_{11} & 0 \\ 0 & 0 & -v_{13}\varepsilon_{11} \end{pmatrix} \tag{5.13}$$

Poisson's ratio, v_{ij}, is the ratio of the transverse strain in the j-direction when stressed in the i-direction, so that

$$v_{ij} = -\frac{\varepsilon_{jj}}{\varepsilon_{ii}} \tag{5.14}$$

for $\sigma_i = \sigma$, and all other stresses are zero. In equation 5.13, $\varepsilon_{22} = -v_{12}\varepsilon_{11}$, or $v_{12} = -\varepsilon_{22}/\varepsilon_{11}$.

The modulus of elasticity in the x_i-direction, E_i, is the ratio of the stress, σ_i, to the strain, ε_i, so that

$$\varepsilon_1 = S_{11}\sigma_1 = \frac{\sigma_1}{E_1}$$

$$\varepsilon_2 = S_{21}\sigma_1 = -v_{12}\varepsilon_1 = -\frac{v_{12}\sigma_1}{E_1}$$

$$\varepsilon_3 = S_{31}\sigma_1 = -v_{13}\varepsilon_1 = -\frac{v_{13}\sigma_1}{E_1}$$

which yield

$$S_{11} = \frac{1}{E_1}, \quad S_{21} = -\frac{v_{12}}{E_1}, \quad S_{31} = -\frac{v_{13}}{E_1}$$

From the similar tensile test in the X_2 direction, we obtain,

$$S_{12} = -\frac{v_{21}}{E_2}, \quad S_{22} = \frac{1}{E_2}, \quad S_{32} = \frac{-v_{23}}{E_2}$$

In the X_3 direction,

$$S_{13} = -\frac{v_{31}}{E_3}, \quad S_{23} = -\frac{v_{32}}{E_3}, \quad S_{33} = \frac{1}{E_3}$$

The compliance matrix components in terms of the engineering constants for an orthotropic material are

$$[S_{ij}] = \begin{bmatrix} 1/E_1 & -v_{21}/E_2 & -v_{31}/E_3 & 0 & 0 & 0 \\ -v_{12}/E_1 & 1/E_2 & -v_{32}/E_3 & 0 & 0 & 0 \\ -v_{13}/E_1 & -v_{23}/E_2 & 1/E_3 & 0 & 0 & 0 \\ 0 & 0 & 0 & 1/G_{23} & 0 & 0 \\ 0 & 0 & 0 & 0 & 1/G_{31} & 0 \\ 0 & 0 & 0 & 0 & 0 & 1/G_{12} \end{bmatrix} \tag{5.15}$$

By equation 5.3, $S_{ij} = S_{ji}$, from which we obtain

$$\frac{v_{ij}}{E_i} = \frac{v_{ji}}{E_j} \tag{5.16}$$

It is emphasized that there is a clear difference between v_{12} and v_{21} for an orthotropic material since $E_1 \neq E_2$.

From equations 5.10 and 5.11, it is seen that the stiffness and compliance matrices are mutually inverse. By simple matrix algebra, we obtain the following relations for orthotropic materials:

$$C_{11} = (S_{22}S_{33} - S_{23}^2)/D, \qquad C_{22} = (S_{33}S_{11} - S_{13}^2)/D$$

$$C_{33} = (S_{11}S_{22} - S_{12}^2)/D, \qquad C_{12} = (S_{13}S_{23} - S_{12}S_{33})/D \tag{5.17}$$

$$C_{13} = (S_{12}S_{23} - S_{13}S_{22})/D, \quad C_{23} = (S_{12}S_{13} - S_{23}S_{11})/D$$

$$C_{44} = 1/S_{44}, \quad C_{55} = 1/S_{55}, \quad C_{66} = 1/S_{66}$$

where $D = S_{11}S_{22}S_{33} - S_{11}S_{23}^2 - S_{22}S_{13}^2 - S_{33}S_{12}^2 + 2S_{12}S_{23}S_{13}$.

For the stiffness matrix, some composite materials engineers use the notation Q_{ij} instead of C_{ij} used by classical theory of elasticity. Up to now C_{ij} has been used to help readers who are familiar with classical theory understand better, but Q_{ij} can also be used.

Similarly, for shear stresses and shear strains the tensor notation can be used, such that

$$\varepsilon_{ij} = \tfrac{1}{2}\gamma_{ij} \quad \text{and} \quad \sigma_{ij} = \tau_{ij}$$

The components of the stiffness matrix C_{ij}, for an orthotropic material can be obtained either by inversion of the compliance matrix S_{ij} (equation 5.15), or by substituting the elements of S_{ij} into equations 5.17.

$$C_{11} = E_{11}(1 - v_{23}v_{32})/\nabla$$

$$C_{22} = E_{22}(1 - v_{31}v_{13})/\nabla$$

$$C_{33} = E_{33}(1 - v_{12}v_{21})/\nabla$$

$$C_{44} = G_{23}, \quad C_{55} = G_{13}, \quad C_{66} = G_{12} \tag{5.18}$$

$$C_{12} = E_{11}(v_{21} + v_{31}v_{23})/\nabla = E_{22}(v_{12} + v_{13}v_{32})/\nabla$$

$$C_{13} = E_{11}(v_{31} + v_{21}v_{32})/\nabla = E_{33}(v_{13} + v_{13}v_{32})/\nabla$$

$$C_{23} = E_{22}(v_{32} + v_{12}v_{31})/\nabla = E_{33}(v_{23} + v_{21}v_{13})/\nabla$$

where

$$\nabla = 1 - v_{12}v_{21} - v_{23}v_{32} - v_{31}v_{13} - 2v_{21}v_{32}v_{13}$$

Note that we need 2 before C_{44}, C_{55} and C_{66} in equation 5.6 if $\varepsilon_{ij}(i \neq j)$ is used instead of $\gamma_{ij}(i \neq j)$.

5.4 Stress–strain relations of plane stress and plane strain problems for uni-directionally reinforced lamina

The lamina shown in Figure 5.1 is in a plane stress state in the 1–2 plane, so that

$$\sigma_3 = 0, \quad \tau_{23} = \sigma_{23} = 0, \quad \tau_{31} = \sigma_{31} = 0$$

Figure 5.1 Typical lamina co-ordinate system.

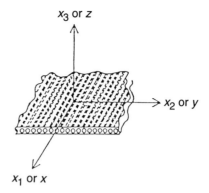

x_3 or z

x_2 or y

x_1 or x

For orthotropic materials, the strains are

$$\varepsilon_3 = S_{13}\sigma_1 + S_{23}\sigma_2$$

$$\gamma_{23} = 2\varepsilon_{23} = 0, \quad \gamma_{31} = 2\varepsilon_{31} = 0$$

The strain–stress relations are

$$\begin{bmatrix} \varepsilon_1 \\ \varepsilon_2 \\ \varepsilon_{12} \end{bmatrix} \begin{bmatrix} S_{11} & S_{12} & 0 \\ S_{12} & S_{22} & 0 \\ 0 & 0 & S_{66}/2 \end{bmatrix} \begin{bmatrix} \sigma_1 \\ \sigma_2 \\ \sigma_{12} \end{bmatrix} \tag{5.19}$$

or

$$[\varepsilon] = [S][\sigma]$$

in which

$$S_{11} = 1/E_1$$

$$S_{12} = -v_{12}/E_1 = -v_{21}/E_2$$

$$S_{22} = 1/E_2$$

$$S_{66} = 1/G_{12}$$

The stress–strain relationship can be obtained by inverting equation 5.19,

$$\begin{bmatrix} \sigma_1 \\ \sigma_2 \\ \sigma_{12} \end{bmatrix} \begin{bmatrix} Q_{11} & Q_{12} & 0 \\ Q_{12} & Q_{22} & 0 \\ 0 & 0 & 2Q_{66} \end{bmatrix} \begin{bmatrix} \varepsilon_1 \\ \varepsilon_2 \\ \varepsilon_{12} \end{bmatrix} \tag{5.20}$$

or

$$[\sigma] = [Q][\varepsilon]$$

in which the Q_{ij}s are called the **reduced stiffnesses** for the plane stress case and are expressed as $Q_{ij} = C_{ij} - C_{i3}C_{j3}/C_{33}$, or

$$Q_{11} = S_{22}/(S_{11}S_{22} - S_{12}^2)$$

$$Q_{12} = -S_{12}/(S_{11}S_{22} - S_{12}^2)$$

$$Q_{22} = S_{11}/(S_{11}S_{22} - S_{12}^2)$$

$$Q_{66} = 1/S_{66}$$

Note that $\sigma_{12} = \sigma_6$ and $[Q] = [S]^{-1}$.

Recalling that $v_{12}/E_1 = v_{21}/E_2$, and replacing the S_{ij} by the engineering constants, we obtain

$$Q_{11} = \frac{E_1}{1 - v_{12}v_{21}}$$

$$Q_{12} = \frac{v_{12}E_2}{1 - v_{12}v_{21}} = \frac{v_{21}E_1}{1 - v_{12}v_{21}} \qquad (5.21)$$

$$Q_{22} = \frac{E_2}{1 - v_{12}v_{21}}$$

$$Q_{66} = G_{12}$$

Tsai [6] uses letter subscripts for the on-axis orthotropic constants to distinguish them from numeric subscripts which he uses for the anisotropic and off-axis orthotropic constants. 'On-axis' means that the structural axis coincides with the material axis. If the numerical subscripts of equations 5.21 represent anisotropic and off-axis orthotropic constants, those equations do not hold true.

For the plane strain case in the 1–2 plane,

$$[\sigma] = [C][\varepsilon] \text{ or } \sigma_i = C_{ij}\varepsilon_j, \ i, j = 1, 2, 6$$

$$[\varepsilon] = [\beta][\sigma] \text{ or } \varepsilon_i = \beta_{ij}\sigma_j, \quad i, j = 1, 2, 6$$

where

$$\beta_{ij} = S_{ij} - \frac{S_{i3}S_{j3}}{S_{33}} \qquad (5.22)$$

β_{ij} is the **reduced compliance** for the plane strain case.

For the plane strain case, the compliance and stiffness matrices have the following relationship:

$$[C] = [\beta]^{-1}$$

$$[\beta] = [C]^{-1} \qquad (5.23)$$

$$[C][\beta] = [I]$$

The following must be clearly kept in mind:

- For the three-dimensional case,

$$[\sigma] = [C][\varepsilon], \quad [\varepsilon] = [S][\sigma]$$

- For the two-dimensional plane stress case,

$$[\sigma] = [Q][\varepsilon], \quad [\varepsilon] = [S][\sigma]$$

- For the two-dimensional plane strain case,

$$[\sigma] = [C][\varepsilon], \quad [\varepsilon] = [\beta][\sigma]$$

Table 5.1 Elastic properties of various composite materials (courtesy of Think Composites)

Type: Fiber/cloth: Matrix:	CFRP T300 N5208	BFRP B(4) N5505	CFRP AS H3501	GFRP E-glass Epoxy	KFRP kev49 Epoxy	CFRTP AS4 PEEK APC2	CFRP IM6 Epoxy	CFRP T300 Fbrt934 4-mil tp	CCRP T300 Fbrt934 13-mil c	CCRP T300 Fbrt934 7-mil c
V_f	0.70	0.50	0.66	0.45	0.60	0.66	0.66	0.60	0.60	0.60
Density	1.60	2.00	1.60	1.80	1.46	1.60	1.60	1.50	1.50	1.50
h_0 (mm)	0.125	0.125	0.125	0.125	0.125	0.125	0.125	0.100	0.325	0.175
h_0 (mil)	4.925	4.925	4.925	4.925	4.925	4.925	4.925	3.94	12.805	6.895
Ply engineering constants and data										
E_x (GPa)	181.0	204.0	138.0	38.6	76.0	134.0	203.0	148.0	74.0	66.0
E_y (GPa)	10.30	18.50	8.96	8.27	5.50	8.90	11.20	9.65	74.00	66.00
v/x	0.28	0.23	0.30	0.26	0.34	0.28	0.32	0.30	0.05	0.04
G (GPa)	7.17	5.59	7.10	4.14	2.30	5.10	8.40	4.55	4.55	4.10
E_x (Msi)	26.25	29.58	20.01	5.60	11.02	19.43	29.44	21.46	10.73	9.57
E_y (Msi)	1.49	2.68	1.30	1.20	0.80	1.29	1.62	1.40	10.73	9.57
G (Msi)	1.04	0.81	1.03	0.60	0.33	0.74	1.22	0.66	0.66	0.59
[Q] 0° SI units (GPa)										
Q_{xx}	181.8	205.0	138.8	39.2	76.6	134.7	204.2	148.9	74.2	66.1
Q_{yy}	10.35	18.59	9.01	8.39	5.55	8.95	11.26	9.71	74.19	66.13
Q_{xy}	2.90	4.28	2.70	2.18	1.89	2.51	3.60	2.91	3.71	2.91
Q_{ss}	7.17	5.59	7.10	4.14	2.30	5.10	8.40	4.55	4.55	4.10
Imperial (Msi)										
Q_{xx}	26.36	29.73	20.13	5.68	11.11	19.53	29.61	21.59	10.76	9.58
Q_{yy}	1.50	2.70	1.31	1.22	0.80	1.30	1.63	1.41	10.76	9.59
Q_{xy}	0.42	0.62	0.39	0.32	0.27	0.36	0.52	0.42	0.54	0.42
Q_{ss}	1.04	0.81	1.03	0.60	0.33	0.74	1.22	0.66	0.66	0.59
[S] 0° SI units (1/TPa)										
S_{xx}	5.5	4.9	7.2	25.9	13.2	7.5	4.9	6.8	13.5	15.2
S_{yy}	97.1	54.1	111.6	120.9	181.8	112.4	89.3	103.6	13.5	15.2
S_{xy}	-1.5	-1.1	-2.2	-6.7	-4.5	-2.1	-1.6	-2.0	-0.7	-0.7
S_{ss}	139.5	178.9	140.8	241.5	434.8	196.1	119.0	219.8	219.8	243

Imperial (1/10^6 psi)										
S_{xx}	37.9	33.8	49.7	178.6	91.0	51.7	33.8	46.9	93.1	104.8
S_{yy}	670	373	770	834	1254	775	616	714	93	105
S_{xy}	−10.3	−7.6	−15.2	−46.2	−31.0	−14.5	−11.0	−13.8	−4.8	−4.8
S_{ss}	962	1234	971	1666	2999	1352	821	1516	1516	1682
Linear combinations of [Q]										
SI units (GPa)										
U_1^*	76.37	87.70	59.66	20.45	32.44	57.04	85.88	62.47	58.84	52.37
U_2	85.73	93.20	64.90	15.39	35.55	62.88	96.44	69.58	0.00	0.00
U_3	19.71	24.08	14.25	3.33	8.65	14.78	21.83	16.82	15.34	13.75
U_4^*	22.61	28.36	16.96	5.51	10.54	17.28	25.43	19.73	19.05	16.66
U_5^*	26.88	29.67	21.35	7.47	10.95	19.88	30.23	21.37	19.89	17.85
Imperial (Msi)										
U_1^*	11.07	12.72	8.65	2.97	4.70	8.27	12.45	9.06	8.53	7.59
U_2	12.43	13.51	9.41	2.23	5.15	9.12	13.98	10.09	0.00	0.00
U_3	2.86	3.49	2.07	0.48	1.25	2.14	3.17	2.44	2.22	1.99
U_4^*	3.28	4.11	2.46	0.80	1.53	2.51	3.69	2.86	2.76	2.42
U_5^*	3.90	4.30	3.10	1.08	1.59	2.88	4.38	3.10	2.88	2.59
Quasi-isotropic constants										
E (GPa)	69.68	78.53	54.84	18.96	29.02	51.81	78.35	56.24	52.67	47.07
ν	0.30	0.32	0.28	0.27	0.32	0.30	0.30	0.32	0.32	0.32
G (GPa)	26.88	29.67	21.35	7.47	10.95	19.88	30.23	21.37	19.89	17.8
E (Msi)	10.10	11.39	7.95	2.75	4.21	7.51	11.36	8.15	7.64	6.83
G (Msi)	3.90	4.30	3.10	1.08	1.59	2.88	4.38	3.10	2.88	2.59

* Invariant.
V_f: fiber volume fraction.
tp: tape.
c: cloth.
CFRP: carbon fiber reinforced thermoset composites.
BF: boron fiber.
KF: Kevlar fiber.
GF: glass fiber.
CC: carbon cloth.
CFRTP: carbon fiber reinforced thermoplastic composites.

For an isotropic material, the strain–stress relationship for the plane stress case is

$$
\begin{bmatrix} \varepsilon_1 \\ \varepsilon_2 \\ \varepsilon_{12} \end{bmatrix} = \begin{bmatrix} S_{11} & S_{12} & 0 \\ S_{12} & S_{11} & 0 \\ 0 & 0 & S_{11} - S_{12} \end{bmatrix} \begin{bmatrix} \sigma_1 \\ \sigma_2 \\ \sigma_{12} \end{bmatrix}
$$

in which $S_{11} = 1/E$, $S_{12} = -v/E$.

The stress–strain relationship is

$$
\begin{bmatrix} \sigma_1 \\ \sigma_2 \\ \sigma_{12} \end{bmatrix} = \begin{bmatrix} Q_{11} & Q_{12} & 0 \\ Q_{12} & Q_{11} & 0 \\ 0 & 0 & 2Q_{66} \end{bmatrix} \begin{bmatrix} \varepsilon_1 \\ \varepsilon_2 \\ \varepsilon_{12} \end{bmatrix}
$$

in which

$$
Q_{11} = \frac{E}{1 - v^2}
$$

$$
Q_{12} = \frac{vE}{1 - v^2}
$$

$$
Q_{66} = \frac{E}{2(1 + v)} = G
$$

These equations are identical to equations 3.10 if we recall $\gamma_{12} = 2\varepsilon_{12}$.

Numerical example

Tsai [6] presents an extensive amount of information on several different kinds of composites. Since the structural engineers working on civil construction will be using a large quantity of glass fiber reinforced composites (GFRP), though the noble composites will also be employed for considerable parts of such structures, such as elements requiring high strength, low deflection, higher fatigue strength, and so on, numerical examples will be given for GFRP.

From Table 5.1, the E-glass/epoxy composite has

- fiber volume fraction (V_f) 0.45
- density   1.80
- unit ply thickness (h_0) 0.125 mm

The engineering constants are

$$E_x = 38.6 \, \text{GPa}, \quad E_y = 8.27 \, \text{GPa}, \quad v_x = 0.26, \quad G_{12} = 4.14 \, \text{GPa}$$

By equation 5.16,

$$v_y = v_x E_y / E_x = 0.26(8.27/38.6) = 0.0557$$

From equation 5.21, for the plane stress case,

$$Q_{xx} = E_x/(1 - v_x v_y) = 38.6/(1 - 0.26 \times 0.0557) = 39.17 \, \text{GPa}$$
$$Q_{yy} = E_y/(1 - v_x v_y) = 8.27/(1 - 0.26 \times 0.0557) = 8.39 \, \text{GPa}$$
$$Q_{xy} = v_x Q_{yy} = v_y Q_{xx} = 0.26(8.39) = 2.18 \, \text{GPa}$$
$$Q_{ss} = G_{12} = 4.14 \, \text{GPa}$$

From equation 5.19,

$$S_{xx} = 1/E_x = 1/38.6 = 25.9 \,(1/\text{TPa})$$
$$S_{yy} = 1/E_y = 1/8.27 = 120.9 \,(1/\text{TPa})$$
$$S_{xy} = -v_x/E_x = -v_y/E_y = -0.26/38.6 = -6.74 \,(1/\text{TPa})$$
$$S_{ss} = 1/G_{12} = 1/4.14 = 241.5 \,(1/\text{TPa})$$

For the plane strain case,

$$E_z = E_y = 8.27 \text{ GPa}$$
$$S_{zz} = 1/E_z = 1/8.27 = 120.9 \,(1/\text{TPa})$$

We assume $v_{yz} = 0.5$ [6], then

$$S_{yz} = -v_{yz}/E_y = -0.5/8.27 = -60.5(1/\text{TPa})$$

From equation 5.22, letting $x = 1$, $y = 2$, $z = 3$, we obtain the reduced compliance matrix as follows. With $S_{xy} = S_{xz}$,

$$\beta_{xx} = S_{xx} - S_{xz}^2/S_{zz} = 25.9 - (-6.74)^2/120.9 = 25.52 \,(1/\text{TPa})$$
$$\beta_{yy} = S_{yy} - S_{yz}^2/S_{zz} = 120.9 - (60.5)^2/120.9 = 90.62 \,(1/\text{TPa})$$
$$\beta_{xy} = S_{xy} - S_{xz}S_{yz}/S_{zz} = (-6.74) - (-6.74)(-60.5)/120.9 = -10.11 \,(1/\text{TPa})$$
$$\beta_{ss} = S_{ss} = 241.5 \,(1/\text{TPa})$$

The stiffness matrix components for a transversely isotropic material are as follows:

$$\nabla = (1 + v_{23})(1 - v_{23} - 2v_{21}v_{12})$$
$$= (1 + 0.5)(1 - 0.5 - 2 \times 0.0557 \times 0.26) = 0.707$$
$$C_{11} = (1 - v_{23}^2)E_1/\nabla = (1 - 0.5^2)(38.6)/0.707 = 40.95 \text{ GPa}$$
$$C_{22} = C_{33} = (1 - v_{21}v_{12})E_2/\nabla = (1 - 0.0557 \times 0.26)(8.27)/0.707 = 11.53 \text{ GPa}$$
$$C_{12} = C_{13} = v_{21}(1 + v_{23})E_1/\nabla = v_{12}(1 + v_{23})E_2/\nabla$$
$$= (2.26)(1 + 0.5)(8.27)/0.707 = 4.562 \text{ GPa}$$
$$C_{23} = (v_{23} + v_{21}v_{12})E_2/\nabla = (0.5 + 0.26 \times 0.0557)(8.27)/0.707 = 6.018 \text{ GPa}$$
$$C_{44} = (1 - v_{23} - 2v_{21}v_{12})E_2/2\nabla = (1 - 0.5 - 2 \times 0.26 \times 0.0557)(8.27)/$$
$$(2 \times 0.707) = 2.755 \text{ GPa}$$
$$C_{55} = C_{66} = G_{12} = 4.14 \text{ GPa}$$

5.5 Transformation equations

In practice, the most of the composite material structural elements are highly directional, i.e. anisotropic. Most of the polymer based composites are made by stacking thin plies made of unidirectional fibers bound by the matrix. Each ply or lamina may have a different orientation to the others and it is necessary to consider the stress and strain in the off-axis or the structural axis orientation.

The equations for stress and strain at a point given by equations 3.13 and 3.22a can be used for this transformation. In this section, we consider the ply material axes to be rotated away from the laminate axes by θ, positive in the **counterclockwise** direction. This means that the laminate axis is at an angle θ **clockwise** from the material axes.

If the x–y coordinate system for a plane stress state is rotated by an angle θ to form a new 1–2 system, we have

$$[\sigma]_{1-2} = [T][\sigma]_{x-y}$$
$$[\sigma]_{x-y} = [T]^{-1}[\sigma]_{1-2}$$
$$[\varepsilon]_{1-2} = [T][\varepsilon]_{x-y}$$
$$[\varepsilon]_{x-y} = [T]^{-1}[\varepsilon]_{1-2}$$

(5.24)

where, with $m = \cos \theta$ and $n = \sin \theta$,

$$[T] = \begin{bmatrix} m^2 & n^2 & +2mn \\ n^2 & m^2 & -2mn \\ -mn & mn & (m^2 - n^2) \end{bmatrix}$$

(5.25)

$$[T]^{-1} = \begin{bmatrix} m^2 & n^2 & -2mn \\ n^2 & m^2 & +2mn \\ mn & -mn & (m^2 - n^2) \end{bmatrix}$$

Note that the second row of $[T]$ can be obtained by setting $\theta = \theta + \pi/2$ in equations 3.13 and 3.22. The third row of $[T]$ is obtained by the second of these equations, using the tensor notation $\varepsilon_{ij}(i \neq j)$. $[T]^{-1}$ can be obtained by setting $\theta = -\theta$. Equation 5.24 is possible because the tensor notation, $\varepsilon_{ij}(i \neq j)$, is used instead of the engineering notation, $\gamma_{ij}(i \neq j)$.

We may write the following detailed forms:

$$\begin{bmatrix} \sigma_1 \\ \sigma_2 \\ \sigma_6 \end{bmatrix} = [T] \begin{bmatrix} \sigma_x \\ \sigma_y \\ \sigma_{xy} \end{bmatrix}$$

$$\begin{bmatrix} \varepsilon_1 \\ \varepsilon_2 \\ \varepsilon_{12} \end{bmatrix} = [T] \begin{bmatrix} \varepsilon_x \\ \varepsilon_y \\ \varepsilon_{xy} \end{bmatrix}$$

(5.26)

The mechanical properties of a composite, transverse to the fiber direction, are provided by weaker matrix material, and the effects of transverse shear deformation may be significant. For such cases, the transformation matrix should include terms related to σ_4, σ_5, ε_4 and ε_5. Since the third axis does not rotate, we can have

$$\begin{bmatrix} \sigma_1 \\ \sigma_2 \\ \sigma_3 \\ \sigma_4 \\ \sigma_5 \\ \sigma_6 \end{bmatrix} = [T] \begin{bmatrix} \sigma_x \\ \sigma_y \\ \sigma_z \\ \sigma_{yz} \\ \sigma_{zx} \\ \sigma_{xy} \end{bmatrix}, \quad \begin{bmatrix} \varepsilon_1 \\ \varepsilon_2 \\ \varepsilon_3 \\ \varepsilon_{23} \\ \varepsilon_{31} \\ \varepsilon_{12} \end{bmatrix} = [T] \begin{bmatrix} \varepsilon_x \\ \varepsilon_y \\ \varepsilon_z \\ \varepsilon_{yz} \\ \varepsilon_{zx} \\ \varepsilon_{xy} \end{bmatrix}$$

(5.27)

in which the 'new' transformation matrix is

$$[T] = \begin{bmatrix} m^2 & n^2 & 0 & 0 & 0 & +2mn \\ n^2 & m^2 & 0 & 0 & 0 & -2mn \\ 0 & 0 & 1 & 0 & 0 & 0 \\ 0 & 0 & 0 & m & -n & 0 \\ 0 & 0 & 0 & n & m & 0 \\ -mn & mn & 0 & 0 & 0 & (m^2 - n^2) \end{bmatrix}$$

and

$$\begin{bmatrix} \sigma_x \\ \sigma_y \\ \sigma_z \\ \sigma_{yz} \\ \sigma_{zx} \\ \sigma_{xy} \end{bmatrix} = [T]^{-1} \begin{bmatrix} \sigma_1 \\ \sigma_2 \\ \sigma_3 \\ \sigma_4 \\ \sigma_5 \\ \sigma_6 \end{bmatrix}, \quad \begin{bmatrix} \varepsilon_x \\ \varepsilon_y \\ \varepsilon_z \\ \varepsilon_{yz} \\ \varepsilon_{zx} \\ \varepsilon_{xy} \end{bmatrix} = [T]^{-1} \begin{bmatrix} \varepsilon_1 \\ \varepsilon_2 \\ \varepsilon_3 \\ \varepsilon_{23} \\ \varepsilon_{31} \\ \varepsilon_{12} \end{bmatrix} \qquad (5.28)$$

in which

$$[T]^{-1} = \begin{bmatrix} m^2 & n^2 & 0 & 0 & 0 & -2mn \\ n^2 & m^2 & 0 & 0 & 0 & +2mn \\ 0 & 0 & 1 & 0 & 0 & 0 \\ 0 & 0 & 0 & m & +n & 0 \\ 0 & 0 & 0 & -n & m & 0 \\ mn & -mn & 0 & 0 & 0 & (m^2 - n^2) \end{bmatrix}$$

The physical meaning may be clearer, for some engineers, when the **engineering shear strain** is used. Then,

$$\begin{bmatrix} \sigma_1 \\ \sigma_2 \\ \sigma_3 \\ \sigma_4 \\ \sigma_5 \\ \sigma_6 \end{bmatrix} = [T] \begin{bmatrix} \sigma_x \\ \sigma_y \\ \sigma_z \\ \sigma_{yz} \\ \sigma_{zz} \\ \sigma_{xy} \end{bmatrix}, \quad \begin{bmatrix} \sigma_x \\ \sigma_y \\ \sigma_y \\ \sigma_{yz} \\ \sigma_{zx} \\ \sigma_{xy} \end{bmatrix} = [T]^{-1} \begin{bmatrix} \sigma_1 \\ \sigma_2 \\ \sigma_3 \\ \sigma_4 \\ \sigma_5 \\ \sigma_6 \end{bmatrix}$$

$$\qquad (5.29)$$

$$\begin{bmatrix} \varepsilon_1 \\ \varepsilon_2 \\ \varepsilon_3 \\ \varepsilon_4 \\ \varepsilon_5 \\ \varepsilon_6 \end{bmatrix} = [T^T]^{-1} \begin{bmatrix} \varepsilon_x \\ \varepsilon_y \\ \varepsilon_z \\ \gamma_{yz} \\ \gamma_{zx} \\ \gamma_{xy} \end{bmatrix}, \quad \begin{bmatrix} \varepsilon_x \\ \varepsilon_y \\ \varepsilon_y \\ \gamma_{yz} \\ \gamma_{zx} \\ \gamma_{xy} \end{bmatrix} = [T^T] \begin{bmatrix} \varepsilon_1 \\ \varepsilon_2 \\ \varepsilon_3 \\ \varepsilon_4 \\ \varepsilon_5 \\ \varepsilon_6 \end{bmatrix}$$

in which

$$[T^T] = \begin{bmatrix} m^2 & n^2 & 0 & 0 & 0 & -mn \\ n^2 & m^2 & 0 & 0 & 0 & mn \\ 0 & 0 & 1 & 0 & 0 & 0 \\ 0 & 0 & 0 & m & n & 0 \\ 0 & 0 & 0 & -n & m & 0 \\ 2mn & -2mn & 0 & 0 & 0 & (m^2 - n^2) \end{bmatrix}$$

$$[T^T]^{-1} = \begin{bmatrix} m^2 & n^2 & 0 & 0 & 0 & mn \\ n^2 & m^2 & 0 & 0 & 0 & -mn \\ 0 & 0 & 1 & 0 & 0 & 0 \\ 0 & 0 & 0 & m & -n & 0 \\ 0 & 0 & 0 & n & m & 0 \\ -2mn & 2mn & 0 & 0 & 0 & (m^2 - n^2) \end{bmatrix}$$

For the plane stress problem, the stress–strain relationship in the principal material axes directions is

$$[\sigma]_i = [Q][\varepsilon]_i$$

and

$$\begin{aligned}
[\varepsilon]_i &= [T^T]^{-1}[\varepsilon]_{xy} \\
&= \begin{bmatrix} m^2 & n^2 & mn \\ n^2 & m^2 & -mn \\ -2mn & 2mn & (m^2 - n^2) \end{bmatrix} \begin{bmatrix} \varepsilon_x \\ \varepsilon_y \\ \gamma_{xy} \end{bmatrix}
\end{aligned} \tag{5.30}$$

Therefore,

$$\begin{aligned}
[\sigma]_{xy} &= [T]^{-1}[\sigma]_i = [T]^{-1}[Q][\varepsilon]_i \\
&= [T]^{-1}[Q][T^T]^{-1}[\varepsilon]_{xy}
\end{aligned}$$

Letting $[\bar{Q}] = [T]^{-1}[Q][T^T]^{-1}$,

$$[\sigma]_{xy} = [\bar{Q}][\varepsilon]_{xy} \tag{5.31}$$

This $\bar{Q}_{ij}$ matrix is called the **transformed reduced stiffness matrix**. This may be written in more detail,

$$\begin{pmatrix} \sigma_x \\ \sigma_y \\ \sigma_z \\ \sigma_{yz} \\ \sigma_{zx} \\ \sigma_{xy} \end{pmatrix} = \begin{bmatrix} \bar{Q}_{11} & \bar{Q}_{12} & \bar{Q}_{13} & 0 & 0 & 2\bar{Q}_{16} \\ \bar{Q}_{12} & \bar{Q}_{22} & \bar{Q}_{23} & 0 & 0 & 2\bar{Q}_{26} \\ \bar{Q}_{13} & \bar{Q}_{23} & \bar{Q}_{33} & 0 & 0 & 2\bar{Q}_{36} \\ 0 & 0 & 0 & 2\bar{Q}_{44} & 2\bar{Q}_{45} & 0 \\ 0 & 0 & 0 & 2\bar{Q}_{45} & 2\bar{Q}_{55} & 0 \\ \bar{Q}_{16} & \bar{Q}_{26} & \bar{Q}_{36} & 0 & 0 & 2\bar{Q}_{66} \end{bmatrix} \begin{pmatrix} \varepsilon_x \\ \varepsilon_y \\ \varepsilon_z \\ \varepsilon_{yz} \\ \varepsilon_{zx} \\ \varepsilon_{xy} \end{pmatrix} \tag{5.32}$$

in which

$$\bar{Q}_{11} = Q_{11}m^4 + 2(Q_{12} + 2Q_{66})m^2n^2 + Q_{22}n^4$$

$$\bar{Q}_{12} = (Q_{11} + Q_{22} - 4Q_{66})m^2n^2 + Q_{12}(m^4 + n^4)$$

$$\bar{Q}_{13} = Q_{13}m^2 + Q_{23}n^2$$

$$\bar{Q}_{16} = -Q_{22}mn^3 + Q_{11}m^3n - (Q_{12} + 2Q_{66})mn(m^2 - n^2)$$

$$\bar{Q}_{22} = Q_{11}n^4 + 2(Q_{12} + 2Q_{66})m^2n^2 + Q_{22}m^4$$

$$\bar{Q}_{23} = (Q_{13}n^2 + Q_{23}m^2$$

$$\bar{Q}_{26} = -Q_{22}m^3n + Q_{11}mn^3 + (Q_{12} + 2Q_{66})mn(m^2 - n^2)$$

$$\bar{Q}_{33} = Q_{33}$$

$$\bar{Q}_{36} = (Q_{13} - Q_{23})mn$$

$$\bar{Q}_{44} = Q_{44}m^2 + Q_{55}n^2$$

$$\bar{Q}_{45} = (Q_{55} - Q_{44})mn$$

$$\bar{Q}_{55} = Q_{55}m^2 + Q_{44}n^2$$

$$\bar{Q}_{66} = (Q_{11} + Q_{22} - 2Q_{12})m^2n^2 + Q_{66}(m^2 - n^2)^2$$

Note that Tsai uses Q_{xy} etc. instead of Q_{12} etc. for the on-axis stiffness, and Q_{11} etc. instead of $\bar{Q}_{11}$ etc., for off-axis quantities. Recall that on-axis means the material principal direction, and the laminate direction (off-axis) is $\theta°$ clockwise from this direction. If the transverse shear effect is neglected,

$$\begin{bmatrix} \sigma_x \\ \sigma_y \\ \sigma_{xy} \end{bmatrix} = \begin{bmatrix} \bar{Q}_{11} & \bar{Q}_{12} & 2\bar{Q}_{16} \\ \bar{Q}_{12} & \bar{Q}_{22} & 2\bar{Q}_{26} \\ \bar{Q}_{16} & \bar{Q}_{26} & 2\bar{Q}_{66} \end{bmatrix} \begin{bmatrix} \varepsilon_x \\ \varepsilon_y \\ \varepsilon_{xy} \end{bmatrix} \tag{5.33}$$

in which $\bar{Q}_{ij}$ are as given above but with Q_{ij} given in equation 5.21.

Using the engineering shear strain, equation 5.7 can be written as

$$\begin{bmatrix} \varepsilon_1 \\ \varepsilon_2 \\ \varepsilon_6 \end{bmatrix} = \begin{bmatrix} S_{11} & S_{12} & 0 \\ S_{12} & S_{22} & 0 \\ 0 & 0 & S_{66} \end{bmatrix} \begin{bmatrix} \sigma_1 \\ \sigma_2 \\ \sigma_6 \end{bmatrix}$$

or

$$[\varepsilon]_i = [S][\sigma]_i \tag{5.34}$$

Then, since $[\sigma]_i = [T][\sigma]_{xy}$,

$$[\varepsilon]_{xy} = [T^T][\varepsilon]_i = [T^T][S][\sigma]_i$$
$$= [T^T][S][T][\sigma]_{xy}$$
$$= [\bar{S}][\sigma]_{xy}$$

or

$$\begin{bmatrix} \varepsilon_x \\ \varepsilon_y \\ 2\varepsilon_{xy} \end{bmatrix} = [\bar{S}] \begin{bmatrix} \sigma_x \\ \sigma_y \\ \sigma_{xy} \end{bmatrix} \tag{5.35}$$

where $[\bar{S}] = [T]^T[S][T]$, in which

$$\bar{S}_{11} = S_{11}m^4 + (2S_{12} + S_{66})m^2n^2 + S_{22}n^4$$

$$\bar{S}_{12} = (S_{11} + S_{22} - S_{66})m^2n^2 + S_{12}(m^4 + n^4)$$

$$\bar{S}_{22} = S_{11}n^4 + (2S_{12} + S_{66})m^2n^2 + S_{22}m^4$$

$$\bar{S}_{16} = (2S_{11} - 2S_{12} - S_{66})m^3n - (2S_{22} - 2S_{12} - S_{66})mn^3$$

$$\bar{S}_{26} = (2S_{11} - 2S_{12} - S_{66})mn^3 - (2S_{22} - 2S_{12} - S_{66})m^3n$$

$$\bar{S}_{66} = 2(2S_{11} + 2S_{22} - 4S_{12} - S_{66})m^2n^2 + S_{66}(m^4 + n^4)$$

and

$$S_{11} = 1/E_1, \quad S_{12} = -v_{12}/E_1 = -v_{21}/E_2$$

$$S_{22} = 1/E_2, \quad S_{66} = 1/G_{12}, \quad S_{16} = \frac{\eta_{12,1}}{E_1} = \frac{\eta_{1,12}}{G_{12}}, \quad S_{26} = \frac{\eta_{12,2}}{E_2} = \frac{\eta_{2,12}}{G_{12}}$$

where $\eta_{1.12} = \varepsilon_1/\gamma_{12}$ for all $\tau_{ij} = 0$ except $ij = 12$, and $\eta_{12.1} = \gamma_{12}/\varepsilon_1$ for all $\tau_{ij} = 0$ except $i = j = 1$. Recall that $[\bar{S}]$ can be obtained by inverting $[\bar{Q}]$ without 2, i.e.

$$[\bar{Q}] = \begin{bmatrix} \bar{Q}_{11} & \bar{Q}_{12} & \bar{Q}_{16} \\ \bar{Q}_{12} & \bar{Q}_{22} & \bar{Q}_{26} \\ \bar{Q}_{16} & \bar{Q}_{26} & \bar{Q}_{66} \end{bmatrix}$$

Note that $\bar{S}$ can be expressed as [4]

$$\frac{1}{E_x} = \bar{S}_{11} = \frac{1}{E_1}\cos^4\theta + \left(\frac{1}{G_{12}} - \frac{2v_{12}}{E_1}\right)\sin^2\theta\cos^2\theta + \frac{1}{E_2}\sin^4\theta$$

$$v_{xy} = E_x\left[\frac{v_{12}}{E_1}(m^4 + n^4) - \left(\frac{1}{E_1} + \frac{1}{E_2} - \frac{1}{G_{12}}\right)m^2n^2\right]$$

$$\frac{1}{E_y} = \bar{S}_{22} = \frac{1}{E_1}n^4 + \left(\frac{1}{G_{12}} - \frac{2v_{12}}{E_1}\right)m^2n^2 + \frac{1}{E_2}m^4$$

$$\frac{1}{G_{xy}} = \bar{S}_{66} = 2\left(\frac{2}{E_1} + \frac{2}{E_2} + \frac{4v_{12}}{E_1} - \frac{1}{G_{12}}\right)m^2n^2 + \frac{1}{G_{12}}(m^4 + n^4)$$

$$\eta_{xy,x} = \bar{S}_{16}E_x = E_x\left[\left(\frac{2}{E_1} + \frac{2v_{21}}{E_2} - \frac{1}{G_{12}}\right)m^3n - \left(\frac{2}{E_2} + \frac{2v_{12}}{E_1} - \frac{1}{G_{12}}\right)mn^3\right]$$

$$\eta_{xy,y} = \bar{S}_{26}E_y = E_y\left[\left(\frac{2}{E_1} + \frac{2v_{21}}{E_2} - \frac{1}{G_{12}}\right)mn^3 - \left(\frac{2}{E_2} + \frac{2v_{12}}{E_1} - \frac{1}{G_{12}}\right)m^3n\right]$$

(5.36)

These are called the **apparent engineering constants** for an orthotropic lamina which is stressed in nonprincipal x–y coordinates.

For some engineers, use of straightforward **tensor notation** may look simpler, mathematically. In this case, we start from equations 5.19 and 5.20 for $[Q]$ and

$[S]$ in $[\sigma]_i = [Q][\varepsilon]_i$ and $[\varepsilon]_i = [S][\sigma]_i$. We have $[\varepsilon]_i = [T][\varepsilon]_{xyz}$, and $[\sigma]_i = [T][\sigma]_{xyz}$, equations 5.24. Therefore,

$$[\sigma]_{xyz} = [T]^{-1}[\sigma]_i = [T]^{-1}[Q][\varepsilon]_i$$
$$= [T]^{-1}[Q][T][\varepsilon]_{xyz}$$
$$= [\bar{Q}]'[\varepsilon]_{xyz} \tag{5.37}$$

$$[\varepsilon]_{xyz} = [T]^{-1}[\varepsilon]_i = [T]^{-1}[S][\sigma]_i$$
$$= [T]^{-1}[S][T][\sigma]_{xyz}$$
$$= [\bar{S}]'[\sigma]_{xyz} \tag{5.38}$$

We note that both $[\bar{Q}]'$ and $[\bar{S}]'$ have similar matrix expressions:

$$[\bar{Q}]' = [T]^{-1}[Q][T]$$
$$[\bar{S}]' = [T]^{-1}[S][T]$$

in which $\bar{Q}'_{ij}$ and $\bar{S}'_{ij}$ are as follows:

$$\bar{Q}'_{11} = Q_{11}m^4 + 2(Q_{12} + 2Q_{66})m^2n^2 + Q_{22}n^4 = \bar{Q}_{11}$$

$$\bar{Q}'_{12} = (Q_{11} + Q_{22} - 4Q_{66})m^2n^2 + Q_{12}(m^4 + n^4) = \bar{Q}_{12}$$

$$\bar{Q}'_{13} = Q_{13}m^2 + Q_{23}n^2 = \bar{Q}_{13}$$

$$\bar{Q}'_{16} = 2[-Q_{22}mn^3 + Q_{11}m^3n - (Q_{12} + 2Q_{66})mn(m^2 - n^2)] = 2\bar{Q}_{16}$$

$$\bar{Q}'_{61} = -Q_{22}mn^3 + Q_{11}m^3n - (Q_{12} + 2Q_{66})mn(m^2 - n^2) = \bar{Q}_{61}$$

$$\bar{Q}'_{22} = Q_{11}n^4 + 2(Q_{12} + 2Q_{66})m^2n^2 + Q_{22}m^4 = \bar{Q}_{22}$$

$$\bar{Q}'_{23} = Q_{13}n^2 + Q_{23}m^2 = \bar{Q}_{23}$$

$$\bar{Q}'_{26} = 2[-Q_{22}m^3n + Q_{11}mn^3 + (Q_{12} + 2Q_{66})mn(m^2 - n^2)] = 2\bar{Q}_{26}$$

$$\bar{Q}'_{62} = -Q_{22}m^3n + Q_{11}mn^3 + (Q_{12} + 2Q_{66})mn(m^2 - n^2) = \bar{Q}_{62}$$

$$\bar{Q}'_{33} = Q_{33} = \bar{Q}_{33}$$

$$\bar{Q}'_{36} = 2(Q_{13} - Q_{23})mn = 2\bar{Q}_{36}$$

$$\bar{Q}'_{63} = (Q_{13} - Q_{23})mn = \bar{Q}_{63}$$

$$\bar{Q}'_{44} = 2(Q_{44}m^2 + Q_{55}n^2) = 2\bar{Q}_{44}$$

$$\bar{Q}'_{45} = 2(Q_{55} - Q_{44})mn = 2\bar{Q}_{45}$$

$$\bar{Q}'_{54} = 2(Q_{55} - Q_{44})mn = 2\bar{Q}_{54}$$

$$\bar{Q}'_{55} = 2(Q_{55}m^2 + Q_{44}n^2) = 2\bar{Q}_{55}$$

$$\bar{Q}'_{66} = 2[(Q_{11} + Q_{22} - 2Q_{12})m^2n^2 + Q_{66}(m^2 - n^2)^2] = 2\bar{Q}_{66}$$

$$\bar{S}'_{11} = S_{11}m^4 + (2S_{12} + S_{66})m^2n^2 + S_{22}n^4 = \bar{S}_{11}$$

$$\bar{S}'_{12} = (S_{11} + S_{22} - S_{66})m^2n^2 + S_{12}(m^4 + n^4) = \bar{S}_{12}$$

$$\bar{S}'_{22} = S_{11}n^4 + (2S_{12} + S_{66})m^2n^2 + S_{22}m^4 = \bar{S}_{22}$$

$$\bar{S}'_{16} = (2S_{11} - 2S_{12} - S_{66})m^3n - (2S_{22} - 2S_{12} - S_{66})mn^3 = \bar{S}_{16}$$

$$\bar{S}'_{61} = (S_{11} - S_{12} - \tfrac{1}{2}S_{66})m^3n - (S_{22} - S_{12} - \tfrac{1}{2}S_{66})mn^3 = \tfrac{1}{2}\bar{S}_{61}$$

$$\bar{S}'_{26} = (2S_{11} - 2S_{12} - S_{66})mn^3 - (2S_{22} - 2S_{12} - S_{66})m^3n = \bar{S}_{26}$$

$$\bar{S}'_{62} = (S_{11} - S_{12} - \tfrac{1}{2}S_{66})mn^3 - (S_{22} - S_{12} - \tfrac{1}{2}S_{66})m^3n = \tfrac{1}{2}\bar{S}_{62}$$

$$\bar{S}'_{66} = (2S_{11} + 2S_{22} - 4S_{12} - S_{66})m^2n^2 + \tfrac{1}{2}S_{66}(m^4 + n^4) = \tfrac{1}{2}\bar{S}_{66}$$

Using $[\bar{Q}]'$, we have

$$\begin{bmatrix} \sigma_x \\ \sigma_y \\ \sigma_z \\ \sigma_{yz} \\ \sigma_{zx} \\ \sigma_{xy} \end{bmatrix} = \begin{bmatrix} \bar{Q}_{11} & \bar{Q}_{12} & \bar{Q}_{13} & 0 & 0 & 2\bar{Q}_{16} \\ \bar{Q}_{12} & \bar{Q}_{22} & \bar{Q}_{23} & 0 & 0 & 2\bar{Q}_{26} \\ \bar{Q}_{13} & \bar{Q}_{23} & \bar{Q}_{33} & 0 & 0 & 2\bar{Q}_{36} \\ 0 & 0 & 0 & 2\bar{Q}_{44} & 2\bar{Q}_{45} & 0 \\ 0 & 0 & 0 & 2\bar{Q}_{45} & 2\bar{Q}_{55} & 0 \\ \bar{Q}_{61} & \bar{Q}_{62} & \bar{Q}_{63} & 0 & 0 & 2\bar{Q}_{66} \end{bmatrix} \begin{bmatrix} \varepsilon_x \\ \varepsilon_y \\ \varepsilon_z \\ \varepsilon_{yz} \\ \varepsilon_{zx} \\ \varepsilon_{xy} \end{bmatrix}$$

which is exactly the same as equation 5.32. For the plane stress problem, we have

$$\begin{bmatrix} \sigma_x \\ \sigma_y \\ \sigma_{xy} \end{bmatrix} = \begin{bmatrix} \bar{Q}_{11} & \bar{Q}_{12} & 2\bar{Q}_{16} \\ \bar{Q}_{12} & \bar{Q}_{22} & 2\bar{Q}_{26} \\ \bar{Q}_{16} & \bar{Q}_{26} & 2\bar{Q}_{66} \end{bmatrix} \begin{bmatrix} \varepsilon_x \\ \varepsilon_y \\ \varepsilon_{xy} \end{bmatrix}$$

which is the same as equation 5.33, and

$$\begin{bmatrix} \varepsilon_x \\ \varepsilon_y \\ \varepsilon_{xy} \end{bmatrix} = \begin{bmatrix} \bar{S}_{11} & \bar{S}_{12} & \bar{S}_{16} \\ \bar{S}_{12} & \bar{S}_{22} & \bar{S}_{26} \\ \tfrac{1}{2}\bar{S}_{16} & \tfrac{1}{2}\bar{S}_{26} & \tfrac{1}{2}\bar{S}_{66} \end{bmatrix} \begin{bmatrix} \sigma_x \\ \sigma_y \\ \sigma_{xy} \end{bmatrix} \tag{5.39}$$

$[Q]'$ and $[S]'$ are asymmetric matrices while $[\bar{Q}]$ and $[\bar{S}]$ are symmetric. However, the off-axis stress–strain relationship can be obtained, straightforwardly, by the same matrix operation for both stiffness and compliance matrices.

Note that $[\bar{Q}]$ and $[\bar{S}]$ are for the case where engineering shear strain is used, and $[\bar{Q}]'$ and $[\bar{S}]'$ are for the case where tensor notation is used for the shear strain.

In equation 3.5, the effects of both temperature and hygrothermal strains were noted. The hygrothermal effect is change in properties due to moisture absorption and temperature change. When temperature changes, a material may expand or contract, and the stiffness and the strength change. Therefore, in addition to considering $\alpha\Delta T$ in equation 3.5, one must use the strengths and the moduli of elasticity of the materials of the operational temperature. When a polymer matrix is used, the combination of high temperature and high humidity causes very serious effects on the structural performance of such composite. This combination of high temperature and moisture results in increased weight and swelling of the matrix. In fact, the hygrothermal effect is similar to accelerated temperature effect.

For an orthotropic material, equation 3.5 may be rewritten as

$$\varepsilon_i = S_{ij}\sigma_i + \alpha_i\Delta T + \beta_i\Delta m \qquad (i = 1, 2, 3)$$

$$\varepsilon_i = S_{ij}\sigma_i \qquad (i = 4, 5, 6) \tag{5.40}$$

Equation 5.6 may be rewritten as

$$
\begin{bmatrix} \sigma_1 \\ \sigma_2 \\ \sigma_3 \\ \sigma_4 \\ \sigma_5 \\ \sigma_6 \end{bmatrix} = \begin{bmatrix} Q_{11} & Q_{12} & Q_{13} & 0 & 0 & 0 \\ Q_{12} & Q_{22} & Q_{23} & 0 & 0 & 0 \\ Q_{13} & Q_{23} & Q_{33} & 0 & 0 & 0 \\ 0 & 0 & 0 & 2Q_{44} & 0 & 0 \\ 0 & 0 & 0 & 0 & 2Q_{55} & 0 \\ 0 & 0 & 0 & 0 & 0 & 2Q_{66} \end{bmatrix} \begin{bmatrix} \varepsilon_1 - \alpha_1 \Delta T - \beta_1 \Delta m \\ \varepsilon_2 - \alpha_2 \Delta T - \beta_2 \Delta m \\ \varepsilon_3 - \alpha_3 \Delta T - \beta_3 \Delta m \\ \varepsilon_{23} \\ \varepsilon_{31} \\ \varepsilon_{12} \end{bmatrix} \quad (5.41)
$$

$[\varepsilon]_{xyz}$ in equation 5.27 may be expressed as

$$
[\varepsilon]_{xyz} = \begin{bmatrix} \varepsilon_x & -\alpha_x \Delta T & -\beta_x \Delta m \\ \varepsilon_y & -\alpha_y \Delta T & -\beta_y \Delta m \\ \varepsilon_z & -\alpha_z \Delta T & -\beta_z \Delta m \\ \varepsilon_{yz} & & \\ \varepsilon_{zx} & & \\ \varepsilon_{xy} & -\tfrac{1}{2}\alpha_{xy}\Delta T & -\tfrac{1}{2}\beta_{xy}\Delta m \end{bmatrix} \quad (5.42)
$$

in which

$$\alpha_x = \alpha_1 m^2 + \alpha_2 n^2 \qquad \beta_x = \beta_1 m^2 + \beta_2 n^2$$

$$\alpha_y = \alpha_2 m^2 + \alpha_1 n^2 \qquad \beta_y = \beta_2 m^2 + \beta_1 n^2$$

$$\alpha_z = \alpha_3 \qquad\qquad\qquad \beta_z = \beta_3$$

$$\alpha_{xy} = (\alpha_1 - \alpha_2)mn \qquad \beta_{xy} = (\beta_1 - \beta_2)mn$$

5.6 Invariants

Tsai [6] introduced an important concept of invariants. Invariants are combinations of stress or strain components that remain constant under coordinate transformation ([6], pp. 5-5, 6-3). The linear invariants are

$$P_\sigma = (\sigma_1 + \sigma_2)/2 = (\sigma_x + \sigma_y)/2$$
$$P_\varepsilon = (\varepsilon_1 + \varepsilon_2)/2 = (\varepsilon_x + \varepsilon_y)/2 \quad (3.21c)$$

The quadratic invariants are

$$R_\sigma^2 = \left(\frac{\sigma_1 - \sigma_2}{2}\right)^2 + \tau_{12}^2 = \left(\frac{\sigma_x - \sigma_y}{2}\right)^2 + \tau_{xy}^2$$

$$R_\varepsilon^2 = \left(\frac{\varepsilon_1 - \varepsilon_2}{2}\right)^2 + \left(\frac{\varepsilon_6}{2}\right)^2 = \left(\frac{\varepsilon_x - \varepsilon_y}{2}\right)^2 + \left(\frac{\gamma_{xy}}{2}\right)^2 \quad (3.20)$$

Note that $\varepsilon_6 = 2\varepsilon_{12}$.

Recall that P_σ ad P_ε are the centers and R_σ and R_ε are the radii of Mohr's circles for the stress and strain, respectively. With the following trigonometric identities,

$$m^4 = (3 + 4\cos 2\theta + \cos 4\theta)/8$$

$$n^4 = (3 - 4\cos 2\theta + \cos 4\theta)/8$$

$$m^2 n^2 = (1 - \cos 4\theta)/8$$

$$m^3 n = (2\sin 2\theta + \sin 4\theta)/8$$

$$mn^3 = (2\sin 2\theta - \sin 4\theta)/8$$

and defining [8, 14]

$$
\begin{aligned}
U_1 &= U_4 + 2U_5 = (3Q_{11} + 3Q_{22} + 2Q_{12} + 4Q_{66})/8 \\
U_2 &= (Q_{11} - Q_{22})/2 \\
U_3 &= (Q_{11} + Q_{22} - 2Q_{12} - 4Q_{66})/8 \\
U_4 &= U_1 - 2U_5 = (Q_{11} + Q_{22} + 6Q_{12} - 4Q_{66})/8 \\
U_5 &= (U_1 - U_4)/2 = (Q_{11} + Q_{22} - 2Q_{12} + 4Q_{66})/8
\end{aligned}
$$

(5.43)

the following result is obtained for $\bar{Q}_{ij}$ in equation 5.33. Recall the Q_{ij} are those in equation 5.21.

$$
\begin{aligned}
\bar{Q}_{11} &= U_1 + U_2 \cos 2\theta + U_3 \cos 4\theta \\
\bar{Q}_{22} &= U_1 - U_2 \cos 2\theta + U_3 \cos 4\theta \\
\bar{Q}_{12} &= U_4 - U_3 \cos 4\theta = \bar{Q}_{21} \\
\bar{Q}_{66} &= U_5 - U_3 \cos 4\theta \\
\bar{Q}_{16} &= \tfrac{1}{2}U_2 \sin 2\theta + U_3 \sin 4\theta = \bar{Q}_{61} \\
\bar{Q}_{26} &= \tfrac{1}{2}U_2 \sin 2\theta - U_3 \sin 4\theta = \bar{Q}_{62}
\end{aligned}
$$

(5.44)

Tsai presents the following stress and strain invariants:

$$|\sigma^{\text{mises}}|^2 = |\sigma|^2 = \sigma_1^2 - \sigma_1\sigma_2 + \sigma_2^2 + 3\sigma_6^2 \tag{5.45}$$

$$|\varepsilon^{\text{iso}}|^2 = [(1 + v^2)/E^2][(\sigma^{\text{mises}})^2$$
$$+ (1 - 4v + v^2)(\sigma_1\sigma_2 - \sigma_6)^2/(1 + v^2)] \tag{5.46}$$

$$|\varepsilon|^2 = \varepsilon_1^2 + \varepsilon_2^2 + \varepsilon_6^2/2 \tag{5.47}$$

$|\sigma|$ in equation 5.45 is called the **von Mises invariant** and can be used as an effective stress or strength of a material under combined stresses. $|\varepsilon|$ in equation 5.47 is an **effective strain invariant** and can be used as a measure of deformation or susceptibility to buckling of a material under combined strains. Use of the effective strain (scalar) as a basis for design is better than using one of the normal strain components (a tensor). We can arrive at an invariant design by designing with invariants. If we design without using the invariants, such as the case of using the maximum normal strain of a laminate, the design will depend on the choice of the coordinate system.

Numerical example

As a numerical example of transformation, we use the glass fiber/epoxy composite used in section 5.4. The on-axis Q_{ij} are

$$[Q] = \begin{bmatrix} 39.17 & 2.18 & 0 \\ 2.18 & 8.39 & 0 \\ 0 & 0 & 4.14 \end{bmatrix}$$

We try to obtain $\bar{Q}_{ij}$ at $45°$ off-axis for the plane stress case. $m = 1/\sqrt{2}$, $n = 1/\sqrt{2}$ for $\theta = 45°$. By equations 5.32,

$$\bar{Q}_{11} = Q_{11}m^4 + 2(Q_{12} + 2Q_{66})m^2n^2 + Q_{22}n^4$$
$$= \tfrac{1}{4}[39.17 + 2(2.18 + 2 \times 4.14) + 8.39] = 17.12 \, \text{GPa}$$
$$\bar{Q}_{26} = -Q_{22}m^3n + Q_{11}mn^3 + (Q_{12} + 2Q_{66})mn(m^2 - n^2)$$
$$= \tfrac{1}{4}(-8.39 + 39.17) = 7.695 \, \text{GPa}$$

If $\theta = -45°$, $\bar{Q}_{11}$ is the same, and $\bar{Q}_{26}$ becomes negative. The other values can be obtained by a similar method.

Table 5.2 shows transformed elastic moduli of $45°$ off-axis laminae of various composites given in Table 5.1.

We may obtain the same result by the use of the invariants (equations 5.44):

$$\bar{Q}_{11} = U_1 + U_2 \cos 2\theta + U_3 \cos 4\theta = U_1 + U_3(-1)$$
$$\bar{Q}_{26} = \tfrac{1}{2}U_2 \sin 2\theta - U_3 \sin 4\theta = \tfrac{1}{2}U_2(1)$$

We have

$$U_1 = \tfrac{1}{8}(3Q_{xx} + 3Q_{yy} + 2Q_{xy} + 4Q_{ss})$$
$$= \tfrac{1}{8}(3 \times 39.17 + 3 \times 8.39 + 2 \times 2.18 + 4 \times 4.14) = 20.45$$
$$U_2 = \tfrac{1}{2}(Q_{xx} - Q_{yy}) = \tfrac{1}{2}(39.17 - 8.39) = 15.39$$
$$U_3 = \tfrac{1}{8}(Q_{xx} + Q_{yy} - 2Q_{xy} - 4Q_{ss})$$
$$= \tfrac{1}{8}(39.17 + 8.39 - 2 \times 2.18 - 4 \times 4.14) = 3.33$$

from which we get

$$\bar{Q}_{11} = U_1 - U_3 = 20.45 - 3.33 = 17.12$$
$$\bar{Q}_{26} = \tfrac{1}{2}U_2 = 15.39/2 = 7.695$$

The remaining $45°$ off-axis quantities can be obtained by equations 5.44:

$$\bar{Q}_{12} = U_4 + U_3 \cos(4 \times 45°) = 8.84 \, \text{GPa} = \bar{Q}_{21}$$
$$\bar{Q}_{16} = \tfrac{1}{2}U_2 \sin(2 \times 45°) - U_3 \sin(4 \times 45°) = 7.69 \, \text{GPa} = \bar{Q}_{26}$$
$$\bar{Q}_{66} = U_5 - U_3 \cos(4 \times 45°) = 10.80 \, \text{GPa}$$

Table 5.2 Transformed elastic moduli of 45° ply orientation of various composite materials (courtesy of Think Composites)

Type: Fiber/cloth: Matrix:	CFRP T300 N5208	BFRP B(4) N5505	CFRP AS H3501	GFRP E-glass Epoxy	KFRP kev49 Epoxy	CFRTP AS4 PEEK	CFRP IM6 Epoxy	CFRP T300 Fbrt934	CCRP T300 Fbrt934	CCRP T300 Fbrt934
[Q] 45° (GPa)										
11 = 22	56.66	63.62	45.41	17.12	23.79	42.26	64.06	45.65	43.50	38.62
12	42.32	52.44	31.21	8.84	19.19	32.06	47.26	36.55	34.40	30.42
66	46.59	53.76	35.60	10.80	19.60	34.66	52.05	38.19	35.24	31.61
16 = 26	42.87	46.60	32.45	7.69	17.77	31.44	48.22	34.79	0.00	0.00
[S] 45° (1/TPa)										
11 = 22	59.7	58.9	63.8	93.7	155.2	77.9	52.5	81.5	61.4	68.2
12	−10.0	−30.5	−6.6	−27.0	−62.2	−20.1	−7.0	−28.4	−48.5	−53.7
66	105.7	61.2	123.2	160.3	203.9	124.0	97.4	114.4	28.4	31.6
16 = 26	−45.8	−24.6	−52.2	−47.5	−84.3	−52.4	−42.2	−48.4	0.00	0.0
Engineering constants at 45°										
E1 = E2 (GPa)	16.74	16.98	15.66	10.67	6.44	12.83	19.04	12.27	16.30	14.6
E6 (GPa)	9.46	16.34	8.12	6.24	4.90	8.06	10.27	8.74	35.24	31.6
v/21	0.17	0.52	0.10	0.29	0.40	0.26	0.13	0.35	0.79	0.79
v/61	−0.77	−0.42	−0.82	−0.51	−0.54	−0.67	−0.80	−0.59	0.00	0.00
v/62	−0.43	−0.40	−0.42	−0.30	−0.41	−0.42	−0.43	−0.42	0.00	0.00

Material types as Table 5.1.

Table 5.3 Stiffness values as a function of ply angle (SI units) (courtesy of Think Composites)

	Ply orientation θ (degrees)							
	0.00	15.00	30.00	45.00	60.00	75.00	90.00	105.00
Off-axis stiffness (GPa)								
Q_{11}	181.81	160.47	109.38	56.66	23.65	11.98	10.35	11.98
Q_{22}	10.35	11.98	23.65	56.66	109.38	160.47	181.81	160.47
$Q_{12} = Q_{21}$	2.90	12.75	32.46	42.32	32.46	12.75	2.90	12.75
Q_{66}	7.17	17.03	36.74	46.59	36.74	17.03	7.17	17.03
$Q_{16} = Q_{61}$	0.00	38.50	54.19	42.87	20.05	4.36	0.00	−4.36
$Q_{26} = Q_{62}$	0.00	4.36	20.05	42.87	54.19	38.50	0.00	−38.50
Off-axis compliance (1/TPa)								
S_{11}	5.52	13.77	34.75	59.75	80.53	93.06	97.09	93.06
S_{22}	97.09	93.06	80.53	59.75	34.75	13.77	5.52	13.77
$S_{12} = S_{21}$	−1.55	−3.66	−7.88	−9.99	−7.88	−3.66	−1.55	−3.66
S_{66}	139.47	131.03	114.15	105.71	114.15	131.03	139.47	131.03
$S_{16} = S_{61}$	0.00	−30.20	−46.96	−45.78	−32.34	−15.58	0.00	15.58
$S_{26} = S_{62}$	0.00	−15.58	−32.34	−45.78	−46.96	−30.20	0.00	30.20
Off-axis engineering constants (GPa)								
E_1	181.00	72.63	28.78	16.74	12.42	10.75	10.30	10.75
E_2	10.30	10.75	12.42	16.74	28.78	72.63	181.00	72.63
E_6	7.17	7.63	8.76	9.46	8.76	7.63	7.17	7.63
Off-axis Poisson, shear and normal couplings								
v_{21}	0.28	0.27	0.23	0.17	0.10	0.04	0.02	0.04
v_{12}	0.02	0.04	0.10	0.17	0.23	0.27	0.28	0.27
v_{61}	0.00	−2.19	−1.35	−0.77	−0.40	−0.17	0.00	0.17
v_{16}	0.00	−0.23	−0.41	−0.43	−0.28	−0.12	0.00	0.12
v_{62}	0.00	−0.17	−0.40	−0.77	−1.35	−2.19	0.00	2.19
v_{26}	0.00	−0.12	−0.28	−0.43	−0.41	−0.23	0.00	0.23
Thermal expansion coefficients (10^{-6}/°C and absolute)								
α_1	0.02	1.53	5.64	11.26	16.68	20.99	22.50	20.99
α_2	22.50	20.99	16.88	11.26	5.64	1.53	0.02	1.53
α_6	0.00	−11.24	−19.47	−22.48	−19.47	−11.24	0.00	11.24

The off-axis compliance matrix is

$$[\bar{S}] = [\bar{Q}]^{-1}$$

with

$$|\bar{Q}| = (\bar{Q}_{11}\bar{Q}_{22} - \bar{Q}_{12}^2)\bar{Q}_{66} + 2\bar{Q}_{12}\bar{Q}_{26}\bar{Q}_{16} - \bar{Q}_{11}\bar{Q}_{26}^2 - \bar{Q}_{22}\bar{Q}_{16}^2$$
$$= (17.12 \times 17.12 - 8.84^2) \times 10.80 + 2 \times 8.84 \times 7.69 \times 7.69$$
$$- 17.12 \times 7.69^2 - 17.12 \times 7.69^2$$
$$= 1342.15$$

$$\bar{S}_{11} = (\bar{Q}_{22}\bar{Q}_{66} - \bar{Q}_{26}^2)/|\bar{Q}|$$
$$= (17.12 \times 10.8 - 7.69^2)/|\bar{Q}| = 93.70\,(1/\text{TPa}) = \bar{S}_{22} = (\bar{Q}_{11}\bar{Q}_{66} - \bar{Q}_{16}^2)/|\bar{Q}|$$

$$\bar{S}_{12} = (-\bar{Q}_{12}\bar{Q}_{66} + \bar{Q}_{16}\bar{Q}_{26})/|\bar{Q}| = -27.05\,(1/\text{TPa})$$

$$\bar{S}_{66} = (\bar{Q}_{11}\bar{Q}_{22} - \bar{Q}_{12}^2)/|\bar{Q}| = 160.30$$

$$\bar{S}_{16} = (\bar{Q}_{12}\bar{Q}_{26} - \bar{Q}_{22}\bar{Q}_{16})/|\bar{Q}| = -47.51\,(1/\text{TPa}) = \bar{S}_{26}$$

For 45° off-axis, the engineering constants are, by equations 5.35,

$$E_1 = 1/\bar{S}_{11} = 1/93.73 = E_2 = 10.67\,\text{GPa}$$
$$E_6 = 1/\bar{S}_{66} = G_1 = G_{12} = 1/160.36 = 6.24\,\text{GPa}$$
$$v_{21} = -\bar{S}_{21}/\bar{S}_{11} = +27.05/93.73 = 0.29 = v_{12}$$
$$v_{61} = \bar{S}_{61}/\bar{S}_{11} = -47.51/93.73 = -0.51 = v_{62}$$
$$v_{16} = \bar{S}_{61}/\bar{S}_{66} = -47.51/160.33 = -0.296 = v_{26}$$

Table 5.3 shows the off-axis properties corresponding to 15°, 30°, 45°, 60°, 75°, 90° and 105°, for the carbon/epoxy composite T300/N5208 given in Table 5.1.

5.7 Laminates

A laminate is made of several plies or laminae bonded together. We consider a laminate composed of n laminae. For the kth lamina of the laminate, equation 5.32 may be written as

$$\begin{pmatrix} \sigma_x \\ \sigma_y \\ \sigma_z \\ \sigma_{yz} \\ \sigma_{zx} \\ \sigma_{xy} \end{pmatrix}_k = [\bar{Q}]_k \begin{pmatrix} \varepsilon_x \\ \varepsilon_y \\ \varepsilon_z \\ \varepsilon_{yz} \\ \varepsilon_{zx} \\ \varepsilon_{xy} \end{pmatrix}_k \qquad (5.48)$$

In this equation, the thermal and hygrothermal effects are neglected.

The strain–displacement relationship for any elastic body, in a Cartesian coordinate system is given by equation 3.2 as

$$\varepsilon_x = \frac{\partial u}{\partial x}, \quad \varepsilon_y = \frac{\partial v}{\partial y}, \quad \varepsilon_z = \frac{\partial w}{\partial z}$$

$$\varepsilon_{yz} = \frac{1}{2}\gamma_{yz} = \frac{1}{2}\left(\frac{\partial v}{\partial z} + \frac{\partial w}{\partial y}\right)$$

$$\varepsilon_{zx} = \frac{1}{2}\gamma_{zx} = \frac{1}{2}\left(\frac{\partial u}{\partial z} + \frac{\partial w}{\partial x}\right) \tag{3.2}$$

$$\varepsilon_{xy} = \frac{1}{2}\gamma_{xy} = \frac{1}{2}\left(\frac{\partial u}{\partial y} + \frac{\partial v}{\partial x}\right)$$

In linear elastic plate theory [7, 8], it is assumed that a linear element through the thickness of a thin plate, perpendicular to the mid-surface before loading, undergoes, upon loading, at most a translation and a rotation with respect to the original coordinate system. Based on such an assumption, the components of the displacement of the plate may be expressed as

$$u(x, y, z) = u_0(x, y) + z\bar{\alpha}(x, y)$$

$$v(x, y, z) = v_0(x, y) + z\bar{\beta}(x, y) \tag{5.49}$$

$$w(x, y, z) = w(x, y)$$

where u_0, v_0 and w are the mid-surface displacements which are the translations of the linear element, and the second terms of u and v are related to the rotations. Substituting equations 5.49 into equation 3.2,

$$\varepsilon_x = \frac{\partial u_0}{\partial x} + z\frac{\partial \bar{\alpha}}{\partial x}, \quad \varepsilon_y = \frac{\partial v_0}{\partial y} + z\frac{\partial \bar{\beta}}{\partial y}, \quad \varepsilon_z = 0$$

$$\varepsilon_{yz} = \frac{1}{2}\left(\bar{\beta} + \frac{\partial w}{\partial y}\right)$$

$$\varepsilon_{zx} = \frac{1}{2}\left(\bar{\alpha} + \frac{\partial w}{\partial x}\right) \tag{5.50}$$

$$\varepsilon_{xy} = \frac{1}{2}\left(\frac{\partial u_0}{\partial y} + \frac{\partial v_0}{\partial x}\right) + \frac{z}{2}\left(\frac{\partial \bar{\alpha}}{\partial y} + \frac{\partial \bar{\beta}}{\partial x}\right)$$

In other words, the laminate strains, ε_{ij}, at any distance z from the mid-surface are given by

$$[\varepsilon] = [\varepsilon_0] + z[\kappa] \tag{5.51}$$

From these equations, we obtain the mid-surface strains and the curvatures as

$$\varepsilon_{x0} = \frac{\partial u_0}{\partial x}, \quad \varepsilon_{y0} = \frac{\partial v_0}{\partial y}$$

$$\varepsilon_{xy0} = \frac{1}{2}\left(\frac{\partial u_0}{\partial y} + \frac{\partial v_0}{\partial x}\right)$$

$$\kappa_x = \frac{\partial \bar{\alpha}}{\partial x}, \quad \kappa_y = \frac{\partial \bar{\beta}}{\partial y} \tag{5.52}$$

$$\kappa_{xy} = \frac{1}{2}\left(\frac{\partial \bar{\alpha}}{\partial y} + \frac{\partial \bar{\beta}}{\partial x}\right)$$

Since all laminae of a composite laminated plate are bonded together, continuity of strains and displacements through the thickness of the laminate exists, regardless of the orientation of individual lamina. However, regardless of the displacement continuity through the thickness of the laminate, the stresses are discontinuous because each lamina may have different orientation resulting in a different stiffness from the others. Equation 5.48 may be written as

$$\begin{pmatrix} \sigma_x \\ \sigma_y \\ \sigma_{yz} \\ \sigma_{zx} \\ \sigma_{xy} \end{pmatrix}_k = [\bar{Q}]_k \begin{pmatrix} \varepsilon_{x0} + z\kappa_x \\ \varepsilon_{y0} + z\kappa_y \\ \varepsilon_{yz} \\ \varepsilon_{zx} \\ \varepsilon_{xy0} + z\kappa_{xy} \end{pmatrix}_k \tag{5.53}$$

Note that an assumption is made such that $\varepsilon_z = 0$, since plate thickening is neglected, and σ_z is negligible for a 'thin' composite plate.

As for the laminate code, there is no standard one. We use the code which goes from the top surface to the bottom surface, used in reference [6]. T and S stand for total and symmetric, respectively. For example,

$$[0_2/90_2/45_2/-45_2/-45_2/45_2/90_2/0_2]_T$$

may be expressed as

$$[0_2/90_2/45_2/-45_4/45_2/90_2/0_2]_T$$

or

$$[0_2/90_2/45_2/-45_2]_S$$

If a laminate is made of repeating sublaminates, for example, if we have four of $[0_2/90_2/45_2/-45_2]$, we use the repeating index of 2, because of symmetry, as follows:

$$[0_2/90_2/45_2/-45_2]_{2S}$$

Figure 5.2 Notations and positive directions of stress resultants and couples.

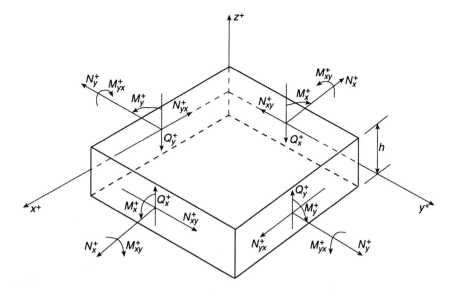

or

$$\{[0_2/90_2/45_2/-45_2]_2\}_s$$

or

$$[0_2/90_2/45_2/-45_2/0_2/90_2/45_2/-45_2/-45_2/45_2/90_2/0_2/-45_2/45_2/90_2/0_2]_T$$

We consider a laminated plate with thickness h made of n laminae. The same notations used in the classical theory of plates for stress resultant N, stress couple M and shear resultant Q, are used here with the sign convention as shown in Figure 5.2. Note that the positive direction of the z-axis is different from that of conventional theory [9]. The stacking sequence is as shown in Figure 5.3. In this figure, h_k is the vectorial distance from the plate mid-plane, $z = 0$, to the upper surface of the kth lamina. Figure 5.3 shows an example.

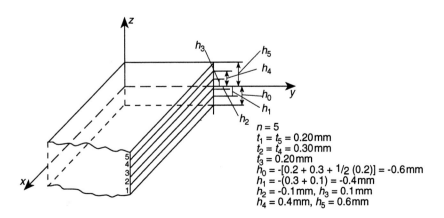

$n = 5$
$t_1 = t_5 = 0.20\text{mm}$
$t_2 = t_4 = 0.30\text{mm}$
$t_3 = 0.20\text{mm}$
$h_0 = -[0.2 + 0.3 + 1/2\,(0.2)] = -0.6\text{mm}$
$h_1 = -(0.3 + 0.1) = -0.4\text{mm}$
$h_2 = -0.1\text{mm}, h_3 = 0.1\text{mm}$
$h_4 = 0.4\text{mm}, h_5 = 0.6\text{mm}$

Figure 5.3 Stacking sequence.

For the whole plate, regardless of the number and the orientation of the laminae, we have

$$\begin{bmatrix} N_x \\ N_y \\ N_{xy} \\ Q_x \\ Q_y \end{bmatrix} = \int_{-h/2}^{h/2} \begin{bmatrix} \sigma_x \\ \sigma_y \\ \sigma_{xy} \\ \sigma_{xz} \\ \sigma_{yz} \end{bmatrix} dz \qquad (5.54)$$

$$\begin{bmatrix} M_x \\ M_y \\ M_{xy} \end{bmatrix} = \int_{-h/2}^{h/2} \begin{bmatrix} \sigma_x \\ \sigma_y \\ \sigma_{xy} \end{bmatrix} z \, dz$$

For a whole laminated plate, the stress components can be integrated across the thickness of each individual lamina, and added together. Substituting equation 5.53 into equation 5.54, we obtain

$$\begin{bmatrix} N_x \\ N_y \\ N_{xy} \end{bmatrix} = \sum_{k=1}^{n} \int_{h_{k-1}}^{h_k} \begin{bmatrix} \sigma_x \\ \sigma_y \\ \sigma_{xy} \end{bmatrix}_k dz$$

$$= \sum_{k=1}^{n} \left(\int_{h_{k-1}}^{h_k} [\bar{Q}]_k \begin{bmatrix} \varepsilon_{x0} \\ \varepsilon_{y0} \\ \varepsilon_{xy0} \end{bmatrix}_k dz + \int_{h_{k-1}}^{h_k} [\bar{Q}]_k \begin{bmatrix} \kappa_x \\ \kappa_y \\ \kappa_{xy} \end{bmatrix}_k z \, dz \right) \qquad (5.55)$$

$$\begin{pmatrix} M_x \\ M_y \\ M_{xy} \end{pmatrix} = \sum_{k=1}^{n} \int_{h_{k-1}}^{h_k} \begin{bmatrix} \sigma_x \\ \sigma_y \\ \sigma_{xy} \end{bmatrix}_k z \, dz$$

$$= \sum_{k=1}^{n} \left(\int_{h_{k-1}}^{h_k} [\bar{Q}]_k \begin{bmatrix} \varepsilon_{x0} \\ \varepsilon_{y0} \\ \varepsilon_{xy0} \end{bmatrix}_k z \, dz + \int_{h_{k-1}}^{h_k} [\bar{Q}]_k \begin{bmatrix} \kappa_x \\ \kappa_y \\ \kappa_{xy} \end{bmatrix}_k z^2 \, dz \right) \qquad (5.56)$$

Note that thermal and hygrothermal effects are neglected in these equations and, if necessary, we can add them. Note also that, naturally, only the related parts of the $[\bar{Q}]_k$ are used.

Since the derivatives of the mid-surface displacements u_0 and v_0, the rotations $\bar{\alpha}$ and $\bar{\beta}$, and the $\bar{Q}$s are independent of z, equations 5.55 and 5.56 can be written as

$$\begin{bmatrix} N_x \\ N_y \\ N_{xy} \end{bmatrix} = \sum_{k=1}^{n} \left([\bar{Q}]_k \begin{bmatrix} \varepsilon_{x0} \\ \varepsilon_{y0} \\ \varepsilon_{xy0} \end{bmatrix}_k \int_{h_{k-1}}^{h_k} dz + [\bar{Q}]_k \begin{bmatrix} \kappa_x \\ \kappa_y \\ \kappa_{xy} \end{bmatrix}_k \int_{h_{k-1}}^{h_k} z \, dz \right) \qquad (5.57)$$

$$\begin{bmatrix} M_x \\ M_y \\ M_{xy} \end{bmatrix} = \sum_{k=1}^{n} \left([\bar{Q}]_k \begin{bmatrix} \varepsilon_{x0} \\ \varepsilon_{y0} \\ \varepsilon_{xy0} \end{bmatrix}_k \int_{h_{k-1}}^{h_k} z \, dz + [\bar{Q}]_k \begin{bmatrix} \kappa_x \\ \kappa_y \\ \kappa_{xy} \end{bmatrix}_k \int_{h_{k-1}}^{h_k} z^2 \, dz \right) \qquad (5.58)$$

Equations 5.57 and 5.58 can be written as

$$[N] = [A][\varepsilon_0] + [B][\kappa] \qquad (5.59a)$$

$$[M] = [B][\varepsilon_0] + [D][\kappa] \qquad (5.60a)$$

If thermal and hygrothermal effects are considered,

$$[N] = [A][\varepsilon_0] + [B][\kappa] - [N]^{\mathrm{T}} - [N]^{\mathrm{m}} \qquad (5.59b)$$

$$[M] = [B][\varepsilon_0] + [D][\kappa] - [M]^{\mathrm{T}} - [M]^{\mathrm{m}} \qquad (5.60b)$$

where

$$A_{ij} = \sum_{k=1}^{n} (\bar{Q}_{ij})_k (h_k - h_{k-1}) \qquad (5.61)$$

$$B_{ij} = \frac{1}{2} \sum_{k=1}^{n} (\bar{Q}_{ij})_k (h_k^2 - h_{k-1}^2) \qquad (5.62)$$

$$D_{ij} = \frac{1}{3} \sum_{k=1}^{n} (\bar{Q}_{ij})_k (h_k^3 - h_{k-1}^3) \qquad (5.63)$$

For the shear resultants, Q_x and Q_y, the transverse shear stresses are assumed to be distributed parabolically across the laminate thickness [10]. Even though there are discontinuities at the interfaces between laminae, a continuous function $f(z)$ is used as a weighting function [7]:

$$f(z) = \frac{5}{4}\left(1 - 4\left(\frac{z}{h}\right)^2\right) \qquad (5.64)$$

From equation 5.37,

$$\sigma_{yz_k} = 2\bar{Q}_{44_k}\varepsilon_{yz} + 2\bar{Q}_{45_k}\varepsilon_{xz},$$

$$\sigma_{xz_k} = 2\bar{Q}_{45_k}\varepsilon_{yz} + 2\bar{Q}_{55_k}\varepsilon_{xz},$$

and

$$Q_y = \sum_{k=1}^{n} \int_{h_{k-1}}^{h_k} \sigma_{yz_k}\, dz$$

$$= (2)\left(\frac{5}{4}\right)\left\{\sum_{k=1}^{n} \bar{Q}_{44}\varepsilon_{yz_k} \int_{h_{k-1}}^{k_k} \left(1 - 4\left(\frac{z}{h}\right)^2\right) dz\right.$$

$$\left. + \sum_{k=1}^{n} \bar{Q}_{45_k}\varepsilon_{xz_k} \int_{h_{k-1}}^{h_k} \left(1 - 4\left(\frac{z}{h}\right)^2\right) dz\right\}$$

from which we get

$$Q_y = 2(A_{44}\varepsilon_{yz} + A_{45}\varepsilon_{xz})$$

In a similar way,

$$Q_x = 2(A_{45}\varepsilon_{yz} + A_{55}\varepsilon_{xz}) \qquad (5.65)$$

where

$$A_{ij} = \frac{5}{4} \sum_{k=1}^{n} (\bar{Q}_{ij})_k \left[h_k - h_{k-1} - \frac{4}{3}(h_k^3 - h_{k-1}^3)\frac{1}{h^2} \right] \qquad (5.66)$$

where h is the thickness of the laminate, and $i = 4, 5, j = 4, 5$.

From equation 5.59, we see that the in-plane stress resultants for a laminate are not only functions of the mid-plane strains as they are in a homogeneous beam, plat   lso functions of the curvatures and twists. Equation 5.60 shows that the stress couples are the functions of the curvatures and twists, as well as the mid-plane strains.

By adding equations 5.59 and 5.60, neglecting the thermal and hygrothermal effects, we obtain

$$
\begin{bmatrix} N_x \\ N_y \\ N_{xy} \\ M_x \\ M_y \\ M_{xy} \end{bmatrix} = \begin{bmatrix} A_{11} & A_{12} & 2A_{16} & B_{11} & B_{12} & 2B_{16} \\ A_{12} & A_{22} & 2A_{26} & B_{12} & B_{22} & 2B_{26} \\ A_{16} & A_{26} & 2A_{66} & B_{16} & B_{26} & 2B_{66} \\ B_{11} & B_{12} & 2B_{16} & D_{11} & D_{12} & 2D_{16} \\ B_{12} & B_{22} & 2B_{26} & D_{12} & D_{22} & 2D_{26} \\ B_{16} & B_{26} & 2B_{66} & D_{16} & D_{26} & 2D_{66} \end{bmatrix} \begin{bmatrix} \varepsilon_{x0} \\ \varepsilon_{y0} \\ \varepsilon_{xy0} \\ \kappa_x \\ \kappa_y \\ \kappa_{xy} \end{bmatrix} \tag{5.67}
$$

In this equation, the $[A]$ matrix is the extensional stiffness matrix expressing the relationship between the in-plane stress resultants N and the mid-surface strains ε_0, and the $[D]$ matrix is the flexural stiffness matrix expressing the relations between the stress couples M and the curvatures κ, while the $[B]$ matrix is the bending–stretching coupling matrix relating Ms to ε_0s and Ns to κs. It is emphasized that a laminated structure can have bending–stretching coupling even when all laminae are isotropic. This situation exists even with metals. For example, a simple thermoset may be made by a cantilever strip of two different metals bonded together. When temperature changes, the two strips with different coefficients of thermal expansion produce different extensions, resulting in bending. Thus, stretching–bending coupling occurs.

When the structure is exactly symmetric about its middle surface, all of the components of the $[B]$ matrix vanish. This symmetry implies laminae properties, orientation and location from the mid-surface. The sufficient condition for a laminate to have $[B] = 0$ is given by Verchery [1] (section 7.10). A_{16} and A_{26} terms are responsible for stretching–shearing coupling. B_{16} and B_{26} cause bending–shearing and stretching–twisting coupling. D_{16} and D_{26} cause bending–twisting coupling. The designer may try to avoid 16 and 26 terms by proper stacking sequences and orientations (section 7.10).

5.8 Micromechanics – mechanical properties of composites

5.8.1 Introduction

A structural engineer, when designing a structure, tends to rely almost entirely on mechanical test results for the properties of the materials to be used. However, since the application of composite materials for general purpose structures is still at an early stage and the range of composite formulation is so diversified, a design engineer must be able to estimate the properties of composites to be used from the properties of the constituent materials, and may also have either to specify the property requirements or direct the manufacturing process to obtain the required properties. Unlike the traditional engineers, new generation engineers may have to be able to design the material as well as the structure. In any case, it is important for a design engineer to be able to obtain expected mechanical properties of a composite chosen for a specific structural element, from given constituent materials. This will enable comparison of overall design concepts, including structural efficiency, material cost, and long term economy of the structure, and so on.

For orthotropic materials in a plane stress state, four engineering constants are necessary to obtain the composite material behavior. These are Young's moduli in the longitudinal direction, E_{11}, and in the transverse direction, E_{22}, Poisson's ratio v_{12} and the shear modulus G_{12} [11].

There are two groups of methods to obtain such properties. One is the empirical method which is often called the rule or law of mixtures, and the other is the 'exact' method. The rule of mixture, which is often sufficiently accurate for many micromechanical problems, is based on the statement that the composite property is the sum of the properties of each constituent multiplied by its volume fraction [12]. The 'exact' method involves the mechanics of materials and theory of elasticity approaches.

The assumptions for both methods are that both the matrix and the fibers are homogeneous, isotropic and linearly elastic, and the fiber is, in addition to the above, regularly spaced and aligned. An additional important assumption is that the fiber, matrix and composite have the same amount of strain in the same direction.

The nomenclature used is as follows:

- E Young's modulus
- v Poisson's ratio
- G shear modulus
- σ stress
- V volume fraction
- v volume
- W weight fraction
- w weight
- ρ density
- A cross-sectional area

For each of the above, the subscripts f, m and c refer to fiber, matrix and composite, respectively.

5.8.2 Density

The composite density can be obtained by the rule of mixture:

$$\rho_c = \rho_f V_f + \rho_m V_m \tag{5.68}$$

or

$$\frac{1}{\rho_c} = \frac{W_f}{\rho_f} + \frac{W_m}{\rho_m}$$

where

$$W_f = \frac{w_f}{w_c} = \frac{\rho_f v_f}{\rho_c v_c} = \frac{\rho_f}{\rho_c} V_f$$

$$W_m = \frac{w_m}{w_c} = \frac{\rho_m v_m}{\rho_c v_c} = \frac{\rho_m}{\rho_c} V_m$$

5.8.3 Longitudinal modulus of elasticity

From the rule of mixture,

$$E_{11} = E_m V_m + E_f V_f \tag{5.69}$$

This can be obtained by considering mechanics of materials. When a unidirectional lamina is acted upon by either a tensile or compression load parallel to the fiber, it can be assumed that the strains on the fiber, matrix and composite in the loading direction are the same. The resultant axial force, P_{11}, of the composite is shared by both fiber and matrix so that

$$P_{11} = P_m + P_f \quad \text{or} \quad \sigma_{11} A_{11} = \sigma_m A_m + \sigma_f A_f$$

which reduces to

$$\sigma_{11} = \sigma_m V_m + \sigma_f V_f \tag{5.70}$$

Since the strains of all phases are assumed to be equal, $\varepsilon_{11} = \varepsilon_m = \varepsilon_f$, when both sides of equation 5.70 are divided by the composite strain, ε_{11}, we obtain

$$\frac{\sigma_{11}}{\varepsilon_{11}} = \frac{\sigma_m}{\varepsilon_m} V_m + \frac{\sigma_f}{\varepsilon_f} V_f$$

which reduces to

$$E_{11} = E_m V_m + E_f V_f \tag{5.69}$$

or

$$E_{11} = (E_f - E_m)V_f + E_m = E_f - (E_f - E_m)V_m \tag{5.71}$$

Since the fiber stiffness is several times the matrix stiffness, the first term in equation 5.69 may be neglected:

$$E_{11} \doteqdot E_f V_f \tag{5.72}$$

5.8.4 Poisson's ratio, v_{12} and v_{21}

Poisson's ratio, v_{12}, is defined as the strain developed in the 2-direction when the laminate is stressed in the 1-direction:

$$v_{12} = -\frac{\varepsilon_{22}}{\varepsilon_{11}} \tag{5.73}$$

The ε_{22} is a function of two components, namely, the fiber component, $-v_f \varepsilon_{11}$, and the matrix component, $-v_m \varepsilon_{11}$. The lateral strain in the 2-direction, ε_{22}, is $-(v_f \varepsilon_{11} V_f + v_m \varepsilon_{11} V_m)$. Therefore

$$v_{12} = -\varepsilon_{22}/\varepsilon_{11} = v_f V_f + v_m V_m \tag{5.74}$$

The other Poisson's ratio, v_{21} can be obtained from

$$v_{21} = v_{12} \frac{E_{22}}{E_{11}} \tag{5.16}$$

5.8.5 Transverse modulus of elasticity, E_{22}

The transverse strains in both fiber and matrix are

$$\varepsilon_f = \sigma_{22}/E_f$$

$$\varepsilon_m = \sigma_{22}/E_m \tag{5.75}$$

Considering the total transverse displacement, the strain is obtained as

$$\varepsilon_{22} = \sigma_{22}\left(\frac{V_f}{E_f} + \frac{V_m}{E_m}\right) \tag{5.76}$$

Since $E_{22} = \sigma_{22}/\varepsilon_{22}$,

$$E_{22} = \frac{E_f E_m}{E_f V_m + E_m V_f} \tag{5.77}$$

5.8.6 Modulus of rigidity, G_{12}

Assuming the shear stresses on the fiber and the matrix are the same, the shear strains are

$$\gamma_m = \frac{\tau}{G_m}$$

$$\gamma_f = \frac{\tau}{G_f} \tag{5.78}$$

$$\gamma_c = \frac{\tau}{G_{12}}$$

Since

$$\gamma_c = \gamma_m V_m + \gamma_f V_f \tag{5.79}$$

$$\frac{\tau}{G_{12}} = \frac{\tau}{G_m} V_m + \frac{\tau}{G_f} V_f \tag{5.80}$$

which yields

$$G_{12} = \frac{G_m G_f}{G_f V_m + G_m V_f} \tag{5.81}$$

Note that

$$G_f = \frac{E_f}{2(1 + \nu_f)}, \quad G_m = \frac{E_m}{2(1 + \nu_m)} \tag{5.82}$$

5.8.7 Modified rule of mixture

The following two equations take into consideration the orientation and the length effects of a short fiber composite [13],

$$\sigma_{11} = \sigma_f V_f\left(1 - \frac{L}{2L^*}\right)C + \sigma_m V_m \tag{5.83}$$

$$E_{11} = E_f V_f\left(1 - \frac{L}{2L^*}\right)C + E_m V_m \tag{5.84}$$

where

- L fiber critical length
- $L*$ average fiber length
- C fiber orientation factor

These equations may be multiplied by the bonding efficiency factor, B, a measure of the lack of perfect interfacial bonding in some fiber resin systems A PPG manual [13] states that, for polypropylene and glass, B is 0.85, and for nylon and for thermoplastic polyester with glass, B is 1.0.

Tsai modified the rule of mixture to take into account imperfections in fiber alignment:

$$E_{11} = K(E_f V_f + E_m V_m) \tag{5.85}$$

The fiber misalignment factor, K, is an experimentally determined constant and is highly dependent on the manufacturing process. K varies from 0.9 to 1.

The fiber critical length is expressed as $L = \sigma_{f.max} d_f/(4\tau_{intf})$ in which d_f is the diameter of the fiber and τ_{intf} is the effective shear strength of the fiber/matrix interface. C is unity when all fibers are oriented undirectionally. Note that $\sigma_{f.max}(1 - L/2L*)$ is the average fiber axial stress. We try to make $L* \geq L$.

5.8.8 Exact approach

Four elastic constants obtained by Tsai [5, 14], using the 'exact' approach are given as follows:

$$E_{11} = K(E_f V_f + E_m V_m) \tag{5.85}$$

$$E_{22} = 2[1 - v_f + (v_f - v_m)V_m]\left[(1 - C)\frac{K_f(2K_m + G_m) - G_m(K_f - K_m)V_m}{(2K_m + G_m) + 2(K_f - K_m)V_m}\right.$$
$$\left. + C\frac{K_f(2K_m + G_f) + G_f(K_m - K_f)V_m}{(2K_m + G_f) - 2(K_m - K_f)V_m}\right] \tag{5.86}$$

$$v_{12} = (1 - C)\frac{K_f v_f(2K_m + G_m)V_f + K_m v_m(2K_f + G_m)V_m}{K_f(2K_m + G_m) - G_m(K_f - K_m)V_m}$$
$$+ C\frac{K_m v_m(2K_f + G_f)V_m + K_f v_f(2K_m + G_f)V_f}{K_f(2K_m + G_m) + G_f(K_m - K_f)V_m} \tag{5.87}$$

$$G_{12} = (1 - C)G_m\frac{2G_f - (G_f - G_m)V_m}{2G_m + (G_f - G_m)V_m}$$
$$+ CG_f\frac{(G_f + G_m) - (G_f - G_m)V_m}{(G_f + G_m) + (G_f - G_m)V_m} \tag{5.88}$$

where K_f and K_m are the bulk moduli for fiber and matrix, respectively, expressed as

$$K_f = \frac{E_f}{2(1 - v_f)}, \quad K_m = \frac{E_m}{2(1 - v_m)} \tag{5.89}$$

and C relates to fiber spacing. $C = 0$ represents the case where each fiber is isolated. $C = 1$ corresponds to the case where all fibers in the matrix are in full contact.

Table 5.4 Elastic constants

	P	P_f	P_m	η
E_{11}	E_{11}	E_{11f}	E_m	1
v_{12}	v_{12}	v_{12f}	v_m	1
G_{12}	$1/G_{12}$	$1/G_{12f}$	$1/G_m$	η_6
G_{23}	$1/G_{23}$	$1/G_{23f}$	$1/G_m$	η_4
K_T	$1/K_T$	$1/K_f$	$1/K_m$	η_k

5.8.9 Alternative formulas

Hahn [15] states that the elastic constants all have the same functional form, as

$$P = \frac{(P_f V_f + \eta P_m V_m)}{(V_f + \eta V_m)} \tag{5.90}$$

where the P_f, P_m and η, for the elastic constant P, are given in Table 5.4. K_T is the plane strain bulk modulus, and

$$\eta_4 = \frac{3 - 4v_m + (G_m/G_{23f})}{4(1 - v_m)}$$

$$\eta_6 = \frac{1 + (G_m/G_{12f})}{2}$$

$$\eta_k = \frac{1 + (G_m/K_f)}{2(1 - v_m)}$$

The transverse moduli of elasticity of the composite, $E_{22} = E_{33}$, are found from

$$E_{22} = E_{33} = \frac{4K_T G_{23}}{K_T + mG_{23}} \tag{5.91}$$

where

$$m = 1 + \frac{4K_T v_{12}^2}{E_{11}}$$

The formulas given above can be used for composites with anisotropic fibers such as carbon and aramid fiber.

If the fibers are isotropic,

$$\eta_k = \frac{1 + [(1 - 2v_f)G_m/G_f]}{2(1 - v_m)}$$

Hahn [15] indicates that $G_m/G_f < 0.05$ for most of the structural composites so that

$$\eta_6 \doteqdot 0.5, \quad \eta_4 \doteqdot \frac{3 - 4v_m}{4(1 - v_m)}, \quad \eta_k \doteqdot \frac{1}{2(1 - v_m)}$$

Table 5.5 Ply stiffness/cash register method (courtesy of Think Composites)

Type: Fiber: Matrix:	CFRP T300 N5208	BFRP B(4) N5505	CFRP AS H3501	GFRP E-glass Epoxy	KFRP kev49 Epoxy	CFRTP AS4 PEEK	CFRP IM6 Epoxy	CFRP T300 Fbrt934	CCRP T300 Fbrt934	CCRP T300 Fbrt934
h_0 (mm)	0.125	0.125	0.125	0.125	0.125	0.125	0.125	0.100	0.325	0.175
h_0 (mil)	5	5	5	5	5	5	5	4	13	7
[$A°$]/0 (unit ply stiffness)										
SI units (MN/m)										
11	22.73	25.62	17.35	4.90	9.58	16.84	25.52	14.89	24.11	11.57
22	1.29	2.32	1.13	1.05	0.69	1.12	1.41	0.97	24.11	11.57
21 = 12	0.36	0.53	0.34	0.27	0.24	0.31	0.45	0.29	1.21	0.51
66	0.90	0.70	0.89	0.52	0.29	0.64	1.05	0.46	1.48	0.72
16 = 26	0.00	0.00	0.00	0.00	0.00	0.00	0.00	0.00	0.00	0.00
Imperial (kip/in)										
11	129.77	146.31	99.08	27.96	54.70	96.14	145.71	85.01	137.67	66.08
22	7.38	13.27	6.43	5.99	3.96	6.39	8.04	5.54	137.67	66.08
21 = 12	2.07	3.05	1.93	1.56	1.35	1.79	2.57	1.66	6.88	2.91
66	5.12	3.99	5.07	2.95	1.64	3.64	6.00	2.60	8.44	4.10
16 = 26	0.00	0.00	0.00	0.00	0.00	0.00	0.00	0.00	0.00	0.00
[$A°$]/90										
SI units (MN/m)										
11	1.29	2.32	1.13	1.05	0.69	1.12	1.41	0.97	24.11	11.57
22	22.73	25.62	17.35	4.90	9.58	16.84	25.52	14.89	24.11	11.57
21 = 12	0.36	0.53	0.34	0.27	0.24	0.31	0.45	0.29	1.21	0.51
66	0.90	0.70	0.89	0.52	0.29	0.64	1.05	0.46	1.48	0.72
16 = 26	0.00	0.00	0.00	0.00	0.00	0.00	0.00	0.00	0.00	0.00

22	129.77	146.31	99.08	27.96	54.70	96.14	145.71	85.01	137.67	66.08
21 = 12	2.07	3.05	1.93	1.56	1.35	1.79	2.57	1.66	6.88	2.91
66	5.12	3.99	5.07	2.95	1.64	3.64	6.00	2.60	8.44	4.10
16 = 26	0.00	0.00	0.00	0.00	0.00	0.00	0.00	0.00	0.00	0.00
$[A°]/45$										
SI units (MN/m)										
11 = 22	7.08	7.95	5.68	2.14	2.97	5.28	8.01	4.57	14.14	6.76
21 = 12	5.29	6.56	3.90	1.11	2.40	4.01	5.91	3.66	11.18	5.32
66	5.82	6.72	4.45	1.35	2.45	4.33	6.51	3.82	11.45	5.53
16 = 26	5.36	5.82	4.06	0.96	2.22	3.93	6.03	3.84	0.00	0.00
Imperial (kip/in)										
11 = 22	40.44	45.41	32.41	12.22	16.98	30.17	45.72	26.07	80.72	38.59
21 = 12	30.20	37.43	22.27	6.31	13.70	22.89	33.73	20.87	63.83	30.40
66	33.25	38.37	25.41	7.71	13.99	24.74	37.15	21.81	65.39	31.59
16 = 26	30.60	33.26	23.16	5.49	12.69	22.44	34.42	19.87	0.00	0.00
$[A°]/-45$										
SI units (MN/m)										
11 = 22	7.08	7.95	5.68	2.14	2.97	5.28	8.01	4.57	14.14	6.76
21 = 12	5.29	6.56	3.90	1.11	2.40	4.01	5.91	3.66	11.18	5.32
66	5.82	6.72	4.45	1.35	2.45	4.33	6.51	3.82	11.45	5.53
16 = 26	−5.36	−5.82	−4.06	−0.96	−2.22	−3.93	−6.03	−3.48	0.00	0.00
Imperial (kip/in)										
11 = 22	40.44	45.41	32.41	12.22	16.98	30.17	45.72	26.07	80.72	38.59
21 = 12	30.20	37.43	22.27	6.31	13.70	22.89	33.73	20.87	63.83	30.40
66	33.25	38.37	25.41	7.71	13.99	24.74	37.15	21.81	65.39	31.59
16 = 26	−30.60	−33.26	−23.16	−5.49	−12.69	−22.44	−34.42	−19.87	0.00	0.00

Material types as Table 5.1.

5.8.10 Notes

The equations given above can be used to estimate the elastic constants for a composite material if the constituent properties and volume fractions are known. However, an engineer should try to confirm the estimated values with the actual tested values. The following example shows the ratio of the predicted to the measured properties of some composites [16].

- PPL/carbon: $E_{22} = 0.79$, $\sigma_{22} = 1.44$;
- PEEK/carbon: $E_{22} = 0.88$, $\sigma_{22} = 1.15$.

5.9 Numerical examples

Numerical examples on micromechanics are given in section 7.8.3 in which the example on plate vibration is explained.

In this section, the methods of obtaining a laminate's properties from ply properties are illustrated. Tsai [6] presents the 'cash register method'. The in-plane laminate stiffness of m ply groups is given by

$$[A] = \sum_{i=1}^{m} [Q]^{(i)} h_0 n^{(i)} = \sum_{i=1}^{m} [A^0]^{(i)} n^{(i)} \tag{5.92}$$

where a ply group is defined as plies with the same angle grouped or banded together, and

- $[Q]^{(i)}$ off-axis stiffness of the ith ply group with angle θ measured from the laminate axes, equivalent to $[\bar{Q}]_i$
- h_0 unit ply thickness
- $n^{(i)}$ ply number in the ith group
- $[A^0]^{(i)}$ the ith unit ply stiffness $= [Q]^{(i)} h_0$

Table 5.5 shows the unit ply data for several composites.

We find $[A]$ for $[0/90/45_2]_s$ for the glass fiber/epoxy composite given in Table 5.5 by the cash register method.

$$h = 8h_0 = 8 \times 0.125 \text{ mm} = 0.001 \text{ m}$$

$$[A^*] = [A]/h = 1000[A]$$

$$A_{ij} = 2(A_{ij}^{0(0)} + A_{ij}^{0(90)} + 2A_{ij}^{0(45)})$$

Using $[A_{ij}^0]$ from Table 5.5,

$$A_{11} = 2(4.9 + 1.05 + 2 \times 2.14) = 20.46 \text{ MN/m}, \quad A_{11}^* = 20.46 \text{ GPa}$$

$$A_{22} = 2(1.05 + 4.90 + 2 \times 2.14) = 20.46 \text{ MN/m}, \quad A_{22}^* = 20.46 \text{ GPa}$$

$$A_{12} = 2(0.27 + 0.27 + 2 \times 1.11) = 5.52 \text{ MN/m}, \quad A_{12}^* = 5.52 \text{ GPa} = A_{21}^*$$

$$A_{66} = 2(0.52 + 0.52 + 2 \times 1.35) = 7.48 \text{ MN/m}, \quad A_{66}^* = 7.48 \text{ GPa}$$

$$A_{16} = 2(0 + 0 + 2 \times 0.96) = 3.84 \text{ MN/m}, \quad A_{16}^* = 3.84 \text{ GPa}$$

Table 5.6 shows the stiffness moduli for the same four ply angles, 0, 90, ± 45. We find $[A^*]$ for $[0/90/45_2]_s$ by the rule of mixture method using this table. We can express

$$[A^*] = [A]/h = \sum_{i=1}^{m} [\bar{Q}]^i h^i/h = \sum_{i=1}^{m} [\bar{Q}]^i v_{\theta i}$$

Table 5.6 [A^*]/rule of mixtures method (courtesy of Think Composites)

Type: Fiber: Matrix:	CFRP T300 N5208	BFRP B(4) N5505	CFRP AS H3501	GFRP E-glass Epoxy	KFRP kev49 Epoxy	CFRTP AS4 PEEK	CFRP IM6 Epoxy	CFRP T300 Fbrt934	CCRP T300 Fbrt934	CCRP T300 Fbrt934
[Q] 0° SI units (GPa)										
Q_{xx}	181.8	205.0	138.8	39.2	76.6	134.7	204.2	148.9	74.2	66.1
Q_{yy}	10.35	18.59	9.01	8.39	5.55	8.95	11.26	9.71	74.19	66.13
Q_{xy}	2.90	4.28	2.70	2.18	1.89	2.51	3.60	2.91	3.71	2.91
Q_{ss}	7.17	5.59	7.10	4.14	2.30	5.10	8.40	4.55	4.55	4.10
Imperial (Msi)										
Q_{xx}	26.36	29.72	20.13	5.68	11.11	19.53	29.60	21.59	10.76	9.59
Q_{yy}	1.50	2.70	1.31	1.22	0.80	1.30	1.63	1.41	10.76	9.59
Q_{xy}	0.42	0.62	0.39	0.32	0.27	0.36	0.52	0.42	0.54	0.42
Q_{ss}	1.04	0.81	1.03	0.60	0.33	0.74	1.22	0.66	0.66	0.59
[Q] 90° SI units (GPa)										
Q_{11}	10.35	18.59	9.01	8.39	5.55	8.95	11.26	9.71	74.19	66.13
Q_{22}	181.8	205.0	138.8	39.2	76.6	134.7	204.2	148.9	74.2	66.1
Q_{12}	2.90	4.28	2.70	2.18	1.89	2.51	3.60	2.91	3.71	2.91
Q_{66}	7.17	5.59	7.10	4.14	2.30	5.10	8.40	4.55	4.55	4.10
$Q_{16} = Q_{26}$	0.00	0.00	0.00	0.00	0.00	0.00	0.00	0.00	0.00	0.00
Imperial (Msi)										
Q_{11}	1.50	2.70	1.31	1.22	0.80	1.30	1.63	1.41	10.76	9.59
Q_{22}	26.36	29.72	20.13	5.68	11.11	19.53	29.60	21.59	10.76	9.59
Q_{12}	0.42	0.62	0.39	0.32	0.27	0.36	0.52	0.42	0.54	0.42
Q_{66}	1.04	0.81	1.03	0.60	0.33	0.74	1.22	0.66	0.66	0.59
$Q_{16} = Q_{26}$	0.00	0.00	0.00	0.00	0.00	0.00	0.00	0.00	0.00	0.00
[Q] 45° SI units (GPa)										
11 = 22	56.66	63.62	45.41	17.12	23.79	42.26	64.06	45.65	43.50	38.62
12	42.32	52.44	31.21	8.84	19.19	32.06	47.26	36.55	34.40	30.42
66	46.59	53.76	35.60	10.80	19.60	34.66	52.05	38.19	35.24	31.61
16 = 26	42.87	46.60	32.45	7.69	17.77	31.44	48.22	34.79	0.00	0.00
Imperial (Msi)										
11 = 22	8.22	9.23	6.58	2.48	3.45	6.13	9.29	6.62	6.31	5.60
12	6.14	7.60	4.53	1.28	2.78	4.65	6.85	5.30	4.99	4.41
66	6.76	7.79	5.16	1.57	2.84	5.03	7.55	5.54	5.11	4.58
16 = 26	6.22	6.76	4.71	1.12	2.58	4.56	6.99	5.04	0.00	0.00
[Q] −45° SI units (GPa)										
11 = 22	56.66	63.62	45.41	17.12	23.79	42.26	64.06	45.65	43.50	38.62
12	42.32	52.44	31.21	8.84	19.19	32.06	47.26	36.55	34.40	30.42
66	46.59	53.76	35.60	10.80	19.60	34.66	52.05	38.19	35.24	31.61
16 = 26	−42.87	−46.60	−32.45	−7.69	−17.77	−31.44	−48.22	−34.79	0.00	0.00
Imperial (Msi)										
11 = 22	8.22	9.23	6.58	2.48	3.45	6.13	9.29	6.62	6.31	5.60
12	6.14	7.60	4.53	1.28	2.78	4.65	6.85	5.30	4.99	4.41
66	6.76	7.79	5.16	1.57	2.84	5.03	7.55	5.54	5.11	4.58
16 = 26	−6.22	−6.76	−4.71	−1.12	−2.58	−4.56	−6.99	−5.04	0.00	0.00

Material types as Table 5.1.

where

$$v_{\theta i} = \frac{\text{volume of } \theta_i^\circ \text{ plies}}{\text{total volume}}$$

$$A_{11}^* = 0.25 \times 39.2 + 0.25 \times 8.39 + 0.5 \times 17.12 = 20.46 \,\text{GPa}$$

$$A_{22}^* = 0.25 \times 8.39 + 0.25 \times 39.2 + 0.5 \times 17.12 = 20.46 \,\text{GPa}$$

$$A_{12}^* = 0.25 \times 2.18 + 0.25 \times 2.18 + 0.5 \times 8.84 = 5.51 \,\text{GPa}$$

$$A_{66}^* = 0.25 \times 4.14 + 0.25 \times 4.14 + 0.5 \times 10.80 = 7.47 \,\text{GPa}$$

We find A_{11}^* of $[0_3/90]$ laminate for the glass fiber/epoxy composite given in Table 5.6, by the rule of mixture method:

$$v_0 = \frac{\text{volume of } 0^\circ \text{ plies}}{\text{total volume}} = \frac{3}{4} = 0.75, \quad v_{90} = 0.25$$

$$A_{11}^* = 0.75 \times 39.17 + 0.25 \times 8.39 = 31.48 \,\text{GPa}$$

We find the same values by the cash register method:

$$h = 4h_0 = 0.005 \,\text{m} \quad [A^*] = [A]/h = 2000[A]$$

$$A_{11} = (3 \times 4.90 + 1 \times 1.05) = 15.75 \,\text{MN/m}$$

$$A_{11}^* = 31.5 \,\text{GPa}$$

References

Note: *ICCM = International Conference on Composite Materials*; SAMPE = Society for the Advancement of Material and Process Engineering.

1. Kandil, N. and Verchery, G. (1989) Some new developments in the design of stacking sequences of laminates, in *Proceedings ICCM 7* (ed. Wu, Y.), Vol. 3, pp. 358–63.
2. Verchery, G. (1990) Designing with anisotropy, in *Textile Composites in Building Construction*, Vol. 3 (eds Hamelin, P. and Verchery, G.), Pluralis, pp. 29–42.
3. Verchery, G. *et al.* (1991) A quantitative study of the influence of anisotropy on the bending deformation of laminates, in *Proceedings ICCM 8* (eds Tsai, S. W. and Springer, G. S.), July, p. 26-J.
4. Jones, R. M. (1975) *Mechanics of Composite Materials*, Scripta Book Co., Washington, D.C.
5. Calcote, L. R. (1969) *The Analysis of Laminated Composite Structures*, Van Nostrand Reinhold, New York.
6. Tsai, S. W. (1988) *Composite Design*, 4th edn, Think Composites, Dayton.
7. Vinson, J. R. and Sierakowski, R. L. (1987) *The Behavior of Structures Composed of Composite Materials*, Martinus Nijhoff, Dordrecht.
8. Hollaway, L. C. (1990) *Polymer and Polymer Composites in Construction*, Thomas Telford, London.
9. Timoshenko, S. and Woinowsky-Krieger, S. (1959) *Theory of Plates and Shells*, 2nd edn, McGraw-Hill, New York.
10. Vinson, J. R. (1974) *Structural Mechanics: The Behavior of Plates and Shells*, Wiley-Interscience, New York.

11. Yang, Y. Y. (1989) Review on micromechanics of fiber-reinforced composites, in *Proceedings ICCM 7* (ed. Wu, Y.), Vol. 3, p. 463.
12. Ashton, J. E. and Whitney, J. M. (1970) *Theory of Laminated Plates*, Technomics.
13. *Fiber Glass Reinforced Plastics ... By Design*, PPG Fiber Glass Reinforcements Market Series, PPG Industries, Inc.
14. Tsai, S. W. (1964) *Structural Behavior of Composite Materials*, NASA CR-71, July.
15. Hahn, H. T. (1980) Simplified formulas for elastic moduli of unidirectional continuous fiber composites. *Composites Technol. Rev.*, Fall.
16. Paul, C. W., Akkapeddi, M. K. and Mares, F. (1988) Poly (pivalolactone) composites; an ultra-crystalline thermoplastic composite prepared by monomer impregnation. *SAMPE J.*, Jan/Feb. 20.

Further reading

Books

Orofine, J. E. (ed.) (1989) *Structural Materials*, Proceedings of sessions related to structural materials at structural congress, ASCE.

Prescott, J. (1924) *Applied Elasticity*, Longman & Green, London.

Sokolnikoff, I. S. (1951) *Tensor Analysis, Theory and Applications*, John Wiley, New York.

Sokolnikoff, I. S. (1956) *Mathematical Theory of Elasticity*, 2nd edn, McGraw-Hill, New York.

Timoshenko, S. and Goodier, J. N. (1951) *Theory of Elasticity*, 2nd edn, McGraw-Hill, New York.

Articles

Fan, T. C. (1993) Strain matrices of composites in magneto-elasticity, in *Proceedings 38th International SAMPE Symposium and Exhibition* (eds Bailey, V. *et al.*), Anaheim, CA, May, p. 868.

Whitney, J. M. (1969) The effect of transverse shear deformation on the bending of laminated plates. *J. Composite Mater.*, **3**, Jul., p. 534.

6 One-dimensional structural elements of composite materials

6.1 Equations for beams and rods

The differential equation for a beam of isotropic material with lateral loading and axial load, derived assuming the transverse shear strains are negligible, is given in equation 3.173. The case of uniform cross-section is given in equation 3.174, so that

$$\frac{d^2}{dx^2}\left(EI\frac{d^2w}{dx^2}\right) + P\frac{d^2w}{dx^2} = q(x) \tag{3.173}$$

or

$$EI\frac{d^4w}{dx^4} + P\frac{d^2w}{dx^2} = q(x) \tag{3.174}$$

The assumption of neglecting the transverse shear strains is valid if thickness of the beam, h, is 'small' relative to the length, L. The difference between a beam and a plate is that the width, b, of the beam is 'small' compared with L.

We call this one-dimensional structural element a beam when it is acted upon by lateral loads. When this element is loaded by an axial loading, we call it a rod if the loading is tensile, and we call it a column if the loading is compression. A combination of such loadings is possible and both equations 3.173 and 3.174 are for the beam under axial and lateral loadings. It is frequently called a beam–column.

We consider a beam as shown in Figure 6.1, under lateral loading acting in the x–y plane. We ignore the thermal and hygrothermal effects, for simplicity. Since a beam is so narrow, all Poisson's ratio effects are neglected, conforming to a classical beam theory. With these assumptions equation 5.67 is reduced to

$$\begin{bmatrix} N_x \\ M_x \end{bmatrix} = \begin{bmatrix} A_{11} & B_{11} \\ B_{11} & D_{11} \end{bmatrix} \begin{bmatrix} \varepsilon_{x0} \\ \kappa_x \end{bmatrix} \tag{6.1}$$

and from equation 5.65,

$$Q_x = 2A_{55}\varepsilon_{xy} \tag{6.2}$$

Note that the y-axis in this equation is equivalent to the z-axis of Figure 5.3, and if we ignore the transverse shear deformation, $\varepsilon_{xy} = 0$.

If the beam has mid-plane symmetry, there is no bending–stretching coupling so that $B_{11} = 0$, and equation 6.1 becomes

$$\begin{aligned} N_x &= A_{11}\varepsilon_{x0} \\ M_x &= D_{11}\kappa_x \end{aligned} \tag{6.3}$$

When the surface shear stresses are zero, we have the following equilibrium equations.

$$\frac{dV}{dx} = -q(x), \quad \frac{dM}{dx} = V \quad \text{and} \quad d\frac{N_x}{dx} = 0 \qquad (3.165)$$

where

$$V = Q_x b \quad \text{and} \quad M = M_x b \qquad (6.4)$$

The additional differential equations of a beam of composite materials with mid-plane symmetry are as follows.

$$P = bA_{11}\varepsilon_{x0} = bA_{11}\frac{du_0}{dx}, \quad \frac{dP}{dx} = 0 \qquad (6.5)$$

where $P = Nb$ and

$$M = bD_{11}\kappa_x = -bD_{11}\frac{d^2w}{dx^2} \qquad (6.6)$$

From equation 6.5,

$$u_0(x) = \left(\frac{P}{bA_{11}}\right)x + C_0 \qquad (6.7)$$

This is the displacement when the rod is loaded by axial load P. Note that when P is a compressive force, we have to consider stability problems which will be discussed later. The stress in each lamina due to P is

$$[\sigma_x]_k = [\bar{Q}_{11}]_k[\varepsilon_{x0}] = [\bar{Q}_{11}]_k\left(\frac{du_0}{dx}\right) \qquad (6.8)$$

Substituting equation 6.6 into equations 3.165, we obtain

$$bD_{11}\frac{d^4w}{dx^4} = q(x) \qquad (6.9)$$

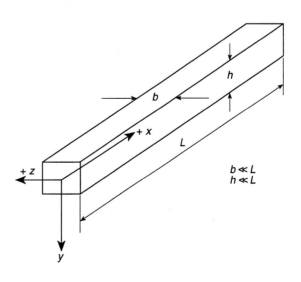

Figure 6.1 Typical beam/rod.

This is exactly the same as equation 3.169, with EI replaced by bD_{11}. For an isotropic beam, $bA_{11} = EA = Ebh$, for a rectangular section.

By integrating equation 6.9 straightforwardly, we can obtain the deflection, w, with four constants, which will be found by the boundary conditions:

$$\text{simple support} \quad w = 0 \quad M = 0 \Rightarrow \frac{d^2w}{dx^2} = 0$$

$$\text{fixed} \qquad\qquad w = 0, \quad \frac{dw}{dx} = 0 \qquad\qquad (6.10)$$

$$\text{free} \qquad\qquad M = 0, \quad V = 0$$

Note that mixed boundaries may happen when we have to solve problems with elastic supports, or structural continuity, and so on.

After the solution of $w(x)$ is found, the stress in each lamina is obtained by

$$[\varepsilon_x]_k = y[\kappa_x] \qquad\qquad (6.11)$$

and

$$[\sigma_x]_k = y[\bar{Q}_{11}]_k[\kappa_x] = -[\bar{Q}_{11}]_k \, y \, \frac{d^2w}{dx^2} \qquad\qquad (6.12)$$

If both in-plane and lateral loads occur simultaneously, the stress in each lamina is

$$[\sigma_x]_k = [\bar{Q}_{11}]_k \frac{du_0}{dx} - [\bar{Q}_{11}]_k \, y \, \frac{d^2w}{dx^2} \qquad\qquad (6.13)$$

Note that this is equivalent to equation (5.53).

As an example, consider a uniformly loaded simple beam (Figure 6.2). The governing differential equation is (equation 6.9):

$$\frac{d^4w}{dx^4} = \frac{q_0}{bD_{11}}$$

Integrating continuously,

$$w(x) = \frac{q_0}{bD_{11}} \frac{x^4}{24} + \frac{C_1 x^3}{6} + \frac{C_2 x^2}{2} + C_3 x + C_4 \qquad\qquad (6.14)$$

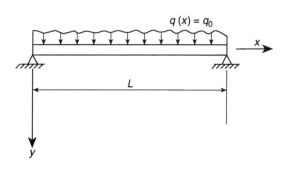

Figure 6.2 Beam under uniform loading.

The boundary conditions are, from equations 6.10,

$$w(0) = 0, \qquad w(L) = 0$$

$$\frac{d^2w}{dx^2}(0) = 0, \qquad \frac{d^2w}{dx^2}(L) = 0$$

from which we get the expression for the deflection as follows:

$$w = \frac{q_0 x}{24bD_{11}}(L^3 - 2Lx^2 + x^3) \tag{6.15}$$

From equations 6.6 and 6.15 the moment is

$$M = \frac{q_0 x}{2}(L - x) \tag{6.16}$$

The maximum moment is

$$M_{\max} = M\left(x = \frac{L}{2}\right) = \frac{q_0 L^2}{8} \tag{6.17}$$

If the beam is of isotropic material with rectangular section, $A = bh$,

$$\sigma_x = \frac{My}{I}$$

where

$$I = \frac{bh^3}{12}$$

and the maximum fiber stress is at $y = \pm h/2$, so that

$$\sigma x_{\max} = \sigma_x(x = L/2, y = \pm h/2) = \pm \frac{6M_{\max}}{bh^2} = \pm \frac{3q_0 L^2}{4bh^2}$$

However, the calculation is not so simple for a beam of composite materials. By equation 6.6,

$$\kappa_{x_{\max}} = \frac{M_{\max}}{bD_{11}} = \frac{q_0 L^2}{8bD_{11}} = -\frac{d^2w}{dx^2} \tag{6.18}$$

Only then can the maximum stress be calculated for each lamina, by equation 6.12

$$[\sigma_x]_{k\max} = [\bar{Q}_{11}]_k[\kappa_x]_{\max} y = -[\bar{Q}_{11}]_k\left(\frac{d^2w}{dx^2}\right)_{\max} y \tag{6.19}$$

This $[\sigma_x]_{k\max}$ must be compared with the allowable strength value which is the failure stress divided by the safety factor.

There could be many types of supports and loading conditions of beams, for which the deflections and moments are available from numerous tables. In any case, the analysis of the beam itself can be carried out by conventional methods.

The calculation of stress requires special attention, as seen from the above example.

To conclude, the determination of the deflection and strength of beams and frames made of composite laminates can be carried out by conventional methods for the homogeneous materials as discussed in Chapter 3. The only difference is that of the material properties resulting in the different stiffness and compliance matrices. When design and analysis involves statically indeterminate structures, special attention is necessary on stress–strain relations.

The stiffness will be discussed in more detail in Chapter 7.

6.2   Beams with hollow cross-sections

Since the structural element made of composite materials has, in general, higher specific strength, the maximum efficiency can be obtained by using hollow cross-sections, resulting in increased stiffnesses. Several types of possible hollow cross-sections are discussed in Chapter 3.

As an example for hollow beams, we consider a box beam as shown in Figure 6.3. This beam may be subjected to axial loads, either tension or compression, in the x-direction, a bending moment with respect to the z-axis, and a torsional moment with respect to the x-axis.

For an isotropic solid rectangular beam, the stiffnesses needed are the extensional stiffness, EA, the flexural stiffness EI and the torsional stiffness, GJ'.

For the present illustration, we assume that the top and bottom panels, ① in the figure, are identical as are the two vertical ones, ②. We also assume that the stacking sequence is made such that no coupling occurs.

The axial force (per unit width) in the x-direction, N_x, is

$$N_x = A_{11}\varepsilon_{x0} \tag{6.3}$$

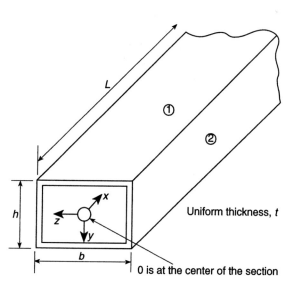

Figure 6.3 Beam with hollow cross-section.

A_{11} is given by equations 5.21, 5.37 and 5.61. The axial load carried by the whole section is the sum of the forces carried by all elements.

$$P = 2N_{x_1}b + 2N_{x_2}h = 2[(A_{11})_1 b + (A_{11})_2 h]\varepsilon_{xo} \qquad (6.20)$$

Therefore, the extensional stiffness for this cross-section is

$$ST_e = 2(A_{11})_1 b + 2(A_{11})_2 h \qquad (6.21)$$

If the box beam is bent in the x–y plane, the bending moment–overall curvature relation is,

$$M = \left[2(D_{11})_1 b + 2(A_{11})_1 b \left(\frac{h}{2}\right)^2 + 2(A_{11})_2 \frac{h^3}{12} \right]\kappa_x \qquad (6.22)$$

Since the top and bottom panels are 'thin' relative to the height of the box, h, we neglect $(D_{11})_1$. Then the bending stiffness of the box is

$$ST_b = 2(A_{11})_1 b \left(\frac{h}{2}\right)^2 + \frac{(A_{11})_2 h^3}{6} \qquad (6.23)$$

If the box beam is acted on by a torque, M_T, this is equivalent to the moment of shear flows with respect to the center of the section, which is the x-axis in our case (equations 3.96 and 3.97). From Figure 6.3, we have

$$M_T = 2N_{xy_1}b(h/2) + 2N_{xy_2}h(b/2) \qquad (6.24)$$

where

$$N_{xy_1} = 2A_{66_1}\varepsilon_{xy0_1}$$

$$N_{xy_2} = 2A_{66_2}\varepsilon_{xy0_2}$$

Note that we may write N_{xz} instead of N_{xy}.

The angle of twist, θ, may be obtained by equation 3.108. For the present problem, we are interested in the stiffness of the section and try to obtain the expression of the shear strains. From Figure 6.3, when the displacements of the centers of each element due to twist are denoted by δ_1 and δ_2, the angle of twist is

$$\theta = \frac{\delta_1}{(h/2)} = \frac{\delta_2}{(b/2)}$$

and

$$\gamma_{xy_1} = 2\varepsilon_{xy_1} = \frac{\delta_1}{L}$$

$$\gamma_{xy_2} = 2\varepsilon_{xy_2} = \frac{\delta_2}{L}$$

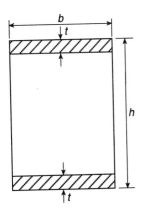

Figure 6.4 Cross-section for sandwich beam.

Then from equation 6.24

$$M_T = \frac{bh}{2L}[(A_{66})_1 h + (A_{66})_2 b]\theta$$

so that the torsional stiffness of this section is

$$ST_t = \frac{bh}{2}[(A_{66})_1 h + (A_{66})_2 b] \tag{6.25}$$

For the section shown in Figure 6.4, which can be the top and bottom skins of a beam with cores, the bending stiffness is

$$ST_b = 2\left(\frac{A_{11}}{t}\right)\cdot bt\left(\frac{h}{2}\right)^2 = \left(\frac{A_{11}}{t}\right)\cdot bt\,\frac{h^2}{2} \tag{6.26}$$

Now, we note that, for a symmetric section, the bending stiffness is

$$ST_b = \frac{A_{11}}{t}I \tag{6.27}$$

In case of a one-dimensional thin structural element, symmetric with respect to the mid-plane,

$$Q_{11} = E_1/(1 - v_{12}v_{21})$$

If we neglect the effect of Poisson's ratio,

$$Q_{11} = E_1$$

therefore

$$A_{11} = E_1 t \tag{6.28}$$

$$ST_b = E_1 I \tag{6.29}$$

and

$$\kappa_1 = M/(E_1 I) \tag{6.30}$$

$$[\varepsilon_x]_{max} = [\kappa]\cdot h/2, \tag{6.31}$$

$$[\sigma]_{max} = [E_1][\varepsilon_x]_{max} \tag{6.32}$$

For the I-section (Figure 6.5),

$$ST_b = \left[2(D_{11})b + 2(A_{11})b\left(\frac{h}{2}\right)^2 + (A_{11})\frac{h^3}{12}\right] \approx A_{11}(6h^2 b + h^3)/12 \tag{6.33}$$

since $D_{11} \approx 0$.

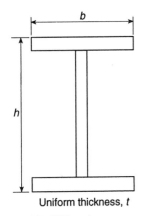

Uniform thickness, t

Figure 6.5 WF section.

Figure 6.6 Beam under axial and transverse loads.

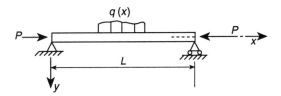

We can now summarize the bending stiffnesses as follows:

- Box section (Figure 6.3) $ST_b = E_1(3h^2b + h^3)t/6$
- Core skins (Figure 6.4) $ST_b = E_1h^2bt/2$
- I-section (Figure 6.5) $ST_b = E_1(6h^2b + h^3)t/12$

For any hollow section other than the above, we can obtain the section moment of inertia from numerous tables.

For thin-walled sections, attention must be given to the buckling of walls. This problem is discussed in Chapter 4. We simply replace E by E_1.

6.3 Eigenvalue problems of beams and frames of anisotropic materials

The stability and vibration problems for one-dimensional structures, which the civil and architectural engineers come to face often are discussed in Chapter 4. The material is assumed to be homogeneous and isotropic. However, the method of approach is basically the same, whether the material is isotropic or anisotropic, as long as the stress–strain relations are not involved in the process of analysis. Even if the deformation is involved, the basic concept is the same for both cases as long as the deformation is calculated by constituent equations used for composite materials.

The governing differential equation for the beam shown in Figure 6.6 is given by equations 3.173 and 3.174.

In the case of a composite beam, we replace EI by bD_{11}. Assuming uniform section along the x-axis and stacking sequence is such that there is no bending–stretching coupling, we have

$$bD_{11}\frac{d^4w}{dx^4} + P\frac{d^2w}{dx^2} = q(x) \tag{6.34}$$

As is discussed in Chapter 4, the buckling load of a column, P_{cr}, is independent of the lateral load, $q(x)$. Equation 6.34 can be written as

$$bD_{11}\frac{d^4w}{dx^4} + P\frac{d^2w}{dx^2} = 0 \tag{6.35}$$

Assuming

$$w = \sum_{n=1}^{\infty} a_n \sin\left(\frac{n\pi x}{L}\right)$$

and substituting into equation 6.35,

$$\sum_{n=1}^{\infty} a_n \left[bD_{11} \frac{n^4\pi^4}{L^4} - P\frac{n^2\pi^2}{L^2} \right] \sin\left(\frac{n\pi x}{L}\right) = 0$$

For a nontrivial solution to exist, $[\cdot] = 0$ and we obtain

$$P_{cr} = \frac{n^2\pi^2}{L^2} bD_{11} \qquad (6.36)$$

This is exactly the same as equation 4.3 with EI replaced by bD_{11}.

The minimum value of P_{cr} is meaningful and is the Euler load for buckling given as

$$P_e = P_{cr(n=1)} = \frac{\pi^2 bD_{11}}{L^2} \qquad (6.37)$$

which is identical to equation 4.4.

If the section is asymmetric with respect to the midplane, the D_{11} in equation 6.36 can be approximately modified as [1]

$$\bar{D}_{11} = \frac{A_{11}D_{11} - B_{11}^2}{A_{11}} \qquad (6.38)$$

If the transverse shear deformation is significant, equation 6.37 may be modified by the basic concept used by Engesser [2],

$$P_{cr} = \frac{\pi^2 bD_{11}}{L^2} \bigg/ \left(1 + \frac{F\pi^2}{L^2} \frac{D_{11}}{A_{55}} \right) \qquad (6.39)$$

In practice, a design engineer selects a structural section considering the maximum stress and the maximum strain, with the stability problem in mind. At the same time, he pays attention to check if the selected member does not get involved with resonances caused by expected external loadings.

Resonance occurs if the applied external loads have the same frequency as one of the natural frequencies of the structure member, and the deflection of this member increases unbounded.

The basic differential equation governing vibration is equation 3.169 with EI replaced by bD_{11}:

$$bD_{11} \frac{d^4w}{dx^4} = q(x) \qquad (6.40)$$

The natural frequencies of the beam depend on the material properties and the geometry of the beam and are independent of the lateral loadings. The force acting on the beam is the dynamic force exerted by the mass on the beam. By

Newton's law,

$$q(x, t) = -\mu \frac{\partial^2 w}{\partial t^2}$$

where μ is the mass per unit length as explained in Chapter 4, so that equation 6.40 becomes

$$bD_{11} \frac{\partial^4 w}{\partial x^4} + \mu \frac{\partial^2 w}{\partial t^2} = 0 \tag{6.41}$$

If the beam is simply supported, we may assume the deflected mode shape as

$$w(x, t) = \sum_{n=1}^{\infty} a_n \sin\left(\frac{n\pi x}{L}\right) \cos \omega t$$

Substituting this into equation 6.41, we obtain

$$\omega^2 = \frac{n^4 \pi^4 b D_{11}}{\mu L^4} \tag{6.42}$$

which is identical to equation 4.26.

For beams or frames other than simply supported, and with variable cross-section, the present method is not effective, if not impossible. In such cases the method presented in section 4.2.11 is very effective even with those structures with composite materials.

As done in the stability discussion, if $B_{11} \neq 0$, the 'reduced' or 'apparent' D given in equation 6.38 may be used.

If the transverse shear deformation is large or in question, the following equation can be used:

$$\omega = \frac{n^2 \pi^2}{L^2} \sqrt{\frac{bD_{11}}{\mu}} \bigg/ \left(1 + \frac{F\pi^2}{L^2} \frac{D_{11}}{A_{55}}\right) \tag{6.43}$$

In equations 6.39 and 6.43, A_{55} is as given by equation 5.66 and F is a factor depending on the shape of the cross-section. For a rectangular section $F = 1.2$, and for a circular cross-section $F = 1.11$. For an I beam, bent in the plane of the flanges, $F = 1.2A/A_f$, and when bent in the plane of the web, $F = A/A_w$, where A, A_f, and A_w are the areas of the total section, of the two flanges and of the web, respectively.

References

1. Calcote, L. R. (1969) *The Analysis of Laminated Composite Structures*, Van Nostrand Reinhold, New York.
2. Timoshenko, S. P. and Gere, J. M. (1961) *Theory of Elastic Stability*, McGraw-Hill, New York.

Further reading

Fukuda, T. *et al.* (1991) Flexural–torsion coupling vibration of fiber composite beams, in *Proceedings ICCM 8* (eds Tsai, S. and Springer, G.), July, p. 26-E.

Xue, Y. *et al.* (1991) A unified approach to modelling the damage process of composite beams under static, impact and cyclic loading, in *Proceedings ICCM 8* (eds Tsai, S. and Springer, G.), July, p. 34-A.

7 Plates and panels of composite materials

7.1 Equilibrium equations

We consider a control element, $dx \, dy \, dz$, in a continuum as shown in Figure 3.31. The three equilibrium equations are obtained as [1–4]

$$\frac{\partial \sigma_x}{\partial x} + \frac{\partial \sigma_{yx}}{\partial y} + \frac{\partial \sigma_{zx}}{\partial z} + X = 0$$

$$\frac{\partial \sigma_{xy}}{\partial x} + \frac{\partial \sigma_y}{\partial y} + \frac{\partial \sigma_{zy}}{\partial z} + Y = 0 \qquad (3.81)$$

$$\frac{\partial \sigma_{xz}}{\partial x} + \frac{\partial \sigma_{yz}}{\partial y} + \frac{\partial \sigma_z}{\partial z} + Z = 0$$

Note that $\sigma_{ij}s(i \neq j)$ are used instead of $\tau_{ij}(i \neq j)$, for shear stresses. Notations and positive directions of stress resultants and couples are as shown in Figure 5.2, to avoid confusion, and follow those used by many composite materials books.

The stress resultants and stress couples are [5–8]

$$\begin{pmatrix} N_x \\ N_y \\ N_{xy} \\ Q_x \\ Q_y \end{pmatrix} = \int_{-h/2}^{h/2} \begin{pmatrix} \sigma_x \\ \sigma_y \\ \sigma_{xy} \\ \sigma_{xz} \\ \sigma_{yz} \end{pmatrix} dz = \sum_{k=1}^{n} \int_{h_{k-1}}^{h_k} \begin{pmatrix} \sigma_x \\ \sigma_y \\ \sigma_{xy} \\ \sigma_{xz} \\ \sigma_{yz} \end{pmatrix}_k dz \qquad \begin{matrix} (5.54) \\ (5.55) \end{matrix}$$

$$\begin{pmatrix} M_x \\ M_y \\ M_{xy} \end{pmatrix} = \int_{-h/2}^{h/2} \begin{pmatrix} \sigma_x \\ \sigma_y \\ \sigma_{xy} \end{pmatrix} z \, dz = \sum_{k=1}^{n} \int_{h_{k-1}}^{h_k} \begin{pmatrix} \sigma_x \\ \sigma_y \\ \sigma_{xy} \end{pmatrix}_k z \, dz \qquad \begin{matrix} (5.54) \\ (5.56) \end{matrix}$$

The first integral of each equation above is for a single layer plate, and the second one, the summation of integrals is for a laminated plate. In a laminated plate, there exist the stress discontinuities since each lamina may have different materials and/or different orientation from the others.

Consider the first of the equilibrium equations (equation 3.81). We neglect the body force X for simplicity. Integrating term by term across each ply and summing up, we obtain

$$\sum_{k=1}^{n} \int_{h_{k-1}}^{h_k} \frac{\partial \sigma_{x_k}}{\partial x} dz + \sum_{k=1}^{n} \int_{h_{k-1}}^{h_k} \frac{\partial \sigma_{yx_k}}{\partial y} dz + \sum_{k=1}^{n} \int_{h_{k-1}}^{h_k} \frac{\partial \sigma_{zx_k}}{\partial z} dz = 0 \qquad (7.1)$$

In the first two terms, we can interchange integration and differentiation,

$$\frac{\partial}{\partial x} \left[\sum_{k=1}^{n} \int_{h_{k-1}}^{h_k} \sigma_{x_k} dz \right] + \frac{\partial}{\partial y} \left[\sum_{k=1}^{n} \int_{h_{k-1}}^{h_k} \sigma_{yx_k} dz \right] + \sum_{k=1}^{n} \sigma_{zx_k} \Big]_{h_{k-1}}^{h_k} = 0 \qquad (7.2)$$

273

By equations 5.54, the first two terms of the above are derivatives of N_x and N_{xy}, respectively. Since the interlamina shear stresses between all plies cancel each other out, the last term is

$$\sigma_{zx}(z = h_n) - \sigma_{zx}(z = h_0) = \tau_x^+ - \tau_x^- \tag{7.3}$$

where

$$\tau_x^+ = \sigma_{zx}(z = h_n), \quad \tau_x^- = \sigma_{zx}(z = h_0)$$

Equation 7.2 can be written as

$$\frac{\partial N_x}{\partial x} + \frac{\partial N_{yx}}{\partial y} + \tau_x^+ - \tau_x^- = 0 \tag{7.4}$$

Proceeding similarly, from the second equilibrium equation (equation 3.81), we obtain

$$\frac{\partial N_{xy}}{\partial x} + \frac{\partial N_y}{\partial y} + \tau_y^+ - \tau_y^- = 0 \tag{7.5}$$

where

$$\tau_y^+ = \sigma_{zy}(z = h_n), \quad \tau_y^- = \sigma_{zy}(z = h_0) \tag{7.6}$$

The third of equations 3.81 yields

$$\frac{\partial Q_x}{\partial x} + \frac{\partial Q_y}{\partial y} + p^+ - p^- = 0 \tag{7.7}$$

where

$$p^+ = \sigma_z(z = h_n), \quad p^- = \sigma_z(z = h_0) \tag{7.8}$$

In addition to the force equilibrium equations as above, we need two equations of moment equilibrium. We multiply each term of the first of equation 3.81 by $z\,dz$, integrate across each ply, and sum up across all laminae to obtain

$$\sum_{k=1}^{n} \int_{h_{k-1}}^{h_k} \frac{\partial \sigma_{x_k}}{\partial x} z \, dz + \sum_{k=1}^{n} \int_{h_{k-1}}^{h_k} \frac{\partial \sigma_{yx_k}}{\partial y} z \, dz + \sum_{k=1}^{n} \int_{h_{k-1}}^{h_k} \frac{\partial \sigma_{zx_k}}{\partial z} z \, dz = 0 \tag{7.9}$$

Recalling that

$$\sum_{k=1}^{n} \int_{h_{k-1}}^{h_k} \frac{\partial \sigma_{zx_k}}{\partial z} z \, dz = \sum_{k=1}^{n} \left[z\sigma_{zx} \big|_{h_{k-1}}^{h_k} - \int_{h_{k-1}}^{h_k} \sigma_{zx} \, dz \right]$$

$$= \frac{h}{2} [\tau_x^+ + \tau_x^-] - Q_x = h_n\tau_x^+ - h_0\tau_x^- - Q_x \tag{7.10}$$

the moment equilibrium equation in the x-direction is

$$\frac{\partial M_x}{\partial x} + \frac{\partial M_{xy}}{\partial y} - Q_x + \frac{h}{2} [\tau_x^+ + \tau_x^-] = 0 \tag{7.11}$$

A similar equation in the y-direction can be obtained as

$$\frac{\partial M_{xy}}{\partial x} + \frac{\partial M_y}{\partial y} - Q_y + \frac{h}{2} [\tau_y^+ + \tau_y^-] = 0 \tag{7.12}$$

Whatever materials are used, there are five equilibrium equations for a rectangular plate, namely, equations 7.4, 7.5, 7.7, 7.11 and 7.12.

Note that the moment equilibrium equations can also be obtained by considering equilibrium conditions about the reference axes.

7.2 Bending of composite plates

To illustrate the method of approach to the problems, we consider a plate of laminated composite material with simplifying assumptions such as

- The plate section is arranged such that $B_{ij} = 0$.
- There is no other coupling term, i.e. $(\cdot)_{16} = (\cdot)_{26} = 0$.
- There are no surface shear stresses, and there are no temperature or hygrothermal effects.

With the above assumptions, the bending equilibrium equations 7.11 and 7.12, and equation 7.7 can be written as

$$\frac{\partial M_x}{\partial x} + \frac{\partial M_{xy}}{\partial y} - Q_x = 0 \qquad (7.13)$$

$$\frac{\partial M_{xy}}{\partial x} + \frac{\partial M_y}{\partial y} - Q_y = 0 \qquad (7.14)$$

$$\frac{\partial Q_x}{\partial x} + \frac{\partial Q_y}{\partial y} + q(x, y) = 0 \qquad (7.15)$$

where $q(x, y) = p^+ - p^-$. Substituting equations 7.13 and 7.14 into 7.15, we obtain

$$\frac{\partial^2 M_x}{\partial x^2} + 2\frac{\partial^2 M_{xy}}{\partial x\,\partial y} + \frac{\partial^2 M_y}{\partial y^2} = -q(x, y) \qquad (7.16)$$

This equation is obtained by considering equilibrium conditions only and is valid for a plate of any material with the assumptions given above.

For a plate of composite material, we need the constitutive relations as given by equation 5.67. With $B_{ij} = 0$, we have

$$M_x = D_{11}\kappa_x + D_{12}\kappa_y \qquad (7.17)$$

$$M_y = D_{12}\kappa_x + D_{22}\kappa_y \qquad (7.18)$$

$$M_{xy} = 2D_{66}\kappa_{xy} \qquad (7.19)$$

in which

$$\kappa_x = \frac{\partial \bar{\alpha}}{\partial x}, \quad \kappa_y = \frac{\partial \bar{\beta}}{\partial y}$$

$$\kappa_{xy} = \frac{1}{2}\left[\frac{\partial \bar{\alpha}}{\partial y} + \frac{\partial \bar{\beta}}{\partial x}\right] \qquad (5.52)$$

The transverse shear deformation $(\varepsilon_{xz}, \varepsilon_{yz})$ has an important role in overall plate behavior. However, in practice, one must 'size' the initial dimensions and decide the stacking sequence of the structure for preliminary design. One can judge the range of such effects on the accuracy of the solution from the abundant literature on such problems. One may adjust the solution obtained by simplified approaches at the final stage of design. The good approach for a designer is to 'compose' the structure in a way such that no complication in analysis and manufacturing is involved unless necessary. In such cases, we can ignore the transverse shear deformation, and from equation 5.50, we have

$$\varepsilon_{xz} = \frac{1}{2}\left(\bar{\alpha} + \frac{\partial w}{\partial x}\right) = 0$$

$$\varepsilon_{yz} = \frac{1}{2}\left(\bar{\beta} + \frac{\partial w}{\partial y}\right) = 0$$

from which we obtain

$$\bar{\alpha} = -\frac{\partial w}{\partial x} \quad \text{and} \quad \bar{\beta} = -\frac{\partial w}{\partial y} \tag{7.20}$$

From these and equations 5.52, we get

$$\kappa_x = -\frac{\partial^2 w}{\partial x^2}$$

$$\kappa_y = -\frac{\partial^2 w}{\partial y^2} \tag{7.21}$$

$$\kappa_{xy} = -\frac{\partial^2 w}{\partial x \, \partial y}$$

Substituting these equations into equations 7.17, 7.18 and 7.19,

$$M_x = -D_{11}\frac{\partial^2 w}{\partial x^2} - D_{12}\frac{\partial^2 w}{\partial y^2} \tag{7.22}$$

$$M_y = -D_{12}\frac{\partial^2 w}{\partial x^2} - D_{22}\frac{\partial^2 w}{\partial y^2} \tag{7.23}$$

$$M_{xy} = -2D_{66}\frac{\partial^2 w}{\partial x \, \partial y} \tag{7.24}$$

Substituting these equations into equation 7.16, we obtain

$$D_{11}\frac{\partial^4 w}{\partial x^4} + 2(D_{12} + 2D_{66})\frac{\partial^4 w}{\partial x^2 \, \partial y^2} + D_{22}\frac{\partial^4 w}{\partial y^4} = q(x, y) \tag{7.25}$$

We may denote

$$D_1 = D_{11}, \quad D_2 = D_{22}, \quad D_3 = (D_{12} + 2D_{66}) \tag{7.26}$$

to express equation 7.25 in a simpler way:

$$D_1\frac{\partial^4 w}{\partial x^4} + 2D_3\frac{\partial^4 w}{\partial x^2 \, \partial y^2} + D_2\frac{\partial^4 w}{\partial y^4} = q(x, y) \tag{7.27}$$

This is the governing differential equation for the bending of a specially orthotropic plate of composite materials acted on by lateral loading, $q(x, y)$.

The assumptions needed are:

- The transverse shear deformation is neglected.
- **Specially orthotropic** layers are arranged so that no coupling terms exist, i.e.

$$B_{ij} = 0, \quad (\cdot)_{16} = (\cdot)_{26} = 0$$

- No temperature or hygrothermal terms exist, i.e.

$$\Delta T = \Delta m = 0$$

Equation 7.27 may be solved by several methods used for isotropic plates. The choice of the method largely depends on the boundary and loading conditions.

7.3 Boundary conditions

The necessary and sufficient number of boundary conditions for solving the special orthotropic plate problems expressed by equation 7.27 is two at each of the boundaries under consideration. We denote the direction normal to the edge by n and the tangent to the edge by t.

1. Simply supported edge: We have

$$w = 0, \quad M_n = 0 \tag{7.28a}$$

   Since $w = 0$, $\partial^2 w/\partial t^2 = 0$. From equation 7.22, $M_n = 0$ implies $\partial^2 w/\partial n^2 = 0$.
2. Clamped edge:

$$w = 0, \quad \frac{\partial w}{\partial n} = 0 \tag{7.29a}$$

3. Free edge: the three boundary conditions, $M_n = 0$, $Q_n = 0$ and $M_{nt} = 0$ can be reduced to two,

$$M_n = 0, \quad V_n = Q_n + \frac{\partial M_{nt}}{\partial t} = 0 \tag{7.30}$$

   where V_n is the Kirchhoff shear, or the 'effective' shear resultant.
4. Many practical problems may have mixed boundaries. The fold lines of folded plate structures and the edges of shear walls may be subject to unknown three-dimensional linear displacements and rotations, and three-dimensional forces and couples [1, 9–11]. Most of such problems can be handled by the methods discussed in Chapters 3 and 4.

The boundary conditions for a general laminated plate will be discussed in section 7.7.

7.4 Navier solution for a simply supported plate of specially orthotropic laminate

If all four edges of a plate are simply supported, the Navier solution is quite efficient. The coordinate system could be either Figure 7.1(a) or 7.1(b), but we use 7.1(a). In this method, we expand both the lateral deflection, $w(x, y)$, and

Figure 7.1 Coordinate systems of the plate.

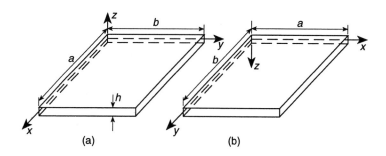

the applied lateral load, $q(x, y)$, into double half-range sine Fourier series which satisfies the boundary conditions given by equation 7.28,

$$w(x, y) = \sum_{m=1}^{\infty} \sum_{n=1}^{\infty} w_{mn} \sin\left(\frac{m\pi x}{a}\right) \sin\left(\frac{n\pi y}{b}\right) \tag{7.31}$$

$$q(x, y) = \sum_{m=1}^{\infty} \sum_{n=1}^{\infty} q_{mn} \sin\left(\frac{m\pi x}{a}\right) \sin\left(\frac{n\pi y}{b}\right) \tag{7.32}$$

Substituting equations 7.31 and 7.32 into equation 7.27, we obtain, for given m and n,

$$w_{mm} = \frac{(q_{mn}/\pi^4)}{D_1(m/a)^4 + 2D_3(m/a)^2(n/b)^2 + D_2(n/b)^4} \tag{7.33}$$

The q_{mn} for given m and n can be found by multiplying both sides of equation 7.32 by $\sin(n'\pi y/b)\, dy$, integrating it from zero to b, multiplying both sides of the resulting equation by $\sin(m'\pi x/a)\, dx$, and integrating it from zero to a:

$$q_{mn} = \frac{4}{ab} \int_0^a \int_0^b q(x, y)\sin\left(\frac{m\pi x}{a}\right)\sin\left(\frac{n\pi y}{b}\right) dx\, dy \tag{7.34}$$

For a given load, $q(x, y)$, we can obtain q_{mn} in equation 7.32 by integrating equation 7.34. By substituting q_{mn} into equation 7.33, we obtain w_{mn}. If we substitute this into equation 7.31, we obtain the lateral deflection of a simply supported plate made of specially orthotropic laminates.

As an example, we consider the plate acted on by a uniformly distributed loading with intensity of $q(x, y) = q_0$. Expanding q_0 into a half-range sine Fourier series, q_{mn} for fixed values of m and n is,

$$q_{mn} = \frac{4q_0}{ab} \int_0^a \int_0^b \sin\frac{m\pi x}{a} \sin\frac{n\pi y}{b}\, dx\, dy = \frac{16q_0}{\pi^2 mn} \tag{7.35}$$

The deflection, $w(x, y)$, is obtained by substituting equation 7.35 into equation 7.33,

$$w(x, y) = \frac{16q_0}{\pi^6} \sum_{m=1,3,5,\ldots}^{\infty} \sum_{n=1,3,5,\ldots}^{\infty} \frac{1}{mn} \sin\left(\frac{m\pi x}{a}\right)\sin\left(\frac{n\pi y}{b}\right) / \text{DEN}$$

where

$$\text{DEN} = D_1\left(\frac{m}{a}\right)^4 + 2D_3\left(\frac{m}{a}\right)^2\left(\frac{n}{b}\right)^2 + D_2\left(\frac{n}{b}\right)^4 \tag{7.36}$$

Since the deflections are known, the stresses are obtained straightforwardly. From equation 5.53,

$$\begin{bmatrix} \sigma_x \\ \sigma_y \\ \sigma_{xy} \end{bmatrix}_k = [\bar{Q}]_k \begin{bmatrix} \varepsilon_{x0} + z\kappa_x \\ \varepsilon_{y0} + z\kappa_y \\ \varepsilon_{xy0} + z\kappa_{xy} \end{bmatrix} \tag{7.37}$$

Since we assumed the transverse shear deformation is negligible, the curvatures are given by equation 7.21. Since we assumed that there is no mid-plane displacement ($B_{ij} = 0$), the stresses at the kth lamina are,

$$\begin{bmatrix} \sigma_x \\ \sigma_y \\ \sigma_{xy} \end{bmatrix}_k = [\bar{Q}]_k \begin{bmatrix} -z\dfrac{\partial^2 w}{\partial x^2} \\[2mm] -z\dfrac{\partial^2 w}{\partial y^2} \\[2mm] -z\dfrac{\partial^2 w}{\partial x\,\partial y} \end{bmatrix} \tag{7.38}$$

which yields

$$\begin{bmatrix} \sigma_x \\ \sigma_y \\ \sigma_{xy} \end{bmatrix}_k = \frac{16q_0 z}{\pi^4} \sum\sum \frac{1}{mn\,\text{DEN}} \begin{bmatrix} \left[\bar{Q}_{11_k}\left(\dfrac{m}{a}\right)^2 + \bar{Q}_{12_k}\left(\dfrac{n}{b}\right)^2\right]\sin\left(\dfrac{m\pi x}{a}\right)\sin\left(\dfrac{n\pi y}{b}\right) \\[3mm] \left[\bar{Q}_{12_k}\left(\dfrac{m}{a}\right)^2 + \bar{Q}_{22_k}\left(\dfrac{n}{b}\right)^2\right]\sin\left(\dfrac{m\pi x}{a}\right)\sin\left(\dfrac{n\pi y}{b}\right) \\[3mm] \left[-2\bar{Q}_{66_k}\left(\dfrac{m}{a}\right)\left(\dfrac{n}{b}\right)\right]\cos\left(\dfrac{m\pi x}{a}\right)\cos\left(\dfrac{n\pi y}{b}\right) \end{bmatrix}_k \tag{7.39}$$

Note that

$$[\bar{Q}]_k = \begin{bmatrix} \bar{Q}_{11} & \bar{Q}_{12} & 0 \\ \bar{Q}_{12} & \bar{Q}_{22} & 0 \\ 0 & 0 & 2\bar{Q}_{66} \end{bmatrix}_k$$

for the special orthotropic plate in a plane stress state.

As another example, we consider a simply supported specially orthotropic laminated plate under partial loading (Figure 7.2). The single load P is uniformly distributed over the rectangular area uv as shown in this figure. By equation 7.34, q_{mn} for fixed values of m and n, is

$$\begin{aligned} q_{mn} &= \frac{4p}{abuv} \int_{\xi-u/2}^{\xi+u/2} \int_{\eta-v/2}^{\eta+v/2} \sin\left(\frac{m\pi x}{a}\right)\sin\left(\frac{n\pi y}{b}\right) \, dy \, dx \\ &= \frac{16p}{\pi^2 mnuv} \sin\left(\frac{m\pi\xi}{a}\right)\sin\left(\frac{n\pi\eta}{b}\right)\sin\left(\frac{m\pi u}{2a}\right)\sin\left(\frac{n\pi v}{2b}\right) \end{aligned} \tag{7.40}$$

Figure 7.2 Plate under partial load.

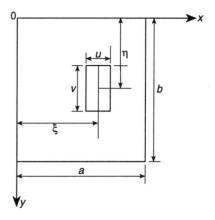

Substituting this q_{mn} into equation 7.33, we obtain w_{mn}, and resulting $w(x, y)$ from equation 7.31. The strains and stresses can be obtained as the previous example.

Another example of practical interest such as obtaining influence coefficients is the case of a single load P, concentrated at any given point, $x = \xi$ and $y = \eta$, of the plate. In the case of a concentrated load, $u \to 0$ and $v \to 0$. Since $\lim_{\theta \to 0} \sin \theta \approx \theta$,

$$\lim_{u \to 0} \sin\left(\frac{m\pi u}{2a}\right) = \frac{m\pi u}{2a}$$

$$\lim_{v \to 0} \sin\left(\frac{n\pi v}{2b}\right) = \frac{n\pi v}{2b}$$

Substituting these results into equation 7.40, we obtain

$$q_{mn} = \frac{4p}{ab} \sin\frac{m\pi\xi}{a} \sin\frac{n\pi\eta}{b} \tag{7.41}$$

Substituting this q_{mn} into equation 7.33, we obtain w_{mn} and the deflection $w(x, y)$ by equation 7.31, as

$$w(x, y) = \frac{4p}{\pi^4 ab} \sum_{m=1}^{\infty} \sum_{n=1}^{\infty} \frac{\sin(m\pi\xi/a)\sin(n\pi\eta/b)}{\text{DEN}} \sin\left(\frac{m\pi x}{a}\right)\sin\left(\frac{n\pi y}{b}\right) \tag{7.42}$$

where DEN is given by equation 7.36.

If $P = 1$, $w(x, y)$ in equation 7.42 represents an influence surface, i.e. the deflection at (x, y) due to a unit load at (ξ, η). We may express this as

$$w(x, y) = \Delta(x, y, \xi, \eta)$$

$$= \frac{4}{\pi^4 ab} \sum_{m=1}^{\infty} \sum_{n=1}^{\infty} \frac{\sin(m\pi\xi/a)\sin(n\pi\eta/b)}{\text{DEN}} \sin\left(\frac{m\pi x}{a}\right)\sin\left(\frac{n\pi y}{b}\right) \tag{7.43}$$

Note that m and n are 'all' integers, unlike the uniform load case where these are odd integers. This $\Delta(x, y, \xi, \eta)$ is sometimes called **Green's function** of the plate with simply supported boundaries. As this equation explicitly shows,

$$\Delta(x, y, \xi, \eta) = \Delta(\xi, \eta, x, y) \tag{7.44}$$

which also follows from the reciprocal theorem of Maxwell.

7.5 Alternative solution for plates of specially orthotropic laminate

For the problems of bending of rectangular plates with two opposite edges simply supported, M. Levy [1] suggested the solution in the form

$$w = \sum_{m=1}^{\infty} Y_m \sin\left(\frac{m\pi x}{a}\right) \tag{7.45}$$

in which Y_m is a function of y only.

We consider a composite plate as shown in Figure 7.1 with boundary conditions $w(0, y) = w(a, y) = 0$, and

$$M_x(0, y) = 0 \quad \text{or} \quad \frac{\partial^2 w(0, y)}{\partial x^2} = 0$$

$$M_x(a, y) = 0 \quad \text{or} \quad \frac{\partial^2 w(a, y)}{\partial x^2} = 0$$

The other two opposite edges are arbitrary. The governing differential equation for the present problem is equation 7.27. A. Nadai [1] used the solution of such an equation, for isotropic materials, in the form

$$w(x, y) = w_1(x) + w_2(x, y) \tag{7.46}$$

where the particular solution w_1 represents the deflection of a strip parallel to the x-axis and w_2 is the homogeneous solution satisfying equation 7.27 with $q(x, y) = 0$. w_2 must be chosen such that $w(x, y)$ in equation 7.46 must satisfy all boundary conditions of the plate.

For our present problem, with $w(x, y)$ in the form given in equation 7.45, we express the lateral loading as

$$q(x, y) = g(x)h(y) \tag{7.47}$$

To match the form of w given by equation 7.45, we assume

$$g(x) = \sum_{m=1}^{\infty} a_m \sin\left(\frac{m\pi x}{a}\right) \tag{7.48}$$

where

$$a_m = \frac{2}{a} \int_0^a g(x)\sin\left(\frac{m\pi x}{a}\right) dx \tag{7.49}$$

Substituting equations 7.45, 7.47 and 7.48 into equation 7.27,

$$D_2 \frac{d^4 Y_m(y)}{dy^4} - 2D_3 \lambda_m^2 \frac{d^2 Y_m(y)}{dy^2} + D_1 \lambda_m^4 Y_m(y) = a_m h_m(y) \tag{7.50}$$

where $\lambda_m = m\pi/a$. The homogeneous solution of equation 7.50 has four constants of integration, and this homogeneous solution together with the particular solution must satisfy the boundary conditions at two edges, $y = 0$ and $y = b$. The homogeneous solution may be obtained from equation 7.50 with the right-hand side equal to zero,

$$\frac{d^4 Y_m(y)}{dy^4} - \frac{2D_3}{D_2} \lambda_m^2 \frac{d^2 Y_m(y)}{dy^2} + \frac{D_1}{D_2} \lambda_m^4 Y_m(y) = 0 \qquad (7.51a)$$

We let $Y_m(y) = e^{Sy}$ and substitute it into the above, and divide the result by e^{Sy} to obtain

$$S^4 - \frac{2D_3}{D_2} \lambda_m^2 S^2 + \frac{D_1}{D_2} \lambda_m^4 = 0 \qquad (7.51b)$$

In the case of an isotropic plate, with uniform load, q, $D_1 = D_2 = D_3 = D$, and S has repeated roots of $\pm\lambda_m$, and

$$Y_m(y) = \frac{qa^4}{D} \left[A_m \cosh\left(\frac{m\pi y}{a}\right) + B_m \left(\frac{m\pi y}{a}\right) \sinh\left(\frac{m\pi y}{a}\right) \right.$$

$$\left. + C_m \sinh\left(\frac{m\pi y}{a}\right) + D_m \left(\frac{m\pi y}{a}\right) \cosh\left(\frac{m\pi y}{a}\right) \right] \qquad (7.52)$$

In the case of the present problem, depending on the relative plate stiffness in various directions, there are three sets of roots. Vinson gives the roots in the form as follows [5]: for

$$\left(\frac{D_1}{D_2}\right)^{1/2} < \frac{D_3}{D_2}$$

$$Y_{m_H}(y) = C_1 \cosh(\lambda_m S_1 y) + C_2 \sinh(\lambda_m S_1 y) + C_3 \cosh(\lambda_m S_2 y)$$

$$+ C_4 \sinh(\lambda_m S_2 y) \qquad (7.53)$$

where

$$S_1 = \left\{ \frac{D_3}{D_2} + \left[\left(\frac{D_3}{D_2}\right)^2 - \frac{D_1}{D_2} \right]^{1/2} \right\}^{1/2}, \quad S_2 = \left\{ \frac{D_3}{D_2} - \left[\left(\frac{D_3}{D_2}\right)^2 - \frac{D_1}{D_2} \right]^{1/2} \right\}^{1/2}$$

For

$$\left(\frac{D_1}{D_2}\right)^{1/2} = \frac{D_3}{D_2}$$

$$Y_{m_H}(y) = (C_5 + C_6 y)\cosh(\lambda_m S_3 y) + (C_7 + C_8 y)\sinh(\lambda_m S_3 y) \qquad (7.54)$$

where

$$S_3 = \pm\left(\frac{D_3}{D_2}\right)^{1/2}$$

For

$$\left(\frac{D_1}{D_2}\right)^{1/2} > \frac{D_3}{D_2}$$

$$Y_{m_H}(y) = (C_9 \cos \lambda_m S_5 y + C_{10} \sin \lambda_m S_5 y)\cosh(\lambda_m S_4 y)$$

$$+ (C_{11} \cos \lambda_m S_5 y + C_{12} \sin \lambda_m S_5 y)\sinh(\lambda_m S_4 y) \qquad (7.55)$$

where

$$S_4 = \left\{ \frac{1}{2} \left[\left(\frac{D_1}{D_2} \right)^{1/2} + \frac{D_3}{D_2} \right] \right\}^{1/2}, \quad S_5 = \left\{ \frac{1}{2} \left[\left(\frac{D_1}{D_2} \right)^{1/2} - \frac{D_3}{D_2} \right] \right\}^{1/2}$$

If $q(x, y)$ is at most linear in y, Vinson shows the particular solution in the form

$$Y_{m_p}(p) = \frac{a_m h_m(y)}{\lambda_m^4 D_1} \tag{7.56}$$

After proper boundary conditions are applied with $Y_m(y) = Y_{m_h}(y) + Y_{m_p}(y)$, resulting $w(x, y)$ can be obtained. The curvatures are given by equation 7.21. The stresses at each lamina are obtained by equation 7.38.

7.6 Review of stiffnesses

In order to avoid possible confusion, a summary of stiffnesses is given in this section.

7.6.1 Single layered plate

Isotropic single layer

$$A_{11} = \frac{Et}{1 - v^2} = A, \quad A_{12} = vA, \quad A_{22} = A, \quad A_{16} = A_{26} = 0$$

$$A_{66} = \frac{Et}{2(1 + v)} = \frac{1 - v}{2} A, \quad B_{ij} = 0$$

$$D_{11} = \frac{Et^3}{12(1 - v^2)} = D, \quad D_{12} = vD, \quad D_{22} = D, \quad D_{16} = D_{26} = 0 \tag{7.57}$$

$$D_{66} = \frac{Et^3}{24(1 + v)} = \frac{1 - v}{2} D$$

where t is the thickness of the layer.

Special orthotropic single layer

$$A_{11} = Q_{11}t, \quad A_{12} = Q_{12}t, \quad A_{22} = Q_{22}t, \quad A_{16} = A_{26} = 0, \quad A_{66} = Q_{66}t,$$

$$B_{ij} = 0, \quad D_{11} = \frac{Q_{11}t^3}{12}, \quad D_{12} = \frac{Q_{12}t^3}{12},$$

$$D_{22} = \frac{Q_{22}t^3}{12}, \quad D_{16} = D_{26} = 0, \quad D_{66} = \frac{Q_{66}t^3}{12} \tag{7.58}$$

Generally orthotropic single layer

$$A_{ij} = \bar{Q}_{ij}t, \quad B_{ij} = 0, \quad D_{ij} = \frac{\bar{Q}_{ij}t^3}{12} \tag{7.59}$$

There is no bending–extension coupling. However, there is extension–shear coupling, and bending–twisting coupling as the following equation shows:

$$
\begin{bmatrix} N_x \\ N_y \\ N_{xy} \\ M_x \\ M_y \\ M_{xy} \end{bmatrix} = \begin{bmatrix} A_{11} & A_{12} & 2A_{16} & 0 & 0 & 0 \\ A_{12} & A_{22} & 2A_{26} & 0 & 0 & 0 \\ A_{16} & A_{26} & 2A_{66} & 0 & 0 & 0 \\ 0 & 0 & 0 & D_{11} & D_{12} & 2D_{16} \\ 0 & 0 & 0 & D_{12} & D_{22} & 2D_{26} \\ 0 & 0 & 0 & D_{16} & D_{26} & 2D_{66} \end{bmatrix} \begin{bmatrix} \varepsilon_{x0} \\ \varepsilon_{y0} \\ \varepsilon_{xy0} \\ \kappa_x \\ \kappa_y \\ \kappa_{xy} \end{bmatrix}
\tag{7.60}
$$

Anisotropic single layer

$$
A_{ij} = Q_{ij}t, \quad B_{ij} = 0, \quad D_{ij} = \frac{Q_{ij}t^3}{12}
\tag{7.61}
$$

Equation 7.60 is applicable for this case. Note that Q_{ij} is used instead of the reduced stiffness, $\bar{Q}_{ij}$.

7.6.2 Symmetrically laminated plate

In design and analysis, an engineer may try to use symmetric sections if possible because:

1. It is much easier for analysis since all bending–extension coupling stiffness terms, B_{ij}, vanish.
2. The laminate does not have an inherent tendency to twist when extended or contracted by any reason including thermal effect.

For the symmetric case, the resultant forces and couplings may be expressed by equation 7.60.

Specially orthotropic laminated plate

If the laminate is made with orthotropic layers which have the principal material directions aligned with the laminate structural axes, $A_{16} = A_{26} = D_{16} = D_{26} = 0$. Since the laminate is symmetric with respect to the mid-surface, $B_{ij} = 0$. The extensional and bending stiffnesses are obtained from equations 5.61 and 5.63 as

$$
A_{ij} = \sum_{k=1}^{n} (\bar{Q}_{ij})_k (h_k - h_{k-1})
$$

$$
D_{ij} = \frac{1}{3} \sum_{k=1}^{n} (\bar{Q}_{ij})_k (h_k^3 - h_{k-1}^3)
$$

where

$$
(\bar{Q}_{11})_k = \left(\frac{E_1}{1 - v_{12}v_{21}} \right)_k
$$

$$
(\bar{Q}_{12})_k = \left(\frac{v_{12}E_1}{1 - v_{12}v_{21}} \right)_k
$$

$$
(\bar{Q}_{22})_k = \left(\frac{E_2}{1 - v_{12}v_{21}} \right)_k
\tag{7.62}
$$

$$
(\bar{Q}_{16})_k = (\bar{Q}_{26})_k = 0
$$

$$
(\bar{Q}_{66})_k = (G_{12})_k
$$

Figure 7.3 Three-layered regular
symmetric cross-ply laminate (all
laminae have the same thickness
and material properties).

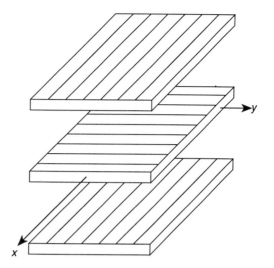

Figure 7.3 Three-layered regular symmetric cross-ply laminate (all laminae have the same thickness and material properties).

Such a laminate may have an orientation of $[0°/90°/0°]$, $[0°/90°/0°/90°/0°]$ and so on. Figure 7.3 shows an exploded view of a typical 'regular' symmetric cross-ply laminate. One may use the term 'regular' if all cross-plied laminae have the same material properties and thicknesses. Note that all thicknesses are not necessarily equal in practice.

Laminated plate with isotropic layers

Each of the laminae may have different material and thickness from others but the arrangement is such that the section is symmetric with respect to the mid-surface. There is no bending–extension coupling. A_{ij} and D_{ij} are found from equations 5.61 and 5.63. Because of the isotropy of each layer, for the kth layer,

$$
\begin{aligned}
(E_{12})_k &= (E_{21})_k \\
(v_{12})_k &= (v_{21})_k
\end{aligned}
\tag{7.63}
$$

and

$$
\begin{aligned}
(\bar{Q}_{11})_k = (\bar{Q}_{22})_k &= \frac{E_k}{1 - (v_k)^2} \\
(\bar{Q}_{12})_k &= \frac{v_k E_k}{1 - (v_k)^2} \\
(\bar{Q}_{66})_k &= \frac{E_k}{2(1 + v_k)}
\end{aligned}
\tag{7.64}
$$

which implies that

$$
\begin{aligned}
A_{11} &= A_{22} \\
D_{11} &= D_{22}
\end{aligned}
\tag{7.65}
$$

The force and coupling resultants are

$$\begin{bmatrix} N_x \\ N_y \\ N_{xy} \end{bmatrix} = \begin{bmatrix} A_{11} & A_{12} & 0 \\ A_{12} & A_{11} & 0 \\ 0 & 0 & 2A_{66} \end{bmatrix} \begin{bmatrix} \varepsilon_{x0} \\ \varepsilon_{y0} \\ \varepsilon_{xy0} \end{bmatrix}$$

$$\begin{bmatrix} M_x \\ M_y \\ M_{xy} \end{bmatrix} = \begin{bmatrix} D_{11} & D_{12} & 0 \\ D_{12} & D_{11} & 0 \\ 0 & 0 & 2D_{66} \end{bmatrix} \begin{bmatrix} \kappa_x \\ \kappa_y \\ \kappa_{xy} \end{bmatrix} \tag{7.66}$$

Laminated plate with multiple general orthotropic layers
An example of such a laminate is $[+\alpha/-\beta/+\theta/-\beta/+\alpha]$ (Figure 7.4). This laminate has mid-surface symmetry and there is no bending–extension coupling. However, A_{16}, A_{26}, D_{16} and D_{26} are not zero, and there are couplings between N_x, N_y and ε_{xy0}, N_{xy} and ε_{x0}, ε_{y0}, M_x, M_y and κ_{xy}, and M_{xy} and κ_x, κ_y (equation 7.60).

When the number of layers of a laminate with a certain ratio α/β increases, the values of A_{16}, A_{26}, D_{16} and D_{26} may decrease compared with those of $(\cdot)_{11}$ and $(\cdot)_{22}$. Recall that

$$A_{ij} = \sum_{k=1}^{n} (\bar{Q}_{ij})_k (h_k - h_{k-1})$$

$$D_{ij} = \frac{1}{3} \sum_{k=1}^{n} (\bar{Q}_{ij})_k (h_k^3 - h_{k-1}^3)$$

and

$$(\bar{Q}_{16})_{+\theta} = -Q_{22}mn^3 + Q_{11}m^3n - (Q_{12} + 2Q_{66})mn(m^2 - n^2) = -(\bar{Q}_{16})_{-\theta} \tag{7.67a}$$

$$(\bar{Q}_{26})_{+\theta} = -Q_{22}m^3n + Q_{11}mn^3 + (Q_{12} + 2Q_{66})mn(m^2 - n^2) = -(\bar{Q}_{26})_{-\theta}$$

so that A_{16}, A_{26}, D_{16} and D_{26} can be the sums of alternating signs.

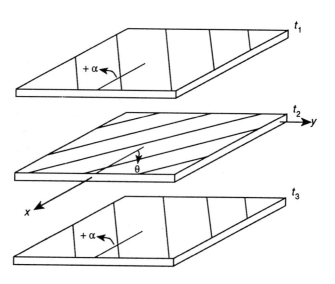

Figure 7.4 Symmetric angle-ply laminate ('regular' if $t_1 = t_2 = t_3$ and $\theta = -\alpha$).

Some layered symmetric angle-ply laminates have the advantage of $B_{ij} = 0$, and small values of A_{16}, A_{26}, D_{16} and D_{26} compared with other terms of A_{ij} and D_{ij}, and higher shear stiffness than cross-ply laminates.

Laminated plate with multiple anisotropic layers
This is the case where $B_{ij} = 0$, but all other stiffnesses must be taken into account.

7.6.3   Antisymmetrically laminated plate

Typical antisymmetric laminates include $[+\alpha/-\beta/+\beta/-\alpha]$ and each pair of laminates has the same thickness. If the central layer has $0°$ or $90°$ orientation, an odd number of layers fits into this category, the central layer being considered as split in two to form two layers of the same orientation. If either one of the central pair has $0°$ or $90°$ $(-\beta)$ orientation, the other one will have $90°$ or $0°(+\beta)$ orientation.

There are the bending–extension coupling stiffnesses, B_{ij}. However, since

$$(\bar{Q}_{16})_{+\theta} = -(\bar{Q}_{16})_{-\theta}$$
$$(\bar{Q}_{26})_{+\theta} = -(\bar{Q}_{26})_{-\theta}$$

(7.67a)

and the layers symmetric about the mid-surface have equal thicknesses,

$$A_{16} = \sum_{k=1}^{n} (\bar{Q}_{16})_k (h_k - h_{k-1}) = 0$$

$$A_{26} = D_{16} = D_{26} = 0$$

The force and coupling resultants are expressed as

$$
\begin{bmatrix} N_x \\ N_y \\ N_{xy} \\ M_x \\ M_y \\ M_{xy} \end{bmatrix}
=
\begin{bmatrix}
A_{11} & A_{12} & 0 & B_{11} & B_{12} & 2B_{16} \\
A_{12} & A_{22} & 0 & B_{12} & B_{22} & 2B_{26} \\
0 & 0 & 2A_{66} & B_{16} & B_{26} & 2B_{66} \\
B_{11} & B_{12} & 2B_{16} & D_{11} & D_{12} & 0 \\
B_{12} & B_{22} & 2B_{26} & D_{12} & D_{22} & 0 \\
B_{16} & B_{26} & 2B_{66} & 0 & 0 & 2D_{66}
\end{bmatrix}
\begin{bmatrix} \varepsilon_{x0} \\ \varepsilon_{y0} \\ \varepsilon_{xy0} \\ \kappa_x \\ \kappa_y \\ \kappa_{xy} \end{bmatrix}
$$

(7.67b)

Note that $\sigma_1 = \sigma_x$, $\sigma_2 = \sigma_y$, $\sigma_3 = \sigma_z$, $\sigma_4 = \sigma_{yz}$, $\sigma_5 = \sigma_{zx}$, $\sigma_6 = \sigma_{xy}$.

Cross-ply laminated plate
$[90°/0°/90°/0°]$ is an example of such an orientation (Figure 7.5). For such a laminate, $A_{16} = A_{26} = D_{16} = D_{26} = 0$. $B_{11} = -B_{22}$, $A_{22} = A_{11}$, $D_{22} = D_{11}$.
The force and coupling resultants are shown as [6, 12]

$$
\begin{bmatrix} N_x \\ N_y \\ N_{xy} \\ M_x \\ M_y \\ M_{xy} \end{bmatrix}
=
\begin{bmatrix}
A_{11} & A_{12} & 0 & B_{11} & 0 & 0 \\
A_{12} & A_{22} & 0 & 0 & -B_{11} & 0 \\
0 & 0 & 2A_{66} & 0 & 0 & 0 \\
B_{11} & 0 & 0 & D_{11} & D_{12} & 0 \\
0 & -B_{11} & 0 & D_{12} & D_{22} & 0 \\
0 & 0 & 0 & 0 & 0 & 2D_{66}
\end{bmatrix}
\begin{bmatrix} \varepsilon_{x0} \\ \varepsilon_{y0} \\ \varepsilon_{xy0} \\ \kappa_x \\ \kappa_y \\ \kappa_{xy} \end{bmatrix}
$$

(7.68)

Figure 7.5 Antisymmetric cross-ply laminate.

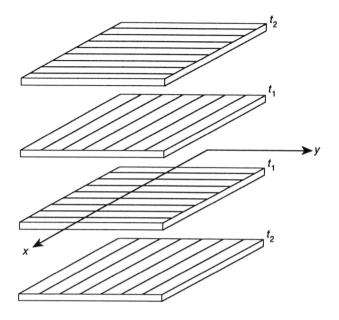

Angle-ply laminated plate

A typical orientation is $[-\alpha/+\beta/-\theta/+\theta/-\beta/+\alpha]$, that is, a lamina oriented at $+\alpha°$ to the laminate coordinate axis on one side of the mid-surface has a lamina of equal thickness oriented at $-\alpha°$ on the other side, at the same distance from the mid-surface. The orientation $[+90°/0°/90°/0°]$ in Figure 7.5 may be replaced by $[+\beta/-\theta/+\theta/-\beta]$. The force and coupling resultants are [6, 12]

$$
\begin{bmatrix} N_x \\ N_y \\ N_{xy} \\ M_x \\ M_y \\ M_{xy} \end{bmatrix} = \begin{bmatrix} A_{11} & A_{12} & 0 & 0 & 0 & 2B_{16} \\ A_{12} & A_{22} & 0 & 0 & 0 & 2B_{26} \\ 0 & 0 & 2A_{66} & B_{16} & B_{26} & 0 \\ 0 & 0 & 2B_{16} & D_{11} & D_{12} & 0 \\ 0 & 0 & 2B_{26} & D_{12} & D_{22} & 0 \\ B_{16} & B_{26} & 0 & 0 & 0 & 2D_{66} \end{bmatrix} \begin{bmatrix} \varepsilon_{x0} \\ \varepsilon_{y0} \\ \varepsilon_{xy0} \\ \kappa_x \\ \kappa_y \\ \kappa_{xy} \end{bmatrix} \quad (7.69)
$$

The B_{16} and B_{26} stiffnesses approach zero as the number of layers in the laminate increases.

7.6.4 Asymmetric laminated plate

Asymmetric laminates with multiple generally orthotropic or multiple anisotropic layers have all stiffnesses present and the force and moment resultants are exactly as shown by equations 5.59 and 5.60.

7.7 Plates with various sections

Consider a symmetric laminated plate with bending–twisting coupling. Because of mid-plane symmetry, $B_{ij} = 0$. The moment–curvature relation is (equation 5.60),

$$M_x = D_{11}\kappa_x + D_{12}\kappa_y + 2D_{16}\kappa_{xy}$$

$$M_y = D_{12}\kappa_x + D_{22}\kappa_y + 2D_{26}\kappa_{xy}$$

$$M_{xy} = D_{16}\kappa_x + D_{26}\kappa_y + 2D_{66}\kappa_{xy}$$

Note that in the case of a symmetric angle-ply plate, $A_{16} = A_{26} \neq 0$, but these two stiffnesses do not affect the bending of the plate (equation 5.60).

If the transverse shear deformation is neglected,

$$M_x = -D_{11}\frac{\partial^2 w}{\partial x^2} - D_{12}\frac{\partial^2 w}{\partial y^2} - 2D_{16}\frac{\partial^2 w}{\partial x\,\partial y}$$

$$M_y = -D_{12}\frac{\partial^2 w}{\partial x^2} - D_{22}\frac{\partial^2 w}{\partial y^2} - 2D_{26}\frac{\partial^2 w}{\partial x\,\partial y} \qquad (7.70)$$

$$M_{xy} = -D_{16}\frac{\partial^2 w}{\partial x^2} - D_{26}\frac{\partial^2 w}{\partial y^2} - 2D_{66}\frac{\partial^2 w}{\partial x\,\partial y}$$

Substituting these moment terms into the equilibrium equation 7.16, we obtain

$$D_{11}\frac{\partial^4 w}{\partial x^4} + 4D_{16}\frac{\partial^4 w}{\partial x^3\,\partial y} + 2(D_{12} + 2D_{66})\frac{\partial^4 w}{\partial x^2\,\partial y^2}$$

$$+ 4D_{26}\frac{\partial^4 w}{\partial x\,\partial y^3} + D_{22}\frac{\partial^4 w}{\partial y^4} = q(x, y) \quad (7.71)$$

Because of terms involving D_{16} and D_{26}, the method of separation of variables, such as the use of Fourier expansion of the deflection w, cannot be used. Farshad and Ahmadi [5] used the perturbation solutions for the case of a plate of one lamina. Ashton [6], by using the Rayleigh–Ritz method, found that the difference between the maximum deflection of a simply supported plate obtained by this approach and the result obtained by assuming the plate to be a special orthotropic laminated one was 24%.

We consider the equilibrium equations for the **general case of laminated plates**. The governing differential equations are the equilibrium equations 7.4, 7.5 and 7.16,

$$\frac{\partial N_x}{\partial x} + \frac{\partial N_{xy}}{\partial y} = 0 \qquad (7.4)$$

$$\frac{\partial N_{xy}}{\partial x} + \frac{\partial N_y}{\partial y} = 0 \qquad (7.5)$$

$$\frac{\partial^2 M_x}{\partial x^2} + 2\frac{\partial M_{xy}}{\partial x\,\partial y} + \frac{\partial^2 M_y}{\partial y^2} = -q(x, y) \qquad (7.16)$$

These equations of classical plate theory are good for plates of any material. Equilibrium equations for plates of composite laminates have different form only when the force–moment resultant equations 5.59 and 5.60 are introduced. Using the expression for curvature, equation 7.21, and for mid-plane strain, with subscript 0 dropped (equation 5.51), we obtain

$$A_{11} \frac{\partial^2 u}{\partial x^2} + 2A_{16} \frac{\partial^2 u}{\partial x \partial y} + A_{66} \frac{\partial^2 u}{\partial y^2} + A_{16} \frac{\partial^2 v}{\partial x^2} (A_{12} + A_{66}) \frac{\partial^2 v}{\partial x \partial y}$$

$$+ A_{26} \frac{\partial^2 v}{\partial y^2} - B_{11} \frac{\partial^3 w}{\partial x^3} - 3B_{16} \frac{\partial^3 w}{\partial x^2 \partial y} - (B_{12} + 2B_{66}) \frac{\partial^3 w}{\partial x \partial y^2}$$

$$- B_{26} \frac{\partial^3 w}{\partial y^3} = 0 \tag{7.72}$$

$$A_{16} \frac{\partial^2 u}{\partial x^2} + (A_{12} + A_{66}) \frac{\partial^2 u}{\partial x \partial y} + A_{26} \frac{\partial^2 u}{\partial y^2} + A_{66} \frac{\partial^2 v}{\partial x^2} + 2A_{26} \frac{\partial^2 v}{\partial x \partial y}$$

$$+ A_{22} \frac{\partial^2 v}{\partial y^2} - B_{16} \frac{\partial^3 w}{\partial x^3} - (B_{12} + 2B_{66}) \frac{\partial^3 w}{\partial x^2 \partial y} - 3B_{26} \frac{\partial^3 w}{\partial x \partial y^2}$$

$$- B_{22} \frac{\partial^3 w}{\partial y^3} = 0 \tag{7.73}$$

$$D_{11} \frac{\partial^4 w}{\partial x^4} + 4D_{16} \frac{\partial^4 w}{\partial x^3 \partial y} + 2(D_{12} + 2D_{66}) \frac{\partial^4 w}{\partial x^2 \partial y^2} + 4D_{26} \frac{\partial^4 w}{\partial x \partial y^3}$$

$$+ D_{22} \frac{\partial^4 w}{\partial y^4} - B_{11} \frac{\partial^3 u}{\partial x^3} - 3B_{16} \frac{\partial^3 u}{\partial x^2 \partial y} - (B_{12} + 2B_{66}) \frac{\partial^3 u}{\partial x \partial y^2}$$

$$- B_{26} \frac{\partial^3 u}{\partial y^3} - B_{16} \frac{\partial^3 v}{\partial x^3} - (B_{12} + 2B_{66}) \frac{\partial^3 v}{\partial x^2 \partial y} - 3B_{26} \frac{\partial^3 v}{\partial x \partial y^2}$$

$$- B_{22} \frac{\partial^3 v}{\partial y^3} = q(x, y) \tag{7.74}$$

The boundary conditions are four kinds such as normal in-plane displacement or normal in-plane force, N_n, tangential in-plane displacement or in-plane shear force, N_{nt}, slope to normal direction or moment to normal direction, lateral deflection or Kirchhoff shear,

$$V_n = \frac{\partial M_{nt}}{\partial t} + Q_n \tag{7.30}$$

There are 12 possible conditions on each of the four edges of a rectangular plate, even if only the simply supported (S), clamped (C) and free edges are considered. The eight possible types of simply supported and clamped edge boundary condition classified by Almroth are as follows [6]:

$$\begin{aligned} &S1: w = 0, \quad M_n = 0, \quad U_n = U_{n,g}, \quad U_t = U_{t,g} \\ &S2: w = 0, \quad M_n = 0, \quad N_n = N_{n,g}, \quad U_t = U_{t,g} \\ &S3: w = 0, \quad M_n = 0, \quad U_n = U_{n,g}, \quad N_{nt} = N_{nt,g} \\ &S4: w = 0, \quad M_n = 0, \quad N_n = N_{n,g}, \quad N_{nt} = N_{nt,g} \end{aligned} \tag{7.28b}$$

$$C1: w = 0, \quad \partial w / \partial n = 0, \quad U_n = U_{n,g}, \quad U_t = U_{t,g}$$

$$C2: w = 0, \quad \partial w / \partial n = 0, \quad N_n = N_{n,g}, \quad U_t = U_{t,g}$$

$$C3: w = 0, \quad \partial w / \partial n = 0, \quad U_n = U_{n,g}, \quad N_{nt} = N_{nt,g}$$

$$C4: w = 0, \quad \partial w / \partial n = 0, \quad N_n = N_{n,g}, \quad N_{nt} = N_{nt,g}$$

(7.29b)

The second subscript, g, indicates the given boundary condition. We have three equations with three unknown dependent variables, u, v and w, and a complete solution is possible.

As seen from the above equations, the problems of plates with composite laminate sections are, generally, very complicated and obtaining analytic solutions is difficult except in a few simple cases.

The general case of asymmetric laminates requires solving the complete set of coupled equations, equations 7.72, 7.73 and 7.74. For such complicated problems, we may resort to some numerical methods which we will discuss in Appendix B.

Considerable simplifications can be made if the laminate is arranged such that $B_{ij} = 0$, or specially orthotropic in which case $B_{ij} = 0$, $A_{16} = A_{26} = D_{16} = D_{26} = 0$. In all these cases equations 7.72 and 7.73 are uncoupled with equation 7.74, and only equation 7.74 needs to be solved to obtain the deflections of the plate.

7.8 Eigenvalue problems for laminated plates

7.8.1 Buckling equations

A plate buckles when the deflection starts to increase without increase of the in-plane loads. Analysis of plate buckling under in-plane loading is concerned with obtaining the solution of an eigenvalue problem, different from the boundary value problem of equilibrium analysis. However, at the moment of the beginning of instability, in other words, just before the start of buckling, the equilibrium must hold.

The governing differential equations are equations 7.4 and 7.5, and equation 7.16 modified for the influence of in-plane forces. When the plate is under the in-plane forces N_x, N_y and N_{xy}, equation 7.16 can be expressed as [13]

$$\frac{\partial^2 M_x}{\partial x^2} + 2 \frac{\partial^2 M_{xy}}{\partial x \, \partial y} + \frac{\partial^2 M_y}{\partial y^2} = \left[q(x, y) + N_x \frac{\partial^2 w}{\partial x^2} + N_y \frac{\partial^2 w}{\partial y^2} + 2 N_{xy} \frac{\partial^2 w}{\partial x \, \partial y} \right]$$

(7.75)

The other two equations are equations 7.4 and 7.5.

Introducing the force–moment resultant equations 5.59 and 5.60, with the expressions of curvature (equation 7.21), and mid-plane strain (equation 5.51), we obtain three equations which are the same as equations 7.72 and 7.73 and

7.74 modified as

$$D_{11} \frac{\partial^4 w}{\partial x^4} + 4D_{16} \frac{\partial^4 w}{\partial x^3 \, \partial y} + 2(D_{12} + 2D_{66}) \frac{\partial^4 w}{\partial x^2 \, \partial y^2} + 4D_{26} \frac{\partial^4 w}{\partial x \, \partial y^3}$$

$$+ D_{22} \frac{\partial^4 w}{\partial y^4} - B_{11} \frac{\partial^3 u}{\partial x^3} - 3B_{16} \frac{\partial^3 u}{\partial x^2 \, \partial y}$$

$$- (B_{12} + 2B_{66}) \frac{\partial^3 u}{\partial x \, \partial y^2} - B_{26} \frac{\partial^3 u}{\partial y^3} - B_{16} \frac{\partial^3 v}{\partial x^3}$$

$$- (B_{12} + 2B_{66}) \frac{\partial^3 v}{\partial x^2 \, \partial y} - 3B_{26} \frac{\partial^3 v}{\partial x \, \partial y^2} - B_{22} \frac{\partial^3 v}{\partial y^3}$$

$$= q(x, y) + N_x \frac{\partial^2 w}{\partial x^2} + N_y \frac{\partial^2 w}{\partial y^2} + 2N_{xy} \frac{\partial^2 w}{\partial x \, \partial y} \tag{7.76}$$

The buckling loads are independent of the lateral loads, and we let $q(x, y) = 0$ in equation 7.76. Generally, the buckling problem of laminated plates involves coupling between bending and extension. However, when the laminate is arranged such that there is no coupling, only equation 7.76 needs to be solved.

Buckling of plates with specially orthotropic laminates
Even though the buckling loads of a structural element are independent of the lateral loads, the combined stresses due to the lateral loads and simultaneous in-plane loads may cause failure before the in-plane buckling load is reached. Existence of initial deformation has a similar effect. In Chapter 4, the cases of simultaneous lateral and in-plane loadings are discussed. In design practice, it is recommended to arrange the section such that $B_{ij} = 0$. Bending–extension coupling may cause premature overstressing before the critical load is reached.

A specially orthotropic laminate has stiffnesses $A_{11}, A_{12}, A_{22}, A_{66}, D_{11}, D_{12}, D_{22}$ and D_{66}. There is neither shear nor twist coupling ('16' and '26' terms), nor bending–stretching coupling ($B_{ij} = 0$). The buckling problem is expressed by one equation, equation 7.76. Under such conditions, equation 7.76 with in-plane forces, reduces to

$$D_{11} \frac{\partial^4 w}{\partial x^4} + 2(D_{12} + 2D_{66}) \frac{\partial^4 w}{\partial x^2 \, \partial y^2} + D_{22} \frac{\partial^4 w}{\partial y^4}$$

$$= q(x, y) + N_x \frac{\partial^2 w}{\partial x^2} + N_y \frac{\partial^2 w}{\partial y^2} + 2N_{xy} \frac{\partial^2 w}{\partial x \, \partial y} \tag{7.77}$$

When this plate is under uniaxial compressive force, N_x (Figure 7.6), using the expression of equation 7.26, equation 7.77 is reduced as

$$D_1 \frac{\partial^4 w}{\partial x^4} + 2D_3 \frac{\partial^4 w}{\partial x^2 \, \partial y^2} + D_2 \frac{\partial^4 w}{\partial y^4} - N_x \frac{\partial^2 w}{\partial x^2} = 0 \tag{7.78}$$

For a simply supported plate, the boundary conditions are, at $x = 0$ and a,

$$w = 0, \quad M_x = -D_{11} \frac{\partial^2 w}{\partial x^2} - D_{12} \frac{\partial^2 w}{\partial y^2} = 0$$

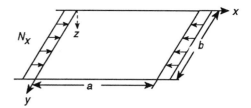

Figure 7.6 Plate under unidirectional axial load. All boundaries are simply supported.

at $y = 0$ and b,

$$w = 0, \quad M_y = -D_{12} \frac{\partial^2 w}{\partial x^2} - D_{22} \frac{\partial^2 w}{\partial y^2} = 0$$

Assuming the buckling mode shape of this plate to be

$$w(x, y) = \sum_{m=1}^{\infty} \sum_{n=1}^{\infty} w_{mn} \sin\left(\frac{m\pi x}{a}\right)\sin\left(\frac{n\pi y}{b}\right)$$

we substitute this expression into equation 7.78 to obtain

$$N_{x_{cr}} = -\frac{\pi^2 a^2}{m^2}\left[D_1\left(\frac{m}{a}\right)^4 + 2D_3\left(\frac{m}{a}\right)^2\left(\frac{n}{b}\right)^2 + D_2\left(\frac{n}{b}\right)^4\right]$$

$$= -\pi^2\left[D_1\left(\frac{m}{a}\right)^2 + 2D_3\left(\frac{n}{b}\right)^2 + D_2\left(\frac{n}{b}\right)^4\left(\frac{a}{m}\right)^2\right] \quad (7.79)$$

The minus sign indicates that $N_{x_{cr}}$ is a compressive force. Buckling occurs at the lowest value of $N_{x_{cr}}$. Obviously $n = 1$, and equation 7.79 reduces to

$$N_{x_{cr}} = -\pi^2\left[D_1\left(\frac{m}{a}\right)^2 + 2D_3 \frac{1}{b^2} + D_2 \frac{1}{b^4}\left(\frac{a}{m}\right)^2\right] \quad (7.80)$$

To find which m provides the smallest $N_{x_{cr}}$ is not simple. The combination of the stiffnesses, D_1, D_2 and D_3, the plate aspect ratio a/b, and m determine $N_{x_{cr}}$.

Symmetric angle-ply laminates are discussed in section 7.6.2. '16' and '26' terms do not vanish but since $B_{ij} = 0$, the governing differential equation can be reduced from equation 7.76 as

$$D_{11} \frac{\partial^4 w}{\partial x^4} + 4D_{16} \frac{\partial^4 w}{\partial x^3 \, \partial y} + 2(D_{12} + 2D_{66}) \frac{\partial^4 w}{\partial x^2 \, \partial y^2}$$

$$+ 4D_{26} \frac{\partial^4 w}{\partial x \, \partial y^3} + D_{22} \frac{\partial^4 w}{\partial y^4} - N_x \frac{\partial^2 w}{\partial x^2} = 0 \quad (7.81)$$

The simply supported boundary conditions are, at $x = 0$ and a, $w = 0$

$$M_x = -D_{11} \frac{\partial^2 w}{\partial x^2} - D_{12} \frac{\partial^2 w}{\partial y^2} - 2D_{16} \frac{\partial^2 w}{\partial x \, \partial y} = 0$$

and at $y = 0$ and b, $w = 0$

$$M_y = -D_{12} \frac{\partial^2 w}{\partial x^2} - D_{22} \frac{\partial^2 w}{\partial y^2} - 2D_{26} \frac{\partial^2 w}{\partial x \, \partial y} = 0$$

Because of the '16' and '26' terms, obtaining a closed-form solution is very difficult, if not impossible. We may resort to the Rayleigh–Ritz solution or numerical approaches.

Antisymmetric cross-ply laminates are discussed in section 7.6.3. For this case, we need to handle all three equations, equations 7.72, 7.73 and 7.74, because of the existence of $B_{11} = -B_{22}$. These equations reduce to

$$A_{11}\frac{\partial^2 u}{\partial x^2} + A_{66}\frac{\partial^2 u}{\partial y^2} + (A_{12}+A_{66})\frac{\partial^2 v}{\partial x\,\partial y} - B_{11}\frac{\partial^3 w}{\partial x^3} = 0 \qquad (7.82)$$

$$(A_{12}+A_{66})\frac{\partial^2 u}{\partial x\,\partial y} + A_{66}\frac{\partial^2 v}{\partial x^2} + A_{11}\frac{\partial^2 v}{\partial y^2} + B_{11}\frac{\partial^3 w}{\partial y^3} = 0 \qquad (7.83)$$

$$D_{11}\left(\frac{\partial^4 w}{\partial x^4} + \frac{\partial^4 w}{\partial y^4}\right) + 2(D_{12}+2D_{66})\frac{\partial^4 w}{\partial x^2\,\partial y^2} - B_{11}\left(\frac{\partial^3 u}{\partial x^3} - \frac{\partial^3 v}{\partial y^3}\right) - N_x\frac{\partial^2 w}{\partial x^2} = 0$$
$$(7.84)$$

Note $D_{22} = D_{11}$ and $A_{22} = A_{11}$.

The problem of simply supported edge boundary conditions of the type S2, at $x = 0$ and a, $w = 0$

$$M_x = B_{11}\frac{\partial u}{\partial x} - D_{11}\frac{\partial^2 w}{\partial x^2} - D_{12}\frac{\partial^2 w}{\partial y^2} = 0$$

$$v = 0$$

$$N_x = A_{11}\frac{\partial u}{\partial x} + A_{12}\frac{\partial v}{\partial y} - B_{11}\frac{\partial^2 w}{\partial x^2} = 0$$

and at $y = 0$ and b, $w = 0$

$$M_y = -B_{11}\frac{\partial v}{\partial y} - D_{12}\frac{\partial^2 w}{\partial x^2} - D_{11}\frac{\partial^2 w}{\partial y^2} = 0$$

$$u = 0$$

$$N_y = A_{12}\frac{\partial u}{\partial x} + A_{11}\frac{\partial v}{\partial y} + B_{11}\frac{\partial^2 w}{\partial y^2} = 0$$

was solved by Jones [6, 14] by using

$$u = \bar{u}\cos\left(\frac{m\pi x}{a}\right)\sin\left(\frac{n\pi y}{b}\right)$$

$$v = \bar{v}\sin\left(\frac{m\pi x}{a}\right)\cos\left(\frac{n\pi y}{b}\right) \qquad (7.85)$$

$$w = \bar{w}\sin\left(\frac{m\pi x}{a}\right)\sin\left(\frac{n\pi y}{b}\right)$$

to get the buckling load as

$$N_x = -\left(\frac{a}{m\pi}\right)^2\left(T_{33} + \frac{2T_{12}T_{23}T_{13} - T_{22}T_{13}^2 - T_{11}T_{23}^2}{T_{11}T_{22} - T_{12}^2}\right) \qquad (7.86)$$

where

$$T_{11} = A_{11}\left(\frac{m\pi}{a}\right)^2 + A_{66}\left(\frac{n\pi}{b}\right)^2$$

$$T_{12} = (A_{12} + A_{66})\left(\frac{m\pi}{a}\right)\left(\frac{n\pi}{b}\right)$$

$$T_{13} = -B_{11}\left(\frac{m\pi}{a}\right)^3$$

$$T_{22} = A_{11}\left(\frac{n\pi}{b}\right)^2 + A_{66}\left(\frac{m\pi}{a}\right)^2 \tag{7.87}$$

$$T_{23} = B_{11}\left(\frac{n\pi}{b}\right)^3$$

$$T_{33} = D_{11}\left[\left(\frac{m\pi}{a}\right)^4 + \left(\frac{n\pi}{b}\right)^4\right] + 2(D_{12} + 2D_{66})\left(\frac{m\pi}{a}\right)^2\left(\frac{n\pi}{b}\right)^2$$

Antisymmetric angle-ply laminates are also discussed in section 7.6.3. We have A_{11}, A_{12}, A_{22}, A_{66}, B_{16}, B_{26}, D_{11}, D_{12}, D_{22} and D_{66}. Because of the bending–stretching coupling stiffnesses, B_{16} and B_{26}, all three equations 7.72, 7.73 and 7.74 are needed. These equations are reduced as follows, describing the present problem:

$$A_{11}\frac{\partial^2 u}{\partial x^2} + A_{66}\frac{\partial^2 u}{\partial y^2} + (A_{12} + A_{66})\frac{\partial^2 v}{\partial x\,\partial y} - 3B_{16}\frac{\partial^3 w}{\partial x^2\,\partial y} - B_{26}\frac{\partial^3 w}{\partial y^3} = 0 \tag{7.88}$$

$$(A_{12} + A_{66})\frac{\partial^2 u}{\partial x\,\partial y} + A_{66}\frac{\partial^2 v}{\partial x^2} + A_{22}\frac{\partial^2 v}{\partial y^2} - B_{16}\frac{\partial^3 w}{\partial x^3} - 3B_{26}\frac{\partial^3 w}{\partial x\,\partial y^2} = 0 \tag{7.89}$$

$$D_{11}\frac{\partial^4 w}{\partial x^4} + 2(D_{12} + 2D_{66})\frac{\partial^4 w}{\partial x^2\,\partial y^2} + D_{22}\frac{\partial^4 w}{\partial y^4}$$
$$- B_{16}\left(3\frac{\partial^3 u}{\partial x^2\,\partial y} + \frac{\partial^3 v}{\partial x^3}\right) - B_{26}\left(\frac{\partial^3 u}{\partial y^3} + 3\frac{\partial^3 v}{\partial x\,\partial y^2}\right) - N_x\frac{\partial^2 w}{\partial x^2} = 0 \tag{7.90}$$

Whitney [6, 15] solved this problem with simply supported boundary conditions of the type S3, at $x = 0$ and a, $w = 0$

$$M_x = B_{16}\left(\frac{\partial v}{\partial x} + \frac{\partial u}{\partial y}\right) - D_{11}\frac{\partial^2 w}{\partial x^2} - D_{12}\frac{\partial^2 w}{\partial y^2} = 0$$

$$u = 0$$

$$N_{xy} = A_{66}\left(\frac{\partial v}{\partial x} + \frac{\partial u}{\partial y}\right) - B_{16}\frac{\partial^2 w}{\partial x^2} - B_{26}\frac{\partial^2 w}{\partial y^2} = 0$$

and at $y = 0$ and b, $w = 0$

$$M_y = B_{26}\left(\frac{\partial v}{\partial x} + \frac{\partial u}{\partial y}\right) - D_{12}\frac{\partial^2 w}{\partial x^2} - D_{22}\frac{\partial^2 w}{\partial y^2} = 0$$

$$v = 0$$

$$N_{xy} = A_{66}\left(\frac{\partial v}{\partial x} + \frac{\partial u}{\partial y}\right) - B_{16}\frac{\partial^2 w}{\partial x^2} - B_{26}\frac{\partial^2 w}{\partial y^2} = 0$$

by employing

$$u = \bar{u} \sin\left(\frac{m\pi x}{a}\right)\cos\left(\frac{n\pi y}{b}\right)$$

$$v = \bar{v} \cos\left(\frac{m\pi x}{a}\right)\sin\left(\frac{n\pi y}{b}\right) \tag{7.91}$$

$$w = \bar{w} \sin\left(\frac{m\pi x}{a}\right)\sin\left(\frac{n\pi y}{b}\right)$$

to obtain the buckling load as

$$N_{x_{cr}} = -\left(\frac{a}{m\pi}\right)^2\left(T_{33} + \frac{2T_{12}T_{23}T_{13} - T_{22}T_{13}^2 - T_{11}T_{23}^2}{T_{11}T_{22} - T_{12}^2}\right) \tag{7.92}$$

where

$$T_{11} = A_{11}\left(\frac{m\pi}{a}\right)^2 + A_{66}\left(\frac{n\pi}{b}\right)^2$$

$$T_{12} = (A_{12} + A_{66})\left(\frac{m\pi}{a}\right)\left(\frac{n\pi}{b}\right)$$

$$T_{13} = -\left[3B_{16}\left(\frac{m\pi}{a}\right)^2 + B_{26}\left(\frac{n\pi}{b}\right)^2\right]\left(\frac{n\pi}{b}\right) \tag{7.93}$$

$$T_{22} = A_{22}\left(\frac{n\pi}{b}\right)^2 + A_{66}\left(\frac{m\pi}{a}\right)^2$$

$$T_{23} = -\left[B_{16}\left(\frac{m\pi}{a}\right)^2 + 3B_{26}\left(\frac{n\pi}{b}\right)^2\right]\left(\frac{m\pi}{a}\right)$$

$$T_{33} = D_{11}\left(\frac{m\pi}{a}\right)^4 + 2(D_{12} + 2D_{66})\left(\frac{m\pi}{a}\right)^2\left(\frac{n\pi}{b}\right)^2 + D_{22}\left(\frac{n\pi}{b}\right)^4$$

7.8.2 Vibration equations for laminated plates

Basic concepts and some methods of vibration analysis for (isotropic) beams and frames are discussed in Chapter 4. As with the buckling problem, the governing equations for vibration are equations 7.4 and 7.5, and equation 7.16 expressed as

$$\frac{\partial^2 M_x}{\partial x^2} + 2\frac{\partial^2 M_{xy}}{\partial x\,\partial y} + \frac{\partial^2 M_y}{\partial y^2}$$

$$= -\left[q(x, y, t) + \rho h\frac{\partial^2 w}{\partial t^2} + N_x\frac{\partial^2 w}{\partial x^2} + N_y\frac{\partial^2 w}{\partial y^2} + 2N_{xy}\frac{\partial^2 w}{\partial x\,\partial y}\right] \tag{7.94}$$

where ρ is the mass density and h is the thickness of the plate. In the case of a plate of a hybrid composite, we may use the average of the mass density across the thickness, so that

$$\rho = \frac{1}{h}\sum_{k=1}^{n} \rho_k(h_k - h_{k-1}) \tag{7.95}$$

When transverse shear effects are neglected, the governing equations are equations 7.72 and 7.73, and 7.74 with the right-hand side replaced by that of equation 7.94 times minus one.

Plates with specially orthotropic laminates

The stiffnesses are A_{11}, A_{12}, A_{22}, A_{66}, D_{11}, D_{12}, D_{22} and D_{66}. A specially orthotropic laminate is symmetric about the mid-plane so that $B_{ij} = 0$, and all '16' and '26' terms vanish.

To investigate the natural vibration problems, we take the forcing function $q(x, y, t)$ and all in-plane forces, N_x, N_y and N_{xy}, to be zero. Since the transverse shear deformation is neglected, the governing equation is obtained directly from equations 7.72, 7.73 and 7.74 as

$$D_{11}\frac{\partial^4 w}{\partial x^4} + 2(D_{12} + 2D_{66})\frac{\partial^4 w}{\partial x^2\,\partial y^2} + D_{22}\frac{\partial^4 w}{\partial y^4} = \rho h \frac{\partial^2 w}{\partial t^2} \qquad (7.96)$$

or

$$D_1\frac{\partial^4 w}{\partial x^4} + 2D_3\frac{\partial^4 w}{\partial x^2\,\partial y^2} + D_2\frac{\partial^4 w}{\partial y^4} = \rho h \frac{\partial^2 w}{\partial t^2} \qquad (7.97)$$

The simply supported boundary conditions are, at $x = 0$ and a, $w = 0$

$$M_x = -D_{11}\frac{\partial^2 w}{\partial x^2} - D_{12}\frac{\partial^2 w}{\partial y^2} = 0$$

and at $y = 0$ and b, $w = 0$

$$M_y = -D_{12}\frac{\partial^2 w}{\partial x^2} - D_{22}\frac{\partial^2 w}{\partial y^2} = 0 \qquad (7.98)$$

Note that

$$D_1 = D_{11}, \quad D_2 = D_{22} \quad \text{and} \quad D_3 = D_{12} + 2D_{66}$$

Let $w(x, y, t) = F(t)G(x, y)$. The free vibration of an elastic body is a harmonic motion. Ashton and Whitney [16] chose the form

$$w(x, y, t) = (A\cos\omega t + B\sin\omega t)G(x, y) \qquad (7.99)$$

We choose $w(x, y, t) = \bar{w}\sin\omega_n t G(x, y)$, as in Chapter 4, so that

$$w(x, y, t) = \sum_{m=1}^{\infty}\sum_{n=1}^{\infty} w_{mn}\sin\left(\frac{m\pi x}{a}\right)\sin\left(\frac{n\pi y}{b}\right)\sin\omega_n t \qquad (7.100)$$

Substituting the mode shape of equation 7.100 for each value of m and n, into equation 7.97, we obtain the natural frequencies as

$$\omega_n^2 = \frac{\pi^4}{\rho h}\left[D_1\left(\frac{m}{a}\right)^4 + 2D_3\left(\frac{m}{a}\right)^2\left(\frac{n}{b}\right)^2 + D_2\left(\frac{n}{b}\right)^4\right] \qquad (7.101)$$

The fundamental frequency, the lowest frequency, occurs when $m = 1$ and $n = 1$ and the mode shape is half sine curves in both directions. Note that the 'maximum' amplitude, w_{mn}, cannot be found from this eigenvalue theory.

Plates with symmetric angle-ply laminates

We have, as discussed in section 7.6.2, full A_{ij} and D_{ij} matrices, but no bending–stretching coupling matrix, i.e. $B_{ij} = 0$. Therefore, equations 7.72 and 7.73 are not coupled with equations 7.74. For our vibration problem, equation 7.74 can be reduced to

$$D_{11} \frac{\partial^4 w}{\partial x^4} + 4D_{16} \frac{\partial^4 w}{\partial x^3 \partial y} + 2(D_{12} + 2D_{66}) \frac{\partial^4 w}{\partial x^2 \partial y^2}$$

$$+ 4D_{26} \frac{\partial^4 w}{\partial x \partial y^3} + D_{22} \frac{\partial^4 w}{\partial y^4} = \rho h \frac{\partial^2 w}{\partial t^2} \quad (7.102)$$

The boundary conditions for simply supported edges are, at $x = 0$ and a, $w = 0$

$$M_x = -D_{11} \frac{\partial^2 w}{\partial x^2} - D_{12} \frac{\partial^2 w}{\partial y^2} - 2D_{16} \frac{\partial^2 w}{\partial x \partial y} = 0$$

and at $y = 0$ and b, $w = 0$

$$M_y = -D_{12} \frac{\partial^2 w}{\partial x^2} - D_{22} \frac{\partial^2 w}{\partial y^2} - 2D_{26} \frac{\partial^2 w}{\partial x \partial y} = 0.$$

The boundary condition, $M_n = 0$ at all edges, and D_{16} and D_{26} terms in the differential equation preclude the use of the separation of variables method. The Rayleigh–Ritz method may be used for such a problem.

Plates with antisymmetric cross-ply laminates

As discussed in section 7.6.3, we have stiffnesses $A_{11} = A_{22}$, A_{12}, A_{66}, $B_{11} = -B_{22}$, $D_{11} = D_{22}$, D_{12} and D_{66}. Because of the appearance of B_{11} and B_{22}, we need all of equations 7.72, 7.73, and 7.74, from which we obtain

$$A_{11} \frac{\partial^2 u}{\partial x^2} + A_{66} \frac{\partial^2 u}{\partial y^2} + (A_{12} + A_{66}) \frac{\partial^2 v}{\partial x \partial y} - B_{11} \frac{\partial^3 w}{\partial x^3} = 0 \quad (7.103)$$

$$(A_{12} + A_{66}) \frac{\partial^2 u}{\partial x \partial y} + A_{66} \frac{\partial^2 v}{\partial x^2} + A_{11} \frac{\partial^2 v}{\partial y^2} + B_{11} \frac{\partial^3 w}{\partial y^3} = 0 \quad (7.104)$$

$$D_{11}\left(\frac{\partial^4 w}{\partial x^4} + \frac{\partial^4 w}{\partial y^4}\right) + 2(D_{12} + 2D_{66}) \frac{\partial^4 w}{\partial x^2 \partial y^2} - B_{11}\left(\frac{\partial^3 u}{\partial x^3} - \frac{\partial^3 v}{\partial y^3}\right) = \rho h \frac{\partial^2 w}{\partial t^2}$$

$$(7.105)$$

Whitney [6, 17] solved this problem with simple support boundary conditions of the type S2, at $x = 0$ and a, $w = 0$

$$M_x = B_{11} \frac{\partial u}{\partial x} - D_{11} \frac{\partial^2 w}{\partial x^2} - D_{12} \frac{\partial^2 w}{\partial y^2} = 0$$

$$v = 0$$

$$N_x = A_{11} \frac{\partial u}{\partial x} + A_{12} \frac{\partial v}{\partial y} - B_{11} \frac{\partial^2 w}{\partial x^2} = 0$$

and at $y = 0$ and b, $w = 0$

$$M_y = -B_{11}\frac{\partial v}{\partial y} - D_{12}\frac{\partial^2 w}{\partial x^2} - D_{11}\frac{\partial^2 w}{\partial y^2} = 0$$

$$u = 0$$

$$N_y = A_{12}\frac{\partial u}{\partial x} + A_{11}\frac{\partial v}{\partial y} + B_{11}\frac{\partial^2 w}{\partial y^2} = 0$$

by employing

$$u(x, y, t) = \bar{u}\cos\left(\frac{m\pi x}{a}\right)\sin\left(\frac{n\pi y}{b}\right)e^{i\omega t},$$

$$v(x, y, t) = \bar{v}\sin\left(\frac{m\pi x}{a}\right)\cos\left(\frac{n\pi y}{b}\right)e^{i\omega t} \qquad (7.106)$$

$$w(x, y, t) = \bar{w}\sin\left(\frac{m\pi x}{a}\right)\sin\left(\frac{n\pi y}{b}\right)e^{i\omega t}$$

to obtain

$$\omega^2 = \frac{\pi^4}{\rho h}\left(T_{33} + \frac{2T_{12}T_{23}T_{13} - T_{22}T_{13}^2 - T_{11}T_{23}^2}{T_{11}T_{22} - T_{12}^2}\right) \qquad (7.107)$$

where

$$T_{11} = A_{11}\left(\frac{m}{a}\right)^2 + A_{66}\left(\frac{n}{b}\right)^2$$

$$T_{12} = (A_{12} + A_{66})\left(\frac{m}{a}\right)\left(\frac{n}{b}\right)$$

$$T_{13} = -B_{11}\left(\frac{m}{a}\right)^3$$

$$T_{22} = A_{11}\left(\frac{n}{b}\right)^2 + A_{66}\left(\frac{m}{a}\right)^2 \qquad (7.108)$$

$$T_{23} = B_{11}\left(\frac{n}{b}\right)^3$$

$$T_{33} = D_{11}\left[\left(\frac{m}{a}\right)^4 + \left(\frac{n}{b}\right)^4\right] + 2(D_{12} + 2D_{66})\left(\frac{m}{a}\right)^2\left(\frac{n}{b}\right)^2$$

No simple conclusion can be made on the values of m and n for the lowest frequency. Equation 7.107 must be treated as a function of the discrete variables m and n.

Plates with antisymmetric angle-ply laminates

There are stiffnesses A_{11}, A_{12}, A_{22}, A_{66}, B_{16}, B_{26}, D_{11}, D_{12}, D_{22} and D_{66}. Equations 7.72, 7.73 and 7.74 are reduced to

$$A_{11}\frac{\partial^2 u}{\partial x^2} + A_{66}\frac{\partial^2 u}{\partial y^2} + (A_{12} + A_{66})\frac{\partial^2 v}{\partial x\,\partial y} - 3B_{16}\frac{\partial^3 w}{\partial x^2\,\partial y} - B_{26}\frac{\partial^3 w}{\partial y^3} = 0 \quad (7.109)$$

$$(A_{12} + A_{66})\frac{\partial^2 u}{\partial x\,\partial y} + A_{66}\frac{\partial^2 v}{\partial x^2} + A_{22}\frac{\partial^2 v}{\partial y^2} - B_{16}\frac{\partial^3 w}{\partial x^3} - 3B_{26}\frac{\partial^3 w}{\partial x\,\partial y^2} = 0 \quad (7.110)$$

$$D_{11}\frac{\partial^4 w}{\partial x^4} + 2(D_{12} + 2D_{66})\frac{\partial^4 w}{\partial x^2\,\partial y^2} + D_{22}\frac{\partial^4 w}{\partial y^4}$$

$$- B_{16}\left(3\frac{\partial^3 u}{\partial x^2\,\partial y} + \frac{\partial^3 v}{\partial x^3}\right) - B_{26}\left(\frac{\partial^3 u}{\partial y^3} + 3\frac{\partial^3 v}{\partial x\,\partial y^2}\right) = \rho h\frac{\partial^2 w}{\partial t^2} \quad (7.111)$$

Again, Whitney [6, 17] solved this problem with simple support boundary conditions of the type S3, at $x = 0$, and a, $w = 0$

$$M_x = B_{16}\left(\frac{\partial v}{\partial x} + \frac{\partial u}{\partial y}\right) - D_{11}\frac{\partial^2 w}{\partial x^2} - D_{12}\frac{\partial^2 w}{\partial y^2} = 0$$

$$u = 0$$

$$N_{xy} = A_{66}\left(\frac{\partial v}{\partial x} + \frac{\partial u}{\partial y}\right) - B_{16}\frac{\partial^2 w}{\partial x^2} - B_{26}\frac{\partial^2 w}{\partial y^2} = 0$$

and at $y = 0$, and b, $w = 0$

$$M_y = B_{26}\left(\frac{\partial v}{\partial x} + \frac{\partial u}{\partial y}\right) - D_{12}\frac{\partial^2 w}{\partial x^2} - D_{22}\frac{\partial^2 w}{\partial y^2} = 0$$

$$v = 0$$

$$N_{xy} = A_{66}\left(\frac{\partial v}{\partial x} + \frac{\partial u}{\partial y}\right) - B_{16}\frac{\partial^2 w}{\partial x^2} - B_{26}\frac{\partial^2 w}{\partial y^2} = 0$$

by using

$$u(x, y, t) = \bar{u}\sin\left(\frac{m\pi x}{a}\right)\cos\left(\frac{n\pi y}{b}\right)e^{i\omega t}$$

$$v(x, y, t) = \bar{v}\cos\left(\frac{m\pi x}{a}\right)\sin\left(\frac{n\pi y}{b}\right)e^{i\omega t} \quad (7.112)$$

$$w(x, y, t) = \bar{w}\sin\left(\frac{m\pi x}{a}\right)\sin\left(\frac{n\pi y}{b}\right)e^{i\omega t}$$

to obtain

$$\omega^2 = \frac{\pi^4}{\rho h}\left(T_{33} + \frac{2T_{12}T_{23}T_{13} - T_{22}T_{13}^2 - T_{11}T_{23}^2}{T_{11}T_{22} - T_{12}^2}\right) \quad (7.113)$$

where

$$T_{11} = A_{11}\left(\frac{m}{a}\right)^2 + A_{66}\left(\frac{n}{b}\right)^2$$

$$T_{12} = (A_{12} + A_{66})\left(\frac{m}{a}\right)\left(\frac{n}{b}\right)$$

$$T_{13} = -\left[3B_{16}\left(\frac{m}{a}\right)^2 + B_{26}\left(\frac{n}{b}\right)^2\right]\left(\frac{n}{b}\right)$$

$$T_{22} = A_{22}\left(\frac{n}{b}\right)^2 + A_{66}\left(\frac{m}{a}\right)^2 \tag{7.114}$$

$$T_{23} = -\left[B_{16}\left(\frac{m}{a}\right)^2 + 3B_{26}\left(\frac{n}{b}\right)^2\right]\left(\frac{m}{a}\right)$$

$$T_{33} = D_{11}\left(\frac{m}{a}\right)^4 + 2(D_{12} + 2D_{66})\left(\frac{m}{a}\right)^2\left(\frac{n}{b}\right)^2 + D_{22}\left(\frac{n}{b}\right)^4$$

7.8.3 Eigenvalue problems of plates with irregular sections and complex boundaries

In practice, it is common to have a structural member with nonuniform section, supported on mixed boundaries.

In section 4.1.5, a procedure useful for buckling analysis of columns with nonuniform cross-sections with other than simply supported ends is discussed. As long as calculation of deflection is possible, this procedure is applicable to any one-dimensional structural element. This method can be extended to two-dimensional problems with some degree of elaborate work.

In section 4.2.11, a method of vibration analysis of irregularly shaped 'one-dimensional' structural elements with any boundary conditions developed by the author is explained. This method has been extended by him to two-dimensional problems including composite laminated plates [18, 19], and to the thick laminated plates with shear deformation [20–22].

The basic concept and method of analysis are as explained in section 4.2.11.

Example 1. Vibration of specially orthotropic laminated plate

As an illustration of this method, consider a simply supported specially orthotropic laminated plate with uniform section as shown in Figure 7.7.

The material properties are

- Matrix modulus $E_m = 3.4\,\text{GPa}$
- Fiber modulus $E_f = 110\,\text{GPa}$
- Matrix Poisson's ratio $v_m = 0.35$
- Fiber Poisson's ratio $v_f = 0.22$
- Matrix volume ratio $V_m = 0.4$
- Fiber volume ratio $V_f = 0.6$

Figure 7.7 Specially orthotropic
laminate.

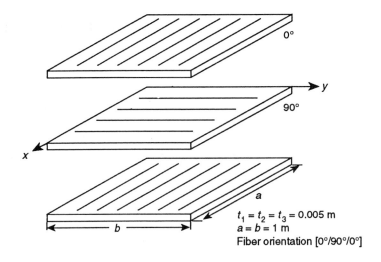

By the rule of mixture, in simpler forms,

$$E_1 = E_f V_f + E_m V_m = 67.36 \text{ GPa} \tag{7.115a}$$

$$E_2 = \frac{E_f E_m}{E_f V_m + E_m V_f} = 8.12 \text{ GPa} \tag{7.115b}$$

$$G_{12} = \frac{G_m G_f}{V_m G_f + V_f G_m} \tag{7.116}$$

$$G_m = \frac{3.4 \text{ GPa}}{2(1 + 0.35)} = 1.2593 \text{ GPa}$$

$$G_f = \frac{110 \text{ GPa}}{2(1 + 0.22)} = 45.0820 \text{ GPa}$$

$$G_{12} = 3.0217 \text{ GPa}$$

$$v_{12} = V_m v_m + V_f v_f = 0.2720 \tag{7.117}$$

$$v_{21} = v_{12} \frac{E_2}{E_1} = 0.0328$$

from equation 5.16. The stiffnesses are

$$A_{ij} = \sum_{k=1}^{n} (\bar{Q}_{ij})_k (h_k - h_{k-1}), \quad \text{in N/m} \tag{5.61}$$

$$B_{ij} = \frac{1}{2} \sum_{k=1}^{n} (\bar{Q}_{ij})_k (h_k^2 - h_{k-1}^2), \quad \text{in N} \tag{5.62}$$

$$D_{ij} = \frac{1}{3} \sum_{k=1}^{n} (\bar{Q}_{ij})_k (h_k^3 - h_{k-1}^3), \quad \text{in Nm} \tag{5.63}$$

$\bar{Q}_{ij}$ are given by equation 5.32; Q_{ij} are obtained by equation 5.21.

$$Q_{11} = E_1/(1 - v_{12}v_{21}) = 67.36/(1 - 0.272 \times 0.0328) = 67.97 \text{ GPa}$$

$$Q_{22} = E_2/(1 - v_{12}v_{21}) = 8.12/(1 - 0.272 \times 0.0328) = 8.194 \text{ GPa}$$

$$Q_{12} = Q_{21} = v_{12}Q_{22} = 0.272 \times 8.194 = 2.229 \text{ GPa}$$
$$= v_{21}Q_{11} = 0.0328 \times 67.97 = 2.229 \text{ GPa}$$

$$Q_{66} = G_{12} = 3.02 \text{ GPa}$$

We obtain the A_{ij}s by the cash register method:

$$\begin{matrix} 0^\circ_2 & 90^\circ \end{matrix}$$

$$A_{11} = 0.005 \text{ m} \times (2 \times 67.97 + 8.194) \text{ GPa} = 720.67 \text{ MN/m}$$

$$A_{22} = 0.005 \text{ m} \times (2 \times 8.194 + 67.97) \text{ GPa} = 421.79 \text{ MN/m}$$

$$A_{12} = 0.005 \text{ m} \times 3 \times 2.229 = 33.4315 \text{ MN/m}$$

$$A_{66} = 0.005 \text{ m} \times 3 \times 3.0217 = 45.325527 \text{ MN/m}$$

The other stiffnesses are obtained in a similar way, and

$$\begin{matrix} A(i, j) \\ (\text{MN/m}) \end{matrix} = \begin{vmatrix} 720.67 & 33.43 & 0 \\ 33.43 & 421.79 & 0 \\ 0 & 0 & 45.33 \end{vmatrix}$$

$$B(i, j) = 0 \text{ (from symmetry)}$$

$$\begin{matrix} D(i, j) = \\ (\text{Nm}) \end{matrix} \begin{vmatrix} 18\,492 & 627 & 0 \\ 627 & 2927 & 0 \\ 0 & 0 & 849 \end{vmatrix}$$

The influence surfaces are calculated by using equation 7.42, with $P = 1$,

$$w(x, y) = \sum_{m=1}^{\infty} \sum_{n=1}^{\infty} w_{mn} \sin\left(\frac{m\pi x}{a}\right)\sin\left(\frac{n\pi y}{b}\right)$$

where

$$w_{mn} = \frac{(P_{mn}/\pi^4)}{D_{11}(m/a)^4 + 2(D_{12} + 2D_{66})(m/a)^2(n/b)^2 + D_{22}(n/b)^4} \tag{7.118}$$

in which

$$P_{mn} = \frac{4(1)}{ab} \sin\left(\frac{m\pi\xi}{a}\right)\sin\left(\frac{n\pi\eta}{b}\right)$$

From equation 4.77,

$$w(i, j)^{(1)} = W(i, j)^{(1)}$$

where W is the maximum amplitude, (i, j) or (x, y) is the point under consideration, and the superscript (1) after (i, j) indicates the first assumed mode shape. We assume

$$W(i, j)^{(1)} = \begin{vmatrix} 10 & 20 & 20 & 10 \\ 20 & 35 & 35 & 20 \\ 20 & 35 & 35 & 20 \\ 10 & 20 & 20 & 10 \end{vmatrix}$$

By equation 4.78, which is

$$F(i, j)^{(1)} = + m(i, j)[\omega(i, j)^{(1)}]^2 w(i, j)^{(1)},$$

where

$$m(i, j) = \text{the mass at point } (i, j) = \rho h(i, j)\Delta x \Delta y,$$

where Δx and Δy are the mesh sizes in the x- and y-directions, respectively, and ρ is the mass density at (i, j), h is the thickness of the plate at (i, j).
$\omega(i, j)^{(1)} = $ the 'first' natural circular frequency at point (i, j), we obtain $F(i, j)^{(1)}$ in terms of $\omega(i, j)^{(1)}$. Substituting $F(i, j)^{(1)}$ into equation 4.79, which is

$$w(i, j)^{(2)} = \sum^{k,l} \Delta(i, j, k, l)F(i, j)^{(1)}$$
$$= \sum^{k,l} \Delta(i, j, k, l)\{+ m(i, j)[\omega(i, j)^{(1)}]^2 w(k, l)^{(1)}\}$$

where $\Delta(i, j, k, l)$ is the influence surface, i.e. the deflection at point (i, j) caused by a unit load at point (k, l), we can obtain $w(i, j)^{(2)}$. Some values of the influence surface are as follows.

$$\Delta(1, 1, 1, 1,) = 0.1696 \times 10^{-7} = \Delta(4, 4, 4, 4)$$

$$\Delta(2, 2, 1, 1) = 0.1694 \times 10^{-7} \quad \Delta(2, 2, 3, 1) = 0.3621 \times 10^{-7}$$

$$\Delta(2, 2, 1, 2) = 0.4078 \times 10^{-7} \quad \Delta(2, 2, 3, 2) = 0.8181 \times 10^{-7}$$

$$\Delta(2, 2, 1, 3) = 0.2505 \times 10^{-7} \quad \Delta(2, 2, 3, 3) = 0.5605 \times 10^{-7}$$

$$\Delta(2, 2, 1, 4) = 0.0662 \times 10^{-7} \quad \Delta(2, 2, 3, 4) = 0.1615 \times 10^{-7}$$

$$\Delta(2, 2, 2, 1) = 0.3913 \times 10^{-7} \quad \Delta(2, 2, 4, 1) = 0.1411 \times 10^{-7}$$

$$\Delta(2, 2, 2, 2) = 0.9876 \times 10^{-7} \quad \Delta(2, 2, 4, 2) = 0.3017 \times 10^{-7}$$

$$\Delta(2, 2, 2, 3) = 0.5872 \times 10^{-7} \quad \Delta(2, 2, 4, 3) = 0.2238 \times 10^{-7}$$

$$\Delta(2, 2, 2, 4) = 0.1602 \times 10^{-7} \quad \Delta(2, 2, 4, 4) = 0.0673 \times 10^{-7}$$

$$\begin{aligned}
w(2, 2)^{(2)} = {}& m(2, 2)[\omega(2, 2)^{(1)}]^2[0.1694(10) + 0.4078(20) + 0.2505(20) \\
& + 0.0662(10) + 0.3913(20) + 0.9867(35) + 0.5872(35) + 0.1602(20) \\
& + 0.3621(20) + 0.8181(35) + 0.5605(35) + 0.1615(20) + 0.1411(10) \\
& + 0.3017(20) + 0.2238(20) + 0.0673(10)] \times 10^{-7} \\
= {}& 152.95 \times 10^{-7} \cdot m(2, 2)[\omega(2, 2)^{(1)}]^2
\end{aligned}$$

From equation 4.80

$$w(i, j)^{(1)}/w(i, j)^{(2)} = 1$$

or

$$35 = 152.95 \times 10^{-7} \cdot m(2, 2)[\omega(2, 2)^{(1)}]^2$$

from which, we obtain,

$$\omega(2, 2)^{(1)} = 1512.68/\sqrt{m(2, 2)}$$

The calculation is carried out at all points (i, j) and we obtain

$$\omega(i, j)^{(1)} = (1913-1512)/\sqrt{m(i, j)}$$

Since the range of $\omega(i, j)^{(1)}$ is too large, we proceed one more cycle. For $W(i, j)^{(2)}$ to be used for $F(i, j)^{(2)}$, the absolute numerics of $w(i, j)^{(2)}$ are used. Then

$$\omega(i, j)^{(2)} = (1586.5-1631.3)/\sqrt{m(i, j)}$$

$$\omega(2, 2)^{(2)} = 1587/\sqrt{m(2, 2)}$$

We proceed further to obtain

$$\omega(i, j)^{(3)} = (1592.5-1598.0)/\sqrt{m(i, j)}$$

$$\omega(i, j)^{(4)} = (1593.6-1594.3)/\sqrt{m(i, j)}$$

$$\omega(2, 2)^{(3)} = 1593/\sqrt{m(2, 2)}$$

$$\omega(2, 2)^{(4)} = 1593.6/\sqrt{m(2, 2)}$$

The result by equation 7.101 is

$$\omega = 1593.7/\sqrt{m}$$

For the second mode, we appoint

$$W(i, j)^{(1)} = \begin{vmatrix} 10 & 20 & 20 & 10 \\ 10 & 20 & 20 & 10 \\ -10 & -20 & -20 & -10 \\ -10 & -20 & -20 & -10 \end{vmatrix}$$

and proceed as the first mode case. The result is obtained as

$$\omega(i, j)^{(1)} = (2906-2658)/\sqrt{m(i, j)}$$

$$\omega(2, 2)^{(1)} = 2658/\sqrt{m(2, 2)}$$

$$\omega(i, j)^{(2)} = (2707-2694)/\sqrt{m(i, j)}$$

$$\omega(2, 2)^{(2)} = 2694.1/\sqrt{m(2, 2)}$$

$$\omega(i, j)^{(3)} = (2696.4-2695.8)/\sqrt{m(i, j)}$$

$$\omega(2, 2)^{(3)} = 2695.8/\sqrt{m(2, 2)}$$

The result by equation 7.101 is

$$\omega = 2695.9/\sqrt{m}$$

Note that the first mode natural frequency obtained by the first cycle at point $(2, 2)$, $\omega(2, 2)^{(1)}$ is only 4.7% off. For the second mode, $\omega(2, 2)^{(1)}$ is 1.4% off.

Example 2. Vibration of antisymmetric angle-ply laminated plate
We consider an antisymmetric laminated plate with $[+45°/-45°/+45°/-45°]$ orientation. The plate geometry and material properties are the same as in the

previous example. The stiffnesses are found as

$$A(i,j) = \begin{vmatrix} 463\,540\,000 & 342\,678\,000 & 0 \\ 342\,678\,000 & 463\,540\,000 & 0 \\ 0 & 0 & 358\,520\,000 \end{vmatrix} \text{N/m}$$

$$B(i,j) = \begin{vmatrix} 0 & 0 & -747\,123 \\ 0 & 0 & -747\,123 \\ -747\,123 & -747\,123 & 0 \end{vmatrix} \text{N}$$

$$D(i,j) = \begin{vmatrix} 15\,451.3 & 11\,422.6 & 0 \\ 11\,422.6 & 15\,451.3 & 0 \\ 0 & 0 & 11\,950.7 \end{vmatrix} \text{N m}$$

For the first mode, it is assumed that

$$W(i,j)^{(1)} = \begin{vmatrix} 10 & 20 & 20 & 10 \\ 20 & 30 & 30 & 20 \\ 20 & 30 & 30 & 20 \\ 10 & 20 & 20 & 10 \end{vmatrix}$$

In order to obtain the influence surfaces, we need to solve equations 7.109, 7.110, and equation 7.111 with the right-hand side replaced by $q(x, y)$, thus

$$A_{11}\frac{\partial^2 u}{\partial x^2} + A_{66}\frac{\partial^2 u}{\partial y^2} + (A_{12} + A_{66})\frac{\partial^2 v}{\partial x\,\partial y} - 3B_{16}\frac{\partial^3 w}{\partial x^2\,\partial y} - B_{26}\frac{\partial^3 w}{\partial y^3} = 0 \quad (7.109)$$

$$(A_{12} + A_{66})\frac{\partial^2 u}{\partial x\,\partial y} + A_{66}\frac{\partial^2 v}{\partial x^2} + A_{22}\frac{\partial^2 v}{\partial y^2} - B_{16}\frac{\partial^3 w}{\partial x^3} - 3B_{26}\frac{\partial^3 w}{\partial x\,\partial y^2} = 0 \quad (7.110)$$

$$D_{11}\frac{\partial^4 w}{\partial x^4} + 2(D_{12} + 2D_{66})\frac{\partial^4 w}{\partial x^2\,\partial y^2} + D_{22}\frac{\partial^4 w}{\partial y^4}$$

$$- B_{16}\left(3\frac{\partial^3 u}{\partial x^2\,\partial y} + \frac{\partial^3 v}{\partial x^3}\right) - B_{26}\left(\frac{\partial^3 u}{\partial y^3} + 3\frac{\partial^3 v}{\partial x\,\partial y^2}\right) = q(x, y) \quad (7.119)$$

Following Whitney's [6, 17] solution, with simple boundary conditions of the type S3, at $x = 0$ and a, $w = 0$

$$M_x = B_{16}\left(\frac{\partial v}{\partial x} + \frac{\partial u}{\partial y}\right) - D_{11}\frac{\partial^2 w}{\partial x^2} - D_{12}\frac{\partial^2 w}{\partial y^2} = 0$$

$$u = 0$$

$$N_{xy} = A_{66}\left(\frac{\partial v}{\partial x} + \frac{\partial u}{\partial y}\right) - B_{16}\frac{\partial^2 w}{\partial x^2} - B_{26}\frac{\partial^2 w}{\partial y^2} = 0$$

and at $y = 0$ and b, $w = 0$

$$M_y = B_{26}\left(\frac{\partial v}{\partial x} + \frac{\partial u}{\partial y}\right) - D_{12}\frac{\partial^2 w}{\partial x^2} - D_{22}\frac{\partial^2 w}{\partial y^2} = 0$$

$$v = 0$$

$$N_{xy} = A_{66}\left(\frac{\partial v}{\partial x} + \frac{\partial u}{\partial y}\right) - B_{16}\frac{\partial^2 w}{\partial x^2} - B_{26}\frac{\partial^2 w}{\partial y^2} = 0$$

we assume

$$u(x, y) = \sum_{m=1}^{\infty} \sum_{n=1}^{\infty} \bar{u}_{mn} \sin\left(\frac{m\pi x}{a}\right)\cos\left(\frac{n\pi y}{b}\right) \qquad (7.120a)$$

$$v(x, y) = \sum_{m=1}^{\infty} \sum_{n=1}^{\infty} \bar{v}_{mn} \cos\left(\frac{m\pi x}{a}\right)\sin\left(\frac{n\pi y}{b}\right) \qquad (7.120b)$$

$$w(x, y) = \sum_{m=1}^{\infty} \sum_{n=1}^{\infty} \bar{w}_{mn} \sin\left(\frac{m\pi x}{a}\right)\sin\left(\frac{n\pi y}{b}\right) \qquad (7.120c)$$

$$q(x, y) = \sum_{m=1}^{\infty} \sum_{n=1}^{\infty} \bar{q}_{mn} \sin\left(\frac{m\pi x}{a}\right)\sin\left(\frac{n\pi y}{b}\right) \qquad (7.120d)$$

In the case of a concentrated load P at $x = \xi$ and $y = \eta$,

$$\bar{q}_{mn} = \frac{4P}{ab} \sin\left(\frac{m\pi\xi}{a}\right)\sin\left(\frac{n\pi\eta}{b}\right) \qquad (7.41)$$

Substituting equations 7.120 into equations 7.109, 7.110, and 7.119, and applying the boundary conditions, we obtain

$$\bar{u}_{mn} = \bar{q}_{mn} \frac{R^3 b^3 n}{\pi^3 (\text{Den})} [(A_{66}m^2 + A_{22}n^2 R^2)(3B_{16}m^2 + B_{26}n^2 R^2)$$

$$- m^2(A_{12} + A_{66})(B_{16}m^2 + 3B_{26}n^2 R^2)] \qquad (7.121)$$

$$\bar{v}_{mn} = \bar{q}_{mn} \frac{R^3 b^3 m}{\pi^3 (\text{Den})} [(A_{11}m^2 + A_{66}n^2 R^2)(B_{16}m^2 + 3B_{26}n^2 R^2)$$

$$- n^2 R^2(A_{12} + A_{66})(3B_{16}m^2 + B_{26}n^2 R^2)] \qquad (7.122)$$

$$\bar{w}_{mn} = \bar{q}_{mn} \frac{R^4 b^4}{\pi^4 (\text{Den})} [(A_{11}m^2 + A_{66}n^2 R^2)(A_{66}m^2 + A_{22}n^2 R^2)$$

$$- (A_{12} + A_{66})^2 m^2 n^2 R^2] \qquad (7.123)$$

where $R = a/b$ and

$$\text{Den} = [(A_{11}m^2 + A_{66}n^2 R^2)(A_{66}m^2 + A_{22}n^2 R^2) - (A_{12} + A_{66})^2 m^2 n^2 R^2]$$

$$\times [D_{11}m^4 + 2(D_{12} + 2D_{66})m^2 n^2 R^2 + D_{22}n^4 R^4]$$

$$+ 2m^2 n^2 R^2(A_{12} + A_{66}) \times (3B_{16}m^2 + B_{26}n^2 R^2)(B_{16}m^2 + 3B_{26}n^2 R^2)$$

$$- n^2 R^2(A_{66}m^2 + A_{22}n^2 R^2) \times (3B_{16}m^2 + B_{26}n^2 R^2)^2$$

$$- m^2(A_{11}m^2 + A_{66}n^2 R^2)(B_{16}m^2 + 3B_{26}n^2 R^2)^2 \qquad (7.124)$$

For the influence surface, we set $p = 1$ and we need only $w(x, y, \xi, \eta) = \Delta(x, y, \xi, \eta)$, that is, we need equation 7.120c with equations 7.41, 7.123 and 7.124. Proceeding as in the previous example,

$$\omega(i, j)^{(1)} = (3707 - 2737)/\sqrt{m(i, j)}$$

$$\omega(2, 2)^{(1)} = 2737/\sqrt{m(2, 2)}$$

$$\omega(i, j)^{(2)} = (3016 - 2937)/\sqrt{m(i, j)}$$

$$\omega(i, j)^{(3)} = (2949.3 - 2954.6)/\sqrt{m(i, j)}$$

If the plate is divided by 5×5 mesh,

$$\omega(i, j)^{(4)} = (2957.8 - 2958.3)/\sqrt{m(i, j)}.$$

The result obtained by equation 7.114 is

$$\omega = 2957.99/\sqrt{m}$$

For the second mode, it is assumed that

$$\omega(i, \lambda)^{(1)} = \begin{vmatrix} 10 & 20 & 20 & 10 \\ 10 & 20 & 20 & 20 \\ -10 & -20 & -20 & -20 \\ -10 & -20 & -20 & -10 \end{vmatrix}$$

Proceeding as in the first mode case,

$$\omega(i, j)^{(2)} = (6826.8 - 6728.3)/\sqrt{m(i, j)}$$

$$\omega(i, j)^{(3)} = (6758.1 - 6740.3)/\sqrt{m(i, j)}$$

$$\omega(i, j)^{(4)} = (6745.8 - 6742.3)/\sqrt{m(i, j)}$$

If the plate is divided by 6×6 mesh,

$$\omega(i, j)^{(5)} = (6867.3 - 6869.2)/\sqrt{m(i, j)}$$

The result by equation 7.113 is

$$\omega = 6868.2/\sqrt{m}$$

Comments on the method

The method explained in this section can be used for a structural member with irregular section and with any boundary conditions, provided that the deflection can be obtained. Any structural design and analysis involve calculating influence coefficients for displacements, moments and shears, to obtain such displacements and tractions under all kinds of anticipated loadings. After materials and cross-sections of each member are designed according to the result of such calculations, vibration and stability problems are analyzed independently. The method explained in this section uses the deflection influence coefficients, the calculation of which is the starting point of any structural analysis and is already used. As seen in examples, the mass $m(i, j)$ is the 'real' value at point (i, j) in the case of an irregular section. Stiffnesses of the irregular sections are already taken into account when $\Delta(i, j, \xi, \eta)$ was calculated.

One may consider that this method is similar to either the 'matrix' method or 'influence coefficient' method. The 'matrix' method, in a narrow sense, is the Vianello method using matrix iteration. In modern practice, almost all calculations involve matrix operation. The 'influence coefficient' method for beams may be closer. In the case of composite plates, the stiffnesses and flexibility matrices are very complex as shown by equations 5.32 to 5.36, 5.67 and others. These equations are different at each point of the plate if the cross-section is not uniform. The method in this section is intended to solve such problems and the concept is directly based on Newton's law.

Figure 7.8 Effect of thickness (first mode).

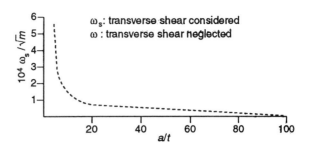

Figure 7.9 Effect of transverse shear (first mode).

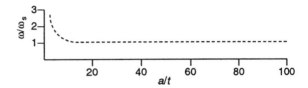

The accuracy of this method is proportional to that of the deflection calculation. It is noted that, in the case of specially orthotropic plates, the first mode natural frequency obtained by the first cycle of calculation at near the center point, $\omega(2, 2)^{(1)}$, is only 4.7% from the 'exact' value. For the second mode, $\omega(2, 2)^{(1)}$ is only 1.4% off.

The relation between the deflection and the accuracy is discussed in section 4.2.11 [23].

This method is used to study the effect of thickness and transverse shear on the natural frequency [20–22, 24, 25]. The result is shown in Figures 7.8 and 7.9.

The numerical procedure for columns explained in section 4.1.5 is a good method for columns with nonuniform cross-section. This method may be readily extended to two-dimensional problems in the near future.

7.9 Simple method of analysis for preliminary design of the composite laminated primary structures for civil construction

7.9.1 Introduction

In Chapter 10, it is recommended to use the quasi-isotropic constants by Tsai for the preliminary design of the composite primary structures for civil construction. Such structures generally require a large number of laminae. This concept greatly simplifies the calculation effort at the early stage of design because

1. The classical mechanics and elasticity theories can be used.
2. There is no coupling between the bending and the mid-plane extension, reducing the three simultaneous fourth-order partial differential equations to one fourth-order partial differential equation.
3. At the preliminary design stage, the orientations of laminae in a laminate are not known. This fact discourages most engineers from the beginning. Use of the quasi-isotropic constants gives a guideline toward a simple and accurate analysis.

We seek simple equations which can predict 'exact' values of the deflections, the natural frequencies of vibration and the critical buckling loads for the specially orthotropic laminates. Many laminates with certain orientations have decreasing values of B_{16}, B_{26}, D_{16} and D_{26} as the number of plies increases. Such laminates can be solved by the same equation as for specially orthotropic laminates. If the quasi-isotropic constants are used, the equations for the isotropic plates can be used. Use of some coefficients representing the anisotropy of the laminates can produce 'exact' values for laminates with such a configuration [26–30]. Most of the structures for civil construction require many layers of plies even though the ratio of the thickness to the length is small, so that the effect of transverse shear deformation can be neglected.

7.9.2 Deflection of laminated plates

Rectangular specially orthotropic laminated plates
If all four edges are simply supported, the expression for the lateral deflection can be obtained by equation 7.31 as

$$w(x, y) = \sum_{m=1}^{\infty} \sum_{n=1}^{\infty} w_{mn} \sin\left(\frac{m\pi x}{a}\right)\sin\left(\frac{n\pi y}{b}\right)$$

where

$$w_{mn} = \frac{q_{mn}}{\pi^4(\text{DEN})} \tag{7.33}$$

in which

$$\text{DEN} = D_1\left(\frac{m}{a}\right)^4 + 2D_3\left(\frac{m}{a}\right)^2\left(\frac{n}{b}\right)^2 + D_2\left(\frac{n}{b}\right)^4 \tag{7.36}$$

where $D_3 = D_{12} + 2D_{66}$, and

$$q_{mn} = \frac{4}{ab} \int_0^a \int_0^b q(x, y)\sin\left(\frac{m\pi x}{a}\right)\sin\left(\frac{n\pi y}{b}\right) dx\, dy \tag{7.34}$$

in which $q(x, y)$ is the applied lateral loading.

Rectangular antisymmetric angle-ply laminated plates with S3 type boundaries
For laminates with such an orientation, $A_{16} = A_{26} = B_{11} = B_{22} = B_{66} = D_{16} = D_{26} = 0$. The equation for the lateral deflection given by Ashton and Whitney (equation 7.120c) is

$$w(x, y) = \sum_{m=1}^{\infty} \sum_{n=1}^{\infty} w_{mn} \sin\left(\frac{m\pi x}{a}\right)\sin\left(\frac{n\pi y}{b}\right)$$

$$w_{mn} = \frac{q_{mn} R^4 b^4}{\pi^4(\text{Den})}[(A_{11}m^2 + A_{66}n^2 R^2)(A_{66}m^2 + A_{22}n^2 R^2)$$

$$- (A_{12} + A_{66})^2 m^2 n^2 R^2] \tag{7.123}$$

where $R = a/b$ and

$$
\begin{aligned}
\text{Den} = {} & [(A_{11}m^2 + A_{66}n^2R^2)(A_{66}m^2 + A_{22}n^2R^2) - (A_{12} + A_{66})^2m^2n^2R^2] \\
& \times [D_{11}m^4 + 2(D_{12} + 2D_{66})m^2n^2R^2 + D_{22}n^4R^4] \\
& + 2m^2n^2R^2(A_{12} + A_{66}) \times (3B_{16}m^2 + B_{26}n^2R^2)(B_{16}m^2 + 3B_{26}n^2R^2) \\
& - n^2R^2(A_{66}m^2 + A_{22}n^2R^2) \times (3B_{16}m^2 + B_{26}n^2R^2)^2 \\
& - m^2(A_{11}m^2 + A_{66}n^2R^2)(B_{16}m^2 + 3B_{26}n^2R^2)^2 \qquad (7.124)
\end{aligned}
$$

If $B_{16} = B_{26} = 0$, we have $u_0 = v_0 = 0$, and equation 7.123 is exactly the same as equation 7.33.

Some orientations with $B_{16} = B_{26} \to 0$
For angle-ply laminates

$$
(B_{16}, B_{26}) = \sum (\bar{Q}_{16}, \bar{Q}_{26})_k(h_k^2 - h_{k-1}^2)
$$

and

$$
(\bar{Q}_{16})_{-\theta} = -(\bar{Q}_{16})_{+\theta}
$$
$$
(\bar{Q}_{26})_{-\theta} = -(\bar{Q}_{26})_{+\theta}
$$

Therefore, for the antisymmetric angle plies, as the number of layers is increased, $B_{16} \to 0$ and $B_{26} \to 0$. Some other laminates with certain orientations will have a similar situation with $D_{16} = D_{26} \to 0$. For such cases, the deflection equation is the same as the case of specially orthotropic laminates.

DEN, the denominator
Consider DEN in equation 7.36:

$$
\begin{aligned}
\text{DEN} &= D_1\left(\frac{m}{a}\right)^4 + 2D_3\left(\frac{m}{a}\right)^2\left(\frac{n}{b}\right)^2 + D_2\left(\frac{n}{b}\right)^4 \\
&= \left(\frac{m}{a}\right)^4\left[D_1 + 2D_3\left(\frac{na}{mb}\right)^2 + D_2\left(\frac{na}{mb}\right)^4\right] \\
&= \left(\frac{m}{a}\right)^4[(D_1 + 2D_3 + D_2 + 2D_3(r^2 - 1) + D_2(r^4 - 1)] \\
&= \left(\frac{m}{a}\right)^4(\text{DENGN})
\end{aligned} \qquad (7.125)
$$

in which

$$
r = na/(mb) \quad \text{and} \quad \text{DENGN} = \text{DEN}/(m/a)^4 \qquad (7.126)
$$

When the normalized stiffnesses are used,

$$
\text{DENNM} = (12/h^3)(\text{DENGN}) = D_1^* + 2D_3^* + D_2^* + 2D_3^*(r^2 - 1) + D_2^*(r^4 - 1), \qquad (7.127)
$$

where $D^* = 12D/h^3$.

Consider the ith ply,

$$D_{1,i}^* = \frac{h_i^3}{12} \frac{12}{h_i^3} [U_1 + U_2 \cos 2\theta_i + U_3 \cos 4\theta_i]$$

$$D_{2,i}^* = U_1 - U_2 \cos 2\theta_i + U_3 \cos 4\theta_i$$

$$2D_{3,i}^* = 2[U_1 - 3U_3 \cos 4\theta_i]$$

where U_is are as given by equations 5.43 and 5.44. Hence

$$D_{1,i}^* + 2D_{3,i}^* + D_{2,i}^* = 4[U_1 - U_3 \cos 4\theta_i] \tag{7.128}$$

$$(\text{DENNM})_i = 4[U_1 - U_3 \cos 4\theta_i] + 2[U_1 - 3U_3 \cos 4\theta_i](r^2 - 1)$$
$$+ [U_1 - U_2 \cos 2\theta_i + U_3 \cos 4\theta_i](r^4 - 1) \tag{7.129}$$

If the ply thicknesses and material properties are equal, we can have

$$\text{DENGN} = (h^3/12)[4(U_1 - U_3(\text{CTH4})) + 2(U_1 - 3U_3(\text{CTH4}))(r^2 - 1)$$
$$+ (U_1 - U_2(\text{CTH2}) + U_3(\text{CTH4}))(r^4 - 1)]$$

where

$$\text{CTH2} = \left(\frac{h_0}{h}\right)^3 \sum_{i=1}^{N} \cos 2\theta_i[i^3 - (i-1)^3)] \tag{7.130}$$

$$\text{CTH4} = \left(\frac{h_0}{h}\right)^3 \sum_{i=1}^{N} \cos 4\theta_i[i^3 - (i-1)^3)] \tag{7.131}$$

in which h_0 is the thickness of each ply.

If $r = 1$,

$$\text{DENR1} = 4(U_1 - U_3(\text{CTH4})) \tag{7.132}$$

In the case of the specially orthotropic laminates,

$$\text{DENHR1} = 4(U_1 - U_3) \tag{7.133}$$

Use of the quasi-isotropic constants

The quasi-isotropic constants, by Tsai, are

$$[Q]^{\text{iso}} = \begin{bmatrix} U_1 & U_4 & 0 \\ U_4 & U_1 & 0 \\ 0 & 0 & U_5 \end{bmatrix} \tag{7.134}$$

When the quasi-isotropic constants are used,

$$D_{11} = D_{22} = D_{12} + 2D_{66} = D_3 = (h^3/12)Q_{11}^{\text{iso}} \tag{7.135}$$

and equation 7.36 reduces to

$$\text{DEN}^{\text{iso}} = D_{11}\left[\left(\frac{m}{a}\right)^4 + 2\left(\frac{m}{a}\right)^2\left(\frac{n}{b}\right)^2 + \left(\frac{n}{b}\right)^4\right] \tag{7.136}$$

which is identical to the case of an isotropic plate.

Equation 7.125 can be written as

$$\text{DEN}^{\text{iso}} = \left(\frac{m}{a}\right)^4 \frac{h^3}{12} U_1[4 + 2(r^2 - 1) + (r^4 - 1)] \tag{7.137a}$$

If $r = 1$,

$$\text{DEN}^{\text{iso}} = \left(\frac{m}{a}\right)^4 \frac{h^3}{12} (4U_1) \qquad (7.137b)$$

One may obtain the initial deflection for the preliminary design by the use of equations 7.31, 7.33 and 7.137.

Use of correction factor to obtain the 'exact' solution

When thick laminates are used for civil construction, a considerable number of orientations will have rapidly decreasing quantities of B_{ij}, D_{16} and D_{26} as the number of layers increases, for which cases equation 7.33 can be used with good accuracy. For such cases one may start to use equation 7.137.

Relatively 'exact' values can be obtained from the preliminary design stage by the use of the formulas given as follows. With

$$w_{mn}^{\text{iso}} = \frac{q_{mn}}{\pi^4 \text{DEN}^{\text{iso}}} \qquad (7.138)$$

we define

$$w_{mn} = w_{mn}^{\text{iso}}/\text{FRC}^2 \qquad (7.139)$$

and

$$\text{MDEN}^{\text{iso}} = \text{DEN}^{\text{iso}}(a/m)^4$$

where

$$\text{FRC}(1)^2 = \frac{[D_1 + 2D_3 + D_2 + 2D_3(r^2 - 1) + D_2(r^4 - 1)]}{(h^3/12)U_1[4 + 2(r^2 - 1) + (r^4 - 1)]} = \frac{\text{DENGN}}{\text{MDEN}^{\text{iso}}} \qquad (7.140)$$

$$\text{FRC}(2)^2 = \frac{[D_1^* + 2D_3^* + D_2^* + 2D_3^*(r^2 - 1) + D_2^*(r^4 - 1)]}{U_1[4 + 2(r^2 - 1) + (r^4 - 1)]} \qquad (7.141)$$

$$\text{FRC}(3)^2 = [4(U_1 - U_3(\text{CTH4})) + 2(U_1 - 3U_3(\text{CTH4}))(r^2 - 1)$$
$$+ (U_1 - U_2(\text{CTH2}) + U_3(\text{CTH4}))(r^4 - 1)]$$
$$\times 1/\{U_1[4 + 2(r^2 - 1) + (r^4 - 1)]\} \qquad (7.142)$$

At the preliminary design stage, the orientation of each ply is not known. In such a case, one may use the invariants only.

$$\text{FRC}(4)^2 = \frac{[4(U_1 - V_3) + 2(U_1 - 3U_3)(r^2 - 1) + (U_1 - U_2 + U_3)(r^4 - 1)]}{U_1[4 + 2(r^2 - 1) + (r^4 - 1)]} \qquad (7.143)$$

Design steps using the correction factors
1. Decide the material properties and obtain the U_is.
2. Obtain w_{mn}^{iso} by equations 7.33 and 7.137 and $w_{mn} = w_{mn}^{\text{iso}}/\text{FRC}(4)^2$ as given by equation 7.139.
3. Proceed to analyze the whole structural system.
4. With the result of 3, design the orientations.
5. Obtain the exact deflection by

$$w_{mn} = w_{mn}^{\text{iso}}/\text{FRC}(1)^2,$$

$$w_{mn} = w_{mn}^{\text{iso}}/\text{FRC}(2)^2 \text{ or}$$

$$w_{mn} = w_{mn}^{\text{iso}}/\text{FRC}(3)^2.$$

In reality, use of FRC(1) at this stage may be simpler.

7.9.3 Eigenvalue problems

The concept developed in the previous section can be extended to the eigenvalue problems of laminated composite primary structures for civil construction as long as $D_{16} = D_{26} \to 0$, and $B_{16} \to 0$, $B_{26} \to 0$ as the number of plies increases.

Specially orthotropic rectangular laminates with simple supports
The natural frequency of vibration and the critical buckling strength are given as

$$\omega_n^2 = \frac{\pi^4}{\rho h} \text{(DEN)} \tag{7.144}$$

$$N_{x_{cr}} = -\left(\frac{\pi a}{m}\right)^2 \text{(DEN)} \tag{7.145}$$

Rectangular antisymmetric angle-ply laminated plates with S3 type boundaries
Consider equations 7.113 and 7.92 which can be expressed as

$$\omega_n^2 = \frac{1}{\rho h} (T_{123}) \tag{7.146}$$

$$N_{x_{cr}} = -\left(\frac{a}{m\pi}\right)^2 (T_{123}) \tag{7.147}$$

where

$$T_{123} = T_{33} + \frac{2T_{12}T_{23}T_{13} - T_{22}T_{13}^2 - T_{11}T_{23}^2}{T_{11}T_{22} - T_{12}^2} \tag{7.148}$$

in which

$$T_{11} = A_{11}\left(\frac{m\pi}{a}\right)^2 + A_{66}\left(\frac{n\pi}{b}\right)^2$$

$$T_{12} = (A_{12} + A_{66})\left(\frac{m\pi}{a}\right)\left(\frac{n\pi}{b}\right)$$

$$T_{13} = -\left[3B_{16}\left(\frac{m\pi}{a}\right)^2 + B_{26}\left(\frac{n\pi}{b}\right)^2\right]\left(\frac{n\pi}{b}\right)$$

$$T_{22} = A_{22}\left(\frac{n\pi}{b}\right)^2 + A_{66}\left(\frac{m\pi}{a}\right)^2 \tag{7.149}$$

$$T_{23} = -\left[B_{16}\left(\frac{m\pi}{a}\right)^2 + 3B_{26}\left(\frac{n\pi}{b}\right)^2\right]\left(\frac{m\pi}{a}\right)$$

$$T_{33} = D_{11}\left(\frac{m\pi}{a}\right)^4 + 2(D_{12} + 2D_{66})\left(\frac{m\pi}{a}\right)^2\left(\frac{n\pi}{b}\right)^2 + D_{22}\left(\frac{n\pi}{b}\right)^4 = \pi^4\text{(DEN)}$$

Design steps by the use of simple method
Laminates with some orientations may have $D_{16} = D_{26} = 0$, and $B_{16} \to 0$ and $B_{26} \to 0$ as the number of plies is increased. For such cases, the solutions for

the eigenvalue problems will be similar to the case of specially orthotropic laminates.

The quasi-isotropic constants given by equation 7.134 are used. DEN^{iso} is as given by equation 7.137. One may obtain the initial values by

$$(\omega_n^{iso})^2 = \frac{\pi^4}{\rho h} (DEN^{iso}) \tag{7.150}$$

$$N_{x_{cr}}^{iso} = -\left(\frac{\pi a}{m}\right)^2 (DEN^{iso}) \tag{7.151}$$

We obtain the 'exact' solutions as

$$(\omega_n)^2 = (\omega_n^{iso})^2 (FRC^2), \tag{7.152}$$

$$N_{x_{cr}} = N_{x_{cr}}^{iso} (FRC^2) \tag{7.153}$$

The FRCs defined in the previous section are also used for the eigenvalue problems. Design steps using 'exact' values from the preliminary stage are the same except that the $(\omega_n^{iso})^2$ and $N_{x_{cr}}^{iso}$ are multiplied by FRC^2. In the case of deflection, w_{mn}^{iso} is divided by FRC^2.

7.9.4 Numerical examples
The material properties and the geometries of the laminates used are as follows:

$$E_1 = 38.6\,GPa$$
$$E_2 = 8.27\,GPa$$
$$v_{12} = 0.26$$
$$v_{21} = 0.0557$$
$$h_0 = 0.00125\,m \quad \text{and} \quad h_0 = 0.000125\,m$$
$$G_{12} = 4.14\,GPa$$
$$a = b = 1\,m$$

Note that in this case, h_0 can be the thickness of a group of plies with the same orientation and thickness. The results of calculation are as shown in Tables 7.1 to 7.12.

7.10 Possibility of using classical mechanics for the preliminary design of laminated composite structures for civil construction

7.10.1 Importance of the subject problem
The highest specific strength and stiffness of composites can be obtained by arranging long fiber reinforcements in a parallel fashion, and forming a laminate made of several laminae. Design and analysis of a laminate is so complicated that a considerable number of structural engineers are simply allergic to composite design. In analysis, even boundary conditions are not so simple as with classical mechanics or elasticity. Even simple and clamped boundaries each have four possible types.

Table 7.1 $[\pm\theta]_r$, $\theta = \pm 15°$, $\omega_n = \omega(\text{real})\sqrt{\rho h}$, $h_0 = 0.00125$ m

	r(N)						
	3(6)	6(12)	9(18)	12(24)	15(30)	21(42)	27(54)
ω_n^{exact}	501	1430	2632	4055	5668	9392	13 694
ω_n^{orth}	507	1434	2635	4058	5671	9395	13 696
ω_n^{iso}	529	1497	2750	4234	5917	9802	14 290
B_{16}^*/D_{11}^*	0.0316	0.0158	0.0104	0.0079	0.0062	0.0045	0.0033
B_{26}^*/D_{11}^*	0.0045	0.0023	0.0015	0.0011	0.0009	0.0006	0.00048
FRC(1)	0.9584	0.9584	0.9584	0.9584	0.9584	0.9584	0.9584
FRC(4)	0.9149	0.9149	0.9149	0.9149	0.9149	0.9149	0.9149
ω_n^{iso} FRC(1)	507	1434	2635	4058	5671	9394	13 696
ω_n^{iso} FRC(4)	484	1369	2516	3874	5413	8968	13 074
$\dfrac{\omega_n^{exact}}{\omega_n^{orth}}$	0.988	0.997	0.998	0.999	0.999	0.999	0.999
$\dfrac{\omega_n^{exact}}{\omega_n^{iso}}$	0.947	0.955	0.957	0.958	0.958	0.958	0.958

Table 7.2 $[ABBAAB]_r$, $A = 15°$, $B = -15°$, $\omega_n = \omega(\text{real})\sqrt{\rho h}$, $h_0 = 0.00125$ m

	r(N)						
	1(6)	2(12)	3(18)	4(24)	5(30)	7(42)	9(54)
ω_n^{exact}	506	1434	2635	4057	5671	9394	13 696
ω_n^{orth}	507	1434	2635	4058	5671	9395	13 696
ω_n^{iso}	529	1497	2750	4234	5917	9802	14 290
B_{16}^*/D_{11}^*	0.0104	0.005	0.0033	0.0026	0.0021	0.0015	0.0011
B_{26}^*/D_{11}^*	0.0015	0.00073	0.00048	0.00036	0.0003	0.00021	0.00016
FRC(1)	0.9584	0.9584	0.9584	0.9584	0.9584	0.9584	0.9584
FRC(4)	0.9149	0.9149	0.9149	0.9149	0.9149	0.9149	0.9149
ω_n^{iso} [FRC(1)]	507	1434	2635	4058	5671	9394	13 694
ω_n^{iso} [FRC(4)]	484	1369	2516	3874	5413	8968	13 074
$\dfrac{\omega_n^{exact}}{\omega_n^{orth}}$	0.998	0.999	0.999	0.999	0.999	0.999	0.999
$\dfrac{\omega_n^{exact}}{\omega_n^{iso}}$	0.956	0.958	0.958	0.958	0.958	0.958	0.958

Table 7.3 $[\pm\theta]_r$, $\theta = \pm 15°$, $h_0 = 0.00125$ m

	$r(N)$						
	3(6)	6(12)	9(18)	12(24)	15(30)	21(42)	27(54)
$N_{\mathrm{xcr}}^{(\mathrm{exact})}$	−25 707	−207 848	−702 858	−1 667 169	−3 257 215	−8 940 249	−19 003 430
$N_{\mathrm{xcr}}^{(\mathrm{orth})}$	−26 072	−208 578	−703 952	−1 668 628	−3 259 040	−8 942 803	−19 006 710
$N_{\mathrm{xcr}}^{(\mathrm{iso})}$	−28 382	−227 062	−766 335	−1 816 493	−3 547 847	−9 735 293	−20 691 040
B_{16}^*/D_{11}^*	0.0316	0.0158	0.0104	0.0079	0.0062	0.0045	0.0033
B_{26}^*/D_{11}^*	0.0045	0.0023	0.0015	0.0011	0.0009	0.0006	0.00048
$\mathrm{FRC}^2(1)$	0.91859	0.91859	0.91859	0.91859	0.91859	0.91859	0.91859
$\mathrm{FRC}^2(4)$	0.8372	0.8372	0.8372	0.8372	0.8372	0.8372	0.8372
$N_{\mathrm{xcr}}^{(\mathrm{iso})}[\mathrm{FRC}^2(1)]$	−26 071	−208 577	−703 948	−1 668 612	−3 259 017	−8 942 743	−19 006 582
$N_{\mathrm{xcr}}^{(\mathrm{iso})}[\mathrm{FRC}^2(4)]$	−23 761	−190 096	−641 575	−1 520 772	−2 970 257	−8 150 389	−17 322 538
$\dfrac{N_{\mathrm{xcr}}^{(\mathrm{exact})}}{N_{\mathrm{xcr}}^{(\mathrm{orth})}}$	0.986	0.996	0.998	0.999	0.999	0.999	0.999
$\dfrac{N_{\mathrm{xcr}}^{(\mathrm{exact})}}{N_{\mathrm{xcr}}^{(\mathrm{iso})}}$	0.901	0.915	0.917	0.918	0.918	0.918	0.918

Table 7.4 $[ABBAAB]_r$, $A = 15°$, $B = -15°$, $h_0 = 0.00125$ m

	$r(N)$						
	1(6)	2(12)	3(18)	4(24)	5(30)	7(42)	9(54)
$N_{\mathrm{xcr}}^{\mathrm{exact}}$	−26 031	−208 431	−703 831	−1 668 466	−3 258 837	−8 942 519	−19 006 350
$N_{\mathrm{xcr}}^{\mathrm{orth}}$	−26 072	−208 579	−703 953	−1 668 628	−3 259 040	−8 942 803	−19 006 710
$N_{\mathrm{xcr}}^{\mathrm{iso}}$	−28 382	−227 062	−766 335	−1 816 493	−3 547 847	−9 735 293	−20 691 040
B_{16}^*/D_{11}^*	0.0104	0.005	0.0033	0.0026	0.0021	0.0015	0.0011
B_{26}^*/D_{11}^*	0.0015	0.00073	0.00048	0.00036	0.0003	0.00021	0.00016
$\mathrm{FRC}^2(1)$	0.91859	0.91859	0.91859	0.91859	0.91859	0.91859	0.91859
$\mathrm{FRC}^2(4)$	0.8372	0.8372	0.8372	0.8372	0.8372	0.8372	0.8372
$N_{\mathrm{xcr}}^{\mathrm{iso}}\mathrm{FRC}^2(1)$	−26 071	−208 577	−703 948	−1 668 623	−3 259 017	−8 942 743	−19 006 582
$N_{\mathrm{xcr}}^{\mathrm{iso}}\mathrm{FRC}^2(4)$	−23 761	−190 096	−641 575	−1 520 772	−2 970 257	−8 150 389	−17 322 538
$\dfrac{N_{\mathrm{xcr}}^{\mathrm{exact}}}{N_{\mathrm{xcr}}^{\mathrm{orth}}}$	0.998	0.999	0.999	0.999	0.999	0.999	0.999
$\dfrac{N_{\mathrm{xcr}}^{\mathrm{exact}}}{N_{\mathrm{xcr}}^{\mathrm{iso}}}$	0.917	0.918	0.918	0.919	0.919	0.919	0.919

Table 7.5 $[\pm\theta]_r$, $\theta = \pm 15°$, $\omega_n = \omega(\text{real})\sqrt{\rho h}$, $h_0 = 0.000125$ m

	r(N)						
	3(6)	6(12)	9(18)	12(24)	15(30)	21(42)	27(54)
ω_n^{exact}	15.93	45.29	83.29	128.3	179.3	297.04	433.08
ω_n^{orth}	16.04	45.37	83.35	128.3	179.3	297.09	433.11
ω_n^{iso}	16.73	47.34	86.97	133.9	187.1	309.97	451.9
B_{16}^*/D_{11}^*	0.0316	0.0158	0.0104	0.0079	0.0062	0.0045	0.0033
B_{26}^*/D_{11}^*	0.0045	0.0023	0.0015	0.0011	0.0009	0.0006	0.00048
FRC(1)	0.9584	0.9584	0.9584	0.9584	0.9584	0.9584	0.9584
FRC(4)	0.9149	0.9149	0.9149	0.9149	0.9149	0.9149	9.9149
ω_n^{iso} FRC(1)	16.034	45.37	83.35	128.3	179.3	297.08	433.10
ω_n^{iso} FRC(4)	15.306	43.311	79.568	122.50	171.17	283.6	413.53
$\dfrac{\omega_n^{\text{exact}}}{\omega_n^{\text{orth}}}$	0.9931	0.9982	0.9993	1.000	1.000	1.000	1.000
$\dfrac{\omega_n^{\text{exact}}}{\omega_n^{\text{iso}}}$	0.9522	0.9567	0.9577	0.9582	0.958	0.958	0.958

Table 7.6 $[ABBAAB]_r$, $A = 15°$, $B = -15°$, $\omega_n = \omega(\text{real})\sqrt{\rho h}$, $h_0 = 0.000125$ m

	r(N)						
	1(6)	2(12)	3(18)	4(24)	5(30)	7(42)	9(54)
ω_n^{exact}	16.03	45.36	83.34	128.3	179.3	297	433
ω_n^{orth}	16.04	45.37	83.35	128.3	179.3	297	433
ω_n^{iso}	16.73	47.34	86.97	133.9	187.1	309.97	451.9
B_{16}^*/D_{11}^*	0.0104	0.005	0.0033	0.0026	0.0021	0.0015	0.0011
B_{26}^*/D_{11}^*	0.0015	0.00073	0.00048	0.00036	0.0003	0.00021	0.00016
FRC(1)	0.9584	0.9584	0.9584	0.9584	0.9584	0.9584	0.9584
FRC(4)	0.9149	0.9149	0.9149	0.9149	0.9149	0.9149	9.9149
ω_n^{iso} FRC(1)	16.034	45.37	83.35	128.3	179.3	297.08	433.1
ω_n^{iso} FRC(4)	15.306	43.311	79.568	122.50	171.17	283.6	413.53
$\dfrac{\omega_n^{\text{exact}}}{\omega_n^{\text{orth}}}$	0.9994	0.9998	0.9999	1.0000	1.0000	1.0000	1.0000
$\dfrac{\omega_n^{\text{exact}}}{\omega_n^{\text{iso}}}$	0.9582	0.9582	0.9582	0.9582	0.9582	0.9582	0.9582

Table 7.7 $[\pm\theta]_T$, $\theta = \pm 15°$, $h_0 = 0.000125$ m

	$r(N)$						
	3(6)	6(12)	9(18)	12(24)	15(30)	21(42)	27(54)
N_{xcr}^{exact}	−25.708	−207.849	−702.858	−1667.169	−3257.22	−8940.250	−19 003.430
N_{xcr}^{orth}	−26.072	−208.579	−703.953	−1668.629	−3259.04	−8942.805	−19 006.710
N_{xcr}^{iso}	−28.383	−227.062	−766.335	−1816.498	−3547.85	−9735.295	−20 691.050
B_{16}^*/D_{11}^*	0.0316	0.0158	0.0104	0.0079	0.0062	0.0045	0.0033
B_{26}^*/D_{11}^*	0.0045	0.0023	0.0015	0.0011	0.0009	0.0006	0.00048
$FRC^2(1)$	0.91859	0.91859	0.91859	0.91859	0.91859	0.91859	0.91859
$FRC^2(4)$	0.8372	0.8372	0.8372	0.8372	0.8372	0.8372	0.8372
$N_{xcr}^{iso}FRC^2(1)$	−26.072	−208.577	−703.948	−1668.612	−3259.02	−8942.743	−19 006.582
$N_{xcr}^{iso}FRC^2(4)$	−23.762	−190.096	−641.575	−1520.772	−2970.26	−8150.389	−17 322.547
$\dfrac{N_{xcr}^{exact}}{N_{xcr}^{orth}}$	0.986	0.996	0.998	0.999	0.999	0.999	0.999
$\dfrac{N_{xcr}^{exact}}{N_{xcr}^{iso}}$	0.901	0.915	0.917	0.918	0.918	0.918	0.918

Table 7.8 $[ABBAAB]_T$, $A = 15°$, $B = -15°$, $h_0 = 0.000125$ m

	$r(N)$						
	1(6)	2(12)	3(18)	4(24)	5(30)	7(42)	9(54)
N_{xcr}^{exact}	−26.032	−208.438	−703.831	−1668.466	−3258.84	−8942.519	−19 006.350
N_{xcr}^{orth}	−26.072	−208.579	−703.953	−1668.628	−3259.04	−8942.803	−19 006.710
N_{xcr}^{iso}	−28.383	−227.062	−766.335	−1816.498	−3547.85	−9735.295	−20 691.050
B_{16}^*/D_{11}^*	0.0104	0.005	0.0033	0.0026	0.0021	0.0015	0.0011
B_{26}^*/D_{11}^*	0.0015	0.00073	0.00048	0.00036	0.0003	0.00021	0.00016
$FRC^2(1)$	0.91859	0.91859	0.91859	0.91859	0.91859	0.91859	0.91859
$FRC^2(4)$	0.8372	0.8372	0.8372	0.8372	0.8372	0.8372	0.8372
$N_{xcr}^{iso}FRC^2(1)$	−26.072	−208.577	−703.952	−1668.623	−3259.02	−8942.743	−19 006.582
$N_{xcr}^{iso}FRC^2(4)$	−23.762	−190.096	−641.575	−1520.772	−2970.26	−8150.389	−17 322.547
$\dfrac{N_{xcr}^{exact}}{N_{xcr}^{orth}}$	0.9984	0.9996	0.9998	0.9999	0.9999	1.0000	1.0000
$\dfrac{N_{xcr}^{exact}}{N_{xcr}^{iso}}$	0.9172	0.9182	0.9184	0.9185	0.9185	0.9186	0.9186

FRC(1): by equation 7.140; FRC(4): by equation 7.143.

Table 7.9 $[\pm\theta]_r$, $\theta = \pm 15°$, $h_0 = 0.00125$ m

			$r(N)$			
	3(6)	6(12)	12(24)	15(30)	21(42)	27(54)
$FRC^2(1)$	0.91859	0.91859	0.91859	0.91859	0.91859	0.91859
$FRC^2(2)$	0.91859	0.91859	0.91859	0.91859	0.91859	0.91859
$FRC^2(3)$	0.91859	0.91859	0.91859	0.91859	0.91859	0.91859
$FRC^2(4)$	0.83719	0.83719	0.83719	0.83719	0.83719	0.83719
w_{mn}^{orth}	0.4858×10^{-10}	0.6072×10^{-11}	0.7570×10^{-12}	0.3886×10^{-12}	0.1416×10^{-12}	0.6664×10^{-13}
w_{mn}^{anti}	0.4927×10^{-10}	0.6094×10^{-11}	0.7597×10^{-12}	0.3888×10^{-12}	0.1417×10^{-12}	0.6665×10^{-13}
w_{mn}^{iso}	0.4462×10^{-10}	0.5578×10^{-11}	0.6972×10^{-12}	0.3570×10^{-12}	0.1301×10^{-12}	0.6121×10^{-13}
$\dfrac{w_{mn}^{iso}(1)}{w_{mn}^{orth}}$	1.0	1.0	1.0	1.0	1.0	1.0
$\dfrac{w_{mn}^{iso}(2)}{w_{mn}^{orth}}$	1.0	1.0	1.0	1.0	1.0	1.0
$\dfrac{w_{mn}^{iso}(3)}{w_{mn}^{orth}}$	0.999	1.0	1.0	1.0	1.0	1.0
$\dfrac{w_{mn}^{iso}(4)}{w_{mn}^{orth}}$	0.91138	0.91138	0.91138	0.91138	0.91138	0.91138
$\dfrac{w_{mn}^{iso}}{w_{mn}^{orth}}$	1.08862	1.08862	1.08862	1.08862	1.08862	1.08862
$\dfrac{w_{mn}^{iso}(1)}{w_{mn}^{anti}}$	1.01419	1.00351	1.000875	1.00056	1.000287	1.000172
$\dfrac{w_{mn}^{iso}(2)}{w_{mn}^{anti}}$	1.01419	1.00351	1.000875	1.00056	1.000287	1.000172
$\dfrac{w_{mn}^{iso}(3)}{w_{mn}^{anti}}$	1.01419	1.00351	1.000875	1.00056	1.000287	1.000172
$\dfrac{w_{mn}^{iso}(4)}{w_{mn}^{anti}}$	0.924318	0.91458	0.91218	0.911893	0.911643	0.91154
$\dfrac{w_{mn}^{iso}}{w_{mn}^{anti}}$	1.10468	1.092439	1.08957	1.089226	1.088928	1.08881

Note: $w_{mn}^{iso}(i) = w_{mn}^{iso}/FRC(i)^2$.

Table 7.10 $[ABBAAB]$, $A = 15°$, $B = -15°$, $h_0 = 0.00125$ m

	$r(N)$					
	1(6)	2(12)	4(24)	5(30)	7(42)	9(54)
$FRC^2(1)$	0.91859	0.91859	0.91859	0.91859	0.91859	0.91859
$FRC^2(2)$	0.91859	0.91859	0.91859	0.91859	0.91859	0.91859
$FRC^2(3)$	0.91859	0.91859	0.91859	0.91859	0.91859	0.91859
$FRC^2(4)$	0.83719	0.83719	0.83719	0.83719	0.83719	0.83719
w_{mn}^{orth}	0.4858×10^{-10}	0.6072×10^{-11}	0.7590×10^{-12}	0.3886×10^{-12}	0.1416×10^{-12}	0.6664×10^{-13}
w_{mn}^{anti}	0.4865×10^{-10}	0.6075×10^{-11}	0.7591×10^{-12}	0.3886×10^{-12}	0.1416×10^{-12}	0.6664×10^{-13}
w_{mn}^{iso}	0.4462×10^{-10}	0.5578×10^{-11}	0.6972×10^{-12}	0.4264×10^{-12}	0.1301×10^{-12}	0.6121×10^{-13}
$\dfrac{w_{mn}^{orth}}{w_{mn}^{iso(1)}}$	1.0	1.0	1.0	1.0	1.0	1.0
$\dfrac{w_{mn}^{orth}}{w_{mn}^{iso(2)}}$	1.0	1.0	1.0	1.0	1.0	1.0
$\dfrac{w_{mn}^{orth}}{w_{mn}^{iso(3)}}$	0.999	1.0	1.0	1.0	1.0	1.0
$\dfrac{w_{mn}^{orth}}{w_{mn}^{iso(4)}}$	0.91138	0.91138	0.91138	0.91138	0.91138	0.91138
$\dfrac{w_{mn}^{orth}}{w_{mn}^{iso}}$	1.08862	1.08862	1.08862	1.08862	1.08862	1.08862
$\dfrac{w_{mn}^{anti}}{w_{mn}^{iso(1)}}$	1.001557	1.000388	1.0000972	1.000062	1.0000317	1.000019
$\dfrac{w_{mn}^{anti}}{w_{mn}^{iso(2)}}$	1.001557	1.000388	1.0000972	1.000062	1.0000317	1.000019
$\dfrac{w_{mn}^{anti}}{w_{mn}^{iso(3)}}$	1.001557	1.000388	1.0000972	1.000062	1.0000317	1.000019
$\dfrac{w_{mn}^{anti}}{w_{mn}^{iso(4)}}$	0.912802	0.91173	0.91147	0.911439	0.911439	0.91140
$\dfrac{w_{mn}^{anti}}{w_{mn}^{iso}}$	1.090312	1.08904	1.0887229	1.088684	1.088651	1.088637

Table 7.11 $[\pm\theta]_r$, $\theta = \pm 15°$, $h_0 = 0.000125$ m

			$r(N)$			
	3(6)	6(12)	12(24)	15(30)	21(42)	27(54)
$FRC^2(1)$	0.918596	0.918596	0.918596	0.918596	0.918596	0.918596
$FRC^2(2)$	0.918596	0.918596	0.918596	0.918596	0.918596	0.918596
$FRC^2(3)$	0.918596	0.918596	0.918596	0.918596	0.918596	0.918596
$FRC^2(4)$	0.83719	0.83719	0.83719	0.83719	0.83719	0.83719
w^{orth}_{mn}	0.4858×10^{-7}	0.6072×10^{-8}	0.7590×10^{-9}	0.3886×10^{-9}	0.1416×10^{-9}	0.6664×10^{-10}
w^{anti}_{mn}	0.4927×10^{-7}	0.6094×10^{-8}	0.7597×10^{-9}	0.3888×10^{-9}	0.1417×10^{-9}	0.6665×10^{-10}
w^{iso}_{mn}	0.4462×10^{-7}	0.5578×10^{-8}	0.6972×10^{-9}	0.3570×10^{-9}	0.1301×10^{-9}	0.6121×10^{-10}
$\dfrac{w^{orth}_{mn}}{w^{iso}_{mn}(1)}$	1.0	1.0	1.0	1.0	1.0	1.0
$\dfrac{w^{orth}_{mn}}{w^{iso}_{mn}(2)}$	1.0	1.0	1.0	1.0	1.0	1.0
$\dfrac{w^{orth}_{mn}}{w^{iso}_{mn}(3)}$	0.999	1.0	1.0	1.0	1.0	1.0
$\dfrac{w^{orth}_{mn}}{w^{iso}_{mn}(4)}$	0.91138	0.91138	0.91138	0.91138	0.91138	0.91138
$\dfrac{w^{orth}_{mn}}{w^{iso}_{mn}}$	1.088617	1.08862	1.088617	1.088617	1.088617	1.088617
$\dfrac{w^{anti}_{mn}}{w^{iso}_{mn}(1)}$	1.01419	1.00351	1.000875	1.00056	1.000286	1.000172
$\dfrac{w^{anti}_{mn}}{w^{iso}_{mn}(2)}$	1.01419	1.00351	1.000875	1.00056	1.000286	1.000172
$\dfrac{w^{anti}_{mn}}{w^{iso}_{mn}(3)}$	1.01419	1.00351	1.000875	1.00056	1.000286	1.000172
$\dfrac{w^{anti}_{mn}}{w^{iso}_{mn}(4)}$	0.924318	0.91458	0.91218	0.911893	0.911643	0.91154
$\dfrac{w^{anti}_{mn}}{w^{iso}_{mn}}$	1.10468	1.092439	1.08957	1.089226	1.088928	1.08881

Table 7.12 $[ABBAAB]$, $A = 15°$, $B = -15°$, $h_0 = 0.000125$ m

	$r(N)$					
	1(6)	2(12)	4(24)	5(30)	7(42)	9(54)
$\mathrm{FRC}^2(1)$	0.918596	0.918596	0.918596	0.918596	0.918596	0.918596
$\mathrm{FRC}^2(2)$	0.918596	0.918596	0.918596	0.918596	0.918596	0.918596
$\mathrm{FRC}^2(3)$	0.918596	0.918596	0.918596	0.918596	0.918596	0.918596
$\mathrm{FRC}^2(4)$	0.83719	0.83719	0.83719	0.83719	0.83719	0.83719
w_{mn}^{orth}	0.4858×10^{-7}	0.6072×10^{-8}	0.7590×10^{-9}	0.3886×10^{-9}	0.1416×10^{-9}	0.6664×10^{-10}
w_{mn}^{anti}	0.4865×10^{-7}	0.6075×10^{-8}	0.7591×10^{-9}	0.3886×10^{-9}	0.1416×10^{-9}	0.6664×10^{-10}
w_{mn}^{iso}	0.4462×10^{-7}	0.5578×10^{-8}	0.6972×10^{-9}	0.4264×10^{-9}	0.1301×10^{-9}	0.6121×10^{-10}
$\dfrac{w_{mn}^{orth}}{w_{mn}^{iso}}(1)$	1.0	1.0	1.0	1.0	1.0	1.0
$\dfrac{w_{mn}^{orth}}{w_{mn}^{iso}}(2)$	1.0	1.0	1.0	1.0	1.0	1.0
$\dfrac{w_{mn}^{orth}}{w_{mn}^{iso}}(3)$	0.999	1.0	1.0	1.0	1.0	1.0
$\dfrac{w_{mn}^{orth}}{w_{mn}^{iso}}(4)$	0.91138	0.91138	0.91138	0.91138	0.91138	0.91138
$\dfrac{w_{mn}^{orth}}{w_{mn}^{anti}}$	1.088617	1.088617	1.088617	1.088617	1.088617	1.088617
$\dfrac{w_{mn}^{anti}}{w_{mn}^{iso}}(1)$	1.001557	1.000388	1.0000972	1.000062	1.0000317	1.000019
$\dfrac{w_{mn}^{anti}}{w_{mn}^{iso}}(2)$	1.001557	1.000388	1.0000972	1.000062	1.0000317	1.000019
$\dfrac{w_{mn}^{anti}}{w_{mn}^{iso}}(3)$	1.001557	1.000388	1.0000972	1.000062	1.0000317	1.000019
$\dfrac{w_{mn}^{anti}}{w_{mn}^{iso}}(4)$	0.912802	0.91173	0.91147	0.911439	0.911439	0.91140
$\dfrac{w_{mn}^{anti}}{w_{mn}^{iso}}$	1.090312	1.08904	1.0887229	1.088684	1.088651	1.088637

Even when the transverse shear deformation, and thermal and hygrothermal effects are neglected, the related equations are three simultaneous fourth-order partial differential equations:

$$A_{11}\frac{\partial^2 u}{\partial x^2} + 2A_{16}\frac{\partial^2 u}{\partial x\,\partial y} + A_{66}\frac{\partial^2 u}{\partial y^2} + A_{16}\frac{\partial^2 v}{\partial x^2}\,(A_{12} + A_{66})\frac{\partial^2 v}{\partial x\,\partial y} + A_{26}\frac{\partial^2 v}{\partial y^2}$$

$$- B_{11}\frac{\partial^3 w}{\partial x^3} - 3B_{16}\frac{\partial^3 w}{\partial x^2\,\partial y} - (B_{12} + 2B_{66})\frac{\partial^3 w}{\partial x\,\partial y^2}$$

$$- B_{26}\frac{\partial^3 w}{\partial y^3} = 0 \tag{7.72}$$

$$A_{16}\frac{\partial^2 u}{\partial x^2} + (A_{12} + A_{66})\frac{\partial^2 u}{\partial x\,\partial y} + A_{26}\frac{\partial^2 u}{\partial y^2} + A_{66}\frac{\partial^2 v}{\partial x^2} + 2A_{26}\frac{\partial^2 v}{\partial x\,\partial y} + A_{22}\frac{\partial^2 v}{\partial y^2}$$

$$- B_{16}\frac{\partial^3 w}{\partial x^3} - (B_{12} + 2B_{66})\frac{\partial^3 w}{\partial x^2\,\partial y} - 3B_{26}\frac{\partial^3 w}{\partial x\,\partial y^2}$$

$$- B_{22}\frac{\partial^3 w}{\partial y^3} = 0 \tag{7.73}$$

$$D_{11}\frac{\partial^4 w}{\partial x^4} + 4D_{16}\frac{\partial^4 w}{\partial x^3\,\partial y} + 2(D_{12} + 2D_{66})\frac{\partial^4 w}{\partial x^2\,\partial y^2} + 4D_{26}\frac{\partial^4 w}{\partial x\,\partial y^3}$$

$$+ D_{22}\frac{\partial^4 w}{\partial y^4} - B_{11}\frac{\partial^3 u}{\partial x^3} - 3B_{16}\frac{\partial^3 u}{\partial x^2\,\partial y} - (B_{12} + 2B_{66})\frac{\partial^3 u}{\partial x\,\partial y^2}$$

$$- B_{26}\frac{\partial^3 u}{\partial y^3} - B_{16}\frac{\partial^3 v}{\partial x^3} - (B_{12} + 2B_{66})\frac{\partial^3 v}{\partial x^2\,\partial y} - 3B_{26}\frac{\partial^3 v}{\partial x\,\partial y^2}$$

$$- B_{22}\frac{\partial^3 v}{\partial y^3} = q(x,\,y) \tag{7.74}$$

Considerable simplification can be made in preliminary analysis [31], if

1. classical mechanics and elasticity theories can be used;
2. the bending–extension coupling matrix, B_{ij}, vanishes so that the related equation becomes one fourth order partial differential equation.

7.10.2 Possibility of simplified approaches

The classical theories and formulas can be used if the normalized extensional stiffness equals the normalized bending stiffness, that is

$$A^* = D^*$$

where

$$A^* = A/h \qquad \text{in GPa}$$
$$B^* = 2B/h^2 \qquad \text{in GPa} \tag{7.154}$$
$$D^* = 12D/h^3 \qquad \text{in GPa}$$

in which

$$A_{ij} = \sum_{k=1}^{n} (\bar{Q}_{ij})_k (h_k - h_{k-1}) \qquad (5.61)$$

$$B_{ij} = \frac{1}{2} \sum_{k=1}^{n} (\bar{Q}_{ij})_k (h_k^2 - h_{k-1}^2) \qquad (5.62)$$

$$D_{ij} = \frac{1}{3} \sum_{k=1}^{n} (\bar{Q}_{ij})_k (h_k^3 - h_{k-1}^3) \qquad (5.63)$$

where h is the thickness of the laminate, and $\bar{Q}_{ij}$ is the off-axis stiffness matrix given as

$$\bar{Q}_{11} = Q_{11}m^4 + 2(Q_{12} + 2Q_{66})m^2n^2 + Q_{22}n^4$$

$$\bar{Q}_{12} = (Q_{11} + Q_{22} - 4Q_{66})m^2n^2 + Q_{12}(m^4 + n^4)$$

$$\bar{Q}_{13} = Q_{13}m^2 + Q_{23}n^2$$

$$\bar{Q}_{16} = -Q_{22}mn^3 + Q_{11}m^3n - (Q_{12} + 2Q_{66})mn(m^2 - n^2)$$

$$\bar{Q}_{22} = Q_{11}n^4 + 2(Q_{12} + 2Q_{66})m^2n^2 + Q_{22}m^4$$

$$\bar{Q}_{23} = Q_{13}n^2 + Q_{23}m^2$$

$$\bar{Q}_{26} = -Q_{22}m^3n + Q_{11}mn^3 + (Q_{12} + 2Q_{66})mn(m^2 - n^2) \qquad (5.32)$$

$$\bar{Q}_{33} = Q_{33}$$

$$\bar{Q}_{36} = (Q_{13} - Q_{23})mn$$

$$\bar{Q}_{44} = Q_{44}m^2 + Q_{55}n^2$$

$$\bar{Q}_{45} = (Q_{55} - Q_{44})mn$$

$$\bar{Q}_{55} = Q_{55}m^2 + Q_{44}n^2$$

$$\bar{Q}_{66} = (Q_{11} + Q_{22} - 2Q_{12})m^2n^2 + Q_{66}(m^2 - n^2)^2$$

in which Q_{ij} is given as

$$Q_{11} = \frac{E_1}{1 - v_{12}v_{21}}$$

$$Q_{12} = \frac{v_{12}E_2}{1 - v_{12}v_{21}} = \frac{v_{21}E_1}{1 - v_{12}v_{21}} = Q_{21} \qquad (5.21)$$

$$Q_{22} = \frac{E_2}{1 - v_{12}v_{21}}$$

$$Q_{66} = G_{12}$$

and $m = \cos \alpha$ and $n = \sin \alpha$, where α is the angle of the transformation. Note that the laminate axes are rotated by α clockwise from the ply material axes.

It is generally known that the bending–extension coupling matrix, $[B]$, vanishes only if the cross-section of a laminate is symmetric, in material, geometry and orientation, with respect to its mid-surface. However, a sufficient condition to eliminate the bending–extension coupling is that the sum of the normalized weighting factors of each group of orientation is equal to zero [32–34]. In addition to such conditions, an increase of the number of layers for certain orientations for such as the thick laminates of the primary structures in civil construction may result in a negligibly small B-matrix.

7.10.3   Quasi-isotropic concept

In Chapter 10, the author proposes to use the quasi-isotropic constants by Tsai for the preliminary design of the composite primary structures for civil construction. This concept is indirectly supported by the recent paper of Verchery *et al.* [32].

Every anisotropic material has quasi-isotropic constants derived from the invariants of coordinate transformation. These constants represent the lower bound of each composite performance, and are given by Tsai [35] as

$$[Q]^{\text{iso}} = \begin{vmatrix} U_1 & U_4 & 0 \\ U_4 & U_1 & 0 \\ 0 & 0 & U_5 \end{vmatrix}$$

where

$$\begin{aligned}
U_1 &= \tfrac{1}{8}(3Q_{xx} + 3Q_{yy} + 2Q_{xy} + 4Q_{ss}) \\
U_4 &= \tfrac{1}{8}(Q_{xx} + Q_{yy} + 6Q_{xy} - 4Q_{ss}) = U_1 - 2U_5 \qquad (5.43) \\
U_5 &= \tfrac{1}{8}(Q_{xx} + Q_{yy} - 2Q_{xy} + 4Q_{ss})
\end{aligned}$$

Note that $Q_{xx} = Q_{11}$, $Q_{yy} = Q_{22}$, $Q_{xy} = Q_{12}$ and $Q_{ss} = Q_{66}$. When quasi-isotropic constants are used we always have $A^* = D^*$, $B^* = 0$.

7.10.4   Numerical examples

Several laminate configurations with different orientations and numbers of layers are examined in order to

1. study the validity of the use of the quasi-isotropic constants;
2. find the laminates with $A^* = D^*$, and $B_{ij} \approx 0$.

The results are rather promising as follows. The material properties used are:

$$E_{\text{m}} = 3.4\,\text{GPa}$$
$$E_{\text{f}} = 110\,\text{GPa}$$
$$v_{\text{m}} = 0.35$$
$$v_{\text{f}} = 0.22$$
$$V_{\text{m}} = 0.4$$
$$V_{\text{f}} = 0.6$$

From these values, we obtain

$$E_1 = 67.36 \, \text{GPa}$$

$$E_2 = 8.12 \, \text{GPa}$$

$$\nu_{12} = 0.272$$

$$\nu_{21} = 0.0328$$

$$G_{12} = 3.02 \, \text{GPa}$$

$$h_0 = 0.005 \, \text{m}$$

Note that in this case, h_0 can be the thickness of each group of plies with the same orientation and thickness.

Quasi-homogeneous laminates $(A^* = D^*)$
1. Quasi-isotropic constants are used, $A^* = D^*$.
2. Angle ply laminates, $[\pm \theta]_r$, $A^* = D^*$ (Table 7.13).

Table 7.13 $[\pm \theta]_r$, $\theta = \pm 15°$

				r			
	3	5	8	11	14	17	20
A_{11}^*/D_{11}^*	1	1	1	1	1	1	1
B_{16}^*/D_{11}^*	0.039	0.023	0.014	0.01	0.008	0.007	0.0056
B_{26}^*/D_{11}^*	0.0026	0.0016	0.001	0.0007	0.0006	0.0005	0.0004
D_{11}/D_{11}^{iso}	1.96	1.96	1.96	1.96	1.96	1.96	1.96

3. Quasi-isotropic orientation, $[90/+45/-45/0]_r$, $A^* \approx D^*$, A_{11}^* is constant (Table 7.14). The differences are 4.4% when $r = 2$ and 0.2% when $r = 9$.

Table 7.14 $[90/+45/-45/0]_r$

				r			
	1	2	3	4	5	6	7	9
A_{11}^*/D_{11}^*	0.845	0.956	0.98	0.99	0.993	0.995	0.996	0.997
B_{11}^*/D_{11}^*	0.31	0.18	0.12	0.09	0.07	0.06	0.053	0.04
D_{11}/D_{11}^{iso}		1.05	1.02	1.01	1.007	1.005	1.0039	1.002

4. Special orthotropic laminates, $[0/90/0]_r$ orientation (Table 7.15). The differences are 27% when $r = 3$ and 3% when $r = 51$.

Table 7.15 $[0/90/0]_r$ orientation

				r			
	3	7	11	15	19	27	51
A_{11}^*/D_{11}^*	0.731	0.835	0.883	0.909	0.926	0.97	0.97

5. Special orthotropic laminates, $[90/0/90]_r$ orientation (Table 7.16). The differences are 45% when $r = 5$ and 3.6% when $r = 45$.

Table 7.16   Special orthotropic laminates, $[90/0/90]_r$ orientation

				r		
	5	9	13	17	21	45
A_{11}^*/D_{11}^*	1.56	1.233	1.146	1.107	1.083	1.036

6. $[ABBCAAB]_r$ orientation with $A = 45°$, $B = -45°$, $C = 0°$ (Table 7.17). This orientation has fairly good quasi-homogeneous characteristics when $r \geq 2$.

Table 7.17 $[ABBCAAB]_r$, $A = 45°$, $B = -45°$, $C = 0°$

			$r(N)$		
	1(7)	2(14)	3(21)	4(28)	5(35)
A_{11}^*/D_{11}^*	1.268	1.056	1.024	1.013	1.008
B_{ij}	$\fallingdotseq 0$	$\fallingdotseq 0$	$\fallingdotseq 0$	$\fallingdotseq 0$	$\fallingdotseq 0$

7. $[ABCCABBCA]_r$ orientation with $A = 45°$, $B = -45°$, $C = 0°$ (Table 7.18). This arrangement has fairly good quasi-homogeneous characteristics when $r \geq 2$.

Table 7.18 $[ABCCABBCA]_r$, $A = 45°$, $B = -45°$, $C = 0°$

			$r(N)$		
	1(9)	2(18)	3(27)	4(36)	5(45)
A_{11}^*/D_{11}^*	1.13	1.03	1.013	1.007	1.003
B_{ij}	$\fallingdotseq 0$	$\fallingdotseq 0$	$\fallingdotseq 0$	$\fallingdotseq 0$	$\fallingdotseq 0$

8. Antisymmetric angle-ply $[ABBAAB]_r$, $A = +15°$, $B = -15°$ (Table 7.19).

Table 7.19 $[ABBAAB]_r$, $A = +15°$, $B = -15°$

				r			
	3	7	11	15	19	27	51
B_{16}^*/D_{11}^*	0.0105	0.0052	0.0035	0.0026	0.0021	0.0015	0.0011
B_{26}^*/D_{11}^*	0.0015	0.00075	0.0005	0.00037	0.00030	0.00021	0.00016

9. Symmetric angle-ply $[[45/90/30/0]_r]_s$ (Table 7.20).

Table 7.20 $[[45/90/30/0]_r]_s$

	r(N)						
	1(8)	2(16)	5(40)	10(80)	14(112)	18(144)	20(160)
D_{16}^*/D_{11}^*	0.3455	0.2578	0.2146	0.2015	0.1976	0.1959	0.1953
D_{26}^*/D_{11}^*	0.3058	0.2000	0.1523	0.1386	0.1348	0.1328	0.1321
A_{11}^*/D_{11}^*	1.433	1.1936	1.073	1.035	1.025	1.019	1.0175

Elimination of the bending–extension coupling stiffness, B_{ij}

1. Quasi-isotropic constants are used.
2. The cross-section is symmetric with respect to the mid-surface of the laminate.
3. $[\pm\theta]_r$ angle ply laminate.
4. $[ABBCAAB]_r$.
5. $[ABCCABBCA]_r$.
6. Antisymmetric angle-ply $[ABBAAB]_r$. $B_{16} \doteqdot 0$, $B_{26} \doteqdot 0$ as r increases.
7. Symmetric angle-ply $[[+45/90/30/0]_r]_s$. $B_{ij} \doteqdot 0$ but $D_{16} \neq 0$, $D_{26} \neq 0$. As r increases, $D_{16}^*/D_{11}^* \to 0.2$, $D_{26}^*/D_{11}^* \to 0.13$.

References

Note: *ICCM = International Conference on Composite Materials; JISSE 1 = Proceedings 1st Japan International SAMPE Symposium; SAMPE = Society for the Advancement of Material and Process Engineering.*

1. Kim, D. H. (1966) Theory of non-prismatic folded plate structures, in Applied Mechanics Seminar, Seoul National University, 13 May. (Also *Mem. Korea Milit. Acad.*, **5**, Sept. 1967.)
2. Prescott, J. (1924) *Applied Elasticity*, Longman S. Green, London.
3. Sokolnikoff, I. S. (1956) *Mathematical Theory of Elasticity*, 2nd edn, McGraw-Hill, New York.
4. Timoshenko, S. and Goodier, J. N. (1951) *Theory of Elasticity*, McGraw-Hill, New York.
5. Vinson, J. R. and Sierakowski, R. L. (1987) *The Behavior of Structures Composed of Composite Materials*, Martinus Nijhoff, Dordrecht.
6. Jones, R. M. (1975) *Mechanics of Composite Materials*, Scripta Book Co., Washington, D.C.
7. Calcote, L. R. (1969) *The Analysis of Laminated Composite Structures*, Van Nostrand Reinhold, New York.
8. Marshall, I. H. (1981) *Composite Structures*, Applied Science Publishers.
9. Goldberg, J. E. and Leve, H. L. (1957) *Theory of Prismatic Folded Plate Structures*, International Association of Bridge and Structural Engineers, Zurich.
10. Craemer, H. (1930) *Theorie der Faltwerke*, Beton und Eisen.
11. Kim, D. H. (1965) Analysis of triangulated folded plate roots of umbrella type. Thesis submitted for partial fulfilment for Ph.D., Purdue University.
12. Tsai, S. W. (1964) Structural behavior of composite materials. NASA CR-71, July.
13. Timoshenko, S. and Woinowsky-Krieger, S. (1959) *Theory of Plates and Shells*, McGraw-Hill, New York.

14. Jones, R. M. (1973) Buckling and vibration of rectangular unsymmetrically laminated cross-ply plates. *AIAA J.*, **11**(2), 1626–32.
15. Whitney, J. M. (1969) Bending–extension coupling in laminated plate under transverse loading. *J. Composite Mater.*, Jan, **3**, 204–28.
16. Ashton, J. E. and Whitney, J. M. (1970) *Theory of Laminated Plates*, Technomic Publishing Co., Westport, VA.
17. Whitney, J. M. and Leissa, A. W. (1969) Analysis of heterogeneous anisotropic plates. *J. Appl. Mech. Am. Soc. Mech. Engrs*, **28**, June, 261.
18. Kim, D. H. (1989) A simple method of vibration analysis of irregularly shaped composite structural elements, in *JISSE 1* (First Japan International SAMPE Symposium and Exhibition), Tokyo, 28 Nov.–1 Dec., pp. 863–68.
19. Kim, D. H. (1990) Vibration analysis of irregularly shaped composite structural members – for higher modes, in *1990 Structural Congress*, American Society of Civil Engineers, Baltimore, MD., USA, May, pp. 63–4.
20. Kim, D. H. (1991) Vibration analysis of irregularly shaped laminated thick composite plates, in *ICCM 8* (eds Tsai, S. and Springer, G.), Honolulu, Hawaii, July, p. 30-J.
21. Kim, D. H. (1991) Vibration analysis of laminated thick composite plates, in *Proceedings Third East Asia–Pacific Conference on Structural Engineering and Construction* (eds Fan, L. C. *et al.*), Shanghai, Apr., pp. 1249–54.
22. Kim, D. H., Park, J. S. and Kim, K. J. (1991) Vibration analysis of irregularly shaped laminated thick composite plates II, in *Second Japan International SAMPE Symposium and Exhibition* (eds Kimpura, I. *et al.*), Dec., pp. 1310–17.
23. Kim, D. H. (1974) A method of vibration analysis of irregularly shaped structural elements, in *International Symposium on Engineering Problems in Creating Coastal Industrial Sites* (ed. ISWACO), Seoul, Korea, pp. 39–63.
24. Whitney, J. M. (1969) The effect of transverse shear deformation on the bending of laminated plates. *J. Composite Mater.*, **3**, July, 534.
25. Whitney, J. M. and Rose, D. H. (1991) Effect of transverse normal stress on the bending of thick laminated plates, in *Proceedings 8th International Conference on Composite Materials* (eds Tsai, S. and Springer, G.), July, p. 30-B.
26. Kim, D. H., Park, J. S., Kim, K. J. and Shim, D. S. (1991) The influence of anisotropy on the natural frequencies of vibration of composite laminated structures. *J. Korean Soc. Civ. Engrs*, Oct., 228–32.
27. Kim, D. H., Shim, D. S. and Kim, K. J. (1991) The influence of anisotropy on the buckling strength of composite laminated structures for civil construction. *J. Korean Soc. Civ. Engrs*, Oct., 233–8.
28. Kim, D. H. (1993) A simple method of analysis for eigenvalue problems of certain composite laminates for civil construction, in *Proceedings Fourth East Asia Pacific Conference on Structural Engineering and Construction* (eds Shin, Y. K. *et al.*), Seoul, Korea, 20–22 Sept., pp. 413–18.
29. Kim, D. H. (1991) Simple method of analysis for preliminary design of the composite laminated primary structures for civil construction. *J. Comput. Struct. Engng Inst. Korea*, Oct.
30. Kim, D. H. (1991) Design of composite material structures, in *Proceedings Korean Society of Civil Engineers Meeting*, Oct., pp. 215–21.
31. Kim, D. H., Kim, K. J. and Shim, D. S. (1991) Possibility of using the classical mechanics for the preliminary design of laminated composite structures for civil construction. *J. Korean Soc. Civ. Engrs*, Oct., 222–7.
32. Verchery, G. *et al.* (1991) A quantitative study of the influence of anisotropy on the bending deformation of laminates, in *Proceedings ICCM 8* (eds Tsai, S. and Springer, G.), July, p. 26-J.
33. Kandil, N. and Verchery, G. (1989) Some new developments in the design of stacking sequences of laminates, in *Proceedings ICCM 7* (ed. Wu, Y.).
34. Verchery, G. (1990) Designing with anisotropy, in *Composites in Building Construction*, Vol. 3, Pluralis.
35. Tsai, S. W. (1988) *Composite Design*, 4th edn, Think Composites, Dayton, OH.

Further reading

Books

Girkmann, K. (1954) *Flachentragwerke*, 3rd edn., Springer, Vienna.

Magnus, W. and Oberbettinger, F. (1954) *Formulas and Theorems for the Functions of Mathematical Physics*, Chelsea, New York.

Sokolnikoff, I. S. (1951) *Tensor Analysis, Theory and Applications*, John Wiley, New York.

Timoshenko, S. P. and Gere, J. M. (1961) *Theory of Elastic Stability*, McGraw-Hill, New York.

Tranter, C. J. (1951) *Integral Transforms in Mathematical Physics*, John Wiley, New York.

Vinson, J. R. (1974) *Structural Mechanics, The Behavior of Plates and Shells*, Wiley-Interscience, New York.

Articles and reports

Bai, J. M. and Sun, C. T. (1993) Optimal design and control of a sandwich beam for maximum frequency and minimum dynamic response, in *Proceedings 38th International SAMPE Symposium and Exhibition* (eds Bailey, V. *et al.*), Anaheim, CA, May, pp. 2128–38.

Campbell, G. A. and Foster, R. M. (1942) Fourier integrals for practical applications. Bell Telephone Laboratories, New York.

Carrier, G. F. (1944) Bending of clamped sectional plate. *J. Appl. Mechs*, **66**, A134.

Chen, P., Elshiekh, A. and DeTeresa, S. (1993) Effect of lay-up on the mechanical properties of laminates, in *Proceedings 38th International SAMPE Symposium and Exhibition* (eds Bailey, V. *et al.*), Anaheim, CA, May, pp. 1152–68.

Ferreira, A. J. M. and Marques, A. T. (1991) Composite structure analysis by the use of flat shell element, in *Proceedings ICCM 8* (eds Tsai, S. and Springer, G.), July, p. 26-C.

Hahn, H. T., Kim, I. and Bakis, C. E. (1991) Laminated plate theory for thick composites: experimental correlation, in *Proceedings ICCM 8* (eds Tsai, S. and Springer, G.), July, p. 30-H.

Hasse, H. R. (1950) The bending of uniformly loaded clamped plate in the form of circular sector. *Q. Mechs Appl. Maths*, **3**, 271.

Heitkamp, R. R. (1993) Cibrrier: A new approach to low FST sandwich structures, in *Proceedings 38th International SAMPE Symposium and Exhibition* (eds Bailey, V. *et al.*), Anaheim, CA, May, pp. 1720–35.

Horvay, G. and Hansen, K. L. (1957) The sector problem. *Trans. Am. Soc. Mech. Engrs J. Appl. Mechs*, Dec., 574.

Jensen, D. W. and Hipp, P. A. (1991) Compressive testing of filament–wound cylinders, in *Proceedings ICCM 8* (eds Tsai, S. and Springer, G.), July, p. 35-F.

Jing, H. S. and Tzeng, K. G. (1993) Bending analysis of thick cross-ply laminated doubly-curved shells, in *Proceedings 38th International SAMPE Symposium and Exhibition* (eds Bailey, V. *et al.*), Anaheim, CA, May, pp. 707–18.

Koo, K. N. and Lee, I. (1991) Finite element analysis of vibration and damping for symmetric composite laminates, in *Proceedings ICCM 8* (eds Tsai, S. and Springer, G.), July, p. 34-C.

Kouri, J. V. and Atluri, S. N. (1991) Analytical modeling of laminated composites, in *Proceedings ICCM 8* (eds Tsai, S. and Springer, G.), July, p. 30-A.

Kumar, V. and Weerth, D. E. (1991) Finite element analysis of thick composite plate structures, in *Proceedings ICCM 8* (eds Tsai, S. and Springer, G.), July, p. 30-C.

Li, S. and Sun, C. T. (1989) A global local method for analysis of thick composite laminates, in *Proceedings ICCM 7* (ed. Wu, Y.), Vol. 3, pp. 223–8.

Liaw, D. G. and Yang, H. T. Y. (1989) Dynamic response and buckling of imperfect laminated thin shells, in *Proceedings ICCM 7* (ed. Wu, Y.), Vol. 3, pp. 219–22.

Love, A. E. H. (1944) *A Treatise on the Mathematical Theory of Elasticity*, 4th edn, Dover Publications, New York.

Michell, J. H. (1899) On the direct determination of stress in elastic solid with application to the theory of plates. *Proc. London Math. Soc.*, **31**, 100.

Michell, J. H. (1900) Some elementary distribution of stress in three dimensions. *London Maths Soc. Proc.*, **32**, 23.

Michell, J. H. (1902) The flexure of a circular plate. *Proc. London Math. Soc.*, **34**, Apr., 223.

Mindlin, R. D. (1951) Influence of rotatory inertia and shear on flexural motions of isotropic, elastic plates. *J. Appl. Mechs*, **18**, 31.

Molodtsov, G. A. (1989) Micro-mechanical analysis of prestressed fiber reinforced composite laminates, in *Proceedings ICCM 7* (ed. Wu, Y.), Vol. 3, pp. 117–122.

Nomachi, S. G. (1960) On one method of solving stress problems in cylindrical coordinates by means of finite Fourier Hankel transforms. *Mem. Muroran Inst. Technol.*, **3**, Jun.

Nuriyev, B. R. (1991) Elastic/viscoelastic wave propagation in the layered composites due to an impact, in *Proceedings ICCM 8* (eds Tsai, S. and Springer, G.), July, p. 32-T.

Raoul, R. A. and Palazotto, A. N. (1991) Nonlinear dynamics of anisotropic shell panels, in *Proceedings ICCM 8* (eds Tsai, S. and Springer, G.), July, p. 32-R.

Reissner, E. (1945) The effect of transverse shear deformation on the bending of elastic plates. *J. Appl. Mechs*, **12**, A69.

Reissner, E. (1947–8) On bending of elastic plates. *Q. Appl. Maths*, **5**, 55.

Scherer, H. (1957) Einflussflachen einer Dreiecks Platte mit Aufpunkt am Freien Rand. *Ingenieur Archiv.*, **25**(4), 255.

Silverman, I. K. (1955) Approximate stress functions for triangular wedges. *Trans. Am. Soc. Mech. Engrs*, **77**, 123.

Tranter, C. J. (1948) The use of Mellin transformation in finding stress distribution in infinite wedge. *Q. J. Mechs Appl. Maths*, **1**, 125, 229.

Verchery, G. (1989) An efficient finite element for thick laminated and sandwich beams, in *Proceedings ICCM 7* (ed. Wu, Y.), Vol. 3, pp. 165–170.

Weissman-Berman, D. (1993) Elastic response of thermoplastic sandwich beams, in *Proceedings 38th International SAMPE Symposium and Exhibition* (eds Bailey, V. *et al.*), Anaheim, CA, May, pp. 2102–15.

Whitney, J. M. (1971) Fourier analysis of clamped anisotropic plates. *Trans. Am. Soc. Mech. Engrs J. Appl. Mechs*, **38**, 530–2.

Whitney, J. M. (1972) Stress analysis of thick laminated composite and sandwich plates. *J. Composite Mater.*, **6**, Oct., 426.

Whitney, J. M. (1989) Cylindrical bending of laminated anisotropic plates including transverse shear deformation, in *Proceedings ICCM 7* (ed. Wu, Y.), Vol. 3, pp. 112–122.

Woinowsky-Krieger, S. (1952) Uber die Anwendung der Mellin Transformation zur Lösung einer Aufgabe der Plattenbiegung. *Ingenieur Archiv.*, 391.

8 Failure of composites

8.1 Introduction

While the failure analysis procedures for metallic structures were well established several decades ago, similar procedures for composite material structures are not well defined. Even though some experience obtained from metals is used for composites, the two material groups are of fundamentally different natures especially when continuous fiber laminated composites are used for primary structures.

8.1.1 Causes of failure

The causes of failure of composites can be grouped into three types, as with metals [1–3].

Design errors

Errors in designing composite structures can be made in both material and structure. The most common errors related to the material may include those in analyzing the nature of anisotropy of the individual ply, inadequate assessment of material damage, and improper prediction of environmental sensitivities.

A laminate is assumed to have strain continuity throughout the thickness. The stress level carried by each ply in the laminate depends on its modulus. This may cause large stress gradients and internal shear stresses between plies which are oriented at considerably large angles to each other (e.g. 90°). If the magnitude of these gradients is large enough, a premature fracture may occur.

Composite materials with a highly anisotropic coefficient of thermal expansion may cause design errors at a material level. If the adjacent plies in a laminate are oriented at large angles to one another, a high level of internal stress may develop as the temperature changes. Such internal stresses are analogous to those caused by externally applied loads. Most high-performance composites are cured at elevated temperatures and may be used in ambient conditions. The temperature changes can induce stresses in the laminate and behave as residual stresses.

As with metallic materials, errors in design include improper treatment of environmental problems, and misjudgement of damage tolerance and fatigue strength of the material. Other common errors are unexpected loading conditions, stress concentrations and instabilities. Most of such errors may be alleviated by thorough study and testing during design.

Fabrication and processing errors

Even though manufacturing control and material inspection tests can prevent the chance of producing structural parts with abnormalities, some errors may still occur.

When continuous fiber reinforced composites are fabricated, usually in the form of laminates, placement and orientation of individual plies are critical

333

in obtaining the required mechanical properties. Suppose a unidirectional T300/N5208 CFRP laminate has an overall fiber misorientation of 15°. From Table 5.3, Q_{11} for this laminate is approximately 88.26% (160.47/181.81) of Q_{xx}. If the misorientation is 30°, Q_{11} is 60% of Q_{xx}. The reality is, though, the misorientation of 15° may generate up to 50% reduction in ultimate strength. [2]. The results of Polygon Co. [3] indicate that fiber misalignment of 15° reduces the tensile strength by 25%, the modulus of elasticity by 40%. Misalignment of 30° reduces both the tensile strength and the modulus of elasticity by 60%. If fiber misalignment, unequal fiber loading, broken fibers due to mishandling, etc., are considered in modeling, the reduction of the tensile strength from that obtained by the simple equation is about 30% (150 ksi *vs* 220 ksi).

Unexpected service conditions

The properties of composites may be considerably reduced by temperature variations, impact damage, and chemical attack in the case of some matrices. Thermosetting matrices may be degraded by the hygrothermal effect, i.e. increased temperature with increased moisture. Some service anomalies include improper operation, faulty maintenance, overloads and environmental incurred damage.

8.2 Fracture

Composite fracture mechanics is rather complex because of its anisotropic nature. The types and modes of failure depend on the nature and direction of the applied load, and on the distribution of reinforcements in the composites.

8.2.1 Fracture in continuous fiber composites

Fractures in continuous fiber reinforced composites may be classified into three basic types [4–11] (Figure 8.1),

- interlaminar
- intralaminar
- translaminar

Interlaminar fracture shows the failure developed between plies. Intralaminar fracture is located inside a ply. Translaminar fracture is oriented transverse to the laminated plane and involves significant fiber fracture. Both interlaminar and intralaminar fractures involve few fiber breaks, if any, because they occur in the laminate plane.

Interlaminar and intralaminar fractures

Both interlaminar and intralaminar fractures occur in a plane parallel to that of the fiber reinforcement. Naturally, the fracture mechanism and appearance are influenced by matrix fracture and fiber matrix separation. Separation of the fiber from the matrix at the interface occurs by either tension (mode I) or by in-plane shear (mode II). Since the matrix used for a composite is usually cohesive, the fracture between fibers is that of the matrix resin. The majority

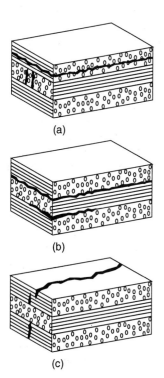

(a)

(b)

(c)

Figure 8.1 Fracture modes in a continuous fiber reinforced laminate: (a) intralaminar; (b) interlaminar; (c) translaminar.

of thermosetting matrices used currently have relatively high tensile strength and are brittle. For such matrices, cohesive matrix failure occurs in a brittle nature.

In practice, the direction of the reinforcement is oriented to various angles to obtain the required specific properties. In such cases, the path of crack propagation may be very complex.

Translaminar fractures

Generally speaking, translaminar fractures have rough fiber dominated features. The macroscopic appearance shows the condition of failure under longitudinal loading. Two distinct translaminar failure mechanisms are tensile and compression microbuckling. One of these, or a combination is responsible for all translaminar fractures.

Tension fractures show extremely rough fracture surfaces, with a large number of fibers protruding from the fracture planes. Delamination near the fracture surface is generally not evident. The primary failure mechanism is the brittle tensile failure of individual fibers. Fiber pull-out, fiber end fracture and matrix fracture are the typical features.

Uniaxial compression induced fractures show gross buckling, extensive delamination and interlocking of the delamination planes. There is a relative post-failure motion between the fractured surfaces in contact. The surface is flatter than that in tension fractures, and no pull-out fibers appear. Fiber buckling, fiber end fracture, matrix fracture and post-fracture damage are typical features of translaminar compression fractures. Compression microbuckling is the primary cause of such a failure mechanism.

8.2.2 Fracture in discontinuous fiber composites

Discontinuous fiber composites such as sheet molding compounds (SMC) or bulk molding compounds have chopped fiber bundles randomly distributed [12]. When pressure is applied during the molding process, some fiber bundles are flattened, and some lose strict planarity. While the matrix of such a composite shows general brittle behavior, sometimes it has ductility. The matrix is generally brittle in tension, but it may not be brittle under complex stresses including superimposed compressive stress. The complex failure mode may result from an effective interweaving of the short fiber bundles in such composites. In addition to such complexities, the failure modes in chopped fiber composites may be caused by other undesirable reasons such as wavy fiber bundles, kinked fiber bundles, matrix rich regions and a proportion of oriented fiber bundles losing random distribution.

Tensile failure

Tensile failure generally occurs with some degree of separation of fiber bundles along the parallel planes, and results in much fiber breakage, intrabundle fiber separation and separation of fiber bundles perpendicular to the fracture plane.

Compressive failure

Compressive failure is often similar to tensile failure as far as separation between the fiber planes is concerned. When a compressive stress is acting in the plane of the sheet, failure begins with delamination, followed by buckling of the resulting layers, made thin by delamination. This type of failure eventually produces sharp kinking of the fiber bundles with occasional nonkinked filaments protruding. If the fibers receive lateral support from a rigid matrix before failure, failure occurs at higher stresses with considerable delamination. If smaller lateral support is given from a softer matrix, the fibers may kink and break locally at lower stresses with less delamination.

Shear failure
The nature of shear failures is similar to that of tensile failures. As discussed in Chapter 3, the principal plane is at 45° to that on which the shear stress acts. Such shear cracks occur upon further loading after the tensile cracks have appeared.

Fibers in a bundle may have relative motion. During fabrication, fiber bundles sometimes come apart and the fibers do not all respond in the same way when the material is deformed. In shear failures, fiber breakage sometimes occurs.

Fatigue failure
Fatigue generally results in failures such as separation, as with static loading. The repeated relative motion between fibers in a bundle, and between bundles, causes resin dust. This can greatly reduce the fatigue life of the material. Compressive fatigure failures tend to be distributed over a larger region than tensile failures, producing a much larger amount of delamination. If the failure criterion is given by the amount of deformation, failure is often defined as the loss of a certain amount of stiffness. In such cases, fatigue failure occurs long before separation. The reduction in stiffness comes mainly from debonding and matrix microcracking. Both of these phenomena are followed by a whitening of the composite.

8.3 Fatigue

8.3.1 Fatigue failure
By definition, fatigue is the failure or decay of mechanical properties after repeated applications of stress. Because of their anisotropic nature, composite materials have very complex failure mechanisms under static and fatigue loadings. In most isotropic brittle materials, fatigue failure mostly results in a single crack. A composite material undergoes fatigue failure with extensive damage throughout the volume of the specimen.

There are four basic failure mechanisms in composites. These are matrix cracking, delamination, fiber breakage and interfacial debonding. Any combination of these is possible and causes fatigue damage. This fatigue damage results in reduced fatigue strength and stiffness. The damage development under fatigue and static loading is similar except that fatigue at a given stress level causes additional damage as a function of cycles [13–16]. (R. Kim [17] explains fatigue failure characteristics in elaborate detail.)

Matrix cracking

If multidirectional laminates are acted upon by in-plane loading, and this loading is either increased or repeated, the failure usually occurs successively from the weakest ply to the strongest. In Chapter 5, the strength of the laminate as a single unit is discussed. However, when failure occurs, the weakest lamina will start to fail. Recall that in the case of the pure bending of an isotropic beam, the external 'fiber' starts to yield. Since the 'fiber' strength of the beam is assumed to be uniform, the one with highest stress starts to yield and this yielding propagates to the next fiber. This process goes on until all fibers above the neutral axis receive an equal amount of strain; eventually failure occurs if the load is continuously increased.

In multidirectional laminates, similar phenomena happen. The difference is that yielding starts from the weakest lamina even though the external stress is equal. As equations 5.25 show, the weakest lamina in any off-axis ply is at 90° orientation, and matrix cracking is initiated by 90° plies. In principle, this cracking propagates to the plies with the next highest degree of orientation, and should continue until failure. However, because composite laminate behavior is rather complex, thorough testing should be carried out to obtain design data. In general, most of the crack multiplication (60–90%) occurs during the first 20% of the fatigue life and a good amount of fatigue life remains after reaching this crack density (Figure 8.2).

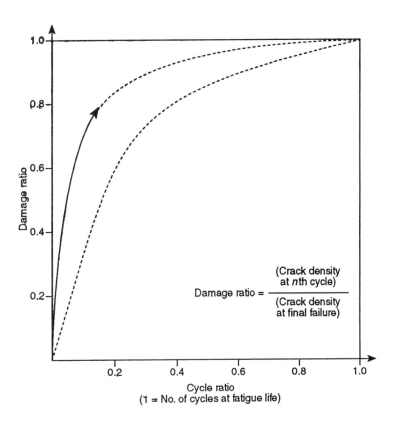

$$\text{Damage ratio} = \frac{\text{(Crack density at } n\text{th cycle)}}{\text{(Crack density at final failure)}}$$

Cycle ratio
(1 = No. of cycles at fatigue life)

Figure 8.2 General tendency of crack increase under fatigue loading for typical carbon–epoxy laminates. (Courtesy of Think Composites.)

It is said that the stacking sequence in a laminate has a significant role in the development of cracks.

Delamination

Delamination is believed to be caused by the interlaminar stresses. Accurate calculation of interlaminar stresses can be made using analytical models. The type of laminate, stacking sequence, properties of the constituent materials and type of loading all influence the interlaminar stresses.

Under fatigue stress, delamination starts at a stress level smaller than the static stress level. This delamination may occur in the very early part of fatigue life and propagates rapidly from the free edge toward the inside of the laminate.

Fiber break and interface debonding

The properties of the constituent materials and defects of the fiber influence fiber break and interface debonding. The tensile strain of the matrix is generally higher than that of the fiber. For most of the advanced composites under fatigue loading, since the matrix failure strain is larger than the composite failure strain, the fatigue damage at the interface is negligible except at the site of fiber breakage.

Reduction in strength and modulus

Fatigue damage often causes a significant reduction in strength and modulus of the composite laminates. The type of laminate, nature of loading, and others influence the degree of the damage to strength. In general, a multidirectional laminate shows a gradual strength reduction until failure, while a unidirectional (0°) laminate shows almost steady strength until immediate failure.

Test results for a typical quasi-isotropic laminate, $[0/45/90/-45]_{2s}$, show [17] that no damage under compression–compression fatigue was detected. Extensive damage is shown by laminates subject to tension–tension and tension–compression fatigue.

The strength reduction seems to be directly related to the amount of fatigue damage. Tension–tension fatigue produces the greatest strength reduction. A significant modulus reduction is observed in tension–tension and tension–compression fatigue, while no modulus change occurs in compression–compression fatigue.

8.3.2 S–N relation

The fatigue behavior of materials is represented by fatigue stress and fatigue life (S–N) cycle to failure relation. The S–N relation is dependent mostly on the constituent material properties (Figure 8.3). Most advanced fibers are very insensitive to fatigue and the composites with such fibers show good fatigue resistance [17]. As the static strength of a laminate is mainly dependent upon the percentage of 0° plies and their strength, the fatigue strength of a multidirectional laminate can be estimated from the 0° fatigue strength unless the laminate undergoes extensive damage during fatigue.

The fatigue stress is composed of a static mean stress, S_m, and a completely reversed cyclic stress. The fatigue strength is expressed by the variable cyclic

Figure 8.3 Typical *S–N* curve. Note the densities of steel, aluminum and glass. Specific strength curves are different from this figure.

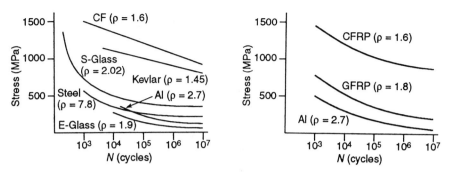

Figure 8.3 Typical *S–N* curve. Note the densities of steel, aluminum and glass. Specific strength curves are different from this figure.

stress, S_r, for failure corresponding to various possible values of mean stress. From mechanics of materials, we have

$$S_m = (S_{max} + S_{min})/2$$
$$S_r = (S_{max} - S_{min})/2$$

(8.1)

where S_{max} and S_{min} are the maximum and minimum fatigue stresses, respectively. The *S–N* behavior of composites seems to be independent of the type of fatigue loading, such as tension–tension, tension–compression or compression–compression fatigue as long as the tensile and compressive strength are equal.

The reduction in fatigue strength resulting from the presence of notches (circular hole and crack) is found to be insignificant [17]. The value of the fatigue notch factor is much smaller than the static stress concentration factor and is close to unity in many cases. The fatigue notch factor is defined as the ratio of the fatigue strength of unnotched specimens to the fatigue strength of notched specimens at *N* cycles. The residual strength after fatigue loading is usually greater than the static notch strength. The damage around the notch tips relaxes the stress concentration, resulting in the excellent fatigue resistance of the notched specimens.

The effect of test frequency on fatigue life is generally negligible in the range from 1 Hz to 30 Hz for most composites with some portion of 0° plies [17]. Matrix controlled laminates are susceptible to test frequency because the matrix is sensitive to loading rate and temperature.

Among several environmental factors, the hygrothermal effect (elevated temperature with simultaneously increased moisture) seems to be most influential on the mechanical performance of composite material. Elevated temperature and high moisture content usually lower the matrix strength, and eventually, the matrix dominant composite failure strength such as transverse and shear strengths, but reduce the residual stresses. The static tensile strength of the laminates with fiber dominant failure modes is relatively insensitive to temperature and moisture. The fatigue strength of angle-ply laminates (matrix dominant failure) slightly decreases with increasing temperature.

The effect of cryogenic (extremely low) temperature is negligible on the fatigue life of boron–epoxy laminates. Moisture does not affect the fatigue life of carbon–epoxy laminates much.

Fatigue behavior of composite materials has been studied extensively. Such studies include fatigue life prediction [17], which is not treated here. Users of composite materials have a great deal of confidence in fatigue and environmental stability.

8.4   Failure theories

8.4.1   Review of failure theories of isotropic materials

The concept of failure of isotropic materials may be extended to composite materials. The concepts include the maximum stress, maximum strain and quadratic criteria [18, 19].

In most cases, the mechanical properties of ductile materials are obtained from tensile tests, while those of brittle materials are usually found from compression tests. For selecting working stresses for the various cases of combined stresses, quite a few strength theories have been developed.

Some scientists such as Lamé and Rankine assumed the maximum principal stress as the criterion of strength. Mathematically, this concept is

$$\sigma_1 = \pm \sigma_{yp}$$
$$\sigma_2 = \pm \sigma_{yp} \tag{8.2}$$
$$\sigma_3 = \pm \sigma_{yp}$$

where σ_{yp} is the yield strength of the material when both tension and compression strength are assumed to be equal. If these are unequal, we have six similar equations, with different yield strengths for tension and compression. For the plane stress case,

$$\sigma_3 = 0$$
$$\sigma_1 = \pm \sigma_{yp} \tag{8.3}$$
$$\sigma_2 = \pm \sigma_{yp}$$

Later the maximum strain theory was generally accepted under the influence of authorities such as Poncelet and Saint Venant. The maximum strain theory can be expressed mathematically by equating the principal strains obtained by Hooke's law to the uniaxial strain at yield. This yields

$$\sigma_1 - v\sigma_2 - v\sigma_3 = \sigma_{yp} \quad \text{(tension or compression)}$$
$$\sigma_2 - v\sigma_3 - v\sigma_1 = \sigma_{yp} \quad \text{(tension or compression)} \tag{8.4}$$
$$\sigma_3 - v\sigma_1 - v\sigma_2 = \sigma_{yp} \quad \text{(tension or compression)}$$

For the plane stress case,

$$\sigma_1 - v\sigma_2 = \sigma_{yp}$$
$$\sigma_2 - v\sigma_1 = \sigma_{yp} \tag{8.5}$$

The maximum shear theory was proposed by Guest in 1900. In fact, this theory is a particular case of Mohr's earlier theory. Before Mohr, Coulomb and Vicat had already assumed that failures are precipitated by shearing stresses. In terms of principal stresses, the maximum shear theory states that

$$\sigma_1 - \sigma_2 = \pm\sigma_{yp}$$
$$\sigma_2 - \sigma_3 = \pm\sigma_{yp} \tag{8.6}$$
$$\sigma_3 - \sigma_1 = \pm\sigma_{yp}$$

For the plane stress case, $\sigma_3 = 0$ and

$$\sigma_1 - \sigma_2 = \pm\sigma_{yp}$$
$$\sigma_2 = \pm\sigma_{yp} \tag{8.7}$$
$$\sigma_1 = \pm\sigma_{yp}$$

Maxwell proposed the use of the expression for strain energy. He showed that the total strain energy per unit volume has two parts, that is, the part of uniform tension or compression, and the strain energy of distortion. He declared that 'when the strain energy of distortion reaches a certain limit, the element will begin to give way'. Maxwell already had the theory of yielding we call the maximum distortion energy theory. Huber in 1904 considered only the energy of distortion in determining the critical state of combined stress as

$$U = \frac{1}{12G}\left[(\sigma_1 - \sigma_2)^2 + (\sigma_1 - \sigma_3)^2 + (\sigma_2 - \sigma_3)^2\right] \tag{8.8}$$

In simple tension, the distortion energy at yield point stress can be obtained by substituting $\sigma_1 = \sigma_{yp}$, $\sigma_2 = \sigma_3 = 0$ into equation 8.8; $U = \sigma_{yp}^2/6G$. Equating this expression to equation 8.8, we obtain

$$(\sigma_1 - \sigma_2)^2 + (\sigma_1 - \sigma_3)^2 + (\sigma_2 - \sigma_3)^2 = 2\sigma_{yp}^2$$

or $\tag{8.9}$

$$\sigma_1^2 + \sigma_2^2 + \sigma_3^2 - \sigma_1\sigma_2 - \sigma_2\sigma_3 - \sigma_3\sigma_1 = \sigma_{yp}^2$$

For a plane stress case, the yielding condition is

$$\sigma_1^2 - \sigma_1\sigma_2 + \sigma_2^2 = \sigma_{yp}^2 \tag{8.10a}$$

or

$$\left(\frac{\sigma_1}{\sigma_{yp}}\right)^2 - \frac{\sigma_1\sigma_2}{\sigma_{yp}^2} + \left(\frac{\sigma_2}{\sigma_{yp}}\right)^2 = 1 \tag{8.10b}$$

For pure shear, $\sigma_1 = -\sigma_2 = \tau_{12}$, and equation 8.10 yields

$$\tau_{yp} = \frac{1}{\sqrt{3}}\sigma_{yp} = 0.5774\sigma_{yp} \tag{8.11}$$

8.4.2 Anisotropic failure criteria

With anisotropic materials, many strength (failure) theories have been developed [19, 20]. Some of the commonly used composite strength theories are presented in this section.

For laminates, a criterion is applied on a ply-by-ply basis and the load-carrying capability of the entire composite is predicted by the laminate theory given in section 5.7. A laminate may be assumed to have failed when the strength criterion of any one of its laminae is reached. However, failure of a single ply need not necessarily lead to total fracture of the structure.

With a laminate, the criterion of an on-axis ply can be determined with relative ease. The off-axis or laminate-axis criterion can be obtained by the coordinate transformation of stress or strain. For failure criteria, ply-by-ply strength analysis can be carried out, taking into consideration the effects of curing stresses, hygrothermal effects and others, if necessary. From such analysis the first-ply failure (FPF), last-ply failure (LPF), limit design, and ultimate design concepts can be developed.

We use **notation for strength** used by Jenkins, and also Tsai [19, 21]:

- X Longitudinal (or uniaxial) tensile strength
- X' Longitudinal (or uniaxial) compressive strength
- Y Transverse tensile strength
- Y' Transverse compressive strength
- S Longitudinal shear strength

These strengths are to be obtained experimentally.

The **strength ratio,** R, is the ratio of the maximum or ultimate strength to the applied stress. It is assumed that:

1. The material is linearly elastic.
2. For each state of combined stresses, there is a corresponding state of combined strains.
3. All components of stress and strain increase by the same proportion.

The definition of the strength ratio indicates that

$$
\{\sigma\}_{\text{max}} = R\{\sigma\}_{\text{applied}}
$$
$$
\{\varepsilon\}_{\text{max}} = R\{\varepsilon\}_{\text{applied}}
$$

(8.12)

R is analogous to the safety factor or the load factor: failure occurs when $R = 1$.

When $R < 1$, the applied stress is larger than the strength by a factor of $1/R$. This is physically impossible but provides useful information for design. For example, one may reduce the applied load by $(R - 1)$.

Maximum strength theory

Jenkins extended the concept of the maximum normal or principal stress theory to predict the strength of planar orthotropic materials such as wood [19]. According to this theory, failure will occur when one or more of the stresses acting in the directions of material symmetry, σ_1, σ_2 and τ_{12}, reaches a respective maximum value, X, Y and S. Mathematically stated, failure will not occur as long as

$$
-X' < \sigma_1 < X
$$
$$
-Y' < \sigma_2 < Y
$$
$$
-S' < \tau_{12} < S
$$

(8.13)

Because of orthotropic symmetry, shear strength is independent of the sign of τ_{12}. There are five independent modes of failure, and there is no interaction among the modes, according to this theory. The reality is that failure processes are highly interacting and far more complex than the values of the stress components [14]. If the stress ratio is used, this criterion can be expressed as

$$R_x = X/\sigma_x, \quad \text{if} \quad \sigma_x > 0 \quad \text{or} \quad R'_x = X'/|\sigma_x| \quad \text{if} \quad \sigma_x < 0$$
$$R_y = Y/\sigma_y, \quad \text{if} \quad \sigma_y > 0 \quad \text{or} \quad R'_y = Y'/|\sigma_y| \quad \text{if} \quad \sigma_y < 0 \qquad (8.14)$$
$$R_s = S/|\sigma_s|$$

Maximum strain theory

The maximum strain theory is an extension of the maximum principal strain theory, promoted by Poncelet and Saint Venant, to anisotropic media. The strain components for an orthotropic lamina are referred to the principal material axes, and there are three strain components in this criterion. Since linear elastic response is assumed to be failure, this criterion can predict strength in terms of loads or stresses. A ply of a laminate is considered to have failed when one of ε_x, ε_y and ε_s reaches the maximum value obtained from simple one-dimensional testing [22, 23]. This maximum strain from each test is either measured or computed from the measured strength divided by the **tangent modulus**

$$\varepsilon_x^* = X/E_x \quad \text{or} \quad \varepsilon_x^{*'} = X'/E_x$$
$$\varepsilon_Y^* = Y/E_Y \quad \text{or} \quad \varepsilon_y^{*'} = Y'/E_y \qquad (8.15a)$$
$$\varepsilon_s^* = S/E_s$$

The minimum common envelope of the superposition of the interaction failure diagrams, for either stress or strain, of all individual plies, becomes the failure diagram for the laminate. Figure 8.5(a) shows the typical envelope on the zero shear strain plane. The maximum stress criterion in stress space would have a similar shape. Each line of the envelope shows the appropriate allowable strain listed in equation 8.15(a).

The strength ratio is expressed by the lowest of three ratios of the maximum strain to the applied strain. Note the similar procedure taken for the maximum stress criterion, equation 8.14.

$$R_x = \varepsilon_x^*/\varepsilon_x \quad \text{if} \quad \varepsilon_x > 0 \quad \text{or} \quad R'_x = \varepsilon_x^{*'}/|\varepsilon_x| \quad \text{if} \quad \varepsilon_x < 0$$
$$R_y = \varepsilon_y^*/\varepsilon_y \quad \text{if} \quad \varepsilon_y > 0 \quad \text{or} \quad R'_y = \varepsilon_y^{*'}/|\varepsilon_y| \quad \text{if} \quad \varepsilon_y < 0 \qquad (8.15b)$$
$$R_s = \varepsilon_s^*/|\varepsilon_s|$$

Both the maximum stress and maximum strain criteria assume no interactions among the possible five modes. Since Poisson's ratio is not zero, there is always coupling between the normal components, and this leads to disagreement between these two criteria regarding the magnitude of the load and the mode for the failure.

For example, consider a unidirectionally reinforced laminate acted upon by uniaxial tension, σ, at some angle θ to the reinforcements. The maximum allowable loading is the smallest of the following equations:

1. From the maximum stress theory,

$$\sigma = \frac{X}{\cos^2 \theta}, \quad \sigma = \frac{Y}{\sin^2 \theta} \quad \text{or} \quad \sigma = \frac{S}{\sin \theta \cos \theta} \tag{8.16}$$

2. From the maximum strain theory,

$$\sigma = \frac{X}{\cos^2 \theta - v_{12} \sin^2 \theta}, \quad \sigma = \frac{Y}{\sin^2 \theta - v_{21} \cos^2 \theta}$$

or

$$\sigma = \frac{S}{\sin \theta \cos \theta} \tag{8.17}$$

The results of the two criteria agree only in the shear plane and along the four lines of constant failure due to uniaxial stresses. Just as the deformation of a body is always coupled by the nonzero Poisson's ratio, failure of a body is also coupled. Because the micromechanics of failure is highly coupled, we should not extend the simple failure modes based on maximum stress or maximum strain components to fiber, matrix and interfacial failure modes.

Quadratic criterion in stress space

According to Tsai [21], an easy way to incorporate a coupled or interacting failure criterion is to use the quadratic criterion. This is a generalization of strain or distortion energy, proposed by Maxwell, and further developed by Huber. By using this criterion, we can recognize failure criteria as useful design tools on fitting available data, instead of depending on failure criteria to define the. modes of failure. Tsai and Wu assume that the criterion in stress space is the sum of linear and quadratic scalar products:

$$F_{ij}\sigma_i\sigma_j + F_i\sigma_i = 1, \quad i,j = 1, 2, 3, 4, 5, 6 \tag{8.18}$$

The F_i and F_{ij} are second and fourth order lamina strength tensors. The linear stress terms account for possible differences in tensile and compressive strengths. The quadratic stress terms describe an ellipsoid in stress space. The F_{ij} ($i \neq j$) terms are new. Off-diagonal terms of the strength tensor represent independent interactions among the stress components.

For a thin orthotropic ply under plane stress status relative to the symmetric axes $x–y$, this failure criterion becomes

$$F_{xx}\sigma_x^2 + F_{yy}\sigma_y^2 + 2F_{xy}\sigma_x\sigma_y + F_{ss}\sigma_s^2 + F_x\sigma_x + F_y\sigma_y + F_s\sigma_s = 1 \tag{8.19}$$

where the strength parameters (Fs) can be obtained from

$$F_{xx} = \frac{1}{XX'}, \quad F_{yy} = \frac{1}{YY'}, \quad F_{ss} = \frac{1}{S^2}$$

$$F_x = \frac{1}{X} - \frac{1}{X'}, \quad F_y = \frac{1}{Y} - \frac{1}{Y'}, \quad F_s = 0 \tag{8.20}$$

The F_{xy}, which is a fourth order tensor term, cannot be determined from any uniaxial test in the principal material directions, but from a biaxial test. Recall that F_{xy} is the coefficient of σ_x and σ_y in equation 8.19. The value of F_{xy} depends on the various engineering strengths and the biaxial tensile failure stresses. Tsai proposes the following empirical formula:

$$F_{xy} = F_{xy}^*(F_{xx}F_{yy})^{1/2}$$

in which F_{xy}^* is the normalized interaction term, and, if reliable biaxial test data are not available, it can be treated as an empirical constant

$$-\tfrac{1}{2} \leq F_{xy}^* \leq 0$$

This numerical value is bounded by the generalized Von Mises criterion of $-1/2$ and zero which yields a result almost identical to Hill's criterion. Since each combination of stress components in equation 8.18 reaches its maximum when the right-hand side reaches unity, we can substitute equation 8.12 in 8.18 to obtain

$$[F_{ij}\cdot\sigma_i\cdot\sigma_j]R^2 + [F_i\cdot\sigma_i]R - 1 = 0 \tag{8.21}$$

The stress component terms in equation 8.21 are those of the applied stresses. For a given material, the strength parameters (Fs) are specified (Table 8.1). For a given state of applied stresses, the σs are known. We need to solve only the quadratic equation in the strength/stress ratio R in equation 8.21. The necessary answer is the positive square root in the quadratic formula in the form

$$aR^2 + bR - 1 = 0 \tag{8.22}$$

where

$$a = F_{ij}\sigma_i\sigma_j, \quad b = F_i\sigma_i$$

and the root is

$$R = -(b/2a) + [(b/2a)^2 + 1/a]^{1/2} \tag{8.23}$$

The absolute value of the conjugate root from the negative square root produces the strength ratio when the sign of all the applied stress components is reversed. In the case of the bending of a symmetric plate, the resulting ply stresses change signs between the positive and negative distance from the mid-plane. Designating the ratios for these stresses R^+ and R^-,

$$\begin{aligned}R^+ &= -(b/2a) + [(b/2a)^2 + 1/a]^{1/2} \\ R^- &= |-(b/2a) - [(b/2a)^2 + 1/a]^{1/2}|\end{aligned} \tag{8.24}$$

Quadratic criterion in strain space

Tsai [14] substitutes the stress–strain relation into the stress criterion equation 8.18, to obtain the plane stress failure criterion in strain space. Even though $\varepsilon_z \neq 0$, it is simply ignored. This is acceptable since all failure criteria are purely empirical. Such criteria are not analytical and cannot be derived from fundamental principles.

Table 8.1 Strength of various composite materials (courtesy of Think Composites)

Type: Fiber: Matrix:	CFRP T300 N5208	BFRP B(4) N5505	CFRP AS H3501	GFRP E-glass Epoxy	KFRP kev49 Epoxy	CFRTP AS4 PEEK APC2	CFRP H-IM6 Epoxy	CFRP T300 Fbrt934 4-mil tp	CCRP T300 Fbrt934 13-mil c	CCRP T300 Fbrt934 7-mil c
Engineering constants SI units										
E_x (GPa)	181.0	204.0	138.0	38.6	76.0	134.0	203.0	148.0	74.0	66.0
E_y (GPa)	10.30	18.50	8.96	8.27	5.50	8.90	11.20	9.65	74.00	66.00
v/x	0.28	0.23	0.30	0.26	0.34	0.28	0.32	0.30	0.05	0.04
G (GPa)	7.17	5.59	7.10	4.14	2.30	5.10	8.40	4.55	4.55	4.10
Imperial										
E_x (Msi)	26.25	29.58	20.01	5.60	11.02	19.43	29.44	21.46	10.73	9.57
E_y (Msi)	1.49	2.68	1.30	1.20	0.80	1.29	1.62	1.40	10.73	9.57
G(Msi)	1.04	0.81	1.03	0.60	0.33	0.74	1.22	0.66	0.66	0.59
Max stress SI units (MPa)										
X	1500	1260	1447	1062	1400	2130	3500	1314	499	375
X'	1500	2500	1447	610	235	1100	1540	1220	352	279
Y	40	61	51.7	31	12	80	56	43	458	368
Y'	246	202	206	118	53	200	150	168	352	278
S	68	67	93	72	34	160	98	48	46	46
Imperial (ksi)										
X	217.5	182.7	209.82	153.99	203	308.85	507.5	190.53	72.355	54.375
X'	217.5	362.5	209.82	88.45	34.075	159.5	223.3	176.9	51.04	40.455
Y	5.8	8.845	7.4965	4.495	1.74	11.6	8.12	6.235	66.41	53.36
Y'	35.67	29.29	29.87	17.11	7.685	29	21.75	24.36	51.04	40.31
S	9.86	9.715	13.485	10.44	4.93	23.2	14.21	6.96	6.67	6.67
Max strain ($\times 10^{-3}$)										
X	8.29	6.18	10.49	27.51	18.42	15.90	17.24	8.88	6.74	5.68
X'	8.29	12.25	10.49	15.80	3.09	8.21	7.59	8.24	4.76	4.23
Y	3.88	3.30	5.77	3.75	2.18	8.99	5.00	4.46	6.19	5.58
Y'	23.88	10.92	22.99	14.27	9.64	22.47	13.39	17.41	4.76	4.21
S	9.48	11.99	13.10	17.39	14.78	31.37	11.67	10.55	10.11	11.22

tp: tape.
c: cloth.

The resulting failure criterion in strain space is

$$G_{ij}\varepsilon_i\varepsilon_j + G_i\varepsilon_i = 1 \qquad (8.25)$$

where

$$G_{xx} = F_{xx}Q_{xx}^2 + 2F_{xy}Q_{xx}Q_{xy} + F_{yy}Q_{xy}^2$$
$$G_{yy} = F_{xx}Q_{xy}^2 + 2F_{xy}Q_{xy}Q_{yy} + F_{yy}Q_{yy}^2$$
$$G_{xy} = F_{xx}Q_{xx}Q_{xy} + F_{xy}(Q_{xx}Q_{yy} + Q_{xy}^2) + F_{yy}Q_{xy}Q_{yy}$$
$$G_{ss} = F_{ss}Q_{ss}^2$$
$$G_x = F_xQ_{xx} + F_yQ_{xy}$$
$$G_y = F_xQ_{xy} + F_yQ_{yy}$$

Since the strength ratios based on both combined stresses and combined strains are assumed to be equal, the strength ratio based on the failure criterion in strain space can be obtained from

$$[G_{ij}\varepsilon_i\varepsilon_j]R^2 + [G_i\varepsilon_i]R - 1 = 0 \qquad (8.26)$$

The solution of this quadratic equation is obtained by setting

$$aR^2 + bR - 1 = 0 \qquad (8.27)$$

where $a = G_{ij}\varepsilon_i\varepsilon_j$, $b = G_i\varepsilon_i$, as

$$R = -(b/2a) + [(b/2a)^2 + 1/a]^{1/2} \qquad (8.28)$$

The constants a and b are invariants and have the same values in stress and strain spaces as long as a linear theory is used.

One may prefer to determine the failure envelopes in strain space (Table 8.2) since strain is at most a linear function of the thickness, in the laminated plate theory. Since failure envelopes are fixed in strain space and independent of other plies with different orientation, they can be regarded as material properties. The stress space failure envelopes of each ply in a multidirectional laminate is dependent on other plies in the laminate and are not material properties.

Example
Consider the E-glass–epoxy composite shown in Table 8.1. The maximum stresses are

$$X = 1062\,\text{MPa}, \quad X' = 610\,\text{MPa}$$
$$Y = 31\,\text{MPa}, \quad Y' = 118\,\text{MPa}$$
$$S = 72\,\text{MPa}$$

Table 8.2 Strength parameters in strain space of various composite materials (courtesy of Think Composites)

Type: Fiber: Matrix:	CFRP T300 N5208	BFRP B(4) N5505	CFRP AS 3501	GFRP E-glass Epoxy	KFRP Kev49 Epoxy	CFRTP AS4 PEEK	CFRP H-IM6 Epoxy	CFRP T300 Fbrt934 4-mil tp	CCRP T300 Fbrt934 13-mil c	CCRP T300 Fbrt934 7-mil c
Strength parameters $F^*_{xy} = -0.5$ (generalized von Mises)										
G_{xx}	12004	10374	7376	1914	13454	6394	5822	10971	29783	40019
G_{yy}	10681	27646	7467	18882	47657	4890	14914	12786	32580	40965
G_{xy}	-3069	-2989	-1746	1712	2069	-1584	-495	-2570	-13120	-17455
G_{ss}	11118	6961	5828	3306	4576	1016	7347	8985	9784	7944
G_x	61	130	39	25	-150	-40	-34	42	-65	-63
G_y	217	214	131	198	351	66	125	168	-52	-61
Strength parameters $F^*_{xy} = 0$ (Modified Hill)										
G_{xx}	15544	14823	9889	3669	23445	8136	9259	14999	31418	41879
G_{yy}	10882	28050	7630	19258	48380	5005	15104	13049	34216	42825
G_{xy}	3280	6728	2467	5137	16885	1545	4938	4183	3273	3720
G_{ss}	11118	6961	5828	3306	4576	1016	7347	8985	9784	7944
G_x	61	130	39	25	-150	-40	-34	42	-65	-63
G_y	217	214	131	198	351	66	125	168	-52	-61

By equation 8.20,

$$F_{xx} = \frac{1}{XX'} = 1.5 \times 10^{-18}$$

$$F_{yy} = \frac{1}{YY'} = 2.7 \times 10^{-16},$$

$$F_{xy}\begin{cases} \text{Generalized Von Mises criterion } (F_{xy}^* = -0.5): \\ \quad F_{xy} = F_{xy}^*(F_{xx}F_{yy})^{1/2} = -10^{-17} \\ \text{Modified Hill criterion } (F_{xy}^* = 0): F_{xy} = 0 \end{cases}$$

$$F_{ss} = \frac{1}{S^2} = 1.93 \times 10^{-16}$$

$$F_x = \frac{1}{X} - \frac{1}{X'} = -7.0 \times 10^{-10}$$

$$F_y = \frac{1}{Y} - \frac{1}{Y'} = 2.4 \times 10^{-8}$$

$$F_s = 0$$

$$G_{xx} = F_{xx}Q_{xx}^2 + 2F_{xy}Q_{xx}Q_{xy} + F_{yy}Q_{xy}^2$$
$$= \begin{cases} 2301 - 1701 + 1283 = 1883 & \text{(Von Mises)} \\ 2301 + 0 + 1283 = 3584 & \text{(Hill)} \end{cases}$$

$$G_{yy} = F_{xx}Q_{xy}^2 + 2F_{xy}Q_{xy}Q_{yy} + F_{yy}Q_{yy}^2$$
$$= \begin{cases} 7.12 - 365.8 + 19\,005 = 18\,646 & \text{(Von Mises)} \\ 7.12 + 0 + 19\,005 = 19\,012 & \text{(Hill)} \end{cases}$$

$$G_{xy} = F_{xx}Q_{xx}Q_{xy} + 2F_{xy}(Q_{xx}Q_{yy} + Q_{xy}^2) + F_{yy}Q_{xy}Q_{yy}$$
$$= \begin{cases} 128 - 3334 + 4938 = 1732 & \text{(Von Mises)} \\ 128 + 0 + 4938 = 5066 & \text{(Hill)} \end{cases}$$

$$G_{ss} = F_{ss}Q_{ss}^2 = 3308$$

$$G_x = F_xQ_{xx} + F_yQ_{xy}$$
$$= -27.4 + 52.32 = 25$$

$$G_y = F_xQ_{xy} + F_yQ_{yy}$$
$$= -1.53 + 201 = 199$$

Failure surface of off-axis plies
In section 3.5, the stresses acting on the inclined face, in terms of double angles, are obtained as

$$\sigma = \tfrac{1}{2}(\sigma_x + \sigma_y) + \tfrac{1}{2}(\sigma_x - \sigma_y)\cos 2\alpha + \tau_{xy} \sin 2\alpha$$
$$\tau = -\tfrac{1}{2}(\sigma_x - \sigma_y)\sin 2\alpha + \tau_{xy} \cos 2\alpha \tag{3.14}$$

The principal stresses are obtained as

$$\sigma_1 = \tfrac{1}{2}(\sigma_x + \sigma_y) + \sqrt{\tfrac{1}{2}(\sigma_x - \sigma_y)]^2 + \tau_{xy}^2}$$
$$\sigma_2 = \tfrac{1}{2}(\sigma_x + \sigma_y) - \sqrt{\tfrac{1}{2}(\sigma_x - \sigma_y)]^2 + \tau_{xy}^2} \tag{3.18}$$

and the maximum and minimum shear stresses are

$$\tau_{\substack{max \\ min}} = \pm \sqrt{[\tfrac{1}{2}(\sigma_x - \sigma_y)]^2 + \tau_{xy}^2} \tag{3.20}$$

Mohr's circle is discussed in section 3.1. Similar equations are developed for strains, and Mohr's circle for strains is also discussed. When Mohr's circle is drawn for strains, the ordinates represent $\gamma_\theta/2$ and the abscissae, ε_θ.

In section 5.5, transformation equations are expressed by the second power of the direction cosines, $m = \cos\theta$, $n = \sin\theta$. Transformations in terms of double angles are simpler than those in terms of single angles because, when double angles are used, the transformation matrices are rigid body rotations with these double angles. Recall that Mohr's circles are constructed by this double angle rotation (Figure 3.8).

With the above review on coordinate transformation and Mohr's circle, we proceed to study the off-axis failure surface.

Tsai defines, using linear combinations of the stress and strain components,

$$p_\sigma = (\sigma_1 + \sigma_2)/2, \quad q_\sigma = (\sigma_1 - \sigma_2)/2, \quad r_\sigma = \sigma_6 \tag{8.29}$$
$$p_\varepsilon = (\varepsilon_1 + \varepsilon_2)/2, \quad q_\varepsilon = (\varepsilon_1 - \varepsilon_2)/2, \quad r_\varepsilon = \varepsilon_6/2 \tag{8.30}$$

where the subscripts σ and ε refer to the stress and strain, respectively, and ε_6 is the engineering shear strain.

Using the same notation as Tsai, the stress and strain transformation equations in matrix form are

$$\{p', q', r'\}_\sigma = [K]\{p, q, r\}_\sigma$$
$$\{p, q, r\}_\sigma = [K]^{-1}\{p', q', r'\}_\sigma \tag{8.31}$$
$$\{p', q', r'\}_\varepsilon = [K^T]^{-1}\{p, q, r\}_\varepsilon$$
$$\{p, q, r\}_\varepsilon = [K^T]\{p', q', r'\}_\varepsilon \tag{8.32}$$

where the Ks are the transformation matrices which can be obtained from equation 3.14 given as

$$[K] = [K^T]^{-1} = \begin{bmatrix} 1 & 0 & 0 \\ 0 & \cos 2\theta & \sin 2\theta \\ 0 & -\sin 2\theta & \cos 2\theta \end{bmatrix}$$

$$\tag{8.33}$$

$$[K]^{-1} = [K^T] = \begin{bmatrix} 1 & 0 & 0 \\ 0 & \cos 2\theta & -\sin 2\theta \\ 0 & \sin 2\theta & \cos 2\theta \end{bmatrix}$$

The principal axes and stresses are

$$\tan 2\theta_0 = \frac{r}{q} = 2\sigma_6/(\sigma_1 - \sigma_2)$$

$$\sigma_{1,pr} = \sigma_I = p + R \tag{8.34}$$

$$\sigma_{2,pr} = \sigma_{II} = p - R$$

$$\sigma_{6,pr} = 0$$

where

$$R = \sqrt{q^2 + r^2}$$

The maximum shear occurs at 45° from the principal axes and the normal stresses on the planes of maximum shear are equal (section 3.6, summary, item 4).

$$\sigma_{1(\text{max shear})} = \sigma_{2(\text{max shear})} = p = \tfrac{1}{2}(\sigma_1 + \sigma_2) = \tfrac{1}{2}(\sigma_x + \sigma_y)$$

$$\sigma_{6(\text{max shear})} = \pm R \tag{8.35}$$

Equations 8.34 and 8.35 are the same as equations 3.15, 3.18, 3.20 and 3.21c. Similar relations for the principal strain can be obtained using the same procedure:

$$\tan 2\theta_0 = \frac{r}{q} = \frac{\varepsilon_6}{(\varepsilon_1 - \varepsilon_2)}$$

$$\varepsilon_{1,pr} = \varepsilon_I = p + R \tag{8.36}$$

$$\varepsilon_{2,pr} = \varepsilon_{II} = p - R$$

$$\varepsilon_{6,pr} = 0$$

where

$$R^2 = q^2 + r^2.$$

At $\theta = \theta_0 \pm 45°$

$$\varepsilon_1(\text{max shear}) = \varepsilon_2(\text{max shear}) = p$$

$$\varepsilon_6(\text{max shear}) = \pm R \tag{8.37}$$

Note that ε_6 is the engineering shear strain. Mohr's circles for stresses are discussed in section 3.6. Mohr's circle for strains is shown in Figure 8.4. The ordinate on the circle, r, in this figure is the tensorial shear strain which is one half of the engineering shear strain.

By equations 8.31, 8.32 and 8.33, we can transform an off-axis stress and strain to an on-axis stress and strain, and vice versa. Failure criteria discussed in the previous section can be applied after such a transformation. The failure stress or strain from an on-axis orientation, which indicates a point on the failure surface, can also be transformed by a similar manner to an off-axis orientation with given specific ply angle. This procedure creates a point on the failure surface of an off-axis ply. As equations 8.33 show, the off-axis surface can be generated from the on-axis surface by a rigid body rotation by an angle equal to twice the ply angle.

Figure 8.4 Mohr's circle for strains.

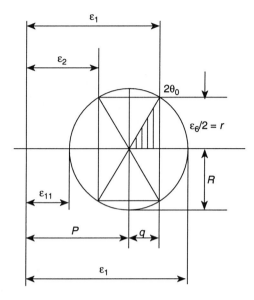

Composite laminates always have off-axis plies and it is shown that the off-axis failure surface can be easily generated from the obtained failure envelope in the symmetry or orthotropic axes of a ply.

Figure 8.5 shows the maximum strain criterion Tsai uses [21, 24]. The box in Figure 8.5(a) is the maximum strain failure envelope of a [0°] orthotropic ply in principal strain space; (b) is the similar envelope for a [90°] ply on the principal strain plane; (c) shows that for [45°] on the principal strain plane. The envelope in (c) is bounded by the allowable values of p and q given as equation 8.30:

$$p = (\varepsilon_1 + \varepsilon_2)/2$$
$$q = (\varepsilon_1 - \varepsilon_2)/2 \qquad\qquad (8.30)$$
$$r = \varepsilon_6/2$$

This state of strains is as shown in Figure 8.4. In Figure 8.5(c), the positive p is the transverse tensile strain while the negative p is the longitudinal compressive strain. In a typical unidirectional ply, the transverse tensile strain is smaller than the longitudinal tensile strain and the lower or controlling strain is used as the positive p. The allowable q is symmetric about the p axis and numerically equal to allowable r.

Figure 8.5(d) is made by overlapping the envelopes in (a), (b) and (c). The inner envelope of all intact plies is the 'first ply failure' (FPF) envelope of a [0°/90°/45°] laminate based on the maximum strain criterion.

Successive ply failures

There are several failure (or strength) criteria other than the maximum stress, maximum strain and quadratic (both in stress and strain space) criteria discussed here. One may use any criterion. However, it must be applied to each ply within a laminate. The ply with the lowest strength ratio fails first (FPF).

Figure 8.5 Maximum strain envelope for (a) [0°]; (b) [90°]; (c) [45°] on principal strain plane. (d) First-ply failure envelope of [0°/90°/+45°] laminate. (Courtesy of Think Composites.)

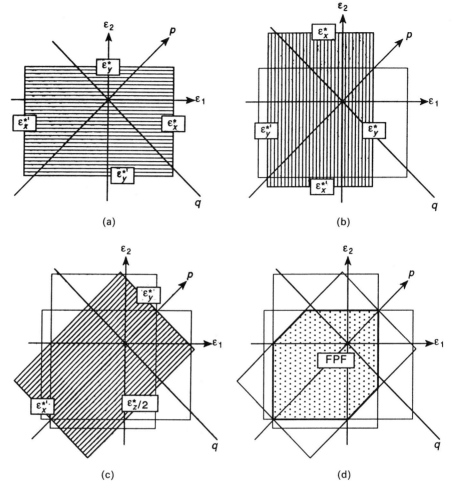

(a)   (b)

(c)   (d)

If the applied stress or strain exceeds this FPF envelope, plies will be degraded by matrix/interface cracks. This will continue until the last-ply failure (LPF). Tsai [24] shows that, by the use of a degradation factor, the FPF can be linked with the LPF through a continuous variation between the intact and degraded plies.

The quadratic criteria are easier to use because each ply is handled by only one strength ratio for each applied load, while the maximum stress or strain criterion requires taking three out of five strength ratios for each ply. As the number of plies increases, the number of ratios required is three times those required for quadratic criteria.

Successive ply failures proceed from the ply with the lowest strength ratio until the ultimate strength of the laminate is reached (LPF).

To conclude, the quadratic criteria in strain space have advantages such as the following.

1.   The presence of ply angles and materials (hybrids included) does not affect the criteria.

2. Laminate failure envelopes are obtained by superposition.
3. The difference between longitudinal and transverse failure strains is less.
4. Obtaining off-axis failure envelopes is done easily by rigid body rotation about the p–q space.
5. The FPF criteria can be extended easily to include LPF (ultimate strength), hygrothermal effects and sensitivity studies.
6. The strain is dimensionless and has the same values for all units.

8.5 Design effort to reduce failure

Some design concepts can reduce failure significantly. As an example, we consider free edge delamination [25–29]. Suppose the two boundaries $y = 0$ and $y = b$ in Figure 7.1(a) are free edges.

There are three **modes** of failure:

- Mode I, peeling or opening mode. It can grow in the direction of either reinforcing fibers or normal to the reinforcing fibers. σ_z stress is responsible for this mode.
- Mode II, shear or forward shearing mode. τ_{yz} is responsible for this type of failure mode.
- Mode III, out-of plane shear, parallel shear or tearing mode. τ_{xz} is responsible for this type.

A combination of the three modes is also possible [30].

Practical reinforcing concepts include putting a U-shaped cap around the free edge, stitching near the free edge to increase the transverse stiffness, inserting interleaved adhesive layers, and modifying the edge to alter the state of the stress by means of ply termination, notching or tapering [27]. Toughening the matrix is also effective. Addition of 20% urethane increased both the mode I and mode II interlaminar fracture toughnesses of CFRP by 70% and 40%, respectively [28, 29].

References

Note: *ICCM = International Conference on Composite Materials*; SAMPE = Society for the *Advancement of Material and Process Engineering*.

1. Masters, J. E. (1987) Failure analysis, in *Composites* (eds Dostal, C. A. *et al.*), Engineering Materials Handbook, ASM International, Metals Park, OH, pp. 765–6.
2. Grove, R. A. (1987) Failure causes, in *Composites* (eds Dostal, C. A. *et al.*), Engineering Materials Handbook, ASM International, Metals Park, OH, pp. 767–9.
3. Dvorak, P. J. (1987) Designing with composites. *Mach. Des.*, 26 Nov., 2.
4. Hayashi, T. (1967) Analytical study of interlaminar shear stresses in a laminated composite plate, in *Proceedings 7th International Symposium on Space Technology and Science* (ed. Kuroda, Y.), Tokyo, pp. 279–86.
5. Soni, S. R. and Kim, R. Y. (1987) Analysis of suppression of free-edge delamination by introducing adhesive layer, in *Proceedings ICCM 6* (eds Matthews, F. L. *et al.*), London, pp. 5-219–5-230.

6. Lagace, P. A. and Brewer, J. C. (1987) Studies of delamination growth and final failure under tensile loading, in *Proceedings ICCM 6* (eds Matthews, F. L. *et al.*), London, pp. 5-262–5-273.

7. Mall, S. and Donaldson, S. L. (1989) Cyclic delamination growth in a graphite/epoxy composite under mode III loading, in *Proceedings ICCM 7* (ed. Wu, Y.), Guangzou, China, pp. 3–8.

8. Allix, O., Daudeville, L. and Ladeveze, P. (1989) A new approach for laminate delamination, in *Proceedings ICCM 7* (ed. Wu, Y.), Guangzou, China, pp. 42–7.

9. Jones, R. M. (1975) *Mechanics of Composite Materials*, McGraw-Hill, New York.

10. Armanios, E. A. and Li, J. (1991) A fracture analysis model for composite laminates under combined loading, in *Proceedings ICCM 8* (eds Tsai, S. and Springer, G.), July, p. 27-A.

11. Chateauminois, A., Chabert, B., Soulier, J. B. and Vincent, L. (1991) Hygrothermal aging effects on viscoelastic and fatigue behavior of glass/epoxy composites, in *Proceedings ICCM 8* (eds Tsai, S. and Springer, G.), July, p. 16-E.

12. Robertson, R. E. (1987) Discontinuous fiber composites, in *Composites* (eds Dostal, C. A. *et al.*), Engineering Materials Handbook, ASM International, Metals Park, OH, pp. 794–7.

13. Kim, R. Y. (1987) Fatigue strength, in *Composites* (eds Dostal, C. A. *et al.*), Engineering Materials Handbook, ASM International, Metals Park, OH, pp. 436–44.

14. Echtermeyer, A. T. *et al.* (1991) Significance of damage caused by fatigue on mechanical properties of composite laminates, in *Proceedings ICCM 8* (eds Tsai, S. and Springer, G.), July, p. 38-A.

15. Beaumont, P. W. R., Spearing, S. M. and Kortschot, M. T. (1991) The mechanics of fatigue damage in structural composite materials, in *Proceedings ICCM 8* (eds Tsai, S. and Springer, G.), July, p. 38-E.

16. Plumtree, A. and Shen, G. (1991) Fatigue damage evolution and life prediction, in *Proceedings ICCM 8* (eds Tsai, S. and Springer, G.), July, p. 38-M.

17. Kim, R. Y. (1988) Fatigue behavior, in *Composite Design* (ed. Tsai, S.), 3rd edn, Think Composites, Dayton, OH, pp. 19-1–19-34.

18. Vinson, J. R. and Sierakowski, R. L. (1987) *The Behavior of Structures Composed of Composite Materials* (ed. Vinson, J. R.), Martinus Nijhoff, Dordrecht, pp. 209–238.

19. Rowlands, R. E. (1985) Strength (failure) theories and their experimental correlation, in *Failure Mechanics of Composites*, Handbook of Composites (eds Sih, G. and Skudra, A.), Vol. 2, Elsevier Science Publishers, Amsterdam, p. 71.

20. Hwang, W. C. *et al.* (1989) Failure analysis of laminated composites, in *Proceedings of 34th International SAMPE Symposium and Exhibition* (ed. Zakrzewski, G. A.), Reno, Nevada, pp. 1369–78.

21. Tsai, S. W. (1988) *Composite Design*, Think Composites, Dayton, OH, pp. 11-1–11.17.

22. Sacharuk, Z., Neale, K. W. and Makinde, A. (1991) A general strain-based failure criterion for fiber-reinforced composites, in *Proceedings ICCM 8* (eds Tsai, S. and Springer, G.), July, p. 31-E.

23. Feng, W. W. and Groves, S. E. (1991) On the finite-strain-invariant failure criterion for composites, in *Proceedings ICCM 8* (eds Tsai, S. and Springer, G.), July, p. 31-G.

24. Tsai, S. W. and Springer, G. S. (1987, 1988, 1989, 1990) Composites design and processing. Tutorial Lecture Notes at SAMPE International Symposium and Exhibition.

25. Pan, J. Z. and Loo, T. T. (1989) A new technique for prevention of free-edge delamination in laminated composites, in *Proceedings ICCM 7* (ed. Wu, Y.), Vol. 3, pp. 61–6.

26. Hong, C. S. and Kim, D. M. (1991) Mixed mode free-edge delamination in composite laminates, in *Proceedings ICCM 8* (eds Tsai, S. and Springer, G.), July, p. 28-D.

27. Jones, R. M. (1991) Delamination-suppression concepts for composite laminate free edges, in *Proceedings ICCM 8* (eds Tsai, S. and Springer, G.), July, p. 28-M.

28. Ozdil, F. and Carlsson, L. A. (1991) Mode I interlaminar toughening of graphite/epoxy through interleaving, in *Proceedings ICCM 8* (eds Tsai, S. and Springer, G.), July, p. 28-N.

29. Messenger, C. R. *et al.* (1991) Effect of resin toughness on delamination growth in composite laminates, in *Proceedings ICCM 8* (eds Tsai, S. and Springer, G.), July, p. 28-V.

30. O'Brien, T. K. (1991) Delamination, durability and damage tolerance of laminated composite materials, in *Proceedings ICCM 8* (eds Tsai, S. and Springer, G.), July, p. 28-A.

Further reading

Book

Timoshenko, S. T. (1953) *History of Strength of Materials*, McGraw-Hill, New York.

Articles and reports

Bogdanovich, A. E. (1991) Dynamic failure analysis of laminated composite plates and cylindrical shells, in *Proceedings ICCM 8* (eds Tsai, S. and Springer, G.), July, p. 30-J.

Crasto, A. S. and Kim, R. Y. (1993) The influence of specimen volume on matrix-dominated composite strength, in *Proceedings 38th International SAMPE Symposium and Exhibition* (eds Bailey, V. *et al.*), Anaheim, CA, May, pp. 759–70.

Groves, S. E., Sanchez, R. J. and Feng, W. W. (1991) Multiaxial failure characterization of composites, in *Proceedings ICCM 8* (eds Yunshu, W. *et al.*), July, p. 37-B.

Hsiao, C. C., Cheng, Y. S., You, S. J. and Yuan, Y. H. (1989) A new damage criterion for composites, in *Proceedings ICCM 7* (eds Yunshu, W. *et al.*), Vol. 3, p. 340.

Kim, R. Y. and Miravete, A. (1991) Effect of residual stresses on the fracture of thermoplastic composites, in *Proceedings ICCM 8* (eds Tsai, S. and Springer, G.), July p. 31-B.

McColskey, J. D. (1993) Compression response of axially loaded E-glass/epoxy tubes, in *Proceedings 38th International SAMPE Symposium and Exhibition* (eds Bailey, V. *et al.*), Anaheim, CA, May, pp. 896–908.

Rubbrecht, P. H. and Verpoest, I. (1993) The development of two new test methods to determine the mode I and mode II fracture toughness for varying fibre orientations at the interface, in *Proceedings 38th International SAMPE Symposium and Exhibition* (eds Bailey, V. *et al.*), Anaheim, CA, May, pp. 875–87.

Shahid, I. S. and Chang, F. K. (1993) Failure and strength of laminated composite plates under multiple in-plane loads, in *Proceedings 38th International SAMPE Symposium and Exhibition* (eds Bailey, V. *et al.*), Anaheim, CA, May, pp. 967–77.

Smith, B. W. (1987) Fractography for continuous fiber composites, in *Composites* (eds Dostal, C. A. *et al.*), Engineering Materials Handbook, ASM International, Metals Park, OH, pp. 767–9.

Tsai, S. W. and Wu, E. M. (1971) A general theory of strength for anisotropic materials. *J. Composite Mater.*, **5**, 58–80.

Wang, C. J. and Jang, B. Z. (1991) Impact performance of polymer composites: deformation process and fracture mechanism, in *Proceedings ICCM 8* (eds Tsai, S. and Springer, G.), July, p. 32-B.

Wang, Z. M. and Mao, T. X. (1989) Load carrying capacity of composite structures, in *Proceedings ICCM 7* (ed. Wu, Y.), Vol. 3, pp. 262–7.

9 Joints

9.1 Introduction

Civil and architectural structures are generally large and are manufactured as units. The use of composite materials significantly reduces the number of component units, but joining is still necessary.

The functional limit of a composite structure is, with very few exceptions, determined by its joints rather than its component units. As with structures of any material, locating and sizing the joints and preparing for unexpected repair work are part of routine design procedure. The design sequence for composites includes optimizing fiber patterns and filling the gaps in between. While the ductility of conventional metals provides yielding, reducing the stress concentration around the bolt or rivet holes and redistributing the stresses at the net section of the joints, most fiber reinforced composites are brittle and have very little stress reduction or redistribution capability even though local debonding reduces the most severe stress concentrations. The techniques for joining metal structures are well developed and diversified, including riveting and bolting, welding, glueing, brazing, soldering, alone or in combination. The methods of joining fiber–matrix composites are

1. chemical joining (adhesive bond);
2. mechanical joining (mechanical fastener);
3. combination of the above.

The highly orthotropic composite units have high load carrying capability along the fiber direction, but the bolted joints in such units have very low, often unacceptably low, capability. On the other hand, the quasi-isotropic laminates such as $[0°/\pm45°/90°]_s$ orientations have almost constant bearing strengths and section strengths, greatly simplifying design procedure. According to the literature [1–3], a $[0°/\pm45°/90°]_s$ quasi-isotropic composite has maximum bearing strength when the orientation has 25% fiber at 0°, 50% at $\pm45°$ and 25% at 90°. The preferred orientation is 37.5% maximum and 12.5% minimum in any one direction. When the structural unit requires a highly orthotropic arrangement, some kind of 'transition area' design is necessary between the main part and connecting part of the unit. A combination of mechanical and adhesive joints may be one solution. Some adhesives become stronger than the adherends after the joints are cured.

The design of mechanical joints in quasi-isotropic laminates is straightforward. The analyses of adhesive bonded joints are very well advanced, even though the mathematics involved is very complicated, requiring 40 equations and 40 unknown functions for a double lap joint [2]. However, design procedures that can be used by ordinary engineers are well established. The fact to remember in designing bolted joints is that the joint strength is always below the local composite strength. It appears that even the most carefully designed bolted joints are at most half as strong as the base laminate [1].

357

9.2 Mechanical joints

9.2.1 General considerations

The design method used for metallic structures may be applied to fiber reinforced composites effectively, provided that consideration is given to several factors, such as tensile and bending stresses, strength and flexibility of the fastener, loss of tensile strength in the adherend caused by fiber cutting due to drilling, shear distribution in the joint, friction between parts, residual stresses, allowable bearing stresses which depend on the fiber orientation, types of fasteners, and fatigue. The anisotropic nature of stiffness and strength, the low interlaminar shear, and low through the thickness tensile strength may cause unexpected failure. Important facts for structural designers to remember are:

1. Design concepts and methods used for a composite system cannot be used for another system unless a careful study is made.
2. Of all parameters influencing the behavior of the joint, the clamping force made by tightening the bolt is most important.

The presence of washers, tightened by the through the thickness clamping force, prevents the laminate from splitting through the thickness on the loaded side of the hole. Such splitting occurs frequently with a pin with no such restraint. According to Matthews [4], the bearing strength of a fully tightened bolt can be up to four times that of a pin joint. Even a 'finger tight' nut can show a strength twice that of a pin joint.

9.2.2 Joint geometry and failure modes

Joint types used for metal structures have been successfully used in composites. It is assumed that readers are familiar with steel design. Some of the common bolted joint types and failure modes are shown in Figures 9.1 and 9.2, respectively.

9.2.3 Design criteria

The bearing strength of composites is the stress that produces 4% elongation of the diameter of the hole carrying the fastener [5]. Figure 9.3 shows the typical geometry of a bolted joint. The geometric factors that influence joint strength are the width w, side distance s, end distance e and thickness t of the laminate, and the bolt diameter D [4]. The stresses are defined in terms of the failure

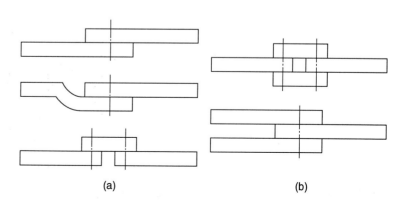

Figure 9.1 Common bolted joint types: (a) single shear joint; (b) double shear joint.

(a)   (b)

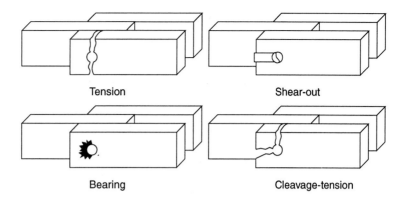

Figure 9.2 Failure modes of bolted joint. (Adapted from reference [4].)

Tension

Shear-out

Bearing

Cleavage-tension

load P, usually taken as the maximum tensile load sustained by the joint [4]. The strength is usually expressed, even when the failure is other than bearing, by the bearing stress

$$\sigma_b = \frac{P}{Dt} \qquad (9.1)$$

for the single bolt case.

The effect of D/t varies with material. Conservative bearing load designs utilize a D/t ratio of 1 for glass fiber reinforced composites with thickness greater than 0.090 in (2.29 mm), while using a slightly larger D/t with reduced allowable bearing strength for thinner laminates.

Dastin [5] gives the typical bearing strengths of glass fiber reinforced composites when $D/t = 1$ (Table 9.1). This table shows the general range of bearing strength and joint design must be verified by subscale element static and fatigue tests, including the environmental effects, for each design configuration.

Dastin [5] also gives the recommended fastener distances for various thicknesses of glass fiber–epoxy composites (Table 9.2).

Gibbs and Cox [6] recommend, for both woven roving and cloth reinforced composites, $e/D = 2.5$, $s/D = 2.5$, and for mat reinforced composites, $e/D = 2.0$,

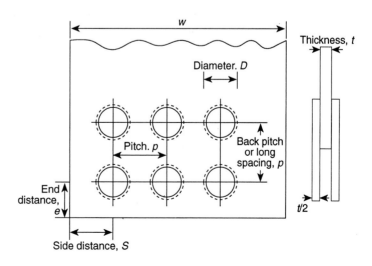

Figure 9.3 Multi-bolt joint.

Table 9.1 Bearing strength of fiberglass reinforced composites [6]

Composite type	Ultimate strength psi (MPa)	Design allowable (4% hole elongation) psi (MPa)
Woven fabric–polyester	43 200(300)	35 000(240)
Mat–polyester	30 000(210)	20 000(140)
Woven fabric–epoxy	46 500(320)	37 000(255)

Table 9.2 Recommended fastener distances for glass fiber/epoxy composites

Thickness of laminate (in. (mm))	Edge distance ratio, e/D	Side distance ratio, s/D
1/8 (3.18) or less	3.0	2.0
1/8–3/16 (3.18–4.76)	2.5	1.50
3/16 (4.76) and greater	2.0	1.25

$s/D = 2.0$. The longitudinal spacings (pitches) between two bolts for both cases are $3D$. All e, s and the longitudinal pitches are the minimum values. If the above distances e, s and p are used, failure of the connection will occur by local laminate crushing (bearing) under the bolt or shearing of the bolt.

The factor of safety (FS) is the ratio of the maximum stress level encountered in service to the allowable design stress. In composite structures, FS values are generally based on ultimate strengths. According to Dastin, a conservative FS for glass fiber reinforced structures is 3, while weight critical designs use 2, with greater quality control requirements. For boron, aramid and carbon reinforced composites, the FS is 1.5 with a practical margin of safety of 15%.

The shear failure criterion is

$$\sigma_s = \frac{P}{2et} \tag{9.2}$$

If fibers are placed at $\pm 45°$, the shear strength of the laminate is high and the stress concentration may not be significant.

The tensile failure stress is

$$\sigma_t = \frac{P}{wt} \text{ on the gross area} \tag{9.3}$$

and

$$\sigma_t = \frac{P}{(w - nD)t} \text{ on the net area} \tag{9.4}$$

where w is the width of the joint.

Dastin [5] gives typical shear properties of a 181 weave glass fiber–epoxy laminate as follows:

- Panel shear (edgewise) strength, psi (MPa) 16 000 (110.2)
- Modulus, Msi (GPa) 0.8 (5.51)
- Tear-out shear, psi (MPa) 8000 (55.1)
- Interlaminar shear, psi (MPa) 4000 (27.6)

9.2.4 Multibolted joints

Dastin [5] gives the expression for the joint efficiency which is defined as the ratio of the strength of the joint members to the strength of the unjoined continuous member of equal size.

For metallic components

$$\text{Efficiency} = \frac{(w - nD)}{w} \tag{9.5}$$

where n is the number of fasteners in a row.

For composites

$$\text{Efficiency} = \frac{L_f}{L_c} \tag{9.6}$$

where L_f and L_c are the loads producing failure in joint and in continuous member, respectively.

The strength per bolt in a row or line joint is less than that of a single bolt unless the pitch, p, is greater than $4D$.

Based on his research experience, Matthews [4] gives the strength of the joint, P, as

$$P = nKP_{\min} \tag{9.7}$$

where n is the number of bolts in a row, K is the strength reduction factor given in Figure 9.4, and $P_{\min}$ may be obtained by equation 9.1 and the 'design allowable' such as in Table 9.1, or, preferably, directly from test results for each laminate design.

In general, the bearing stresses of single lap joints are about 15% lower than those of double lap, single bolt configurations on which most of the test results are based.

9.2.5 Review of studies

Vinson gives an extensive review of joints [2]. According to Vinson, Lehman and his team found that a combination of bolted–bonded joints performed better than bolted joints or bonded joints alone. They also concluded that linear discrete element analysis did not predict load–deformation characteristics of bolted joints, and that extremely sophisticated nonlinear analysis is required for such purposes. Although more difficult to fabricate, the shimmed joint produces a very compact high strength joint. The thickened end design is good

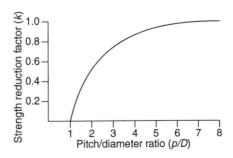

Figure 9.4 Strength reduction factor, k. (Courtesy of Think Composites.)

for general purpose applications. The use of whiskers as a resin additive did not increase the strength of the joint. Dastin [5] strongly recommends the use of bonded threaded inserts, especially for application of multiple assembly/ disassembly. If high installation loads are expected, either threaded or un-threaded metallic spacers should be used to preclude compression or bearing premature failures.

The important conclusions of Kutscha and Hofer [2] attract attention from structural engineers who used to employ simple rule-of-thumb empirical equations in both design office and field. The semi-empirical methods for bolted joint design are the most effective, if the materials are isotropic. For composites, however, the failure modes are so different that a complete stress analysis and failure envelope in four dimensions are required to develop even the semi-empirical procedure to design a given class of joints.

Some important conclusions by Lehman *et al.* are: $e/D = 4.5$ is required to achieve a balance between shear-out and bearing strengths. $s/D = 2$ precludes the tensile failures through the fastener holes. $t/D = 0.8$ results in the maximum bearing strength. Maximum shear-out stresses are obtained in a laminate with $0°$ and $\pm 45°$ orientation in which 2/3 of the laminae are $\pm 45°$ to the load axis. The bearing strength increases until e/D reaches 5. The bolted joint strength increases linearly with e/D ratio, up to $e/D = 4$. Beyond this, increasing e/D does not increase joint strength appreciably regardless of t/D. All these experiments were done with the bolt clamping friction effects included.

Van Siclen found that the allowable shear-out strength for a $[0°/\pm 45°/90°]$ carbon–epoxy laminate is significantly reduced as e/D is increased, contradictory to Lehman's result. For $e/D > 4$, bearing at the hole becomes the failure mode. This fact indicates that shear-out strength is not constant for a particular laminate orientation.

Stockdale and Matthews [7] reported that ultimate bearing loads are increased 40% to 100% in glass fiber reinforced laminates by the clamping pressure of a bolt.

Quinn and Matthews [8] reported that placing $90°$ layers on or next to the outer surface increases the bearing strength. Regarding stacking sequence, $[90°/\pm 45°/0°]_s$ had the highest bearing strength, about 30% higher than the $[0°/90°/\pm 45°]_s$ orientation.

9.2.6 Bolted joint strength

$P_{\min}$ in equation 9.7 can be obtained from the test results with consideration of appropriate end distances. This may be calculated from empirical formulas proposed by Lehman or Van Siclen.

Lehman [2] proposed the following empirical equations for determining the bolted joint strength:

$$\frac{P}{Dt} = K_m\left(\frac{e}{D}\right)(1 - e^{-1.6(s/D - 0.5)})(1 - e^{-3.2(t/D)}) \quad \text{for} \quad 0 < \frac{e}{D} < 4 \quad (9.8)$$

where K_m is an experimentally determined coefficient, and

$$\frac{P}{Dt} = K_n F_{BRU} \quad \text{for} \quad \frac{e}{D} > 4 \quad (9.9)$$

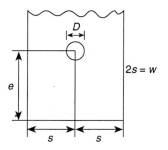

Figure 9.5 Geometry of single bolt joint.

where K_n is a coefficient and F_{BRU} is the ultimate bearing strength of the material. Both equations are based on experiments for $s/D > 2$ and laminates of $0°$ and $\pm 45°$ only. The experiments produced no tension failures but only the shear-out and bearing failures.

Van Siclen [2] presented an empirical formula to determine the net tension stress concentration factor as

$$K_{tc} = 1 + A\left[\left(\frac{D}{2s}\right)^{-0.55} - 1\right]\left(\frac{e}{2s}\right)^{-0.5} \tag{9.10}$$

where A is a constant to be determined experimentally for a particular laminate. For geometry, see Figure 9.5. Assuming the ultimate tensile strength of the laminate, F_x^{tu}, is known, the value of the ultimate tensile strength of the laminate with a bolted connection, F_{nt}, is expressed as

$$F_{nt} = \frac{F_x^{tu}}{K_{tc}} \tag{9.11}$$

The net tensile load a given joint can carry is thus

$$
\begin{aligned}
P_{nt} &= F_{nt}(2s - D)t \\
&= F_{nt}(w - D)t
\end{aligned} \tag{9.12a}
$$

where $2s$ is the specimen width, w. This P_{nt} may be used as P_{min} in equation 9.7. For multiple bolts in a row,

$$P_{nt} = F_{nt}(w - nD)t \tag{9.12b}$$

and

$$P = KP_{min} \tag{9.12c}$$

when $p \le 4D$. The Van Siclen equation for the shear-out strength F_{so}, is [2]

$$F_{so} = A_1\left(\frac{e}{D}\right) + A_2 \tag{9.13}$$

where A_1 and A_2 are constants to be determined for a specific laminate orientation. The allowable load on a fastener at shear-out failure, P_{so}, is

$$P_{so} = F_{so}Dt\left(\frac{2s}{D} - 1\right) = F_{so}(2s - D)t \tag{9.14a}$$

For a multibolted joint,

$$P_{so} = F_{so}(w - nD)t \tag{9.14b}$$

As with P_{nt}, this P_{so} may be used as P_{min} in equation 9.7 without the n term, that is

$$P = KP_{min} \tag{9.14c}$$

The allowable bearing load on the joint, P_{BR}, is

$$P_{BR} = F_{BRU}Dt \tag{9.15}$$

where F_{BRU} is the ultimate bearing strength for a given laminate. This may be used as $P_{\min}$ in equation 9.7.

Hoffman's modified distortion energy criterion for layer-by-layer failure considerations is given as [2]

$$\frac{\sigma_L^2 - \sigma_L\sigma_T}{S_{Lc}S_{Lt}} + \frac{\sigma_T^2}{S_{Tc}S_{Lt}} + \frac{S_{Lc} - S_{Lt}}{S_{Lc}S_{Lt}}\sigma_L + \frac{S_{Tc} - S_{Tt}}{S_{Tc}S_{Tt}}\sigma_t + \frac{\tau_{LT}^2}{T_{LT}^2} = 1 \quad (9.16)$$

where

- σ normal stress
- τ shear stress
- S allowable normal stress
- T allowable shear stress

with subscripts

- L fiber direction
- T normal to fiber direction
- t stress in tension
- c stress in compression

9.2.7 Design approach

Because of the complexities of mechanical fastening joints in composites, obtaining a rational analytical solution is not feasible at present or in the near future.

Vinson [2] considers the design approach of Van Siclen is the best from the design standpoint. His equations are easy to use with the particular tests required.

In structural design, the material system for both reinforcements and matrices, laminate orientation and thickness are determined before joint design is considered. For the chosen laminate, tests should be performed to determine the strengths F_{nt} (net tensile), F_{so} (shear-out) and F_{BR} (bearing). The actual bolts to be used should be tested. Countersunk and protruded head fasteners have different strengths. The normal pressure due to bolt torque can increase the bearing strength by 40% to 100%.

The design approaches may be summarized as follows.

Bearing strength only
The bearing stress is

$$\sigma_b = \frac{P}{Dt} \quad (9.1)$$

for a single bolt. σ_b allowable is the stress that produces 4% elongation of the diameter of the fastener hole.

Start with $e/D = s/D = 2.5$, and the longitudinal pitch $p = 3D$. With such a configuration, the failure is usually of bearing type. The shear and tensile failures can be checked by

$$\sigma_s = \frac{P}{2et} \qquad (9.2)$$

$$\sigma_t = \frac{P}{wt} \qquad (9.3)$$

on the gross area and

$$\sigma_t = \frac{P}{(w - nD)t} \qquad (9.4)$$

on the net area. For multibolted joints with pitch, $p \le 4D$, use equation 9.7. K may be obtained from Figure 9.4.

Van Siclen concept
For each laminate configuration, perform tests to obtain the results in the form of Figures 9.6 to 9.8. F_{nt}, F_{so} and F_{BRU} can be obtained from these figures.
The **net tensile load** this joint can carry is

$$P_{nt} = F_{nt}(w - D)t \qquad (9.12a)$$

for a single bolt and

$$P_{nt} = F_{nt}(w - nD)t \qquad (9.12b)$$

for multiple bolts.

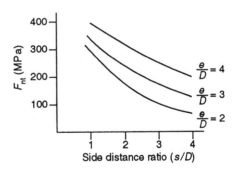

Figure 9.6 Typical shape for net tension strength curve.

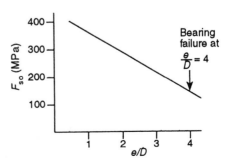

Figure 9.7 Typical shear-out strength curve.

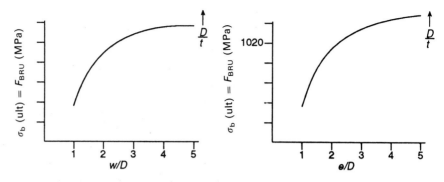

Figure 9.8 Typical shape for bearing stress curve.

If $p \leq 4D$, the strength of the joint is

$$P_{nt} = KF_{nt}(w - nD)t \qquad (9.12c)$$

where n is the number of bolts in a row.

The allowable load on a fastener at **shear-out failure** is

$$P_{so} = F_{so}(w - D)t \qquad (9.14a)$$

for a single bolt and

$$P_{so} = F_{so}(w - nD)t \qquad (9.14b)$$

for multiple bolts.

If the pitch $p \leq 4D$, the strength of the joint is

$$P_{so} = KF_{so}(w - nD)t \qquad (9.14c)$$

The allowable **bearing load** on the joint is

$$P_{BR} = F_{BRU}Dt \qquad (9.15)$$

for a single bolt and

$$P_{BR} = nF_{BRU}Dt \qquad (9.17)$$

for multiple bolts.

If the pitch $p \leq 4D$,

$$P_{BR} = KnF_{BRU}Dt \qquad (9.18)$$

The K value can be obtained from Figure 9.4.

Design, analysis and optimization can be made using the above equations.

9.3   Adhesive bonded joints

9.3.1   General considerations

Adhesives appear to have been used since the dawn of history. The early glues were all natural substances. Modern adhesives are synthetic resins and polymers, and some are capable of bonding metals. Adhesive bonding is also called chemical joining, in contrast to mechanical joining.

Advantages

The advantages of adhesive bonding are as follows.

1. Since the bond is continuous, the stress distribution under loadings over the bonded area is uniform while the mechanically fastened joints and spot welded joints will have a local stress concentration. A continuous welded joint may also be uniformly stressed but the adherend in the heated zone will undergo a change in strength. This lack of stress concentration at bonded joints contributes to thinner element thickness (Figure 9.9).

2. The bonded joint results in a structure with higher stiffness due to shorter unsupported length when subjected to a compression load. For bonded structures, loading may be increased up to 100% before buckling occurs, for example, when $l_1/l_2 = \sqrt{2}$ (Figure 9.10).

3. Fatigue resistance is superior. According to Kuno [2], the fatigue life of adhesive assemblies is 20 times better than riveted or spot welded structures of identical parts. In a bonded structure, fatigue cracks are less likely

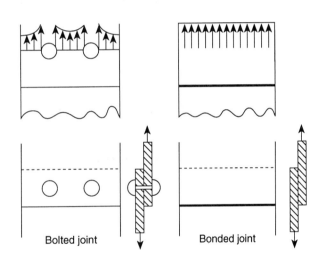

Figure 9.9 Stress distribution at joint.

Bolted joint   Bonded joint

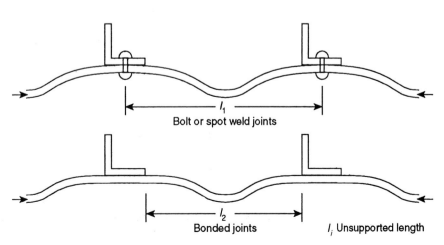

Bolt or spot weld joints

Bonded joints l_i Unsupported length

Figure 9.10 Unsupported length of bolted and bonded joints.

to occur, and if they do, propagate more slowly than in a mechanically fastened structure. The bond lines act as a crack stopper.

4. Vibration damping and noise reduction are superior.
5. Complex assemblies, such as composite sandwich structures, which cannot be joined together by any other method, are possible with adhesives, resulting in a reduced number of parts and thus simplified design.
6. In the case of a light gage structural element, savings in weight and cost are possible. For example [2], 1.30 mm thick aluminum sheets required for riveting could be replaced by 0.508 mm thick sheets when adhesive bonding is used.
7. Because of their flexibility, adhesives can join materials with different moduli, composition, coefficients of expansion (thermal etc.), or thickness.
8. The continuous adhesive bond forms a leakproof seal with resulting corrosion prevention.
9. Heat sensitive materials can be joined without damage.
10. Electrical and thermal insulation are excellent.

All these advantages contribute to beneficial results in terms of improved design, easier assembly, weight and cost reduction, enhanced reliability, flexibility of use and longer service life.

Regardless of all these advantages, adhesive bonding has not been widely introduced in civil and architectural engineering. The ability to design and analyze these joints subject to general mechanical, thermal, and hygrothermal loads must be developed. It has to be borne in mind that confidence in the reliability of this technology is well established among composite engineers.

Limitations

Since adhesives are made from polymers, they have the limitations of this group. In general, there is a prejudicial belief that polymers are not as strong as metals. In fact, such problems can be taken care of by the increased contact surface area provided by the bonded joints. With increased temperature, the bond strength decreases and the strain begins to have plastic properties. This transition temperature depends on the particular adhesive and is in the range 70–220°C.

The environmental resistance of bonded joints depends on the properties of the base polymer. When selecting the adhesive type to use, the anticipated environmental exposure of the bonded structure must be carefully studied.

Unlike mechanical fastening or welding, most of the adhesives require a certain length of time to acquire the design bond strength, necessitating that the assembled unit be supported for such a period.

The quality of the bond depends on the bonding procedure, which includes surface preparation, mixing the constituent materials properly and wetting the bond surface well.

Good quality control is necessary. A badly made joint is often impossible to correct. Finally, bonded structures are difficult to dismantle for in-service repair.

9.3.2 Adhesive types

A good summary of types and characteristics of adhesives are given in a manufacturer's manual [9]. The following is adapted from this reference.

- **Anaerobics.** These adhesives harden in contact with metal when air is excluded. Often known as 'locking compounds' or 'sealants', they are used to seal and retain turned or threaded bolts. They are based on acrylics.
- **Cyanoacrylates.** These adhesives belong to the acrylic group and cure through reaction with moisture held on the surfaces to be bonded. They usually solidify in seconds and are suited to small plastic parts and to rubber.
- **Toughened acrylics.** Being a modified type of acrylic, these adhesives are fast curing and offer high strength and toughness. Resin and catalyst are supplied as two parts. They are usually applied by separate application, resin to one bond surface, catalyst to the other. They tolerate minimum surface preparation and bond well to a wide range of materials.
- **Epoxies.** These adhesives consist of two parts: an epoxy resin and a hardener. They allow great versatility in formulation since there are many varieties of resins and hardeners. They form extremely strong durable bonds with most materials and are available in one-part, two-part and film form.
- **Polyurethanes.** These adhesives are usually two-part and fast-curing. They provide strong resilient joints which are resistant to impact. They are useful for bonding glass fiber reinforced composites. The fast cure usually necessitates applying the adhesives by machine. Polyurethanes are often used as primers.
- **Modified phenolics.** These are the first adhesives for metals and have a long history of successful use for making high-strength metal-to-metal and metal-to-wood joints, and for bonding metal to brakelining materials. They need heat and pressure for the curing process.

The above types set by chemical reaction. Types that are less strong, but important to industry are as follows.

- **Hot melts.** Sealing wax is one of the oldest forms of adhesive. Today's industrial hot melts are based on modern polymers and are used for the fast assembly of structures designed for light loading.
- **Plastisols.** These adhesives are modified PVC dispensions and require heat to harden. The resulting joints are often resilient and tough.
- **Rubber adhesives.** These are based on solutions of latexes and solidify through loss of the solvent or water medium. They are not suitable for sustained loading.
- **Polyvinyl acetates (PVAs).** Vinyl acetate is the principal constituent of the PVA emulsion adhesives. They are suited to the bonding of porous materials, such as paper or wood, and general packaging work.
- **Pressure-sensitive adhesives.** These are suited to use on tapes and labels and do not solidify but are often able to withstand adverse environments. They are not suitable for sustained loading.

9.3.3 Failure modes of adhesive bonded joints

Bonded joints may be subjected to tension, compression, shear, or peel stresses, or any combination of the above. Figure 9.11 shows schematic diagrams of the basic ways such stresses act. Adhesives are strong in shear, compression, and tension but perform poorly under peel and cleavage loadings. A perspective

Figure 9.11 Failure modes of
bonded joints.

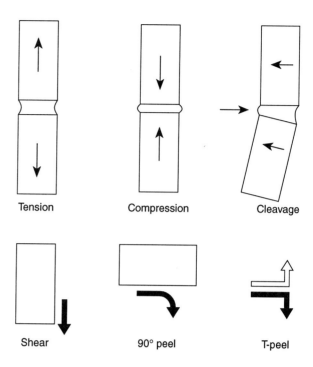

Figure 9.11 Failure modes of
bonded joints.

view of an adhesive bonded joint with all probable failure modes is given in
Figure 9.12.

There are three distinct failure modes in adhesively bonded joints.

1. Failure of adherend including delamination due to excessive interlaminar
   shear in the case of a fibrous laminate, longitudinal tensile failure, and/or
   failure due to transverse stresses.
2. Shear strength failure of the adhesive.
3. Failure of the adhesive under a peel load.

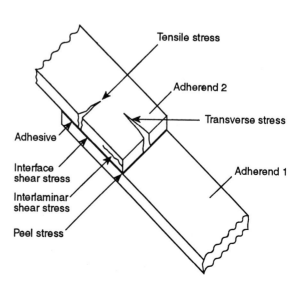

Figure 9.12 Perspective view of an
adhesive bonded joint with probable failure modes.

Figure 9.13 Peel stress failure mode of thick composite joints.

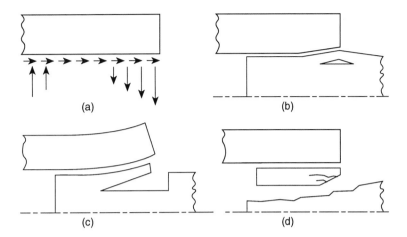

(a)   (b)

(c)   (d)

Type 1 failure occurs when the applied load is 100% of the strength of the adherend. The strength of the joint is linearly proportional to the thickness of the laminate.

Type 2 is an adhesive failure of the bond between the adherend and the adhesive. The shear strength of the bonded joint is proportional to the square root of the laminate thickness. This type of failure is different from 'cohesive failure', which means a failure entirely within the adhesive material. Adhesive failure means the failure of the bond between the adhesive and adherend.

The peel load of type 3 causes a tensile stress, the maximum value of which is near the free end of a double or single lap joint (Figure 9.13(a)). The stress is developed in association with the shear stresses. High peel stresses are developed in single lap joints, and can be severe even in thick double lap joints.

The low interlaminar tension strength of laminate limits the thickness of the adherends which are bonded efficiently by lap joints. If the inner laminae of the adherend split away under the full stresses (Figure 9.13(b)), the shear-transfer capacity between the inner and outer laminae is destroyed. This results in overloading of the outer filaments, which will break in tension (Figure 9.13(c)), and the failure progresses through the adherend thickness (Figure 9.13(d)). In general the interlaminar tension strength is far less than the peel strength of typical structural adhesives. Concrete is another good example. If the concrete is improperly bonded by some polymers, the surface of the concrete simply peels off.

9.3.4 Characteristics of bonded joints

Single lap joint
Figure 9.14 shows the single and tapered lap joints. The eccentricity of the load causes out-of-plane bending moments which may produce high peel stresses. The tapered single lap joint can prevent the premature failure of the single lap joint. For such purposes, Hart-Smith [1] suggests tapering the edges of the splice plates as shown in Figure 9.15. This scheme does not increase the shear strength of the joint. The shear strength can be increased by using a tapered double strap joint (discussed below and shown in Figure 9.18(c)). Another

Figure 9.14 Single and tapered lap joint.

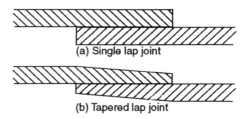

(a) Single lap joint

(b) Tapered lap joint

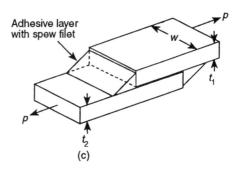

Adhesive layer
with spew filet

(c)

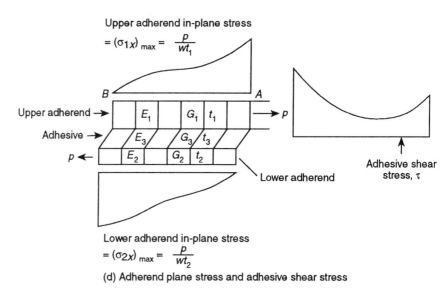

Upper adherend in-plane stress

$= (\sigma_{1x})_{max} = \dfrac{p}{wt_1}$

Upper adherend →

Adhesive →

Lower adherend

Adhesive shear
stress, τ

Lower adherend in-plane stress

$= (\sigma_{2x})_{max} = \dfrac{p}{wt_2}$

(d) Adherend plane stress and adhesive shear stress

method of preventing premature failure is to increase the length of the bond line, even though there is a certain limit on this length for efficiency.

The adherends and adhesive react to the applied force. At the interface, the stresses have the shear component along the plane of the interface and the peel stress components normal to the shear. These stresses are at a maximum at the edges of the bond line (Figures 9.14(d) and 9.16(c) [10]).

Adams and Harris [11] report that if a spew fillet is formed at the ends of single or double lap joint adherends, the shear stress is considerably reduced compared with the square-ended adhesive layer. They conclude that significant increases in single lap joint strength may be achieved in practice by including

Figure 9.15 Tapered splice edge to reduce the peel stresses. (Adapted from reference [1].)

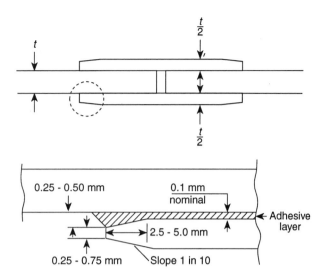

an adhesive fillet at the edges of the overlap and further increases may be achieved by additionally rounding the ends of the adherends.

The simplest analysis considers the adherends to be rigid and the adhesive to deform only in shear [10]. In this case, the shear stress is expressed as

$$\tau = \frac{P}{wl} \tag{9.19}$$

Figure 9.16 Deformations and stresses in loaded single lap joint. (Adapted from reference [10].)

This is shown in Figure 9.16(a). The adherend tensile stress decreases linearly to zero over the joint length, from A to B for the upper adherend and from B to A for the lower adherend (Figure 9.14(d)). If the adherends are elastic, the

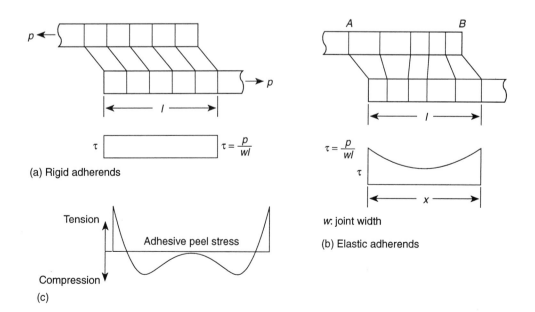

shear stress variation looks like Figure 9.16(b). The distorted shape in Figure 9.16(b) is called 'differential shear'. This was analyzed by Volkerson [10].

The eccentricity of applied load in a single lap joint causes bending of the adherends and a rotation of the bonded region. The adhesive is acted upon by both adherend stretching and bending strains, and will have the complex shaped distribution of shear and peel stresses as shown in Figures 9.16(b) and (c). Volkerson theory does not consider this eccentricity problem.

Vinson [2] presents an extensive review on this subject. After thorough study of the existing literature, Kutscha and Hofer, in 1969 [2], concluded that no rational design procedure for all bonded joints exists. They state that the best design technique is empirical: developing a joint shear strength parameter, using this as a guide for joint design, then building a joint and testing it. They made interesting conclusions such as:

1. Adhesive stresses decrease with increasing specimen width, up to about 10 cm (4 in) width. Beyond this, the stresses remain constant.
2. The maximum stresses do not decrease significantly as the bonded area increases.

Lehman and Hawley [2] found that

1. The maximum joint strength occurs when the extensional stiffnesses of both adherends are the same.
2. Fatigue runout occurs when the maximum shear stress in the adhesive is below the proportional limit shear stress in the adhesive. The fatigue runout is the number of cycles N at which testing is terminated with no fracture prior to this number.
3. For bonded joints of epoxy matrix composites with a length to thickness ratio of about 25, the interlaminar strength of the laminate is the limiting strength.

Grimes [2] makes an important statement: since the predicted mean strength and the type of failure of even complex joints is reasonably accurate compared with experimental tests, design allowable calculation using mean strength should be accurate. The use of a factor of safety of 1.5 on the design limit loads to obtain design ultimate loads should be sufficient for static loadings at room temperature.

Hart-Smith [2] found that fracture of $0°$ filaments close to the adhesives usually initiates an interlaminar shear failure within the laminated adherends. For thicker adherends, the dominant failure modes are peel tensile stresses in the adhesive and the related interlaminar tension stresses in the laminated adherends.

Vinson and colleagues made thorough studies on adhesive bonded joints considering in-plane static and dynamic loads, transverse shear deformation, transverse normal strain and temperature and hygrothermal effects [2, 12–14].

The concept of an **ineffective length**, defined as the lap length beyond which an increase in this dimension is ineffective in reducing peak adhesive shear and peel stresses, is used by Oplinger [2].

As for the efficiency of bonding, a simple lap joint is 'good' while a tapered lap joint is 'very good'.

Double lap joint and double strap (doubler) joint

These are shown in Figures 9.17 and 9.18(b), respectively. The double strap (or doubler) joint is sometimes called a double lap joint.

Extensive studies were carried out by several researchers. Some of the findings are summarized here.

At a symmetric double lap joint, unlike a single lap joint, there is no apparent bending moment because of symmetry. However, the load is applied through the adhesive to the adherend plates away from their neutral axes. This causes internal bending of the double lap joint (Figure 9.19). In a symmetric double lap joint, the center adherend is free from the net bending moment, but the outer adherends do bend, causing tensile stresses across the adhesive layer at the free end of the overlap and compressive stresses at the end of loading direction (Figure 9.19).

Because of symmetry, the double lap joints are more than twice as strong as single lap joints of the same lap length. For practical design, lap length to thickness ratio of 30 may be used.

Figure 9.20 shows the effect of lengthening the overlap. There is still a large, low stress region in the middle of the joint. The ineffective length is defined above.

The tapered double strap (doubler) joint has considerably high shear strength.

The joint efficiencies of both double lap and double strap joints are 'very good', and that of the tapered double strap joint is 'excellent'.

Other types of bonded joints

The single strap (doubler) joint (Figure 9.18(a)) can be important in repair work. The geometry is asymmetric and this type of joint is subject to high peel stresses as with a single lap joint. This joint is 'fair' in performance.

Figure 9.17 Double lap joint.

Figure 9.18 (a) Strap (single doubler) joint. (b) Double strap (double doubler) joint. (c) Tapered double strap (tapered double doubler) joint.

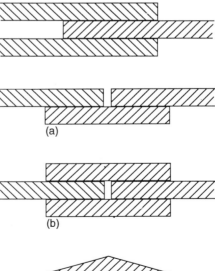
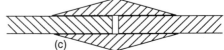

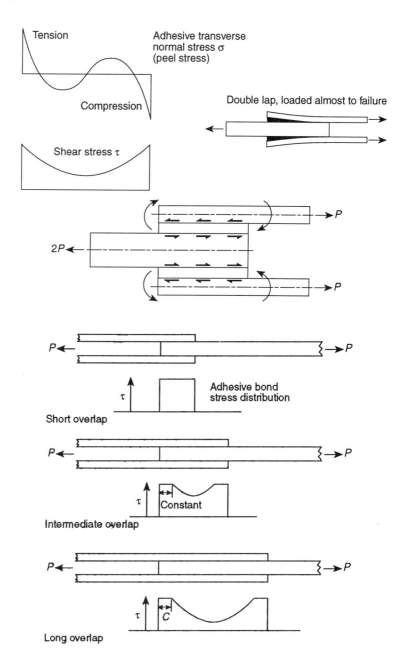

Figure 9.19 Bending moments caused by outer adherends. (After Adams and Wake [10]).

Figure 9.20 Influence of lap length on bond stress. (Adapted from reference [10]).

The scarf joint is shown in Figure 9.21. Because of the machine work required, it may be more useful for metallic adherends than those of composite materials. The scarf joint is 'excellent' in performance and has the capability to transmit increased loads with increased lap length. It was found that the scarf joint could sustain fatigue stress levels 3.5 times greater than those of double lap joints [2]. This type of joint has ideal strain compatibility in the adherends and uniform stress in the adhesive. As the overlap length increases, the peak adhesive stresses are reduced and the full overlap length is efficiently used. There is no 'ineffective length' bound in scarf joints.

Figure 9.21 Scarf joint.

Figure 9.22 Stepped lap joint.

Figure 9.22 shows a stepped lap joint. Stepped lap joints have higher average shear strength than the scarf joint [2]. The strength is not sensitive to the number of steps as long as the total lap length is kept constant. Both scarf and stepped lap joints are lighter in weight than any other lap joints at all load levels. The strain compatibility of the stepped lap joint resembles that of the scarf joint. There is no 'ineffective length' bound.

In practical structures, two or more basic types may be combined for use. Several schemes such as combined use with mechanical fasteners, and insertion and bonding can be used (Figure 9.23). For tubular joints, Adams and Wake [10] give a good review of the theory involved.

9.3.5 Load transfer through adhesive bond

In 1938, O. Volkerson [1] established that the load transfer through adhesive bonds between uniformly thick adherends is not uniform but peaks at each end of the overlap (Figure 9.16(b)). A few years later M. Goland and E. Reissner analyzed the peel stress distribution in the adhesive layer in a single lap joint. The study was made for two cases.

1. The adhesive is assumed to be very thin and relatively stiff, and applicable to the bonding of wood, or similar low modulus adherends.
2. The flexibility of the joint is due mainly to that of the adhesive layer, and is applicable to the bonding of metal and fiber reinforced adherends.

From strain–energy considerations, the criterion for the range of validity of each case is given as:

1. $t_3/E_3 < t_1/10E_1$, or $t_3/G_3 < t_1/10G_1$.
2. $t_1/E_1 < t_3/10E_1$, or $t_1/G_1 < t_3/10G_3$.

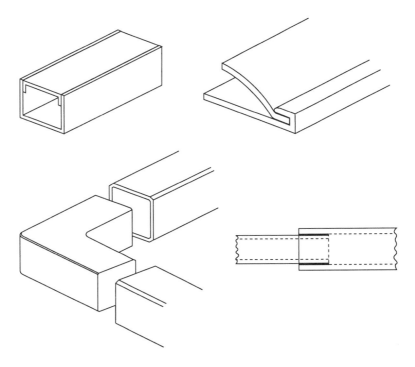

Figure 9.23 Typical insertion and bonding.

The notations are as shown in Figure 9.14(d). However, by assuming the bending moments in the adherends at the ends of the overlap as $M_0 = kPt/2$, a physically unrealistic and mathematically unnecessary simplification is made. The derivation of the bending moment factor k allows for the rotation of the overlap region, but the internal stresses in the joint are derived assuming no rotation. The solutions for the adhesive shear and the tensile stresses are accurate only for $k = 1$, that is, for very low loads. Such classical early work was limited because the peel and shear stresses were assumed constant across the adhesive thickness, the shear was a maximum – not zero – at the overlap end, and the shear deformation of the adherends was neglected.

Adams and Peppiatt [10] showed that there existed significant stresses across the width of an adhesive joint. They considered the existence of shear stresses in the adhesive layer and direct stresses in the adherends acting at right angles to the direction of the applied load, these stresses being caused by Poisson's ratio strains in the adherends. They neglected the effects of bending so that the results are more applicable to double lap joints. The geometry and stresses are as shown in Figure 9.24.

Considering compatibility in the x–y plane, and equilibrium in the x–y and the y–z planes, where z is the coordinate in the width direction, the following two equations are obtained:

$$\frac{\partial^2 \sigma_{1x}}{\partial x^2} = K_a \sigma_{1x} - K_b \sigma_{1z} + C_a \tag{9.20}$$

$$\frac{\partial^2 \sigma_{1z}}{\partial z^2} = K_a \sigma_{1z} - K_b \sigma_{1x} + C_b \tag{9.21}$$

where

$$K_a = 2G_1 G_2 G_3 (E_1 t_1 + E_2 t_2)/(\text{DEN1})$$

$$K_b = 2G_1 G_2 G_3 (v_2 E_1 t_1 + v_1 E_2 T_2)/(\text{DEN1})$$

$$C_a = -2P G_1 G_2 G_3/(\text{DEN2})$$

$$C_b = 2v_2 P G_1 G_2 G_3/(\text{DEN2})$$

in which

$$\text{DEN1} = E_1 t_1 E_2 t_2 (t_1 G_2 G_3 + t_2 G_1 G_3 + 2t_3 G_1 G_2)$$

$$\text{DEN2} = wt_1 t_2 E_2 (t_1 G_2 G_3 + t_2 G_1 G_3 + 2t_3 G_1 G_2)$$

The boundary conditions are

$$\text{at } x = 0, \quad \sigma_{1x} = \frac{P}{wt_1}, \quad \sigma_{2x} = 0$$

$$\text{at } x = l, \quad \sigma_{2x} = \frac{P}{wt_2}, \quad \sigma_{1x} = 0 \tag{9.22}$$

$$\text{at } z = \pm \frac{w}{2}, \quad \sigma_{1z} = 0, \quad \sigma_{2x} = 0$$

Figure 9.24 Geometry and stress at double lap joint. (Adapted from reference [10]).

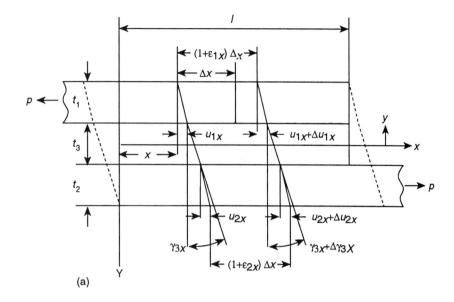

(a)

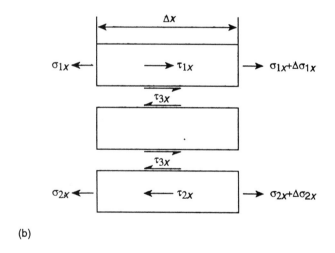

(b)

The solution of two simultaneous partial differential equations 9.20 and 9.21 can be obtained with good accuracy by finite differences.

Approximate simplification can be made by assuming σ_{1x} is constant with respect to z and by neglecting the $K_b\sigma_{1z}$ term which is small compared with $K_a\sigma_{1x}$. The resulting equations are exact at the boundaries and are given as follows.

$$\sigma_{1x} = \frac{P}{wt_1}\left\{1 - \varphi(1 - \cosh \alpha x) - \frac{[1 - \varphi(1 - \cosh \alpha l)]\sinh \alpha x}{\sinh \alpha l}\right\} \quad (9.23)$$

where

$$\alpha = \sqrt{K_a}, \quad \varphi = \frac{E_2 t_2}{E_1 t_1 + E_2 t_2}$$

$$\tau_{3x} = \frac{P\alpha}{w} \left\{ \frac{[1 - \varphi(1 - \cosh \alpha l)]\cosh \alpha x}{\sinh \alpha l} - \varphi \sinh \alpha x \right\} \tag{9.24}$$

$$\sigma_{1z} = t_2(v_1 \sigma_{1x} E_2 - v_2 \sigma_{2x} E_1)\left[\cosh\left(\frac{\alpha w}{2}\right) - \cosh \alpha z\right] \Big/$$

$$\left[(t_1 E_1 + t_2 E_2)\cosh\left(\frac{\alpha w}{2}\right)\right] \tag{9.25}$$

$$\tau_{3z} = \frac{t_1 t_2 \alpha (v_1 \sigma_{1x} E_2 - v_2 \sigma_{2x} E_1)\sinh \alpha z}{(t_1 E_1 + t_2 E_2)\cosh(\alpha w/2)} \tag{9.26}$$

The notations are as shown in Figure 9.24. These equations are exact at the boundaries:

$$\text{at } z = \pm w/2, \quad \sigma_{1z} = 0$$

$$\text{at } x = 0, \quad \sigma_{1x} = P/wt_1, \quad \sigma_{2x} = 0$$

Adams and Peppiatt solved equations 9.20 and 9.21 by finite differences and the results compared well with those of approximate analytical solutions.

Delale *et al.* [15] present a closed form solution to the general plane strain problem of adhesively bonded joints consisting of two different orthotropic adherends. The adherends are treated as plates. The transverse shear effects in the adherends and the in-plane normal strain in the adhesive are taken into account. The problem is reduced to a system of differential equations which is solved in closed form. A single lap joint and a stiffened plate under various loadings are considered as examples. The plate theory used in the analysis not only predicts the correct trend for the adhesive stresses but also gives accurate results. The solution is based on the assumption of linear stress–strain relations for the adhesive.

Through the continuum mechanics approach, it is possible to predict the stresses in simple lap joints by closed form solution. However, two problems still remain to be solved to obtain the real strength of the joints. They are the end effects and the material nonlinearity. Considerable work has been done by Adams on the first subject, Hart-Smith worked on the latter subject, the details of which will be discussed in the next section.

9.3.6 *Elasto-plastic analysis*

Modern adhesives, particularly those such as the rubber-modified epoxies, have a large plastic strain to failure. These new adhesives are so strong that the adherends may be forced to yield. Even with the old brittle adhesives, the adherends in single lap joints often yield plastically in bending [10].

The linearly elastic analysis of bonded joints has been far too conservative for the strong ductile adhesives [1]. The design philosophy behind Hart-Smith's work is that the adhesive should never be the weak line. If peel stresses are likely to occur, these stresses should be alleviated by tapering the adherends or by locally thickening the adhesive layer. Hart-Smith's continuum mechanics approach has the advantage of allowing a parametric study on the effects of bond-line thickness, joint length, etc. At the same time, investigation of adherend and adhesive mechanical properties can be carried out.

Hart-Smith chose an elasto-plastic model to characterize the adhesive, such that the total area under the stress–strain curve is equal to that under the true stress–strain curve. If the maximum stress is less than yield strength the true elastic curve may be used. For a peak stress intermediate between yield and failure, a different and more accurate model is chosen (Figure 9.25). This bilinear model is closer to the true adhesive characteristics over the entire range of loads, so that a single model can be used to calculate the stresses without having to calculate these to establish which intermediate elasto-plastic model should be used.

The analysis for the double lap joints using a bilinear model, given in Figure 9.25, shows that the predicted joint strength is the same for any two straight-line adhesive models having the same strain energy. The advantage of the bilinear model is that a single model will work for all load levels.

In practice, it is sufficient to perform two analyses.

1. A linearly elastic analysis for limit load using the actual adhesive shear modulus.
2. An elasto-plastic model to predict the ultimate joint strength. This elasto-plastic model is inappropriate for low load levels since the initial shear modulus is too low and the elastic strain energy is too high [1].

The ultimate strength of a long overlap bonded joint between uniform adherends should be defined by the strain energy of the adhesive layer in shear, not by any individual properties such as peak shear stress. The strengths of practical bonded joints are not very sensitive to the environment, provided that the temperature is kept below the glass transition temperature of each adhesive [1].

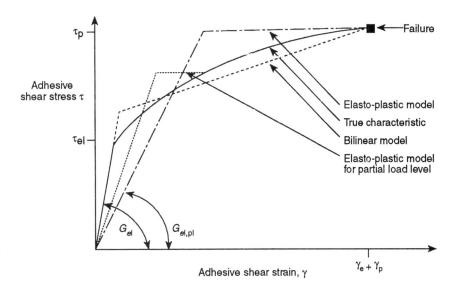

Figure 9.25 Adhesive nonlinear shear behavior. (After Adams and Wake [10].)

Figure 9.26 Shear stress and strain distribution due to increased loads, 1 → 2 → 3. (After Hart-Smith [1].)

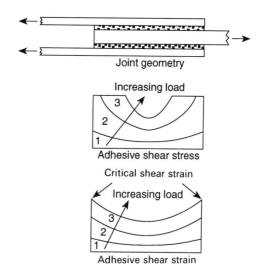

Joint geometry

Increasing load

Adhesive shear stress

Critical shear strain

Increasing load

Adhesive shear strain

Hart-Smith says that failure occurs when the adhesive strain reaches its limiting value (Figure 9.26). If the overlap is short, the minimum adhesive shear stress and strain are almost as high as the maximum values. For the long overlap joint, the overlap length keeps the adhesive shear stress and strain as low as is required (Figure 9.20), and the 'ineffective length' bound appears as the length is continuously increased. This 'ineffective length' contributes to the strength of the joint. There is still a large, low stress region in the middle of the joint. This region is not inefficient but essential to overcome the creep at the ends if low cycle creep/fatigue loading is applied [10]. The creep in the adhesive at the ends of the long overlap cannot accumulate because the stiff adherends push the adhesive back to its original position when the joint is unloaded. This memory action of the adhesive is the key to a durable and successful bonded structure [1]. Figure 9.27 shows Hart-Smith's design criteria for double lap joints. The overlap protects against the extremes of moisture and temperature, etc. For such extreme cases, the plastic zones must be long enough to carry the ultimate load, the elastic region in the middle must be long enough to prevent creep, and the minimum operating stress (elastic minimum) must be kept below 10% of the adhesive shear strength. The width of the elastic trough is adjusted so that the minimum stress is 10% of the maximum [1]. This value is obtained when the elastic trough has a total length of $6/\lambda$, where λ is the exponent of the elastic adhesive shear stress distribution.

In order to transfer a load at least equal to the entire strength of the adherends with the adhesive stressed to the maximum shear strength, a sufficient plastic zone must be added at each end of the elastic overlap.

Hart-Smith [1] says that the static strength of bonded joints between uniform adherends is insensitive to the precise long overlap. This insensitivity of the joint strength to the total bonded area is important. Many structural engineers designing joints have the old notion that the bond strength is equal to the product of the bond area and some fictitious uniform 'allowable' shear strength.

The basic Hart-Smith approach neglects the normal or 'peel' stresses acting across the bond line. The rationale for this is that in order to improve structural

Figure 9.27 Double lap joints.
(After Adams and Wake [10].)

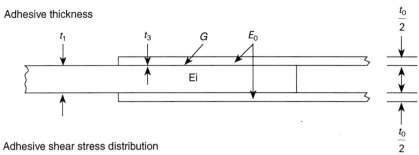

Adhesive thickness

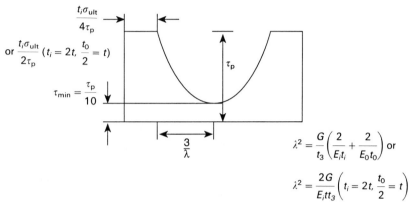

Adhesive shear stress distribution

$$\lambda^2 = \frac{G}{t_3}\left(\frac{2}{E_i t_i} + \frac{2}{E_0 t_0}\right) \text{ or}$$

$$\lambda^2 = \frac{2G}{E_i t t_3}\left(t_i = 2t, \frac{t_0}{2} = t\right)$$

efficiency, the peel stresses should be removed from the structure by simple modifications in design detail rather than including these in a complicated failure criterion [1]. In practice, however, the peel stresses may be significantly serious even for double lap joints. Hart-Smith recognizes this fact and gives for a double lap joint the peak peel stress σ_p as a function of the peak shear stress τ_p as [13]

$$\sigma_p = \tau_p\left[\frac{3E_3(1 - v^2)t_1}{E_1 t_3}\right]^{1/4} \tag{9.27}$$

where t_1 is the adherend thickness, t_3 is the adhesive thickness, E_3 is the effective transverse Young's modulus for the adhesive, and E_1 is Young's modulus for the adherend.

An extensive review on failure criteria of adhesives is presented by Adams and Wake [10], which is not treated here.

9.3.7 Design considerations

Theories and practical applications of adhesive bonded joints have been developed mostly by aerospace engineers. The basic theories and experiences obtained can be used for other branches of engineering. General guidelines of designing adhesive bonded joints are given as follows.

1. Maximum shear and normal stresses can be reduced by making D_{11} and A_{11} of the adherends as high as possible [2].
2. The values of D_{11} and A_{11} of the adherends on each side of the bond should be equal to maintain symmetry of the adhesive shear and normal stresses about the mid-length ($l/2$).

3. For a double lap joint, the lap length is the sum of the plastic zone length and the elastic trough length and is given as

$$2C + \frac{6}{\lambda} \tag{9.28}$$

where

$$C = \frac{t_i \sigma_{ult}}{4\tau_p}$$

$$\lambda^2 = \frac{G}{t_3}\left(\frac{2}{E_i t_i} + \frac{2}{E_0 t_0}\right)$$

The notations are as shown in Figure 9.27. C is obtained from

$$t_i \sigma_{ult} = C \times (2 \text{ faces}) \times (2 \text{ ends})\tau_p.$$

4. For $[0°/45°/90°/-45°]$ orientation, the overlap length can be made 30 times the central adherend thickness for a double lap joint.
5. For adherends thicker than about 3.2 mm ($\frac{1}{8}$ in), a stepped lap joint is appropriate.
6. By making the tips of the adherends thin and flexible, the peel stresses can be made negligible.
7. Ductile adhesives are generally limited to service temperature less than about 70°C (160°F).
8. The stresses from loading must be directed along the lines of the adhesive's greatest strength.
9. In practice, a bonded structure must handle a combination of forces. For maximum strength, cleavage and peel stresses must be designed as far as possible out of the joints.

References

1. Hart-Smith, L. J. (1987) Joints, in *Composites* (eds Dostal, C. A. *et al.*), Engineering Materials Handbook, Vol. 1, ASM International, Metals Park, OH, pp. 479–495.
2. Vinson, J. R. and Sierakowski, R. L. (1987) *The Behavior of Structures Composed of Composite Materials*, Martinus Nijhoff, Dordrecht.
3. Van Siclen, R. C. (1974) Evaluation of bolted joints in graphite/epoxy, in *Proceedings of the Army Symposium on Solid Mechanics*, Sep.
4. Matthews, F. L. (1988) Bolted joints, in *Composite Design*, 4th edn, Think Composites, Dayton, OH.
5. Dastin, S. J. (1982) Joining and machining techniques, in *Handbook of Composites* (ed. Lubin, G.), Van Nostrand Reinhold, p. 603.
6. Gibbs and Cox (1960) *Marine Design Manual for Fiber Glass Reinforced Plastics*, McGraw-Hill, New York.
7. Stockdale, J. H. and Matthews, F. L. (1976) The effect of clamping pressure on bolt bearing loads in glass fiber-reinforced plastics. *Composites*, **7**, Jan., 34–8.
8. Quinn, W. J. and Matthews, F. L. (1977) The effect of stacking sequence on the pin-bearing strength of glass fiber reinforced plastics. *J. Composite Mater.*, **11**, Apr., 139–45.
9. Guide to Araldite bonding. Engineering data book, Ciba-Geigy, Cambridge, UK.
10. Adams, R. D. and Wake, W. C. (1984) *Structural Adhesive Joints in Engineering*, Elsevier Applied Science, New York.

11. Adams, R. D. and Harris, J. A. (1987) The influence of local geometry on the strength of adhesive joints. *Int. J. Adhes. Adhesiv.*, **7**(2), April, 69–80.
12. Renton, W. J. and Vinson, J. R. (1975) On the behavior of bonded joints in composite materials structures. *J. Engng Fract. Mechs*, **7**, 41–60.
13. Renton, W. J. and Vinson, J. R. (1975) The efficient design of adhesive bonded joints. *J. Adhes.*, **7**(3), 175–93.
14. Vinson, J. R. and Zumsteg, J. R. (1979) Analysis of bonded joints in composite materials structures including hygrothermal effects. AIAA Paper 79-0798.
15. Delale, F., Erdogan, F. and Aydivoglu, M. N. (1981) Stresses in adhesively bonded joints: A closed-form solution. *J. Composite Mater.*, **15**, May, 249.

Further reading

Articles and reports

Anon. (1988) A new generation of adhesives that stay tougher longer. *Insights*, **5**(1), Mar. Company note.

Anon. (1987) Adhesives in action, bonded aluminum chassis for new vehicle concept. *Int. J. Adhes. Adhesiv.*, Apr., **7**(2), 68.

Anon. (1987) Cyanamid develops rapid cure adhesives for naval aircraft field repair. *Insights*, **4**(2), Dec. Company note.

Anon. (1987) Fastener technology catches up. *Aviat. Wk Space Technol.*, **127**(24), 14 Dec., 89–93.

Browne, J. (1993) Aerospace adhesives in the 90's, in *Proceedings 38th International SAMPE Symposium and Exhibition* (eds Bailey, V. *et al.*), Anaheim, CA, May, pp. 771–84.

Chon, T. C. (1980) Analysis of tubular lap joint in torsion. *J. Composite Mater.*, **16**, Jan., 266.

De Jong, T. D. (1977) Stresses around pin-loaded holes in elastically orthotropic or isotropic plates. *J. Composite Mater.*, **11**, Jul., 313.

Hart-Smith, L. J. (1981) Further developments in the design and analysis of adhesive bonded structure joints, in *Joining of Composite Materials*, ASTM STP 749 (ed. Kedward, K. T.), American Society for Testing and Materials, pp. 3–31.

Hart-Smith, L. J. (1981) Stress analysis; continuum mechanics approach, in *Developments in Adhesives* (ed. Kinloch, A. J.), Applied Science Publishers.

Hyer, M. W. and Lightfoot, M. C. (1979) Ultimate strength of high-load-capacity composite bolted joints, in ASTM STP 674 (ed. Tsai, S. W.), American Society for Testing and Materials, pp. 118–36.

Juang, S. J. and Chen, M. F. (1991) Design of multi-pin joints in composite laminated plate, in *Proceedings ICCM 8* (eds Tsai, S. and Springer, G.), July, p. 9-A.

Kim, R. Y. and Whitney, J. M. (1976) The effect of temperature and moisture on pin bearing strength of composite laminates. *J. Composite Mater.*, **10**, Apr., 149–55.

Lessard, L. B. (1991) Bearing failure of composite pinned joints using progressive damage modeling, in *Proceedings ICCM 8* (eds Tsai, S. and Springer, G.), July, p. 9-E.

Marinelli, J. M. and Lambing, C. L. T. (1993) A study of surface treatments for adhesive bonding of composite materials, in *Proceedings 38th International SAMPE Symposium and Exhibition* (eds Bailey, V. *et al.*), Anaheim, CA, May, 1196–210.

Miaosheng, W. *et al.* (1991) Optimum design of adhesive bonding of composites, in *Proceedings ICCM 8* (eds Tsai, S. and Springer, G.), July.

Nakamura, K., Maruno, T. and Sasaki, S. (1987) Theory for the decay of the wet shear strength of adhesion and its application to metal/epoxy/metal joints. *Int. J. Adhes. Adhesiv.*, **7**(2), Apr., 97–102.

Sable, W. W. and Sharifi, P. (1991) Structural analysis of bonded joints using the finite element method, in *Proceedings ICCM 8* (eds Tsai, S. and Springer, G.), July, p. 9-F.

Soni, S. R. (1981) Stress and strength analysis of bolted joints in composite laminates, in *Composite Structures* (ed. Marshall, I. H.), Applied Science Publishers.

Tang, S. (1981) Failure of composite joints under combined tension and bolt loads. *J. Composite Mater.*, **15**(4), 329–35.

Wegman, R. F. (1993) The evaluation of the properties in adhesive bonds by nondestructive evaluation, in *Proceedings 38th International SAMPE Symposium and Exhibition* (eds Bailey, V. *et al.*), Anaheim, CA, May.

Zang, K. D. and Ueng, C. E. S. (1984) Stresses around a pin-loaded hole in orthotropic plates. *J. Composite Mater.*, **18**(5), 432–46.

Manufacturers' data

Adhesives and resins. Selector guide, Furane Products Co. Company note.

AF-563 Structural Adhesive. Technical data, 3M.

Araldite, Arathane, Aravite. Structural adhesives, selection guide, Giba-Geigy.

Bighead bonding fasteners. Technical data, Bighead Co.

Bonded structures: high-technology products for aerospace and down-to-earth uses. Resin aspects, Ciba-Geigy.

Cybond structural adhesives. Selection guide, Cyanamid.

Dexter aerospace adhesive products. Technical data, Hysol.

Fiberlite composite fasteners. Technical pamphlet, Fiberlite.

FM 87 adhesive film. Technical data, Cyanamid.

Huck Aerospace ASP fasteners. Technical data, Huck Co.

Huck Automatic fastening systems. Technical data, Huck Co.

Huck Fastening System, Design guide. Technical manual, Huck, July, 1991.

LGP lockbolt fastening system. Technical data book, Huck Co.

Microdot aerospace fastening systems. Technical pamphlet, Microdot Inc.

PR 2701 Permapol polymer-based elastomeric adhesive. Technical data, PRC, April, 1987.

Reactive products, selection guide. Technical data, Adhesive System Div., B. F. Goodrich.

Redux adhesives. Technical data, Ciba-Geigy.

Rivnut-Plusnut engineered fastners. Design guide, engineering data book, BF Goodrich-Aerospace and Defense Div.

Structural adhesive film AF-163-2. Aerospace technical data, 3M, Issue No. 1, March 1986.

Structural adhesives selector guide, Cyanamid.

Unimatic blind bolt fastening system. Technical data, Huck Co.

Z-Axis interconnect film. Technical data, Sheldahl.

10 Design

10.1 Introduction

The basic principles involved in the design of structures made of composite materials are the same as those of isotropic materials such as steel. If we consider the theories related to composite structures for the general cases, the theories for steel belong to a special case of homogeneity and isotropy of materials.

In Chapters 3 and 4, an extensive review of classical mechanics and elasticity is given. These classical theories and methods of analysis can be used for composite structures as long as the constitutive, i.e. stress–strain, relationship takes into account the characteristics of material anisotropy.

It is assumed that the reader has a background in structural design with metals such as steel. The same basic design knowledge and technique used for other materials can be applied to composite structures; reinforced concrete is a good example. However, a design engineer working for the first time with composites often treats them in the same way as metals, which may have disastrous results. The implementation of accurate design methods for steel structures has required a century of research and experience. The range of composite materials is very wide, and new noble materials are reported almost day by day, so that the designer of the future will have to be involved with material specifications to control formulation, form and dimensions of the element, and the manufacturing process. In fact, the concrete design engineer has been doing such things for a long time.

10.1.1 Basic rules for design

The basic rules for designing with composites, most of which are discussed in previous chapters, can be summarized as follows.

As with any material, it is important to know the loads as accurately as possible. Steel has uniform stiffness in all directions. In a composite, the maximum stiffness and strength are obtained in the longitudinal fiber direction, i.e. on-axis direction. The matrix binds the fibers and bonds the laminae. The lower limit of the transverse stiffness is close to that of the matrix. The shear strength is less than the transverse strength (Table 5.1 gives values for Es and Qs). A composite structural member designed for tension cannot be loaded by torsion. In Chapter 3, thin walled structures are discussed. After resolving the loads into in-plane stresses, the optimum laminate configuration has to be designed. It is advised that the normal to the plane shear loading should be avoided. A composite element designed to replace a metal element should not have an identical shape. Composite element shapes are, in general, significantly different from those of metal elements. This is the reason why the author declares that composite structures belong to the fifth basic concept of structures. There is good flexibility in designing with composites. For each type of load, such as tension, compression, shear or torsion, we can simply add features to take care of such kinds of loads. We can select the parameters which control the mechanical behavior. For metals, for example, Young's modulus is the same

387

into all directions, and cannot be changed even in the direction of least stress. With composites, we can change the mechanical properties, even without changing the geometry of the structural element, by simply changing the fiber type, volume and orientation, and the matrix.

There are endless lists of both matrices and reinforcements. However, there are some guidelines which are discussed in Chapter 2. Use glass fibers for low-cost strength and large structures. If stiffness is important, use carbon fibers. For impact resistance, use aramid fibers. If the cost of carbon fiber decreases, it will be used in large quantities. Regarding the matrix, unsaturated polyester is widely used. For a corrosion-critical structure, use vinylester. For higher temperature and better strength, use epoxy. Of course, there are many noble materials deserving the design engineer's careful attention (Chapter 2). A cement matrix may be used in large scale structures if certain required properties are improved.

Some guidelines to designing with composites are as follows.

1. Consult the composite specialists at the early stage of design, on concepts including material and manufacturing process selection. The mechanical property of the composite depends on both of these.
2. The initial design effort should be concentrated on the most effective shape suited to the properties of the material.
3. Never try to develop a composite structural element based on a design used for another material, such as steel. Many commercial composite structural shapes are not well suited to the mechanical properties of the composites.
4. A basic knowledge on manufacturing, fabrication and quality control of composite structural elements must be maintained in order to integrate the design to the manufacturing process. There is a good article by Wilson on this subject [1]. The manufacturing and quality control processes are not treated in this book. However, these are covered in references [1] to [114].
5. Reduction of stiffness and strength after long-term loading must be considered. Both viscoelasticity and fatigue effects are important for some polymer matrix composites.
6. The maximum operational temperature should be decided. The service temperature of composites in nonaggressive environments should be limited to the material's T_g (glass transition temperature) minus 10°C to 38°C (50° to 100°F) depending on the factor of safety, loading and expected service life [115].
7. As with any material, the dimensional tolerances, especially due to temperature changes, must be specified.
8. The combined stress effects on the strengths of composites, under expected service environment must be considered, as discussed in Chapter 8.
9. As discussed in Chapters 8 and 9, lack of ductility requires detailed, thorough, local stress analysis. Design details should be made such that the stress concentrations due to notches, sharp corners, and restraints to deformation, etc. are eliminated.

10.1.2 Steps in the design process

As with any material, proper design procedure is very important for successful accomplishment of the project. Some examples of such processes can be found from references [115–117]. Experienced engineers may have their own ideas and it is assumed again that the reader has a basic knowledge of conventional design methods. Figure 10.1 shows a design process sequence by which it will be explained how the theories involved can be applied to composite design of large scale structures.

After all performance requirements are established (step 2), structural concepts such as structural configuration, material and manufacturing method, are developed with consideration of economy based on the total concept. Steel pipe of a certain diameter may be cheaper than composite pipe on a market price basis. The welding of 1 m diameter steel pipe, however, may take a whole day, while connection of certain composite pipes can be achieved in minutes. If problems such as traffic disruption are considered, not to mention the maintenance required through corrosion of steel, the initial cost of pipes may not be important. If the bridge decks and building walls are made of composite panels, the inertia force, in the case of an earthquake, is far less than those of

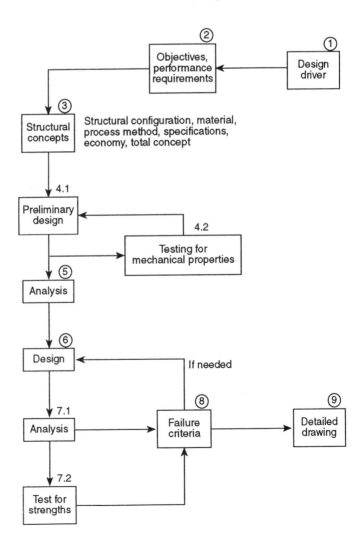

Figure 10.1 Typical design process.

conventional materials, and maintenance will be much easier, because of the light weight of composites. We must consider both short-term and long-term economy. The structural frames may be made as simple as possible. In fact, most of the civil architectural structures can be considered as frames of one-dimensional beams and columns. Even two-dimensional plates can be analyzed as a collection of beams [118]. Three-dimensional shells are often analyzed as a collection of two-dimensional plates or one-dimensional frames [119]. If this is done based on the advice of an experienced engineer, the result of such simplified analysis is not far from that of the 'exact' solution. If too many assumptions are made, obtaining a solution may be easy but the result may be too far out. If too few simplifications are made, the problem may turn out to be impossible to solve. One must trade off practicality and accuracy. Again, experience can make this gap closer.

With chosen material and manufacturing method, the range of material strengths can be estimated. With strength data, the initial dimensions of this framework can be obtained. One may start by assuming the element is either quasi-isotropic or unidirectional to the structural axis. Experienced engineers can assume even variable sections to obtain good results at an early stage. Whether it is quasi-isotropic or unidirectional, the mechanics discussed in Chapters 3 and 4 can be directly applied. A general guideline is to use a quasi-isotropic concept for two-dimensional problems and a unidirectional concept for one-dimensional frameworks. This concept is indirectly supported by the recent papers of Verchery *et al.* [120–122]. At this stage, tests are performed to obtain the mechanical properties, i.e. E_x, E_y, v_x and E_s, of the material chosen.

With this framework and the material properties, an analysis is carried out to obtain the maximum stresses and strains, dynamic responses and stability conditions (step 5). A relatively 'exact' solution can be obtained at this early stage of design if the correction factors (FRC) which the author proposed to use in section 7.9, are used. The possibility of using such factors is discussed in section 7.10. Then we proceed to step 6 to revise the design, including joints. We may have 'new' stresses at this stage, and we modify the sections to take care of such 'new' stresses. With these new sections and frameworks, analysis is carried out again to get 'exact' stresses and strains. Tests are made to obtain the strengths of the material by means of loading the test coupons. With the stresses of the framework and the material strengths, the failure (or strength) criterion, discussed in Chapter 8, is applied. Either limit or ultimate strengths are used for this purpose. If any section turns out to be too weak, we go back to step 6 to reinforce it. We may do the same in the case of over-design. Otherwise, we go to step 9 and make a detailed drawing.

With civil architectural structures, testing of the prototype is almost impossible. These structures are huge and, in most cases, one design is different from the other. We have to be careful in design and analysis of both structures and materials. However, the science of composites is well established and we should have good confidence in huge structures made of composite materials, if enough care is taken.

From the long history of human civilization, four basic concepts of structures have evolved and developed. These are beam and column, masonry arch, wooden truss, and modern steel truss and frame. These concepts were made possible by available materials and applicable technologies of each age. Modern science and engineering have produced numerous noble materials and technologies, and it is necessary to have a new concept which the author wishes to call the fifth concept of structures. Composite structures belong to this concept, which is very much diversified.

The classical four concepts make use of steel I, WF, T, L and other sections. Using composite materials for such elements may not be too efficient structurally. However, other reasons such as corrosion and electro-magnetic problems require the use of composite materials for such sections. Their design will be discussed in the next section and elements based on the new concept will be covered in section 10.3.

10.2 Design and analysis of structures of the four basic concepts

The first concept, beam–column, is different from the modern beam–column concept often used in framed structures, which belongs to the fourth concept. In the first concept, the beam is subjected to bending and shear, and the column is acted on by an axial load only. The second concept, even though the Roman arches were subjected to axial loads only, will include modern arches which are acted on by axial loads, shear, bending and even torsion. The third concept, trusses, are handled by considering axial loads, and moments if necessary. The fourth concept may include modern steel frames and isotropic plates and shells. However the isotropic plates and shells are considered as special cases of anisotropic plates and shells, and therefore, will be treated as the fifth concept.

With the above classification, structural elements for all four concepts are covered by the theories discussed in Chapter 6. The expressions for the stiffnesses are given in that chapter. With these stiffnesses, the methods of analysis in Chapters 3 and 4 can be used to obtain the forces and moments, with consideration of vibration and buckling problems. Then, the methods of calculating the stresses and strains of structural members made of composite materials, given in Chapter 6, can be used before proceeding to apply the strength theory, given in Chapter 8, with proper joints explained in Chapter 9, to conclude the design.

There are several manufacturers producing standard structural elements. Some of them present design guides. These design methods are essentially the same as those of steel structures. It is strongly recommended that the methods for composite materials should be used: stiffnesses, and stresses and strains (Chapter 6), failure criteria (Chapter 8), and joint design (Chapter 9).

Depending on the purpose of application and geological location, we may have different codes and specifications. Any engineer must be able to design and construct structures made of composite materials without such codes and specifications, if these are not available. Confidence in the theories and the engineer's own ability is very important for professional achievement. A large proportion (probably almost 50%) of over 200 structures that the author has designed and built mostly with conventional materials were executed when no code was available. Some structures were constructed with only a fraction of the cost required by the conventional design, and these are still sound and good after several decades.

There is one thing which bothers the engineer most when the code is not available. It is the concept of the factor of safety (FS) or load factor. The FS is the ratio of the ultimate strength to the allowable working stress.

Some manufacturers' manuals give values for the FS. Some of these are too conservative, but given as follows, can be used as a guideline for independent design work.

- Columns   FS = 3.0
- Beams:
  bending   FS = 2.0 (buckling considered)
  shear   FS = 3.0
  deflection   FS = 1.0
- Frame connections FS = 4.0

When the codes are not available, the following facts must be considered to determine the FS:

1. accuracy of assumed loadings;
2. accuracy of theories and methods used for analysis;
3. low ductility of composites prevents the stress relief at stress concentration areas;
4. stiffness and strength greatly depend on environmental condition;
5. influence of loading condition on the ultimate strength.

For example, a design guide suggests FS as follows.

- Static load   FS = minimum 2.0
- Fatigue or repeated load FS = minimum 4.0
- Impact load   FS = minimum 4.0

The selection of the FS is ultimately the responsibility of the designer.

10.3 Design of structures of the fifth concept

10.3.1 General

The structural elements belonging to the fifth concept are very much diversified. These elements have been used extensively by the other branches of the engineering profession. Many civil engineers, including building engineers, however, have had some unjustifiable prejudices against the use of such elements. It is true that composites in the past were used mostly for low volume, high performance applications with little consideration of cost. With present cost of some materials and manufacturing methods, design plays a vital role in determining cost of the product. Even with conventional materials, the design method can bring the construction cost down profoundly. Some of the author's design and construction experiences with structures made of conventional materials are as follows.

1. An elevated expressway with a length of 3.1 km in Seoul was designed in 1967. New design concepts including composite action (which was not new then), grid analysis considering whole bridge deck and girders, welding and use of high-tension bolts, hybrid structure concept, and others resulted in an economical structure. The ratio of the total steel tonnage used to that of the conventional design was 30/80 [123].

2. A tower for use in communications and as a tourist attraction with a height of 270 m from the foundation was designed and built in 1970 [124]. After thorough analysis and careful study of construction methods, it was built at about half the cost originally estimated.

3. Military structures north of Seoul were designed and built based on entirely new and independently developed concepts. Several of these structures were built at 3/25 of the initially estimated cost.

All of the structures mentioned above are structurally sound and perform perfectly after two decades in service.

A very large portion of civil structures can be analyzed by considering them as frameworks of one-dimensional elements. For such structures, the methods presented in Chapters 3 and 4 are good enough. Composite materials are, generally, strong in tension. When an element is designed based on tension load, it will have a thin section which is weak against any loading type other than in-plane on- axis tension load. This requires the section modulus to be increased by means of thin-walled structure or sandwich panels. Even the one-dimensional element, after the frame is analyzed, requires additional study by the methods explained for thin-walled sections. The thin panels of such sections are weak against the loads normal to these panels. The longitudinal stringers are added between transverse diaphragms to take care of such loads. The diaphragms transmit the loads from stringers to the walls of the beam by means of in-plane shear. The problems of thin-walled sections with longerons (stringers) are extensively discussed in Chapter 3. The mechanism and reduced one-dimensional frameworks for a thin walled section are shown in Figures 10.2 and 10.3.

10.3.2 Quasi-isotropic constants

The initial analysis of any composite structural element, including one-dimensional, may be started by using the concept of quasi-isotropic constants. Every anisotropic material has quasi-isotropic constants derived from the invariants of coordinate transformation. These constants represent the lower bound of each composite performance. Whatever ply orientation is selected for an applied load the laminate performance is at least equal to the quasi-isotropic laminate. These quantities are invariants and are better design parameters than any

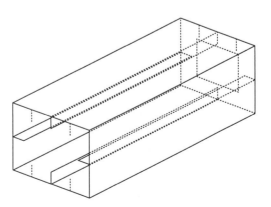

Figure 10.2 Thin-walled box type beam with longitudinal stringers supported between two diaphragms.

DESIGN

Figure 10.3 Box girder with stringer and diaphragms. (a) The walls are similar to a frame with unit width, supported by the stringers and vertical members. (b) The stringer is similar to a continuous beam supported by diaphragms. (c) The diaphragm transmits the load to the walls by in-plane shear.

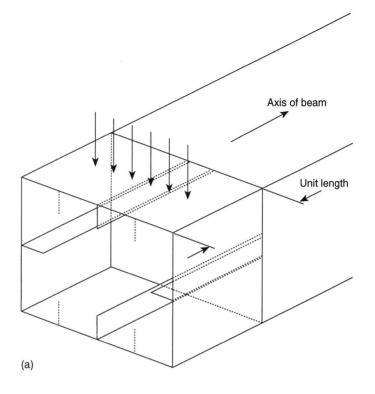

(a)

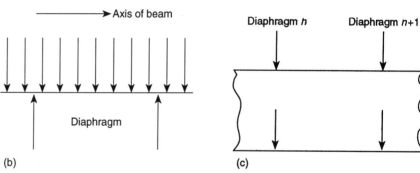

(b)   (c)

stiffness component. Tsai [125] gives these parameters as

$$[\boldsymbol{Q}]^{\text{iso}} = \begin{bmatrix} U_1 & U_4 & 0 \\ U_4 & U_1 & 0 \\ 0 & 0 & U_5 \end{bmatrix}$$

(10.1)

where

$$U_1 = \tfrac{1}{8}(3Q_{xx} + 3Q_{yy} + 2Q_{xy} + 4Q_{ss}),$$
$$U_4 = \tfrac{1}{8}(Q_{xx} + Q_{yy} + 6Q_{xy} - 4Q_{ss}) = U_1 - 2U_5$$
$$U_5 = \tfrac{1}{8}(Q_{xx} + Q_{yy} - 2Q_{xy} + 4Q_{ss})$$

(10.2)

Note that Tsai uses the letter subscripts for on-axis quantities.

$$[S]^{iso} = \begin{bmatrix} U_1/D & -U_4/D & 0 \\ -U_4/D & U_1D & 0 \\ 0 & 0 & 1/U_5 \end{bmatrix} \tag{10.3}$$

where $D = U_1^2 - U_4^2$, and

$$E^{iso} = D/U_1 = (U_1^2 - U_4^2)/U_1 = U_1[1 - (\nu^{iso})^2]$$

$$\nu^{iso} = U_4/U_1 \tag{10.4}$$

$$G^{iso} = U_5$$

When quasi-isotropic constants are used,

$$D_{11} = D_{22} = D_{12} + 2D_{66} = (h^3/12)Q_{11}^{iso}$$

$$A_{11} = A_{22} = A_{12} + 2A_{66} = hQ_{11}^{iso} \tag{10.5}$$

With these values, the preliminary analysis can be carried out. The equations and formulas obtained from textbooks on classical strength of materials theory can be used to analyze the behavior of the structure. We can use the correction factors, FRC, given in section 7.9, to obtain 'exact' values at this early stage of design. The orientations of the laminae may be decided according to this result. Then the off-axis quantities are obtained by equations 5.44. If necessary, we modify the section and calculate stiffnesses for this 'new' section to proceed to 'refined' analysis.

Example
Consider the GFRP given in Table 5.1.

$$[Q] = \begin{bmatrix} 39.17 & 2.18 & 0 \\ 2.18 & 8.39 & 0 \\ 0 & 0 & 4.14 \end{bmatrix} \quad \text{(on-axis)}$$

$$U_1 = \tfrac{1}{8}(3Q_{xx} + 3Q_{yy} + 2Q_{xy} + 4Q_{ss}) = 20.45 \text{ GPa}$$

Similarly

$$U_4 = 5.51 \text{ GPa}$$

$$U_5 = 7.47 \text{ GPa}$$

We may get other Us,

$$U_2 = \tfrac{1}{2}(Q_{xx} - Q_{yy}) = 15.39 \text{ GPa}$$

$$U_3 = \tfrac{1}{8}(Q_{xx} + Q_{yy} - 2_{xy} - 4Q_{ss}) = 3.33 \text{ GPa}$$

From these,

$$[Q]^{iso} = \begin{bmatrix} 20.45 & 5.51 & 0 \\ 5.51 & 20.45 & 0 \\ 0 & 0 & 7.47 \end{bmatrix}$$

$$\nu^{iso} = U_4/U_1 = 5.51/20.45 = 0.27$$

$$G^{iso} = U_5 = 7.47 \text{ GPa}$$

$$E^{iso} = U_1[1 - (\nu^{iso})^2] = 20.45(1 - 0.27^2) = 18.96 \text{ GPa}$$

With the above values and correction factors, FRC, given in section 7.9, a preliminary analysis can be carried out and laminae orientations can be decided on the results of this analysis.

10.3.3 Ultimate strengths of laminates

In Chapter 8, the first-ply failure (FPF) envelopes are discussed. These can be obtained by direct application of laminate theory with chosen failure criteria. Some conservative engineers may consider this FPF as the criterion of the ultimate strengths of the laminated composites. In fact, the failure envelopes based on intact plies are valid only up to the FPF [125]. As long as we do not go beyond the FPF, which defines the limit of intact plies, we can keep on loading and unloading without incurring irreversible effects. A structure may experience an unexpected loading and, if the damage is not serious, it can function as usual so long as the loads are under a certain limit.

If the load is increased beyond the FPF envelope, cracks begin to form within the matrix and at the fiber–matrix interface, parallel to the fibers in unidirectional plies. As the load keeps on increasing, more cracks are formed until a saturation level is reached just before the ultimate failure of the laminate. As long as the ply is still embedded in the laminate, it will continue to contribute to the stiffness of the laminate. Tsai proposes [125] replacing the cracked plies with a continuum of lower stiffness and applying conventional stress analysis. According to his experiments, if the matrix reduction is 40%, the transverse stiffness reduction is 55% and the shear modulus reduction 45%. The conventional arbitrary practice of setting both transverse and shear moduli to nearly zero is not recommended. It is assumed that the reduction of the major Poisson's ratio is by the same percentage as that of the matrix modulus.

It is necessary to calculate only the first- and last-ply failures. If the applied load is increased monotonically, the stress–strain curve moves from the origin to the FPF point on the 'intact' line, then deviates to the LPF point on the 'totally degraded' line. The 'jump' from the 'intact' to the 'degraded' stress–strain curve is not as big as shown in Figure 10.4. For practical laminates, the loss of laminate stiffness due to matrix degradation is less than 10%. The FPF and LPF points appear to be on the same stress–strain curve [125].

Figure 10.4 Simplified prediction of the last-ply failure. (Courtesy of Think Composites.)

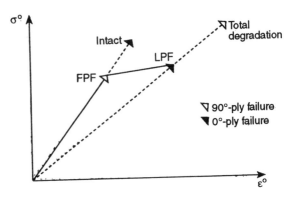

The correct concept of fundamental theories is important. As an example, Tsai considers that the netting analysis for the pressure vessel subjected to internal pressure is not a theory. It is derived from consideration of the equilibrium of a balanced angle ply construction and does not satisfy the strain compatibility. It does not depend on the stress–strain relationship of the material. The notion that the fiber carries all loads in the uniaxial tension of a unidirectional composite cannot be extended to the off-axis plies without laminate plate theory.

LPF is usually greater than FPF in the tension–tension and tension–shear domains. LPF may be less than FPF in compression–compression and tension–compression domains, i.e. FPF is the ultimate strength. If there is alternate loading and unloading, or fatigue loading, the failure modes become complicated.

When a balanced angle ply laminate is subject to a combination of normal stresses with no shear, or when a laminate is subjected to hydrostatic tension or compression, the strains of both LPF and FPF are equal for all fiber orientations, and the laminate will undergo simultaneous failure. This situation, however, does not mean the optimum laminate configuration.

The mode of failure depends on the loading history.

Tsai recommends the use of the matrix degradation model, Figure 10.4, as a means of predicting the ultimate load of laminated composites [125]. The failure envelopes he presents are based on

- stress interaction term   −0.5
- matrix degradation   0.3
- safety factor   1.5

The ultimate strength is defined as the larger of the FPF and LPF (Figure 10.5).

The safety margin (safety factor) is defined as the ratio of the load-carrying capacity of a structure to the limit load. For the design of composite materials this margin is often 1.5. The limit strength is defined as the strength induced by the maximum load which a structure is expected to encounter during its lifetime.

Figure 10.5 Ultimate strength envelope. (Courtesy of Think Composites.)

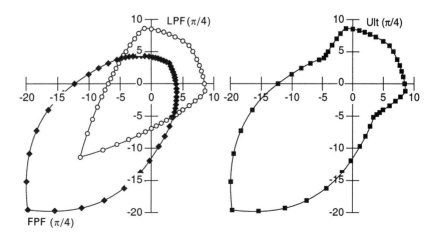

Tsai defines

● the ultimate envelope

$$\text{'ultimate'} = \max(\text{FPF, LPF}) \tag{10.6}$$

or the larger of FPF and LPF forms the ultimate envelope;
● the ultimate based limit envelope

$$\text{'limit*'} = \text{ultimate/FS} \tag{10.7}$$

where FS = factor of safety or safety margin;
● the design limit envelope,

$$\text{'limit'} = \min(\text{FPF, ultimate}) \tag{10.8}$$

There are two FS relations,

$$\text{FS} = \text{'ultimate'/'limit*'} \tag{10.9}$$

$$\text{FS} \geq \text{'ultimate'/'limit'} \tag{10.10}$$

In the case of a filament wound pressure vessel, leakage starts at FPF. If the leakage is prevented by a seal liner, burst occurs when the fiber fails. The fiber strength of this vessel is fully utilized beyond the FPF envelope.

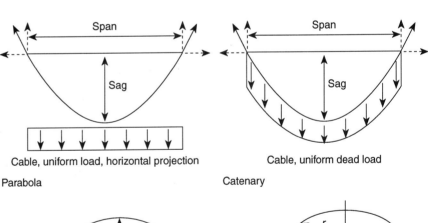

Cable, uniform load, horizontal projection

Parabola

Cable, uniform dead load

Catenary

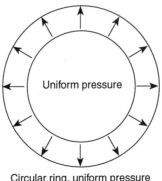

Circular ring, uniform pressure

Circle

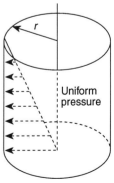

Vertical cylindrical shell under fluid pressure

Circle

Figure 10.6 (a) Typical tension structures. (b) Tension structure. (Courtesy of Ferrari, France.) (c) Arches.

(a)

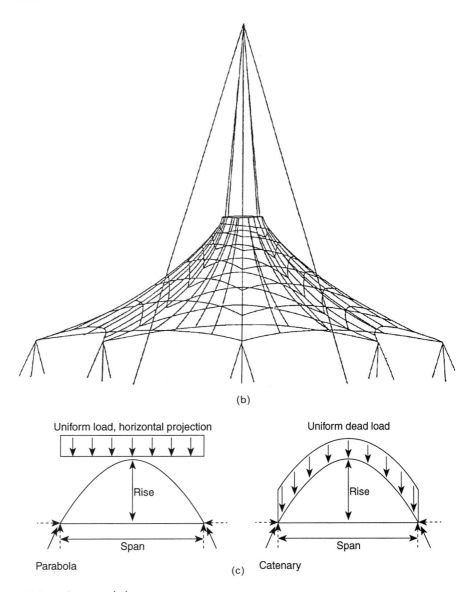

(b)

Uniform load, horizontal projection

Rise

Span

Parabola

(c)

Uniform dead load

Rise

Span

Catenary

10.3.4 Structural shapes

The long fiber reinforced composites generally have high specific strength and stiffness. The maximum structural efficiency can be obtained when such elements are under tension and the shape of the section is thin. This results in stability problems if the loading is other than tension. The cross-section of an element made of composite materials, except when such an element is always in tension, will have shapes such that the moment of inertia of the section is increased. We will review some structural shapes of this fifth concept and briefly discuss the method of analysis.

Tension member

Some of the tension members are shown in Figures 10.6(a) and (b). All configurations in Figure 10.6(a) are the funicular thrust lines of the applied loads. Such shapes are most efficient as long as the type of the loading does not change. If the uniform weight of the cable is the only loading, the best structural shape is catenary. If the load on the cable has uniform horizontal projection, the best configuration is a parabola. The circular cylinder is most effective against uniform hydrostatic pressure. The problems of such structures, taut strings, cables and guyed towers can be solved by the theories in sections 3.20.1, 3.20.2, 3.36, 3.39, 3.40, 4.2.9 (vibrating string) and other elementary mechanics.

Arches, curved girders

The arches in Figure 10.6(c) are under in-plane loading. As with cable, the best shape of the arch under its own uniform weight is a catenary and is a parabola if the horizontal projection of the load is uniform. In civil structures, the loading is always movable and such loadings must be considered by the use of influence coefficients. Such structures under both in-plane and out-of-plane loadings can be solved by the methods given in sections 3.17, 3.33, 3.34 and 3.35. If the sections of such structures are made of thin walled configurations, we have answers in additional sections 3.26, 3.27 and 3.30. If the thin walled section has longitudinal stringers for added strength, we have additional methods of analysis in sections 3.28, 3.29, 3.31 and 3.32.

Structural elements for concentrated loads

The concentrated loads must be transferred to the structural element by means of distributed stresses. For example, the box girder should have a transverse diaphragm for vertical columns, or a bracket to take care of the longitudinal load. From the macroscopic point of view, the plate, shell or whole girder can be under concentrated loads. Such problems can be solved by the methods given in section 3.18.

Flexural members (Figure 10.7)

The solid beam of Figure 10.7(a) is the first basic concept. The W (formerly WF (wide flange)) beam (Figure 10.7(b)) belongs to the fourth basic concept and has considerably increased flexural strength. The sandwich configurations (Figures 10.7(c) to (e)), have excellent flexural strength and belong to the fifth basic concept. This shape will be discussed below. The shapes in Figure 10.8 also belong to the fifth concept. The methods of analysis for such cross-sections are given in previous chapters. For example, the methods for the rectangular tube are the same as for curved girders. The stiffnesses are given in Chapter 6. Figure 10.10 shows some panels made of braided tube cores.

Beam–columns

In Chapters 5, 6 and 7, we considered the coupling problems between bending and in-plane forces, and twisting and so on. The beam–column is only a special case of such problems. We can use the given theories, without difficulty, to solve this type of problem.

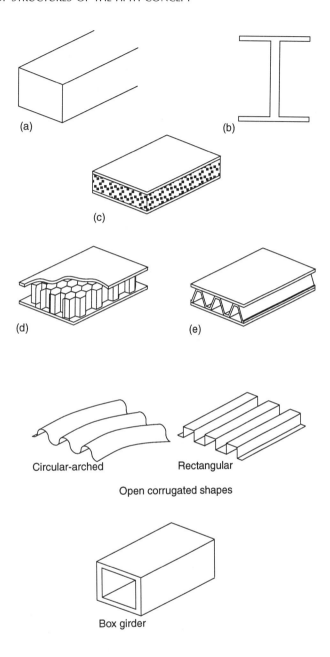

Figure 10.7 Flexural members: (a) solid beam; (b) W beam; (c) foam core; (d) honeycomb core; (e) corrugated core.

Figure 10.8 Typical cross-sections for increased stiffness.

Shells, folded plates

These structures (Figure 10.9) belong to the fifth concept. This subject is rather too bulky to be handled in this book. However, any curved surface can be considered as continuations of certain types of triangular plates. Therefore, the theory of nonprismatic folded plates can be applied to any type of shell structures. This theory will be briefly discussed below.

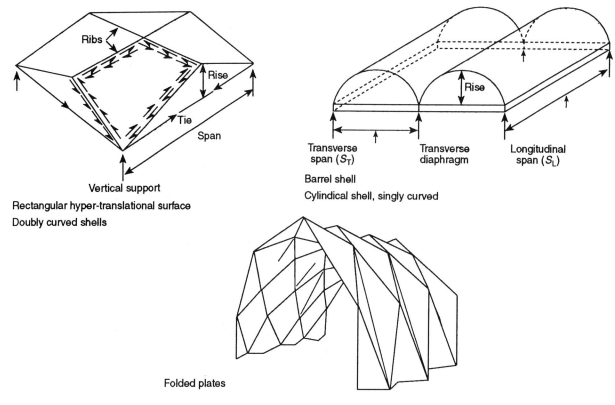

Figure 10.9   Shells, folded plates.

10.3.5 Sandwich panels

From the structural function viewpoint, sandwich panels can be called cored panels. The core can be a foam (Figure 10.7(c)), or a honeycomb (Figure 10.7(d)), corrugated (Figure 10.7(e)) or braided element (Figure 10.10 [126]) construction. Core materials include paper and aluminum. Other core types are extrusions, sheet and stringer, and plywood, shown in Table 10.1 which compares the

Table 10.1   Types of sandwich panels (Adapted from reference [127])

	Design	Relative strength (%)	Relative stiffness (%)
1	Honeycomb sandwich	100	100
2	Foam sandwich	26	68
3	Structural extrusion	62	99
4	Sheet and stringer	64	86
5	Plywood	3	17

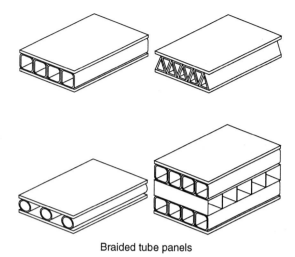

Braided tube panels

Useful cross-sectional shapes for braided tube cores

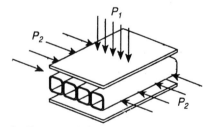

P_1, P_2: consolidation pressures (generally $P_1 > P_2$)

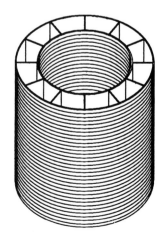

Filament wound cylinder with sandwich
braided tube cores

Figure 10.10 Elements made by braiding. (Adapted from reference [126].)

Table 10.2 The effect of increased thickness of honeycomb sandwich on stiffness (adapted from reference [127])

Relative stiffness	1	7.4	39
Relative strength	1	3.5	9.25
Relative weight	1	1.03	1.06

relative strengths and stiffnesses of the above types [127]. The relative stiffnesses of sections with different thickness are as shown in Table 10.2. If the thickness of the section is increased by four times, the stiffness increase is 39 times, with a weight increase of 6%. Figure 10.11 shows typical sandwich panels of both honeycomb and foam cores. Figure 10.12 shows the nomenclature used in our discussion.

Vinson made an extensive study on this subject [128] with honeycomb core. Consider the panel under axial load (Figure 10.12). We assume that the load is carried by the faces only. Then, the stress on the face in the x-direction is

$$\sigma_{Fx} = \frac{N_x}{2t_F}$$

where t_F is the thickness of the face. If the load is in-plane compression, the panel may fail in various ways such as (Figure 10.13): general buckling; core shear instability; face wrinkling instability; face dimpling or monocell buckling.

For the transverse load, shear is taken care of by the core. The faces take up both in-plane tension and compression due to bending (Figure 10.14). The stiffness can be obtained from section 6.2.

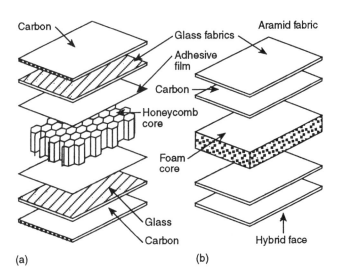

Figure 10.11 Typical hybrid sandwich panels: (a) honeycomb core; (b) foam core.

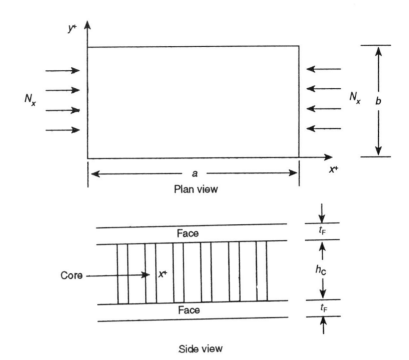

Figure 10.12 Notations for a typical sandwich plate.

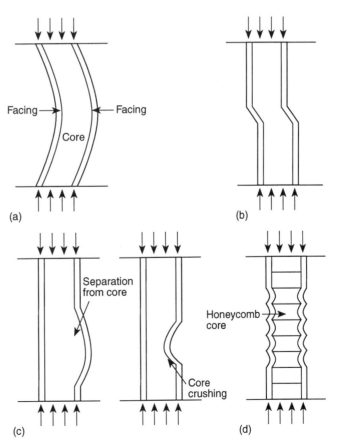

Figure 10.13 Possible modes of failure of sandwich construction under edgewise loads: (a) general buckling; (b) core shear instability; (c) face wrinkling instability; (d) face dimpling.

Figure 10.14 Transverse load and stresses in the element of a cored structure.

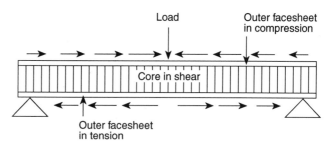

Load

Outer facesheet in compression

Core in shear

Outer facesheet in tension

10.3.6   Theory of nonprismatic folded plate structures

Any three-dimensional structural configuration can be represented by the theory of nonprismatic folded plates, which is composed of sectorial plates. Any sectorial plate problem, both in and normal to the plane, can be solved by the finite difference method, finite element method, and the method explained in section 3.42.

The problem then reduces to that of boundaries of two adjoining sectors. Each sector may be inclined. Figures 10.15 and 10.16 show typical both upper and lower fold lines.

D. H. Kim worked on this problem by finite difference and influence coefficient methods [119, 129, 130].

The joint forces and displacements at the nth fold line can be transformed to a system common to both adjoining plates.

1. When n is at an upper fold line:

$$F_{x(n,n+1)} = -N_{t(n,n+1)} \sin \phi_{(n,n+1)} + V_{t(n,n+1)} \cos \phi_{(n,n+1)}$$

$$F_{y(n,n+1)} = -N_{t(n,n+1)} \cos \phi_{(n,n+1)} - V_{t(n,n+1)} \sin \phi_{(n,n+1)}$$

$$F_{x(n,n-1)} = +N_{t(n,n-1)} \sin \phi_{(n,n-1)} + V_{t(n,n-1)} \cos \phi_{(n,n-1)}$$

$$F_{y(n,n-1)} = -N_{t(n,n-1)} \cos \phi_{(n,n-1)} + V_{t(n,n-1)} \sin \phi_{(n,n-1)}$$

$$D_{x(n,n+1)} = v_{(n,n+1)} \sin \phi_{(n,n+1)} + w_{(n,n+1)} \cos \phi_{(n,n+1)} \qquad (10.11)$$

$$D_{y(n,n+1)} = v_{(n,n+1)} \cos \phi_{(n,n+1)} - w_{(n,n+1)} \sin \phi_{(n,n+1)}$$

$$D_{x(n,n-1)} = +v_{t(n,n-1)} \sin \phi_{(n,n-1)} + w_{(n,n-1)} \cos \phi_{(n,n-1)}$$

$$D_{y(n,n-1)} = -v_{(n,n-1)} \cos \phi_{(n,n-1)} + w_{(n,n-1)} \sin \phi_{(n,n-1)}$$

where $N_t = \sigma_t h$.

2. When n is at a lower fold line:

$$F_{x(n,n+1)} = -N_{t(n,n+1)} \sin \phi_{(n,n+1)} - V_{t(n,n+1)} \cos \phi_{(n,n+1)}$$

$$F_{y(n,n+1)} = N_{t(n,n+1)} \cos \phi_{(n,n+1)} - V_{t(n,n+1)} \sin \phi_{(n,n+1)}$$

$$F_{x(n,n-1)} = N_{t(n,n-1)} \sin \phi_{(n,n-1)} - V_{t(n,n-1)} \cos \phi_{(n,n-1)}$$

$$F_{y(n,n-1)} = N_{t(n,n-1)} \cos \phi_{(n,n-1)} + V_{t(n,n-1)} \sin \phi_{(n,n-1)}$$

$$D_{x(n,n+1)} = v_{(n,n+1)} \sin \phi_{(n,n+1)} + w_{(n,n+1)} \cos \phi_{(n,n+1)} \qquad (10.12)$$

$$D_{y(n,n+1)} = -v_{(n,n+1)} \cos \phi_{(n,n+1)} + w_{(n,n+1)} \sin \phi_{(n,n+1)}$$

$$D_{x(n,n-1)} = v_{(n,n-1)} \sin \phi_{(n,n-1)} - w_{(n,n-1)} \cos \phi_{(n,n-1)}$$

$$D_{y(n,n-1)} = v_{(n,n-1)} \cos \phi_{(n,n-1)} + w_{(n,n-1)} \sin \phi_{(n,n-1)}$$

Figure 10.15 Notations and sign conventions at the upper fold line.

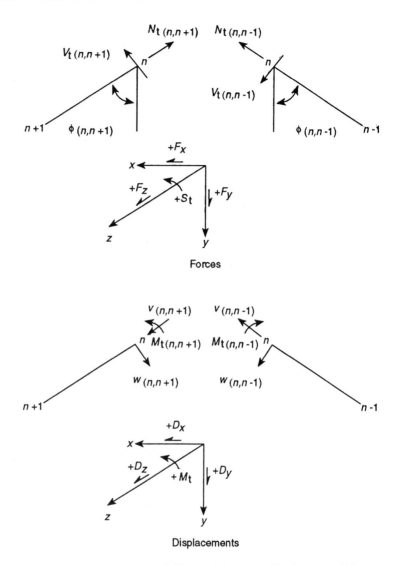

Forces

Displacements

The types of the joint compatibility and joint equilibrium conditions depend on which dependent variables we choose. If the transverse bending moment, M_t, and the three displacement components, u, v and w, are taken as unknowns, it is necessary to satisfy the slope compatibility condition and the three force equilibrium conditions at each fold as follows.

$$S_{t(n,n+1)} - S_{t(n,n-1)} = 0$$
$$F_{xy(n,n+1)} + F_{xy(n,n-1)} = 0$$
$$F_{x(n,n+1)} + F_{x(n,n-1)} = 0$$
$$F_{y(n,n+1)} + F_{y(n,n-1)} = 0$$

$$(10.13)$$

where $F_{xy} = \tau_{r\theta}h = N_{rt}$.

Since these 'force' expressions are to be written in terms of displacements, the compatibility conditions are automatically satisfied. At each fold line, these

Figure 10.16 Notations and sign conventions at the lower fold line.

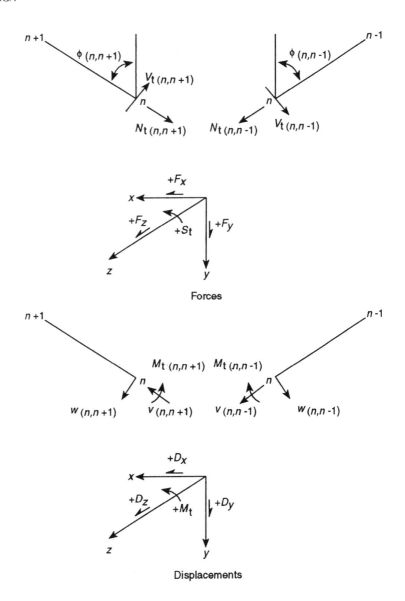

Forces

Displacements

conditions must be satisfied when the governing differential equations are integrated.

If the finite difference method is used to integrate the differential equations, some elaborate work is necessary. A method of solving such a problem was reported by D. H. Kim [129]. The process of calculation is straightforward. A very high degree of accuracy can be obtained by this method [119, 129].

10.3.7 Nonlinear analysis of underground laminated composite pipes

When underground pipes are acted on by a vertical load, the displacements and forces on a segment of the pipe or conduit can be as shown in Figure 10.17. K is the subgrade reaction coefficient or the modulus of the foundation. The magnitude of the vertical load is the weight of the backfill soil, modified by the

Figure 10.17 Definition of loading and sign convention for conduit.

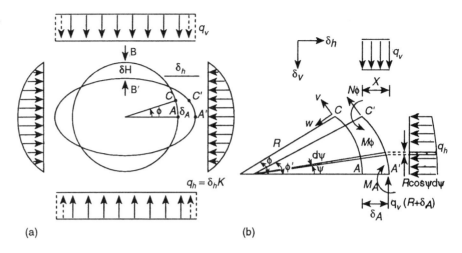

(a) (b)

effect of shear stress due to settlement of backfill, plus the pressure due to any live load present, usually calculated by the Boussinesq solution given in section 3.18. The force of the soil against the pipe side wall tends to prevent the deformation caused by the vertical force. The assumption generally made is that the horizontal force is proportional to the horizontal pipe deflection.

The problem of this type of conduit made of composite laminates is solved and reported by D. Kim [131]. It is assumed that the cross-section is symmetric with respect to the midplane, so that there is no bending–stretching coupling. The moment equilibrium at c (Figure 10.17) yields

$$-M_\phi + M_A - \frac{q_v}{2}\left[R(1 - \cos\varphi) + \delta_A + v\sin\varphi + w\cos\phi\right]$$

$$\times \left[R(1 + \cos\varphi) + \delta_A - v\sin\phi - w\cos\phi\right]$$

$$+ \int_0^\phi \delta_h K R^2(\sin\varphi - \sin\phi)\cos\varphi \, d\varphi = 0 \tag{10.14}$$

Galerkin's method can be used to solve the equation.

In practice, simple elastic analysis is sufficient. Rigorous analysis confirms the recommendation that

$$\max \delta_B \le 0.05R \tag{10.15}$$

10.4 Optimization of composite material structures

10.4.1 Introduction

Optimization of composite material structures involves considering structural aspects, material selection, manufacturing and other factors. The term 'composite' covers a wide range of materials. Both reinforcement and matrix can be a metal, polymer or ceramic, alone or in combination. Thermosetting epoxies and carbon fibers can comprise hundreds of different formulations and combinations. The manufacturing method plays a vital role in determining the properties of a composite. Within any manufacturing method, there are very many controlling factors. As an example, consider pultrusion. In addition to the internal dynamics of the pultrusion process, there are problems of the hydraulic pressure as a result of thermal expansion, and pressure loss as the result of shrinkage. This shrinkage may be caused by curing of the resin matrix in addition to the contraction due to cooling. The shrinkage rate can be controlled by the cure rate of the resin. The cured resin properties largely depend on the choice of curing agent. Over-use of accelerators may compromise pot life; if we want extended pot life, the cure procedure has to be compromised. These and other factors, such as viscosity, have to be optimized. After all chemical problems are solved, the material, type, volume and orientation of reinforcements control the mechanical properties of the composite.

There are too many different kinds of composite structural forms/concepts to choose from which will be best suited for a given design purpose. Even with the sandwich panel, one of the commonly used composite structural forms, the core can be foam, honeycomb, truss, hat type, or some other. The foam core sandwich panel is evolving to a new type called 'syncore'.

In order to design a structure, one must understand the failure/strength mechanism clearly. The causes of failure can be 'design errors', 'fabrication and processing errors', 'unexpected service condition', or any combination of these. The fracture in a continuous fiber composite can be interlaminar, intralaminar, translaminar or any combination.

Fatigue in a laminate may cause matrix cracking, delamination, fiber breakage, interfacial debonding or any combination of these. Fatigue often causes a significant reduction in strength and modulus of the laminate. The type of laminate, nature of loading and other factors influence the degree of damage to strength. Selection of proper failure theory itself is one of the difficult problems.

The failure modes of a composite are tension and compression of fibers, tension and shear of matrix/interface, matrix compression, and interaction of any combination of these modes.

Delamination is caused by a single stress component or any combination, by transverse impact and by nonuniform cure. Matrix degradation, nonmechanical stresses, traction-free expansions, residual strains and other factors cause failure of a composite. Micro–macro and elasto-plastic considerations may be necessary for failure analysis.

The cost of using anisotropic composite materials may still be high; this is mainly because the design may be far from optimal, even though increased computing capability has allowed us to move from closed end solutions suitable for use in designing with isotropic materials to finite element analysis required for these new systems. Because of the complexity and the anisotropic nature of such structures, one must know more precisely the nature of the dynamic stress field this structure will have. This knowledge can be very critical since properties in the plane or through the thickness can differ by an order of magnitude.

Regardless of the many variables and difficult problems as above, the structure still has to be designed for minimum weight with minimum cost; that is, optimization is the necessary procedure in design of composites. The key to the design of an efficient laminate is to make it resist both the magnitude and directional nature of the loads without over-design. The laminate must be tailored to meet the exact specific requirements. In general, isotropic materials are inefficient because unnecessary excessive strength and stiffness are inevitably available in some directions. A laminate can be constructed of individual laminae in such a manner to resist just such loads without excess of strength or stiffness. Use of the invariant concept is best suited to use in the optimization techniques described by Schmit [132]. Since the subject of optimization of composite material structures is critically important for the increased application of such materials for the large sized civil and architectural structures, this subject is explained in detail below.

10.4.2 Optimization of sandwich panels

Vinson worked on optimization of flat honeycomb sandwich plates under in-plane compressive forces and in-plane shear loads [133–135], truss or corrugated core sandwich panels subjected to in-plane compressive loads [136] and in-plane shear loads, web core sandwich panels under in-plane shear loads, solid or foam core sandwich plates under in-plane compressive loads, and truss core sandwich plates subjected to combined uniaxial compression and in-plane shear loads [137].

If the honeycomb core sandwich panel as shown in Figure 10.12 is acted on by a compressive load N_x, the stress (compressive) in the face is given by

$$\sigma_{Fx} = \frac{N_x}{2t_F} \tag{10.16}$$

When the stresses are increased, the plate may fail by any one, or any combination, of the following modes (Figure 10.13):

1. overall elastic instability (buckling);
2. core shear instability;
3. face wrinkling instability;
4. face dimpling or monocell buckling.

Vinson and Shore's study is based on the assumption that the faces are orthotropic plates, that is we have D_x, D_y and D_{xy}, and $B_{ij} = 0$, $(\ )_{16} = (\ )_{26} = (\ )_{45} = 0$, and $E_{Fx}/v_{xy} = E_{Fy}/v_{yx}$. The critical stress for overall buckling can be shown to be

$$\sigma_{cr} = -\frac{\pi^2}{4(1 - v_{xy}v_{yx})}\sqrt{E_{Fx}E_{Fy}}\left(\frac{hc}{b}\right)^2 K_m \tag{10.17}$$

in which K_m is a (complex) function of D_x, D_y, D_{xy}, v_{xy}, the effective transverse shear stiffnesses of the honeycomb core in the x- and y-directions G_{cx}, and G_{cy}, shear modulus of the core itself G_c, and boundary conditions.

For mode 2,

$$\sigma_{cr} = -\frac{2}{3}\left(\frac{t_c}{d}\right)\left(\frac{h_c}{t_F}\right)G_c \tag{10.18}$$

For mode 3,

$$\sigma_{cr} = -\left[\frac{16}{9}\left(\frac{t_F}{h_c}\right)\left(\frac{t_c}{d}\right)\frac{E_c\sqrt{E_{Fx}E_{Fy}}}{(1-v_{xy}v_{yx})}\right]^{1/2} \tag{10.19}$$

For mode 4,

$$\sigma_{cr} = -\frac{2\sqrt{E_{Fx}E_{Fy}}}{(1-v_{xy}v_{yx})}\left(\frac{t_F}{d}\right)^2 \tag{10.20}$$

t_c and d are thickness and the size (i.e. the diameter of the inner circle of the monocell of the honeycomb), respectively. The optimum structure is possible when σ_{cr} for any buckling mode is the same as for all others. When the unloaded edges are simply supported, the minimum weight honeycomb core sandwich plate can be obtained if

$$\frac{h_c}{b} = \frac{\sqrt{10}}{\pi}(1-v_{xy}v_{yx})^{1/2}\left(\frac{\sigma_{Fx}}{E_{Fx}}\right)^{1/2} \tag{10.21}$$

$$\frac{d}{b} = \frac{1}{\pi}\left[\frac{15}{2}(1-v_{xy}v_{yx})\right]^{1/2}\left(\frac{G_c}{E_c}\right)^{1/2}\left(\frac{\sigma_{Fx}}{E_{Fx}}\right)^{1/2} \tag{10.22}$$

$$\frac{t_c}{b} = \frac{9\sqrt{5}}{8\pi}(1-v_{xy}v_{yx})\frac{\sigma_{Fx}^2}{E_c E_{Fx}^{3/4}E_{Fy}^{1/4}} \tag{10.23}$$

$$\frac{t_F}{b} = \frac{\sqrt{15}}{2}\frac{(1-v_{xy}v_{yx})}{\pi}\left(\frac{G_c}{E_c}\right)^{1/2}\frac{\sigma_{Fx}}{E_{Fx}^{3/4}E_{Fy}^{1/4}} \tag{10.24}$$

The minimum panel weight per unit area is

$$\frac{W-W_{ad}}{b} = \sqrt{15}\frac{(1-v_{xy}v_{yx})}{\pi}\left(\frac{G_c}{E_c}\right)^{1/2}\frac{\sigma_{Fx}}{E_{Fx}^{3/4}E_{Fy}^{1/4}}\times\left[\rho_F + 2\rho_c\frac{\sigma_{Fx}}{G_c}\right] \tag{10.25}$$

in which W and W_{ad} are the weights per unit area of the whole panel and the adhesive, respectively, and ρ_F and ρ_c are the unit weight of the face and the core materials, respectively.

Since for the panel with unloaded edge fixed, $W - W_{ad}$ is 7.45% less than in equation 10.25, the panel with any arbitrary unloaded edge condition is at most 7.45% lighter. Vinson reports that the truss core sandwich panel is competitive with the honeycomb core, web core and solid core panels subject to simultaneous uniaxial compression and in-plane shear loads, while for uniaxial load alone, the truss core sandwich panel is a few percent heavier than the honeycomb core panel and significantly heavier when subjected to in-plane shear alone. His study included selecting the best materials from numerous alternatives and the best stacking sequences for these competing material systems. He considered all four failure modes.

10.4.3 Optimization of other composite structures

Optimization of composite structures is a relatively recent subject. Both optimization methods and composite structures have been developed during the last three decades and the combination of both is more recent. Morz, in 1970, tried to obtain the optimal reinforcement using strength criteria. Brizgalin and others, in 1972, used a stiffness criterion for an optimal design of uniform thickness composites. A direct search procedure was used by Lai and Abenbach in 1973 to obtain an optimal design for minimum tensile stress at the interface in a layered structure subjected to time harmonic and transient loads. Taking into account multiple loading conditions and displacement constraints on the structure, Khot, in 1973, suggested an efficient optimization technique based on strain energy distribution and a numerical search for the minimum weight design of structures. The minimum weight design of symmetric composite laminates subject to multiple in-plane loading was presented by Schmit in 1977. Optimization of the fiber volume fractions for columns and of orientation angle for plates and cylinders for buckling strength was presented by Hayashi in 1974.

The problem of stress calculation of anisotropic materials structures is quite easily solved by several analytical means such as the finite element method (FEM). The remaining two main problems are the elaboration of adequate failure criteria, which is still far from being solved perfectly, and the optimization of design and analysis to obtain the lightest solution. (Sometimes, weight can be taken as a constraint, while the maximization of the safety margin is the objective.) When design constraints due to flexibility are involved as eigenfunctions, aerodistortion and flutter, the 'full stress design process' (FSD) is neither able to optimize nor efficient, even with metallic structures.

Dassault developed an optimization for composite design [138] based on mathematical optimization techniques [139]. Design variables include manufacturing constraints and the number of plies in each direction for each group. The number of design variables often exceeds 500 which may have to be handled simultaneously by several analysis models. Constraints include various failure criteria, local buckling criteria, displacement limitation, aeroelastic variation of aerodynamic derivatives, dynamic natural frequencies, flutter speed and aeroelastic dynamic damping, minimum values of design variables, the limitation of the thickness variation between adjacent design variables and others. The constraints considered during the same optimization can appear from each of several analysis models such as the symmetric and anti-symmetric FE aircraft model, local buckling analysis by the Rayleigh–Ritz method, local refined FE analysis, different external load configurations for dynamics and flutter, variation of shape due to control surface deflection and others. Often, over 5000 constraints are handled simultaneously. Mathematical optimization is achieved by using nonlinear approximation of the constraints in terms of design variables, and by changing the variables to minimize a homographic function (weight). Tsai–Hill failure criteria are used inside the optimization loop.

$$C = \left[\left(\frac{\sigma_x}{\sigma_{xa}} \right)^2 + S_1 \left(\frac{\sigma_y}{\sigma_{ya}} \right)^2 + S_2 \left(\frac{\tau_{xy}}{\tau_{xya}} \right)^2 + S_3 \left(\frac{\sigma_x \sigma_y}{\sigma_{xa} \sigma_{ya}} \right) \right]^{1/2} \quad (10.26)$$

Arguments of criteria are adapted to each situation (tension, compression, bending, holed panel, etc.). The constraints must be handled for all potential failure modes simultaneously.

Kolkailah [140] reported the application of FE analysis to an optimum design for multilayered composite cylindrical pressure vessels. By this optimum design technique, a designer can calculate readily the stresses and displacements in each layer during the fabrication process or during the use of the cylinders.

Jun [141] reported his study results on optimization of bidirectional reinforced laminated plates under shear as follows.

1. The optimum structure of laminated plates under shear for buckling load must be symmetric, and the number of different optimum ply angles, including $0°$ and $90°$, is no more than three.
2. As the length to width ratio R of the plate increases, both the optimum ply angle and the buckling load increase.
3. If we exchange the length a and width b (i.e. a/b of two plates are reciprocal to each other), the optimum ply angles of the two plates are complementary and the buckling loads are equal to each other.

In order to increase the operational speed of an energy storage flywheel (momentum wheel) the problems such as the deformation compatibility between the inner disk, which delivers the inertia moment, and the outer filament wound case, the connection of the case with the driving shaft, and the delamination in the disk must be solved. In order to reduce the delamination problem, the disk has to be built up of several subrings each having a higher density and a lower stiffness than its adjacent outer subring. Cuntze and Zaun reported an analytical, optimum design method to determine the subring thickness and required material for the disk, to minimize the delamination [142].

Liu and Qiao report the optimal design procedure for the wing type structures under vibration and flutter [143]. The FE model is formulated using the variable-linking technique. Both Kirchhoff and shear- deformable elements are used. The optimal design of a composite lifting surface under the fundamental frequency constraint is studied. The optimum design of a cantilever wing subjected to the critical flutter speed is investigated. A method of feasible direction is used for optimization.

Other works on optimization include Chao (1975) [144] on the optimum fiber orientation for a symmetric orthotropic composite laminate with in-plane loading; Bert (1977) [144] on the optimal laminate design for a thin plate consisting of multiple layers of equal thickness laminae, with design criteria of maximization of fundamental frequency; Hirano (1979) [144] on the optimum buckling load of laminated plates under uniaxial and biaxial compression; Joshi and Iyengar (1982) [144] on the minimum weight design of composite plates under in-plane and transverse loads, with the fiber orientation and thickness of plies as variables; Soni and Iyengar (1983) [144] on the optimum design of clamped laminated plates, and optimum design under in-plane loads with multiple design variables; Mıravete [144] on the optimum laminate configurations for rectangular plates under uniform transverse load treating the laminate thickness and fiber orientation as design variables (FEM analysis considering in-plane and interlaminar strains is used); Miki and Sugiyama [145] on optimum design of fibrous laminated hybrid composites with required flexural stiffness as variables; Romeo and Baracco [146] on the minimum mass optimization of composite stiffened, unstiffened and sandwich curved panels subjected to combined longitudinal and transverse compression and shear loading; Gao and Mai [147] on optimal design of reinforced materials assuming that the matrix contains microcracks normal to the fibers; by using the sequential quadratic dual programming (SQDP) method, the optimal solution of each subproblem is presented analytically. Ding *et al.* [148] report their efforts to develop a system to develop analysis and optimal design of composite structures. For this purpose, a 'large scale composite structural analysis and synthesis system' (COMPASS) and 'aircraft multi-constraint optimal design system' (YIDOYU-1) are developed.

10.4.4 *Optimization of processing and other factors*

The properties of all polymer matrix composites are dependent on the processing used during fabrication. Factors such as inadequate or excessive resin flow, void formation and the degree of cure influence the properties of the completed structure. Development of cure cycles to minimize such problems has been achieved through trial and error. A Cycom 919 model is used to simulate and optimize cure cycles for the prepreg using the heat history of the part being evaluated, at McDonnell Douglas Helicopter Co. A similar model is used for Hexel F584, 177°C curing resin. The kinetic submodel, viscosity submodel, flow submodel, heat transfer submodel and void submodel have been studied and the result is reported by Frank-Susich *et al.* [149].

Trivisano *et al.* report an intelligent system for the control and optimization of the evolution of the processing variables during autoclave fabrication of high performance composite laminates. The system allows the computation of heat transfer coefficients for each selected tool in real time, prediction of temperature changes as a function of the programmed air temperature and optimization of the cure cycle minimizing the difference between the actual temperature and the value given by process specification [150].

Miaosheng *et al.* report optimum design of adhesive bonding of single lap joints of composites [151]. This report includes stress analysis of lap joints, optimum selection of the adhesive, design for the adherend (selection of modulus, thickness and surface treatment of adherends, and joint form), selection of bonding length, selection of thickness of the adhesive layer and safety coefficient.

Fiber–matrix adhesion is now accepted as a variable to be optimized for any composite material. No quantitative algorithm is available for interface optimization. However, various thermodynamic principles and recently obtained experimental data can be used to design the interphase qualitatively. This includes selection of surface treatments for surface structural and chemical modification, the use of surface finishes and/or sizes to insure thorough wetting and protection of the fiber, creation of interphases with desirable stiffness, toughness and failure modes, and adhesion levels compatible with the structural environment and constituent materials. Drzal reported his study results on optimizing the surface treatment of fibers, such as glass fibers, carbon fibers and polymeric fibers, the interphase design for processing and the interphase design for performance [152].

The analysis of thermal stresses in viscoelastic materials must take into account the temperature dependent properties. Many analysts have idealized such properties by assuming inviscid behavior above and elastic behavior below the critical temperature called the glass transition temperature (T_g) of an amorphous polymer. Such models neglect the important stress relaxation property of viscoelastic materials. During the cooling of a fiber reinforced viscoelastic part, internal stresses develop due to not only the temperature gradient in the part but also the differential thermal contraction between the fibers and the matrix. Cracks and warping may develop if the stresses are too high. In order to reduce the residual stresses, the cooling process must be very slow to achieve a small temperature gradient and maximum stress relaxation. Liou and Suh [153] report the optimum cooling process of such composites. Euler–Lagrange equations are solved with first gradient algorithm to iterate the optimal solution. Static mechanical strength alone cannot fully characterize a composite material. The fatigue properties of advanced composites are related to several fundamental properties such as fiber alignment and matrix wetting at the polymer/fiber interface. Currently available materials systems are not fully utilized to extract their optimum performance output. Once a material system is finalized regarding the matrix and reinforcing fibers, an optimum process method must be developed for such a purpose. While several choices are available for thermoset composites, there are only a few choices with thermoplastic composites. BASF developed a process optimization scheme for thermoplastic composites, and the use of commingled tow is one of the solutions [154].

ARALL laminates consist of thin layers of aramid/epoxy prepreg sandwiched between aluminum and have both the characteristics of high strength of the aramid and high stiffness of the metal, good damage tolerance and high fatigue crack growth resistance. ARALL-4 is made with aluminum alloy 2024-T8 sheets and aramid/epoxy prepreg (50/50). The maximum operating temperature is 160°C. Since the magnitude of creep depends on both the applied stress and the state of residual stress in aluminum and aramid/epoxy layers, it is possible to minimize the creep effects by altering the state of residual stress by mechanical prestressing of the laminate above the yield stress. Pindera *et al.* report [155] the results of their investigation to develop a methodology of optimizing the creep response of ARALL-4 by mechanical prestressing.

References

Note: *ICCM = International Conference on Composite Materials*; SAMPE = Society for the Advancement of Material and Process Engineering.

1. Wilson, B. A. (1987) Design/tooling/manufacturing interfaces, in *Composites* (eds Dostal, C. A. *et al.*), Engineering Materials Handbook, Vol. 1, ASM International, Metals Park, OH, pp. 428–31.
2. Winkel, J. D. and Hurdle, J. R. (1988) Mechanical properties of filament wound polyphenylene sulfide thermoplastic composite tubular structures, in *Proceedings 33rd SAMPE Symposium* (eds Dostal, C. A. *et al.*), pp. 816–28.
3. Epon resins for fiberglass reinforced plastics. Technical bulletin, Shell Chemical Co.
4. Epon/Eponol/Eponex resins/Epon curing agents. Technical bulletin, Shell Chemical Co.
5. Dion, corrosion guide. Technical bulletin, Koppers.

6. Chemical resistance fiber glass reinforced piping systems. Technical bulletin, Smith Fiberglass.
7. Derakane, vinyl ester resins, chemical resistance guide. Technical bulletin, Dow Chemical.
8. Hetron and Aropol, resin selection guide. Technical bulletin, Ashland Chemical Co.
9. Corrosion resistant polyester resins. Technical guidebook, ICI Americas Inc.
10. Dow liquid epoxy resins. Technical data, Dow Chemical, USA.
11. Epon resin 9400/Epon curing agent 9450, epoxy resin system for high performance composite parts fabricated by RTM or wet filament winding. Technical bulletin, Shell Chemical Co.
12. Compimide 65 FWR, Heat curable maleimide type resin designed for processing by wet filament winding and resin transfer molding. Technical bulletin, Shell Chemical Co.
13. Breitigam, W. V. and Stenzenberger, H. D. (1988) Bismaleimide resin formulation for filament winding, in *Proceedings 33rd International SAMPE Symposium* (eds Dostal, C. A. *et al.*), pp. 1229–33.
14. Anon. (1985) Filament winding. *Int. Reinf. Plast. Ind.*, **4**(4), 4–14.
15. Wilhelm, G. F. and Sehab, H. W. (1977) Glass reinforced plastic piping for shipboard applications. *Naval Engng J.*, **189**(2), 139–60.
16. Uberti, G. A. (1976) Fiberglass reinforced piping for shipboard system. National Shipbuilding Research Program.
17. Marshall, S. P. and Brandt, J. L. (1974) Installed cost of corrosion resistant piping. *Chem. Engng*, 28 Oct., 94–7.
18. Tarnopol'skii, Y. M. (1983) Problems of the mechanics of composite filament winding, in *Handbook of Composites* (eds Kelly, A. and Milleiko, Y. N.), Vol. 8, Elsevier Science Publishers, pp. 45–108.
19. Winkel, J. D. and Hurdle, J. R. (1988) Filament wound polyphenylene sulfide thermoplastic composite structures, in *43rd SPI Composites Institute Proceedings* p. 3E. Reprint by Phillips 66(X).
20. Egerton, M. W. and Gurber, M. B. (1988) Thermoplastic filament winding demonstrating economics and properties via in-situ consolidation, in *Proceedings 33rd International SAMPE Symposium* (eds Carrillo, G. and Newell, E. P.), Mar., pp. 35–46.
21. Calius, E. P., Kidron, M., Lee, S. Y. and Springer, G. S. (1988) Manufacturing stresses and strains in filament wound cylinders. *SAMPE J.*, May/Jun., pp. 347–51.
22. Giant RP pipe joints-Y-shape (1989) *Mod. Plast. Int.*, **19**(8), Aug.
23. Flow factors for fiberglass flowtite pipe. Test report by Ohio State University for Owens-Corning Fiberglass Corp.
24. Flowtite pipe. Technical bulletin, Veroc.
25. Servo fiber tension controller STC. Filament winding machines technical bulletin, Josef Baer.
26. Venus Gusmer. Products bulletin.
27. Technical bulletins for equipment series, Pultrex Ltd.
28. Pullwound tubes. Technical bulletin, Pultrex Ltd.
29. Winding machine series. Technical bulletin, Engineering Technology, Inc.
30. Roser, R. R. (1984) Computers remove the burden of programming the filament winding machine, in *Proceedings 29th National SAMPE Symposium*, pp. 73–9.
31. Roser, R. R. (1985) Computer graphics streamline the programming of the filament winding machine, in *Proceedings 30th National SAMPE Symposium*, pp. 1231–7.
32. Roser, R. R. *et al.* (1986) New generation computer controlled filament winding, in *Proceedings 31st International SAMPE Symposium*, pp. 810–21.
33. GRP pipes. Technical bulletin, Vetroresina.
34. Filament-winding machine series. Vetroresina.
35. Red mud corrugate sheet/pipe manufacturing equipment. Technical bulletin, Rai Hsing Plastics Machinery Works Co., Ltd.
36. Filament wound equipment series. Technical bulletin, DuraWound.
37. Downstream equipment for the extrusion of pipes. Technical bulletin, Battenfeld.

38. Pultrusion machine series. Technical bulletin, Pultrex Ltd.
39. Cascalite, corrugated sheets. Technical bulletin, Laminated Profiles, Ltd.
40. Lampro, corrugated and flat sheetings. Technical data, Laminated Profiles, Ltd.
41. Rooflight sheet machinery. Technical bulletin, Laminated Profiles, Ltd.
42. Automatic reciprocator, sheet machine. Technical bulletin, Venus Products, Inc.
43. Corrugated panel machine. Technical bulletin, Venus Products, Inc.
44. Dreger, D. R. (1987) The challenge of manufacturing composites. *Mach. Des.*, 22 Oct., p. 18.
45. Korane, K. J. (1987) Spotting flaws in advanced composites. *Mach. Des.*, 10 Dec., p. 26.
46. Smoluk, G. R. (1988) Downstream hardware keeps up with pipe, tubing, and profile extrusion. *Mod. Plast. Int.*, **18**(12), 40–53.
47. Shaw-Stewart, D. E. (1988) Pullwinding, Pultrex Ltd., UK.
48. Extrusion lines for pipes and profiles. Technical bulletin, McNeil-Akron Repiquet.
49. Design considerations and general properties of filament wound structures. DuraWound.
50. Filament wound fiberglass pipe and pipe fittings. Technical pamphlet, Corrosion Controllers, Inc.
51. FRP pipe for water and sewerage systems throughout the world. Technical pamphlet, Owens-Corning Fiberglass Corp.
52. Tanks for underground petroleum storage. Technical pamphlet, Owens-Corning Fiberglass Corp.
53. Power desalination plant development in Saudi Arabia. Case history, Amianit-Fiberglass.
54. Guideline specification – glass fiber reinforced pipe. Veroc.
55. Double-wall tanks. Technical pamphlet, Owens-Corning Fiberglass Corp.
56. Pipe handling and buried installation instructions. Technical manual, Veroc.
57. McCarvill, W. T. (1987) Filament-winding resins, in *Composites* (eds Dostal, C. A. *et al.*), Engineering Materials Handbook, ASM International, Metals Park, OH, pp. 135–8.
58. McCarvill, W. T. (1987) Prepreg resins, in *Composites* (eds Dostal, C. A. *et al.*), Engineering Materials Handbook, ASM International, Metals Park, OH, pp. 139–142.
59. McCluskey, J. J. and Doherty, F. W. (1987) Sheet molding compounds, in *Composites* (eds Dostal, C. A. *et al.*), Engineering Materials Handbook, ASM International, Metals Park, OH, pp. 159–60.
60. Colclough, W. G. and Dalenburg, D. P. (1987) Bulk molding compounds, in *Composites* (eds Dostal, C. A. *et al.*), Engineering Materials Handbook, ASM International, Metals Park, OH, pp. 161–3.
61. Meyer, F. J. (1987) Injection molding compounds, in *Composites* (eds Dostal, C. A. *et al.*), Engineering Materials Handbook, ASM International, Metals Park, OH, pp. 164–7.
62. Stark, E. B. and Breitigam, W. V. (1987) Resin transfer molding materials, in *Composites* (eds Dostal, C. A. *et al.*), Engineering Materials Handbook, ASM International, Metals Park, OH, pp. 168–76.
63. Hinrichs, R. J. (1987) Quality control, in *Composites* (eds Dostal, C. A. *et al.*), Engineering Materials Handbook, ASM International, Metals Park, OH, pp. 729–66.
64. Extren (fiberglass structural shapes), Fiberbolt (fiberglass studs and nuts). Technical pamphlet, Morrison Molded Fiber Glass Co., VA.
65. Extren chemical resistance guide. Technical pamphlet, Morrison Molded Fiber Glass Co., VA.
66. Application profiles. Technical pamphlet, Morrison Molded Fiber Glass Co., VA.
67. Dura Shield, fiberglass foam core building panels. Technical pamphlet, Morrison Molded Fiber Glass Co., VA.
68. Dura Deck, high strength fiberglass grating. Technical pamphlet, Morrison Molded Fiber Glass Co., VA.

69. Extren, fiberglass structural shapes. Engineering manual, Morrison Molded Fiber Glass Co., VA.
70. Coextrusion sheet lines. ER-WE-PA Maschinen Fabrik and Eisengiesserei GmbH.
71. Fiberforce composites. Technical pamphlet, Fiberfore Composites Ltd.
72. Processing of thermoplastics, injection/compression/extrusion, PEEK, PAI, PI, PES, PEI, PFA. Technical pamphlet, Nief Plastics.
73. Glaspul. Technical data, Pultrusion Corp.
74. Fiberglass products. Technical bulletin, IKG Borden.
75. Fiberglass structures. Technical bulletin, IKG Borden.
76. Lampro continuous process. Technical data, Laminated Profiles Dev. Ltd.
77. Plamix, pultruded profiles. Technical pamphlet, Sanshyo Ind., Japan.
78. SCI advanced composite hardware. Product data, Structural Composites Ltd.
79. Product technical data. Advanced Composite Technology Inc.
80. Macroboard: FRP-plywood composite panels. Product data, Macroboard Inc.
81. Makrolon/Longlife/Plus. Product data, Rohm GmbH.
82. Structural panels and elements. Product data, Albany International.
83. The history of pultrusions. The Polygram, Polygon Co.
84. Ryulex, styrenic copolymer. Dainippon Ink & Chemicals.
85. Commitment to aerospace. Technical pamphlet, The Dexter Corp.
86. Polyglass, structural shapes, rods, electrical shapes. Product guide, Westinghouse Electric Corp.
87. RYTON-PPS pultruded type composite shapes. Technical data, Phillips Petroleum Co.
88. Composite product technical data, Polygon Co.
89. Tuf-Loc, self-lubricating bearings. Technical data, Polygon Co.
90. Product technical pamphlet, Molded Fiber Glass Co.
91. Andreason, K. R. FRP/plywood tests. Test report for fiber-tech industries, Inc., Spokane, Washington.
92. Product technical data for FRP/plywood composites. Fiber Tech Ind., Inc.
93. Panelite. Technical data, Laminated Profiles Development Ltd.
94. Modular filament winding and fiber placement systems. Technical bulletin, Composite Machines Co., 1993.
95. Product technical data for pressure vessels, pipes, and other filament wound composites. Advanced Structures, Inc.
96. Carbon fiber filament wound long roll. Technical data, Shin-Nippon Steel Co.
97. Pipe, tube, extrusion line. Technical pamphlet, Maillefer, Switzerland.
98. AS roller (anti-abrasion HDPE). Technical pamphlet, Asahi Engineering Inc.
99. D.C. pipes. Technical pamphlet, Lucky Chemical Ltd.
100. Advanced reinforced tubing. Product data, J. Kennedy Fisher, Inc.
101. Amalgon, cylinder tubing. Technical pamphlet, Amalga Corp.
102. Custom composite components. Technical pamphlet, Amalga Corp.
103. Commercial filament wound pressure vessels for military and aerospace applications. Technical data, Structural Composites Ind.
104. Epoxy fiberglass rods and tubes. Technical pamphlet, Randolph Co.
105. Product data, filament wound pressure vessels, Brunswick Defense.
106. Fiberglass underground storage tanks – single and double wall. Product data, Xerxes.
107. New generation HDPEs increase pressure pipes' performance range. *Mod. Plast. Int.*, Oct. 1989, **19**(10), 36–7.
108. Ooshiro, Y. (1987) Sewage pipe materials. *Civ. Engng Technol.*, **43**(4), 62 (in Japanese).
109. Smith Fiberglass (1987) General installation instructions, fiberglass reinforced piping system. Manual No. 9474, Smith Fiberglass Products Inc.
110. Anon. (1988) Coupling method hikes pipe rating. *Mod. Plast. Int.*, **18**(3), 87.
111. Wilson, B. A. (1989) Filament winding – past, present and future, in *Proceedings 34th International SAMPE Symposium* (eds Zakrzewski, G. A. *et al.*), May, pp. 242–9.

112. Peters, S. T. and Spencer B. E. (1989) Composite fabrication by filament winding, a tutorial, in *Proceedings 34th International SAMPE Symposium*, May.

113. Ray, H. (1983) Dynamic instability of suddenly heated angle-ply laminated composite cylindrical shells. *Comput. Struct.*, **16**(1–4), pp. 119–24.

114. Shaw-Stewart, D. (1985) Filament winding-materials and engineering. *Mater. Des.*, **6**(3).

115. PPG (1981) Fiber glass reinforced plastics – by design. PPG Fiber Glass Reinforcements Market Series, PPG Ind. Inc.

116. Hollaway, L. C. (1990) *Polymers and Polymer Composites in Construction.* Thomas Telford, London.

117. ASCE (1984) *Structural Plastics Design Manual*, ASCE Manuals and Reports on Engineering Practice No. 63, American Society of Civil Engineers, New York.

118. Kim, D. H. and Chang, S. Y. (1967) A simplified method of analysis of a plate problem, in *Annual Symposium, 22nd Japan Society of Civil Engineers*, Hiroshima, Japan, May, pp. 2–49.

119. Goldberg, J. E. and Kim, D. H. (1966) Analysis of triangularly folded plate roofs of umbrella type, in *16th General Congress of Applied Mechanics*, Tokyo, Japan, Oct., p. 280.

120. Verchery, G. *et al.* (1991) A quantitative study of the influence of anisotropy on the bending deformation of laminates, in *Proceedings ICCM 8* (eds Tsai, S. and Springer, G.), July, p. 26-J.

121. Kandil, N. and Verchery, G. (1989) Some new developments in the design of stacking sequences of laminates, in *Proceedings ICCM 7* (ed. Wu, Y.).

122. Verchery, G. (1990) Designing with anisotropy, in *Textile Composites in Building Construction*, (eds Hamelin, P. and Verchery, G.), Vol. 3, Pruralis, pp. 29–42.

123. Kim, D. H. (1968) Design of welded composite high strength plate girder bridges by grid analysis. *J. Korean Soc. Civ. Engrs*, **16**, p. 26.

124. Kim, D. H. (1971) Report on design and construction of Seoul Tower. Special report, Annual General Meeting, Korean Society of Civil Engineers, May.

125. Tsai, S. W. (1988) *Composite Design*, 4th edn, Think Composites, Dayton, OH.

126. Williams, D. J. and Ajibade, F. (1988) High performance sandwich panels made from braided tubes, in *Proceedings 33rd International SAMPE Symposium*, Mar.

127. Hong, Y. S. *et al.* (1983) Development of honeycomb sandwich construction for aircraft structural applications. Research report, Korea Aerospace Technology Research Institute.

128. Vinson, J. R. and Sierakowski, R. L. (1987) *The Behavior of Structures Composed of Composite Materials*, Martinus Nijhoff Publishers, Dordrecht.

129. Kim, D. H. (1967) Theory of non-prismatic folded plate structures. *Trans. Korea Military Academy* (ed. Lee, S. H.), **5**, (Aug.), 182–268.

130. Kim, D. H. (1966) Matrix analysis of multiple shells, *Proc. Korean Soc. Civ. Engrs.*, **13**(4), 9.

131. Kim, D. H., Lee, I. W. and Byeon, M. J. (1989) Geometric nonlinear analysis of underground laminated composite pipes, in *Proceedings ICCM 7* (ed. Wu, Y.), Guangzou, pp. 177–82.

132. Schmit, L. A. (1967) The structural synthesis concept and its potential role in design with composites, in *Proceedings 5th Symposium Naval Structural Mechanics*.

133. Vinson, J. R. and Shore, S. (1965) Design procedures for the structural optimization of flat sandwich panels. Technical report NAEC-ASL-1084, US Naval Air Engineering Center.

134. Vinson, J. R. (1986) Optimum design of composite hex-cell and square cell honeycomb sandwich panels subjected to uniaxial compression. *AIAA J.*, Aug., **24**, 1690–6.

135. Vinson, J. R. (1985) Minimum weight composite material honeycomb sandwich panels under uniaxial compression, in *Transaction, 1st ECCM*, Bordeax, Sep.

136. Vinson, J. R. and Shore, S. (1968) Minimum weight corrugated core sandwich panels subjected to uniaxial compression. *Int. J. Fiber Sci. Technol.*

137. Vinson, J. R. (1987) Minimum weight composite truss core sandwich panels subjected to combined uniaxial compression and in-plane shear loads, in *Proceedings ICCM 6*, London, July.

138. Petiau, C. (1989) Optimization of aircraft structure in composite material, in *Proceedings ICCM 7* (eds Yunshu, W. *et al.*), p. 33.

139. Petiau, C. and Lecina, G. (1982) Optimization of aircraft structure, in *Foundation of Structural Optimization Approach*, John Wiley.

140. Kolkailah, F. A. (1991) A finite element presentation of an optimum design for composite multilayered cylindrical pressure vessels, in *Proceedings 2nd Japan International SAMPE Symposium (JISSE 2)* (ed. Kimpura, I.), Dec., pp. 455–65.

141. Jun, T. (1989) On optimum analysis of laminated plates under shear, *Proceedings ICCM 7* (ed. Wu, Y.), Vol. 4 Nov., p. 202.

142. Cuntze, R. G. and Zaun, J. (1989) Delamination optimization of hoop wound composite flywheels, *Proceedings ICCM 7* (ed. Wu, Y.), Vol. 3, Nov., p. 16.

143. Liu, S. and Qiao, X. (1987) Frequency and flutter analysis of wing type structures and the relevant optimal design, in *Proceedings ICCM 6*, London, July, pp. 5-144-5-154.

144. Miravete, A. (1989) Optimization of symmetrically laminated composite rectangular plates, in *Proceedings ICCM 7* (ed. Wu, Y.), Vol. 3, pp. 289–294.

145. Miki, M. and Sugiyama, Y. (1989) Optimum design of fibrous laminated hybrid composites with required flexural stiffness, in *Proceedings ICCM 7* (ed. Wu, Y.), Vol. 3, p. 295.

146. Romeo, G. and Baracco, A. (1989) Minimum-mass optimization of composite stiffened, unstiffened and sandwich curved panels subjected to combined longitudinal and transverse compression and shear loading, in *Proceedings ICCM 7* (ed. Wu, Y.), Vol. 3, p. 364.

147. Gao, Y. and Mai, Y. (1989) Optimal design of fiber reinforced materials, in *Proceedings ICCM 7* (ed. Wu, Y.), Vol. 3, p. 377.

148. Ding, H. *et al.* (1989) Analysis and optimal design of composite structures, in *Proceedings ICCM 7* (ed. Wu, Y.), Vol. 3, p. 411.

149. Frank-Susich, D. *et al.* (1992) Computer aided cure optimization, in *Proceedings 37th SAMPE Symposium* (ed. Grimes, G. C.), Mar., pp. 1075–88.

150. Trivisano, A. *et al.* (1992) Control and optimization of autoclave processing of high performance composites, in *Proceedings 37th SAMPE Symposium* (ed. Grimes, G. C.) Mar., pp. 1104–16.

151. Miaosheng, W. *et al.* (1991) Optimum design of adhesive bonding of composites, in *Proceedings ICCM 8* (eds Tsai, S. and Springer, G.), July, p. 9-H.

152. Drzal, L. T. (1991) Optimum design of the fiber–matrix interphase in composite materials. *Proceedings ICCM 8* (eds Tsai, S. and Springer, G.), July, p. 1-E.

153. Liou, M. J. and Suh, N. P. (1989) Optimal cooling for minimum residual stresses, in *Proceedings ICCM 7* (ed. Wu, Y.), Vol. 3, pp. 64–9.

154. Devanathan, D. *et al.* (1992) Process optimization of commingled tow composites, in *Proceedings 37th SAMPE Symposium* (ed. Grimes, G. C.), Mar., pp. 1465–79.

155. Pindra, M. J. *et al.* (1989) Creep optimization of ARALL-4 laminates, in *Proceedings ICCM 7* (ed. Wu, Y.), pp. 64–9.

Further reading

Books
Bruhn, E. F. (1973) *Analysis and Design of Flight Vehicle Structures*, S.R. Jacobs, USA.

Bureau of Reclamation (1981) *Pipe Bedding and Backfill*, Geotechnical Branch Training Manual No. 7.

Caprino, G. and Teti, R. (1989) *Sandwich Structures Handbook*, Il Prato, Padua.

Loken, H. and Hollmann, M. *Designing with Core*. Technical Manual, Du Pont.

Lubin, G. (1987) *Handbook of Composites*, Van Nostrand Reinhold, New York.

Spangler, M. G. and Handy, R. L. (1982) *Soil Engineering*, 4th edn, Harper and Row, New York.

Zoutendijk, G. G. (1960) *Method of Feasible Direction*, Elsevier, Amsterdam.

Articles and reports

Banuk, R. (1993) Design considerations for foam and honeycomb core structure, in *Proceedings 38th International SAMPE Symposium and Exhibition* (eds Bailey, V. *et al.*), Anaheim, CA, May, pp. 1751–61.

Bitzer, T. N. and Castillo, J. I. (1988) Graphite honeycomb, in *Proceedings 33rd International SAMPE Symposium* (eds Carrillo, G. *et al.*), Mar., pp. 73–7.

Bockstedt, R. J. and Sajna, J. L. (1993) Low cost composites based on long carbon fiber thermoplastics, in *Proceedings 38th International SAMPE Symposium and Exhibition* (eds Bailey, V. *et al.*), Anaheim, CA, May, pp. 2011–20.

Borris, P. W. *et al.* (1989) Recent honeycomb developments, in *Proceedings 34th International SAMPE Symposium* (eds Zakrzewski, G. A. *et al.*), pp. 849–60.

Brockenbrough, R. L. (1968) Influence of wall stiffness on corrugated metal culvert design. *HRB Bull.* No. 56.

Brown, R. T. (1985) Through-the-thickness braiding technology, in *Proceedings 30th National SAMPE Symposium*, (eds Newsoh, N. and Brown, W. D.), Mar., p. 1509.

Bucci, R. J., Mueller, L. N., Schultz, R. W., Prohaska, J. L. (1987) ARALL laminates, results from a cooperative test program, in *Proceedings 32nd International SAMPE Symposium* (eds Carrillo, G. *et al.*), Apr., pp. 902–16.

Bucci, R. J., Mueller, L. B., Vogelsang, J. W. and Gunnink, J. W. (1988) ARALL laminates: properties and design updated, in *Proceedings 33rd International SAMPE Symposium* (eds Carson, R. *et al.*), Mar., pp. 902–16.

Burnside, P. *et al.* (1993) Design optimization of an ALL-FRP bridge, in *Proceedings 38th International SAMPE Symposium and Exhibition* (eds Bailey, V. *et al.*), Anaheim, CA, May, pp. 1789–99.

Chao, C. C., Wang, C. C. and Chan, C. Y. (1981) Vibration of web-stiffened foam sandwich panel structures, in *Composite materials* (ed. Marshall), Applied Science Publishers.

Chin, H. B., Prevorsek, D. C. and Li, H. L. (1989) Composite design for structural application, in *Proceedings 34th International SAMPE Symposium* (eds Zakrzewski, G. A. *et al.*), pp. 678–85.

Chisholm, J. M., Kallas, M. N. and Hahn, H. T. (1991) The effect of sea water absorption in pultruded composite rods, in *Proceedings ICCM 8* (Tsai, S. and Springer, G.), July, p. 16-A.

Corden, J. L. and Bitzer, T. N. (1987) Honeycomb materials and applications, in *Proceedings 32nd International SAMPE Symposium and Exhibition* (eds Carson, R. *et al.*), Apr., pp. 68–78.

Costes, N. C. (1956) Factors affecting vertical loads on underground ducts due to arching. *HRB Bull.* 125.

Costin, K. S., Chan, W. S. and Wang, B. P. (1991) Optimum ply layup for delamination resistance in laminates with a hole, in *Proceedings ICCM 8* (eds Tsai, S. and Springer G.), July, p. 1-I.

Demir, H. H. (1965) Cylindrical shells under ring load. *J. Struct. Div. Proc. ASCE*, Jun. Vol. 91, No. ST3, pp. 71–98.

Drzal, L. T. (1991) Optimum design of the fiber-matrix interfaces in composite materials, in *Proceedings ICCM 8* (eds Tsai, S. and Springer, G.), July, p. 1-E.

Dvorak, P. J. (1987) Designing with composites. *Mach. Des.*, 26 Nov.

Ferreira, A. J. M. *et al.* (1991) Finite element analysis of sandwich structures, in *Proceedings ICCM 8* (eds Tsai, S. and Springer, G.), July, p. 3-A.

Ferreira, A. J. M. and Marques, A. T. (1991) Composite structure analysis by the use of flat shell element, in *Proceedings ICCM 8* (eds Tsai, S. and Springer, G.), July, p. 26-C.

Fischer, D. R. (1991) Knowledge-based material selection, in *Proceedings ICCM 8* (eds Tsai, S. and Springer, G.), July, p. 1-A.

Fishman, S. G. (1991) Control of interfacial behavior in structural composites, in *Proceedings ICCM 8* (eds Tsai, S. and Springer, G.), July, p. 19-A.

Florentine, R. A. (1988) 3-D braiding adapted to air foil shapes. Net shape contour preforms, in *Proceedings 33rd SAMPE International Symposium* (eds Carrillo, G. *et al.*), Apr., pp. 921–32.

Folle, G. M. (1971) Stiffness matrix for sandwich folded plates. *J. Struct. Div. Proc. ASCE*, Feb., Vol. 97, ST2, pp. 603–17.

Galili, N. and Shmulevich, I. (1981) A refined elastic model for soil-pipe interaction, in *Proceedings of International Conference on Underground Plastic Pipe*, (ed. Schrock, B. J.), New Orleans, pp. 213–26.

Gao, Y. C. and Mai, Y. W. (1989) Optimal design of fiber-reinforced materials, in *Proceedings ICCM 7* (ed. Wu, Y.), Vol. 3, p. 377.

Geoghegan and Lees, J. K. (1989) Self designed system – the future state, in *Proceedings ICCM 7* (ed. Wu, Y.), Vol. 3, p. 225.

Gordaninejad, F. (1991) Thermally conductive fiber-reinforced composite materials, in *Proceedings ICCM 8* (eds Tsai, S. and Springer, G.), July, p. 29-K.

Ha, K. H., Hussein, P. and Fazio, P. (1982) Analytic solution for continuous sandwich plates. *EM2 ASCE*, Apr., 228–41.

Hahn, H. T. (1991) Design for manufacturability (DFM) for composites, in *Proceedings ICCM 8* (eds Tsai, S. and Springer, G.), July, p. 1-P.

Han, Y. M. and Hahn, H. T. (1989) Design of composite laminates with ply failure, in *Proceedings 34th International SAMPE Symposium and Exhibition* (eds Zakrzewski, G. A. *et al.*), pp. 529–38.

Hasson, D. F. and Crowe, C. R. (1987) Flexural fatigue behavior of aramid reinforced aluminum 7075 laminate (ARALL-L) and AL 7075 alloy sheet in air and in salt ladened humid air, in *Proceedings ICCM 6*, (eds Mathews, F. L. *et al.*) July, p. 2.138.

Hong, C. S. and Kim, D. M. (1989) Stacking method of thick composite laminates considering interlaminar normal stress, in *Proceedings 34th International SAMPE Symposium* (eds Zakrzewski, G. A. *et al.*), pp. 1010–18.

Iyer, S. L. (1993) First composite cable prestressed bridge in the USA, in *Proceedings 38th International SAMPE Symposium and Exhibition* (eds Bailey, V. *et al.*), Anaheim, CA, May, pp. 1766–74.

Jia, J. and Rogers, C. A. (1989) Shells with embedded shape memory alloy actuators, in *Proceedings ICCM 7* (ed. Wu, Y.), Vol. 3, p. 202.

Jones, R. M. (1991) Delamination-suppression concepts for composite laminate free edges, in *Proceedings ICCM 8* (eds Tsai, S. and Springer, G.), July, p. 28-M.

Jones, R. M. and Morgan, H. S. (1974) Buckling and vibration of cross-ply laminated circular cylindrical shells, in *AIAA 12th Aerospace Sciences Meeting*, Washington, DC, AIAA Paper No. 74-33.

Jones, W. K. and Leach, D. C. (1993) Cost effective thermoplastic composites, in *Proceedings 38th International SAMPE Symposium and Exhibition* (eds Bailey, V. *et al.*), Anaheim, CA, May, pp. 1993–99.

Kao, J. S. (1970) Bending of two-layer sandwich plates. *EM2 ASCE*, Apr., 201–5.

Kassaimih, S. A. *et al.* (1993) Optimization of fiber orientations for laminated plates under axial compression, in *Proceedings 38th International SAMPE Symposium and Exhibition* (eds Bailey, V. *et al.*), Anaheim, CA, May, pp. 888–95.

Kawashima, T. and Yamamoto, Y. S. (1981) Development of graphite/epoxy tube truss for satellite, in *Proceedings Japan–US Conference*, (eds Kawate, K. and Akasaka, T.) Tokyo p. 453.

Kemmochi, K., Akasaka, T., Hayashi, R. and Ishwata, K. (1980) Shear-lag effect in sandwich panels with stiffeners under three-point bending. *J. Appl. Mechs.*, **47**, Jun., 383–8.

Kim, D. H. (1987) Application of new materials on structural engineering, in *Meeting of the Structures Committee of Korean Society of Civil Engineers*, Oct.

Kim, D. H. (1992) Optimization of composite material structures – the state of the art, in *Structural Optimization, Proceedings of the Korea–Japan Joint Seminar* (eds Choi, C. K. *et al.*), Seoul, Korea, May, pp. 219–28.

Ko, F., Fang, P. and Chu, H. (1988) 3-D braided commingled carbon fiber/peek composites, in *Proceedings 33rd SAMPE International Symposium* (eds Carrillo, G. *et al.*), Apr., pp. 899–911.

Ko, F., Soebroto, H. B. and Lei, C. (1988) 3-D net shaped composites by the 2-step braiding process, in *Proceedings 33rd SAMPE International Symposium* (eds Carrillo, G. *et al.*), Apr., pp. 912–21.

Lemmer, L. and Lonsinger, H. (1991) Design, analysis and testing of an integrally stiffened composite center fuselage skin for future fighter aircraft, in *Proceedings ICCM 8* (eds Tsai, S. and Springer, G.), July, p. 2-D.

Li, W. and Shiekh, A. E. (1988) The effect of processes and processing parameters on 3-D braided preforms for composites, in *Proceedings 33rd SAMPE International Symposium* (eds Carrillo, G. *et al.*), Apr., pp. 104–115.

Liang, C. and Rogers, C. A. (1989) Behavior of shape memory alloy actuators embedded in composites, in *Proceedings ICCM 7* (ed. Wu, Y.), pp. 475–80.

Manera, M. (1977) Elastic properties of randomly oriented short fiber-glass composites. *J. Composite Mater.*, **11**, Apr., 235–47.

Manso, J. J. *et al.* (1991) Design and manufacturing of 3D composite continuous structures, in *Proceedings ICCM 8* (eds Tsai, S. and Springer, G.), July, p. 2–H.

Marshall, A. (1983) Sandwich construction, *Handbook of Composites* (ed. Lubin, G.), Van Nostrand Reinhold, New York, pp. 557–601.

Meyerhof, G. G. (1963) Strength of steel culvert sheets bearing against compacted sand backfill. *HRB Bull.* No. 30.

Miki, M. and Sugiyama, Y. (1989) Optimum design of fibrous laminated hybrid composites with required flexural stiffeners, in *Proceedings ICCM 7* (ed. Wu, Y.), Vol. 3, p. 295.

Misovec, A. P. and Kempner, J. (1970) Approximate elasticity solution for orthotropic cylinder under hydrostatic pressure and band loads. *J. Appl. Mechs*, Mar., Paper No. 69-APM EE, 101–8.

Molin, J. (1981) Flexible pipes buried in clay, in *Proceedings of International Conference on Underground Plastic Pipe*, (ed. Schrock, B. J.), New Orleans, pp. 322–37.

Noton, B. R. (1987) Cost drivers in design and manufacture of composite structures, in *Composites* (eds Dostal, C. A. *et al.*), Engineering Materials Handbook, Vol. 1, ASM International, Metals Park, OH, pp. 419–29.

Owens, A. D. and Henshaw, J. M. (1991) Recycling, composites, and design, in *Proceedings ICCM 8* (eds Tsai, S. and Springer, G.), July, p. 16-P.

Persad, C. *et al.* (1991) Processing of metal matrix composites, in *Proceedings ICCM 8* (eds Tsai, S. and Springer, G.), July, p. 17-A.

Popper, P. and McConnell, R. (1987) A new 3D braid for integrated parts manufacture and improved delamination resistance – the 2 STEP process, in *Proceedings 32nd SAMPE International Symposium* (eds Carson, R. *et al.*), Apr., pp. 92–115.

Prucz, J., Sivan, J. and Upadhyay, P. C. (1991) On the optimum design of composite ducts and cylinders under combined loading, in *Proceedings ICCM 8* (eds Tsai, S. and Springer, G.), July, p. 1-J.

Rao, K. D. (1988) Buckling of sandwich/stiffened panels, in *Composite Design* (ed. Tsai, S. W.), 4th edn, Think Composites, pp. 21.1–24.15.

Renter, Jr., R. C. (1973) Prediction and control of macroscopic fabrication stresses in hoop wound fiberglass rings, in *ASTM STP521*.

Renter, Jr., R. C. and Guess, T. R. (1974) Analysis, testing, and design of filament wound carbon–carbon burst tubes, in *Composite Materials, Testing and Design*, ASTM STP546.

Richards, R. (1973) Stresses on shallow circular pipe by transformed section. *J. Geotech. Engng Div. ASCE*, **100**(GT6), 2.

Roy, A. K. and Tsai, S. W. (1988) Pressure vessels, in *Composite Design* (ed. Tsai, S. W.), 4th edn, Think Composites, p. 23.

Rubin, C. (1975) Buried pipe deflections using elastic theory. *J. Struct. Div. ASCE*, **101**(12), 2685.

Salamon, N. J. (1981) Stress distribution in sandwich beams in uniform bending, in *ASTM STP734*, pp. 166–77.

Shilbley, A. M. (1982) Filament winding, in *Handbook of Composites* (ed. Lubin, G.), Van Nostrand Reinhold, New York.

Stuart, M. J. (1981) An evaluation of the sandwich beam compression test method for composites, in *ASTM STP734*, pp. 152–65.

Szyszkowski, W. and Glockner, P. G. (1987) Large deformation and collapse behavior of underground aluminum piping, in *Thin-Walled Structures 5* (ed. Rhodes, J.), Elsevier Applied Science, pp. 56–73.

Takagishi, S. K. (1993) Aerospace technology in NGV fuel containers, in *Proceedings 38th International SAMPE Symposium and Exhibition* (eds Bailey, V. *et al.*), Anaheim, CA, May, pp. 2050–4.

Takehana, M., Kimpara, I. and Funatogawa, O. (1981) Effect of core materials on deformation of hat-shaped composite stiffeners, in *Composite Materials* (eds Kawata, K. and Akasaka, T.), Proceedings Japan–US Conference, Tokyo, pp. 400–409.

Timmers, J. H. (1956) Load study of flexible pipes under high fills, HRB Bull. No. 125.

Trott, J. J. and Gaunt, J. (1976) Experimental pipelines under a major road: performance during and after road construction. Transport and Road Research Laboratory Report No. 692, Crownthorne, Berkshire, UK.

Van Vuure, A. W. *et al.* (1993) Survey of intermediate pile length 3D sandwich fabric composites, in *Proceedings 38th International SAMPE Symposium and Exhibition* (eds Bailey, V. *et al.*), Anaheim, CA, May, pp. 1710–19.

Vinson, J. R. (1986) Minimum weight triangulated core sandwich panels subjected to in-plane shear loads, in *Transactions 3rd US–Japan Conference on Composite Materials*, Tokyo, June.

Vinson, J. R. (1986) Minimum weight web-core composite sandwich panels subjected to in-plane compression loads, in *Transactions International Symposium on Composite Materials and Structures*, Beijing, June.

Vinson, J. R. and Shore, S. (1971) Minimum weight web core sandwich panels subjected to uniaxial compression. *AIAA J. of Aircraft*, **8**(11), Nov., 843–7.

Watkins, R. K. (1966) Structural design of buried circular conduits. *HRB Bull.* No. 145.

Weeks, C. A. and Sun, C. T. (1993) Design and characterization of multi-core composite laminates, in *Proceedings 38th International SAMPE Symposium and Exhibition* (eds Bailey, V. *et al.*), Anaheim, CA, May, pp. 1736–50.

Weeks, C. A. and Sun, C. T. (1994) Multi-core composite laminates. *J. Advan. Mater.*, **25**(3), 28–44.

Wu, H. F. (1988) Statistical aspects of tensile strength of ARALL laminates, in *Proceedings 33rd International SAMPE Symposium* (eds Carrillo, G. *et al.*), Apr., pp. 1249–59.

Manufacturers' data

Aeroweb honeycomb sandwich design. Instruction Sheet No. AGC 33a, Ciba-Geigy, 1980.

Aluminum honeycomb. Product Data, Unicel Corp.

Aramid paper honeycomb. Product Data, Unicel Corp.

Divinycell HT Series foam core for aerospace applications. Technical data book, DIAB-Barracuda.

Fabrication guide polyester foam, the lupufoam system. Technical pamphlet, Pennwalt.

Fiberelam. Technical pamphlet, Ciba-Geigy.

Fiberply and fiberfoam building panels. FiberTech Ind., Inc.

Last A Foam series. Product Data, General Plastic Manufacturing Co.

Low-temperature, polyurethane flexible foam. Technical pamphlet, PRC.

PAA-CORE, Aluminum honeycomb. Technical data, Cyanamid.

Rohacell, PMI rigid foam. Technical data ROHM Tech, Inc.

Syn Core. Technical pamphlet, Hysol-Grafil.

Appendix A. Units, symbols and conversion factors

Table A.1 Units and symbols

Measurement	Unit	Symbol
Base units		
Length	meter	m
Mass	kilogram[a]	kg
Thermodynamic temperature	kelvin	K
Time	second	s
Supplementary units		
Plane angle	radian	rad
Derived units		
Acceleration	meter per second squared	m/s^2
Angular acceleration	radian per second squared	rad/s^2
Angular velocity	radian per second	rad/s
Area	square meter	m^2
Density, mass	kilogram per cubic meter	kg/m^3
Energy density	joule per cubic meter	J/m^3
Force	newton	N
Frequency	hertz	Hz
Moment of force	newton meter	N m
Specific heat capacity	joule per kilogram kelvin	J/kg K
Specific energy	joule per kilogram	J/kg
Specific volume	cubic meter per kilogram	m^3/kg
Surface tension	newton per meter	N/m
Velocity	meter per second	m/s
Viscosity, dynamic	pascal[b] second	Pa s
Volume	cubic meter	m^3

[a] 1000 kg = 1 metric ton.
[b] $1 \text{ Pa} = 1 \text{ N/m}^2$.

Table A.2 Conversion factors

To convert from	to	multiply by
Area		
in^2	mm^2	6.451 600 E +02
in^2	cm^2	6.451 600 E +00
in^2	m^2	6.451 600 E −04
ft^2	m^2	9.290 304 E −02
Bending moment or torque		
lbf in	N m	1.129 848 E −01
lbf ft	N m	1.355 818 E +00
kgf m	N m	9.806 650 E +00
ozf in	N m	7.061 552 E −03
Bending moment or torque per unit length		
lbf in/in	N m/m	4.448 222 E +00
lbf ft/in	N m/m	5.337 866 E +01

426

Table A.2 (*Continued*)

To convert from	to	multiply by
Energy (impact, other)		
ft lbf	J	1.355 818 E +00
kW h	J	3.600 000 E +06
W h	J	3.600 000 E +03
Force		
lbf	N	4.448 222 E +00
kip(1000 lbf)	N	4.448 222 E +03
tonf	kN	8.896 443 E +00
kgf	N	9.806 650 E +00
Force per unit length		
lbf/ft	N/m	1.459 390 E +01
lbf/in	N/m	1.751 268 E +02
kip/in	N/m	1.751 268 E +05
Fracture toughness		
ksi$\sqrt{\text{in}}$	MPa$\sqrt{\text{m}}$	1.098 800 E +00
Impact energy per unit area		
ft lbf/ft^2	J/m^2	1.459 002 E +01
Length		
Å	nm	1.000 000 E −01
μin	μm	2.540 000 E −02
mil	μm	2.540 000 E +01
in	mm	2.540 000 E +01
in	cm	2.540 000 E +00
ft	m	3.048 000 E −01
yd	m	9.144 000 E −01
mile	km	1.609 300 E +00
Length per unit mass		
in/lb	m/kg	5.599 740 E −02
yd/lb	m/kg	2.015 907 E +00
Mass		
oz	kg	2.834 952 E −02
lb	kg	4.535 924 E −01
ton (short, 2000 lb)	kg	9.071 847 E +02
ton (short, 2000 lb)	kg × 10^3	9.071 847 E −01
ton (long, 2240 lb)	kg	1.016 047 E +03
Mass per unit area		
oz/in^2	kg/m^2	4.395 000 E +01
oz/ft^2	kg/m^2	3.051 517 E −01
oz/yd^2	kg/m^2	3.390 575 E −02
lb/ft^2	kg/m^2	4.882 428 E +00
Mass per unit length		
lb/ft	kg/m	1.488 164 E +00
lb/in	kg/m	1.785 797 E +01
denier	kg/m	1.111 111 E −07
tex	kg/m	1.000 000 E −06
Mass per unit time		
lb/h	kg/s	1.259 979 E −04
lb/min	kg/s	7.559 873 E −03
lb/s	kg/s	4.535 924 E −01

Table A.2 (*Continued*)

To convert from	to	multiply by
Mass per unit volume		
g/cm^3	kg/m^3	1.000 000 E +03
lb/ft^3	g/cm^3	1.601 846 E −02
lb/ft^3	kg/m^3	1.601 846 E +01
lb/in^3	g/cm^3	2.767 990 E +01
lb/in^3	kg/m^3	2.767 990 E +04
oz/in^3	kg/m^3	1.729 994 E +03
Pressure (fluid)		
atm (standard)	Pa	1.013 250 E +05
bar	Pa	1.000 000 E +05
in Hg(32 °F)	Pa	3.386 380 E +03
in Hg(60 °F)	Pa	3.376 850 E +03
lbf/in^2(psi)	Pa	6.894 757 E +03
torr(mm Hg, 0 °C)	Pa	1.333 220 E +02
Specific area		
ft^2/lb	m^2/kg	2.048 161 E −01
Stress		
$tonf/in^2$(tsi)	MPa	1.378 951 E + 01
kgf/mm^2	MPa	9.806 650 E + 00
ksi	MPa	6.894 757 E + 00
lbf/in^2(psi)	MPa	6.894 757 E −03
MN/m^2	MPa	1.000 000 E + 00
Temperature		
°F	°C	$(5/9)(°F − 32)$
°R	K	5/9
°F	K	$(5/9)(°F + 459.67)$
°C	K	°C + 273.15
Temperature interval		
°F	°C	5/9
Thermal expansion		
$\mu in/in$ °C	$10^{-6}/K$	1.000 000 E + 00
$\mu in/in$ °F	$10^{-6}/K$	1.800 000 E + 00
Velocity		
ft/h	m/s	8.466 667 E −05
ft/min	m/s	5.080 000 E −03
ft/s	m/s	3.048 000 E −01
in/s	m/s	2.540 000 E −02
km/h	m/s	2.777 778 E −01
mph	km/h	1.609 344 E + 00
Viscosity (dynamic and kinematic)		
poise (P)	Pa s	1.000 000 E −01
cP	Pa s	1.000 000 E −03
$lbf\ s/in^2$	Pa s	6.894 757 E + 03
ft^2/s	m^2/s	9.290 304 E −02
in^2/s	mm^2/s	6.451 600 E + 02
Volume		
in^3	m^3	1.638 706 E −05
ft^3	m^3	2.831 685 E −02
fluid oz	m^3	2.957 353 E −05
gal (US liquid)	m^3	3.785 412 E −03

Table A.2 (*Continued*)

To convert from	to	multiply by
Volume per unit time		
ft^3/min	m^3/s	4.719 474 E $-$04
ft^3/s	m^3/s	2.831 685 E $-$02
in^3/min	m^3/s	2.731 177 E $-$07
Wavelength		
Å	nm	1.000 000 E $-$01

Table A.3 Exponential expressions, names and symbols

Exponential expression	Multiplication factor	Name	Symbol
10^{18}	1 000 000 000 000 000 000	exa	E
10^{15}	1 000 000 000 000 000	peta	P
10^{12}	1 000 000 000 000	tera	T
10^9	1 000 000 000	giga	G
10^6	1 000 000	mega	M
10^3	1 000	kilo	k
10^2	100	hecto(a)	h
10^1	10	deka(a)	da
10^0	1	Base unit	
10^{-1}	0.1	deci(a)	d
10^{-2}	0.01	centi(a)	c
10^{-3}	0.001	milli	m
10^{-6}	0.000 001	micro	μ
10^{-9}	0.000 000 001	nano	n
10^{-12}	0.000 000 000 001	pico	p
10^{-15}	0.000 000 000 000 001	femto	f
10^{-18}	0.000 000 000 000 000 001	atto	a

Appendix B. Numerical analysis of structures

B.1 Finite-element method and finite-difference method

B.1.1 Introduction

The finite-difference (FD) approach evolved by replacing the governing differential equation of the continuum by its equations at discrete points. Most of the elasticity problems require solving the governing partial differential equations with certain boundary conditions. The drawback of closed-form elasticity analysis is that only simple boundary conditions can be easily handled. On the other hand, complex boundary value problems can be analyzed by the use of the FD approach. Since the first (known) successful application of the FD approach in elasticity for the torsional problems by Runge, this method has been successfully applied to beams, columns, plates and shells. It was proven to be an efficient method to analyze complex problems with mixed boundaries such as nonprismatic folded plates by the author [1, 2].

The finite-element (FE) method is an outgrowth of beam–frame networks and airframe structural analyses. Structural engineers had been using matrix analysis methods to analyze two- and three-dimensional beam–frame networks, considering each beam as a discrete element. Aircraft designers, such as Turner *et al.* in 1956, adapting the concept of the stiffness methods which were being used by civil engineers in analyzing frame and truss networks, developed the concept of a triangular FE stiffness matrix based on an assumed displacement field. The methods of variational calculus were later adapted, and the FE method used today evolved.

Structural analysts must be able to use both the FD and FE methods at will. There are certain rules to select the method of analysis depending on the nature of the problem. The FD method becomes complicated if nonuniform meshes are required to define an irregular geometry or if a three-dimensional continuum is involved. The FD solutions are more easily solved when Dirichlet boundary conditions rather than Cauchy boundary conditions are defined. FD methods have been used very successfully for beams, columns, plates and two-dimensional shells with variable thicknesses and properties, and are well suited for composite laminate analysis. The advantage of the FD method is that once the equations are set up, the solution of the problem can be obtained considerably faster than the FE method. If a specific structure has to be solved with geometry (thickness) and material property changes, the FD approach is best suited.

The FE method is better for problems with irregular geometry, nonuniform meshes, three-dimensional cases, and Cauchy-type boundary conditions. The basic geometry of the element can be fitted into the desired shape. The element geometry is an integral part of the stiffness formulation for each element. The analyst can choose any of the one-, two- and three-dimensional elements within the same analysis, with compatibility satisfied between the different analyses.

- The Dirichlet boundary condition specifies the solution $\Phi = f(x, y, z)$ on the boundary.
- The Cauchy boundary condition specifies the outward normal derivatives $\partial\Phi/\partial n = g(x, y, z)$ on the boundary.

For the last three decades, extensive application of the FE method has been made and a vast amount of literature is available. Some FE methods are misused, especially for composites, but there are more than enough references available, while the FD method, even though it is efficient for many composite problems, is not well introduced. Therefore the FD method is explained here in more detail. As for the FE method, some comments are necessary for application to composite analysis, as given in the next section.

B.1.2 Use of the finite-element method

Engineers have used both linear and nonlinear methods for composite analysis. When composites are used for load bearing applications, nonlinear analysis yields more accurate results for certain problems such as both buckling and post-buckling behavior, and temperature dependent cases. Laminates have both material and geometric nonlinearity characteristics. Accurate analysis of non-linearity of a laminate requires 3-D elements.

The fiber orientation must be modeled for each ply, defining the material coordinate with three angles at each point in the material. The failure/strength criteria must also be studied for each ply.

When FE analysis of laminates is performed with 2-D shell elements, the results are plane stresses and strains, regardless of the chosen element type. For the analysis of regions of high stress or strain, such as the corners or free edges of laminates, and of reinforced concrete, use of 3-D solid elements yields better results.

B.1.3 Finite differences

In mathematical analysis, two groups of functions can be distinguished. The first group consists of functions in which the dependent variable is continuous, and these functions belong to the domain of **infinitesimal calculus**. The second group consists of functions in which the variable is defined at discrete points. To such functions, the methods of infinitesimal calculus are not applicable. The **calculus of finite differences** deals especially with such functions, but it may be applied to both categories. In fact, the use of a high-performance computer enables us to make the increments of the variable small enough such that the functions of the continuous variable can be well approximated by finite differences. For example, the sectorial plate solution by the use of the FD method has differences less than 0.01% compared with the 'exact' solution result. The mesh size used is 1/10 of the radius [3].

Method of finite differences

Let us suppose that we have a smooth curve passing through three points, sufficiently close together and equally spaced (Figure B.1). The slope at the point mid-way between $i - 1$ and i is

$$\frac{dy}{dx}\bigg|_{i-1/2} \approx \frac{y_i - y_{i-1}}{h} \tag{B.1}$$

In the same manner, the slope at the point mid-way between i and $i + 1$ is

$$\frac{dy}{dx}\bigg|_{i+1/2} \approx \frac{y_{i+1} - y_i}{h} \tag{B.2}$$

If we had two values of a function f, specified at the points $i - \frac{1}{2}$ and $i + \frac{1}{2}$, then the approximate value of the first derivative of this function at i would be

$$\frac{df}{dx}\bigg|_i \approx \frac{f_{i+1/2} - f_{i-1/2}}{h} \tag{B.3}$$

Let us take $f = dy/dx$, that is, the first derivative of the original function. Then we interpret df/dx as

$$\frac{d}{dx}\left(\frac{dy}{dx}\right) = \frac{d^2y}{dx^2}$$

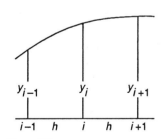

Figure B.1

Now, since

$$f_{i+1/2} = \frac{dy}{dx}\Big|_{i+1/2} \approx \frac{y_{i+1} - y_i}{h}$$

$$f_{i-1/2} = \frac{dy}{dx}\Big|_{i-1/2} \approx \frac{y_i - y_{i-1}}{h} \tag{B.4}$$

it follows that

$$\frac{d^2y}{dx^2} \approx \frac{1}{h^2}\left[y_{i+1} - 2y_i + y_{i-1}\right] \tag{B.5}$$

This process may be continued to obtain approximate expressions for the values of derivatives of higher order.

We see therefore that a differential term may be replaced approximately by an algebraic expression. The set of differential equations are replaced by a set of algebraic equations which can readily be solved by a simple matrix operation. This sometimes permits us to handle certain types of problems which are formulated as differential equations and thus to obtain at least an approximate solution even when the rigorous solution is not known or is not readily obtainable.

Application to the buckling of a column

Let us apply the basic FD technique to the problem of the buckling of a uniform strut which we assume is going to buckle into a shape corresponding to the lowest mode. The differential equation is

$$\frac{d^2y}{dx^2} + \frac{P}{EI}\, y = 0 \tag{B.6}$$

Let us divide the strut into two equal segments. We have shown that

$$\frac{d^2y}{dx^2}\Big|_i \approx \frac{1}{h^2}\left[y_{i+1} - 2y_i + y_{i-1}\right]$$

At $i = 1$,

$$\frac{d^2y}{dx^2}\Big|_{x=L/2} \approx \frac{1}{(L/2)^2}[0 - 2y_i + y_{i-1}] = \frac{-8y_1}{L^2} \tag{B.7}$$

Substituting into the differential equation, we obtain

$$-\frac{8y_1}{L^2} + \frac{Py_1}{EI} = 0$$

and solving for $P = P_{cr}$, $P_{cr} = 8EI/L^2$, which is 19% too low.

Let us try a solution with three equal spaces (Figure B.2). The boundary conditions are

$$y(0) = y_0 = 0$$

$$y(L) = y_3 = 0 \quad \text{(Dirichlet problem)}$$

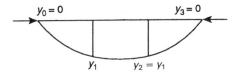

At point 1,

$$\frac{d^2y}{dx^2}\bigg|_{x=L/3} \approx \frac{1}{(L/3)^2}[y_0 - 2y_i + y_2] = \frac{9}{L^2}(y_1) \tag{B.8}$$

Substituting into the differential equation, we obtain $P_{cr} = 9EI/L^2$, which is 9% too low.

When the strut is divided by four segments, if the number and the boundary conditions are symmetric about the mid-point, then $y_3 = y_1$ for the fundamental mode, and there are only two essential unknowns remaining, namely y_1 and y_2. The differential equation is transformed, through the use of the FD approximation, to the following set of algebraic equations:

$$y_{i+1} + \left(\frac{Ph^2}{EI} - 2\right)y_i + y_{i-1} = 0$$

Thus taking i successively to be 1 and 2, and using both the boundary condition at 0 and the symmetry condition, $y_3 = y_1$, we obtain

$$\left(\frac{Ph^2}{EI} - 2\right)y_1 + y_2 = 0$$

$$2y_1 + \left(\frac{Ph^2}{EI} - 2\right)y_2 = 0 \tag{B.9}$$

Since these are a pair of simultaneous homogeneous equations, they have a nontrivial solution if the determinant of the coefficients vanishes. Hence the stability equation (for this mathematical approximation to the actual member) is the determinant equation

$$\begin{vmatrix} (k-2) & 1 \\ 2 & (k-2) \end{vmatrix} = 0 \tag{B.10}$$

where we set $k = Ph^2/EI$, for convenience, since EI is constant. Upon expanding the determinant, the characteristic equation becomes $(k-2)^2 = 2$ which yields $k = \pm\sqrt{2} + 2 = 0.586, 3.414$.

Taking the lower of these values, and recalling the value of k, we obtain $P_{cr} = 9.3715EI/L^2$, which is 6% too low.

Use of the method of interpolating polynomials
To fit the solution at three specified abscissae (called the pivot or pivotal points) (Figure B.3) the approximating polynomials must have the form

$$y = Ax^2 + Bx + C \tag{B.11}$$

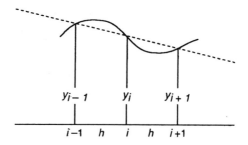

If we take the abscissa at point i to be the origin for the interpolating polynomial,

$$C = y_i$$
$$B = y'_i \tag{B.12}$$
$$2A = y''_i$$

To give the desired values at the adjacent pivoted point, the coefficients must satisfy the following equations:

$$y_{i-1} = A(-h)^2 + B(-h) + C$$
$$y_{i+1} = A(h)^2 + B(h) + C$$

or   (B.13)

$$Ah^2 - Bh = y_{i-1} - y_i$$
$$Ah^2 + Bh^2 = y_{i+1} - y_i$$

From these we obtain

$$2A = \frac{1}{h^2}(y_{i+1} - 2y_i + y_{i-1})$$
$$\tag{B.14}$$
$$B = \frac{1}{2h}(y_{i+1} - y_{i-1})$$

Hence, the approximate formulas for the derivatives are

$$y'_i = \frac{1}{2h}(y_{i+1} - y_{i-1}), \quad y''_i = \frac{1}{h^2}(y_{i+1} - 2y_i + y_{i-1}) \tag{B.15}$$

which agree with the formulas obtained previously.

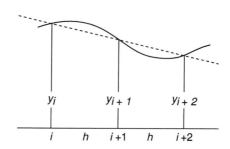

At the boundary points, one-sided expressions which are approximations to the relevant derivatives may be of value. Let i be the boundary point, and $i + 1$ and $i + 2$ be two succeeding or adjacent interior points (Figure B.4). Again without loss of generality, we may take i to be the origin for the FD formula. The interpolating polynomial still has the form

$$y = Ax^2 + By + C$$

and we also have

$$C = y_i$$
$$B = y'_i$$
$$2A = y''_i$$

To give the proper values of the function at the two adjacent points requires that the coefficients satisfy the equations

$$y_{i+1} = A(h)^2 + Bh + C$$
$$y_{i+2} = A(2h)^2 + B(2h) + C \tag{B.16}$$

or

$$Ah^2 + Bh = y_{i+1} - y_i$$
$$4Ah^2 + 2Bh = y_{i+2} - y_i$$

The solution of these equations is

$$2A = \frac{1}{h^2}(y_i - 2y_{i+1} + y_{i+2})$$
$$B = \frac{1}{2h}(-3y_i + 4y_{i+1} + y_{i+2}) \tag{B.17}$$

Hence the one-sided formulas for the first and second derivatives are

$$y'_i = \frac{1}{2h}(-3y_i + 4y_{i+1} - y_{i+2})$$
$$y''_i = \frac{1}{h^2}(y_i - 2y_{i+1} + y_{i+2}) \tag{B.18}$$

Exercise
1. Given four evenly spaced values of y_0, y_1, y_2, y_3, determine the following derivatives: (a) y'_2, (b) y'_1, (c) y'_0, (d) y''_0, (e) y''_2, (f) y''_3.
2. Obtain the expression for the first derivative at i,

$$y'_i = \frac{1}{2(\alpha + 1)h}[y_r + y_i(\alpha^2 - 1) - \alpha^2 y_l]$$

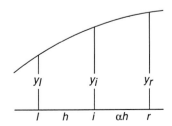

3. Obtain the expression for y_i''',

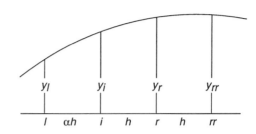

$$y_i'' = \frac{3}{\alpha(\alpha + 1)(\alpha + 2)h^3}$$

$$\times [-2y_l + (\alpha + 1)(\alpha + 2)y_i - 2\alpha(\alpha + 2)y_r + \alpha(\alpha + 1)y_{rr}]$$

B.1.4   Use of Taylor series to obtain finite difference formulas

In terms of the value of a function and its derivatives at x, the value of the function at a neighboring pivotal point $x + h$ (Figure B.5) is given by the Taylor series expansion about x:

$$y(x + h) = y(x) + \left(h\mathscr{D} + \frac{h^2\mathscr{D}^2}{2!} + \frac{h^3\mathscr{D}^3}{3!} + \cdots \right) y(x) \qquad \text{(B.19)}$$

in which $\mathscr{D}$ is the differential operator,

$$\mathscr{D} = \frac{d}{dx}, \ldots, \mathscr{D}^n = \frac{d^n}{dx^n} \qquad \text{(B.20)}$$

and in which the expression

$$1 + h\mathscr{D} + \frac{h^2\mathscr{D}^2}{2!} + \cdots$$

is a linear operator, operating upon y at x.

Alternatively we may write

$$y(x + h) = y(x) + hy'(x) + \frac{h^2}{2!} y''(x) + \frac{h^3}{3!} y'''(x) + \cdots \qquad \text{(B.21)}$$

We may also write, by analogy,

$$y(x - h) = y(x) - hy'(x) + \frac{h^2}{2!} y''(x) - \frac{h^3}{3!} y'''(x) + \cdots \qquad \text{(B.22)}$$

We may eliminate $y''(x)$ between equations B.21 and B.22 and we obtain

$$y(x + h) - y(x - h) = 2hy'(x) + \frac{2h^3}{3!} y'''(x) + \frac{2h^5}{5!} y''''''(x) + \cdots \qquad \text{(B.23)}$$

Solving for $y'(x)$, we obtain

$$y'(x) = \frac{1}{2h} [y(x + h) - y(x - h)] - \frac{h^2}{3!} y'''(x) - \frac{h^4}{5!} y''''''(x) \ldots \qquad \text{(B.24)}$$

Figure B.5

which leads to the formula

$$y'(x) = -\frac{1}{2h}(y_r + y_l) - O(h^2) \qquad \text{(B.25)}$$

where the last term indicates that the 'error' is of order h^2, that is the lowest power of h (the spacing) in the error is the second.

By eliminating $y'(x)$ between equations B.21 and B.22, we get

$$y(x+h) - y(x-h) = 2y(x) + \frac{2h^2}{2!}y''(x) + \frac{2h^4}{4!}y'''(x) + \frac{2h^6}{6!}y''''''(x)\cdots \qquad \text{(B.26)}$$

or solving for $y''(x) = y_i''$, and using the notation $y(x+h) = y_r$, $y(x-h) = y_1$,

$$y_i'' = \frac{1}{h^2}(y_r - 2y_i + y_l) - \frac{h^2}{12}y_i'''' - \frac{h^4}{360}y''''''\cdots \qquad \text{(B.27)}$$

Therefore,

$$y_i'' = \frac{1}{h^2}(y_r - 2y_i + y_l) + O(h^2) \qquad \text{(B.28)}$$

By a similar procedure, we can obtain

$$\mathscr{D}^3 y_i = \frac{1}{2h^3}[y_{i+2} - 2y_{i+1} + 2y_{i-1} - y_{i-2}] + O(h^2)$$

$$\mathscr{D}^4 y_i = \frac{1}{h^4}[y_{i+2} - 4y_{i+1} + 6y_i - 4y_{i-1} + y_{i-2}] + O(h^2) \qquad \text{(B.29)}$$

$$\mathscr{D} y_i = \frac{1}{6h}[-y_{i+1} + 8y_{i+1/2} - 8y_{i-1/2} - y_{i-1}] + O(h^4)$$

The method of interpolating polynomials gives no information regarding the error. The use of the Taylor series allows us to determine the order of error.

Let ∇y_i denote the **backward difference** of the function, y, at i, that is

$$y_i - y_l = \nabla y_i$$

where $y_l = y_{i-1}$. Note that this implies $y_l - y_{ll} = \nabla y_l$. Consequently

$$\nabla(\nabla y_i) = (y_i - y_l) - (y_l - y_{ll}) = y_i - 2y_l + y_{ll}$$

We may obtain the first **forward difference** denoted by

$$\Delta y_i = y_r - y_i$$

where $y_r = y_{i+1}$. The symbols ∇ and Δ represent operators.

By the result of the above,

$$\nabla X_i = X_i - X_l \qquad \text{(B.30)}$$

and if we let $X_i = \nabla y_i$,

$$\nabla(\nabla y_i) = \nabla y_i - \nabla y_l = (y_i - y_l) - (y_l - y_{ll}), \qquad \text{(B.31)}$$

or

$$\nabla^2 y_i = y_i - 2y_l + y_{ll}$$

In a similar manner, we obtain

$$\nabla^3 y_i = y_i - 3y_l + 3y_{ll} - y_{lll}$$
$$\nabla^4 y_i = y_i - 4y_l + 6y_{ll} - 4y_{lll} + y_{llll}$$

(B.32)

One may observe, and readily verify, that the coefficients in the expansion of the nth backward difference are identical with the coefficients of the binomial expansion of $(a - b)^n$.

We may also note that

$$\nabla^n y_i = \nabla(\nabla^{n-1} y_i)$$

(B.33)

The first few forward differences are

$$\Delta y_i = y_r - y_i$$
$$\Delta^2 y_i = y_{rr} - 2y_r + y_i$$
$$\Delta^3 y_i = y_{rrr} - 3y_{rr} + 3y_r - y_i$$
$$\Delta^4 y_i = y_{rrrr} - 4y_{rrr} + 6y_{rr} - 4y_r + y_i$$

(B.34)

Recall that the Taylor series expansion yields

$$y(x + h) = y_i + h\mathscr{D}y_i + \frac{h^2}{2!}\mathscr{D}^2 y_i + \frac{h^3}{3!}\mathscr{D}^3 y_i + \cdots$$

where $\mathscr{D}$ is the differential operator, i.e.

$$\mathscr{D} = \frac{d}{dx} \quad \text{and} \quad \mathscr{D}^n = \frac{d^n}{dx^n}$$

The last result may be written in operator form as

$$y_r = \left[1 + \frac{h\mathscr{D}}{1!} + \frac{h^2 \mathscr{D}^2}{2!} + \frac{h^3 \mathscr{D}^3}{3!} + \cdots \right] y_i$$

(B.35)

Replacing h by $-h$, we also obtain

$$y(x - h) = y_l = \left[1 - \frac{h\mathscr{D}}{1!} + \frac{h^2 \mathscr{D}^2}{2!} - \frac{h^3 \mathscr{D}^3}{3!} + \cdots \right] y_i$$

(B.36)

We now recall the Taylor series for e^z,

$$e^z = 1 + \frac{Z}{1!} + \frac{Z^2}{2!} + \frac{Z^3}{3!} + \cdots$$
$$e^{-z} = 1 - \frac{Z}{1!} + \frac{Z^2}{2!} + \frac{Z^3}{3!} + \cdots$$

(B.37)

We may therefore write, symbolically,

$$e^{h\mathscr{D}} = 1 + \frac{h\mathscr{D}}{1!} + \frac{h^2 \mathscr{D}^2}{2!} + \frac{h^3 \mathscr{D}^3}{3!} + \cdots$$
$$e^{-h\mathscr{D}} = 1 - \frac{h\mathscr{D}}{1!} + \frac{h^2 \mathscr{D}^2}{2!} - \frac{h^3 \mathscr{D}^3}{3!} + \cdots$$

(B.38)

Substituting the last two results into the Taylor series for the functions at $x + h$ and $x - h$, we obtain

$$y_r = e^{h\mathscr{D}} y_i$$
$$y_l = e^{-h\mathscr{D}} y_i$$

(B.39)

In view of the foregoing, we may express the first forward difference and the first backward difference as follows

$$\Delta y_i = y_r - y_i = (e^{h\mathscr{D}} - 1)y_i$$
$$\nabla y_i = y_i - y_l = (1 - e^{-h\mathscr{D}})y_i$$

(B.40)

By means of the last result, we may write

$$\nabla^2 y_i = (1 - e^{-h\mathscr{D}})^2 y_i = (1 - 2 e^{-h\mathscr{D}} + e^{-2h\mathscr{D}})y_i$$

(B.41)

Now, since $\nabla^2 y_i = y_i - 2y_l + y_{ll}$, the above may be solved for $\mathscr{D}^2 y_i$ to yield

$$\mathscr{D}^2 y_i = \frac{1}{h^2} (y_i - 2y_l + y_{ll}) + h\mathscr{D}^3 y_i - \frac{7}{12} h^4 \mathscr{D}^4 y_i + \cdots$$

(B.42)

or

$$\mathscr{D}^2 y_i = \frac{1}{h^2} (y_i - 2y_l + y_{ll}) + O(h)$$

(B.43)

We have

$$\Delta^2 y_i = (e^{h\mathscr{D}} - 1)^2 y_i = (e^{2h\mathscr{D}} - 2 e^{h\mathscr{D}} + 1)y_i$$

(B.44)

Since $\Delta^2 y_i = y_{rr} - 2y_r + y_i$, this yields

$$\mathscr{D}^2 y_i = \frac{1}{h^2} (y_{rr} - 2y_r + y_i) - h\mathscr{D}^3 y_i - \frac{7}{12} h^4 \mathscr{D}^4 y_i + \cdots$$

(B.45)

or

$$\mathscr{D}^2 y_i = \frac{1}{h^2} (y_{rr} - 2y_r + y_i) - O(h)$$

(B.46)

The approximate expressions for the various derivatives, based upon backward differences and forward differences, are frequently given in the form of mathematical molecules as shown in Tables B.1 and B.2.

Table B.1 Backward difference operators

	$i-5$	$i-4$	$i-3$	$i-2$	$i-1$	i	
$h\mathscr{D} =$					-1	1	
$h^2\mathscr{D}^2 =$				1	-2	1	$e = O(h)$
$h^3\mathscr{D}^3 =$			-1	3	-3	1	
$h^4\mathscr{D}^4 =$		1	-4	6	-4	1	
$2h\mathscr{D} =$					-4	3	
$h^2\mathscr{D}^2 =$				1	-5	2	$e = O(h^2)$
$2h^3\mathscr{D}^3 =$			-1	4	-18	5	
$h^4\mathscr{D}^4 =$	-2	3	-14	24	-14	3	

Table B.2 Forward difference operators

	i	$i+1$	$i+2$	$i+3$	$i+4$	$i+5$	
$h\mathscr{D} =$	-1	1					
$h^2\mathscr{D}^2 =$	1	-2	1				$e = O(h)$
$h^3\mathscr{D}^3 =$	-1	3	-3	1			
$h^4\mathscr{D}^4 =$	1	-4	6	-4	1		
$2h\mathscr{D} =$	-3	4	-1				
$h^2\mathscr{D}^2 =$	2	-5	4	-1			$e = O(h^2)$
$2h^3\mathscr{D}^3 =$	-5	18	-24	14	-3		
$h^4\mathscr{D}^4 =$	3	-14	26	-24	11	-2	

The Taylor series expansion for $y(x)$ leads to

$$y(a \pm \eta h) = y(a) \pm \frac{\eta h}{1!} y'(a) + \frac{\eta^2 h^2}{2!} y''(a) \pm \frac{\eta^3 h^3}{3!} y'''(a) + \cdots \qquad \text{(B.47)}$$

where h is the interval spacing and η is any real (positive) number. Symbolically, we may also write

$$y(a + \eta h) = e^{\eta h \mathscr{D}} y(a) = (e^{h\mathscr{D}})^{\eta} y(a) \qquad \text{(B.48)}$$

Furthermore, we have shown that $e^{h\mathscr{D}} = 1 + \Delta$, and therefore

$$(e^{h\mathscr{D}})^{\eta} = (1 + \Delta)^{\eta} \qquad \text{(B.49)}$$

so that symbolically,

$$y(a + \eta h) = (1 + \Delta)^{\eta} y(a) \qquad \text{(B.50)}$$

If we now expand this last expression using the binomial formula, we obtain

$$y(a + \eta h) = \left[1 + \eta\Delta + \frac{\eta(\eta - 1)}{2!} \Delta^2 + \frac{\eta(\eta - 1)(\eta - 2)}{3!} \Delta^3 + \cdots \right] y(a) \quad \text{(B.51)}$$

This is the **Gregory–Newton forward interpolation formula**.

Replacing Δ by $-\nabla$, we obtain the **Gregory–Newton backward interpolation formula**. These formulas may be used for extrapolation (with, of course, diminishing accuracy).

The central difference formulas are more accurate than unilateral formulas, when the same number of terms are used. To define the central differences, we must, temporarily, assume additional points which are located at the mid-points of the intervals. The first central difference, denoted by δy_i, is defined as

$$\delta y_i = y_{i+1/2} - y_{i-1/2} \qquad \text{(B.52)}$$

By this rule, we have

$$\delta y_{i+1/2} = y_{i+1} - y_i$$
$$\delta y_{i-1/2} = y_i - y_{i-1} \qquad \text{(B.53)}$$

The second central difference at i is

$$\delta^2 y_i = \delta(\delta y_i) = \delta y_{i+1/2} - \delta y_{i-1/2} = y_{i+1} - 2y_i + y_{i-1} \qquad \text{(B.54)}$$

Similarly,

$$\delta^3 y_i = y_{i+3/2} - 3y_{i+1/2} + 3y_{i-1/2} - y_{i-3/2}$$

$$\delta^4 y_i = y_{i+2} - 4y_{i+1} + 6y_i - 4y_{i-1} + y_{i-2}$$

(B.55)

Note that, as in the unilateral difference formulas, the coefficients are the same as those of the binomial expansion of $(a - b)^n$, with, however, the additional detail that the odd differences involve the values of the function only at the mid-points of the regular intervals, whereas the even differences involve only the values at the regular pivotal points. To eliminate the use of the values at the mid-points, we introduce the **averaging operator**, μ, defined as

$$\mu f_i = \tfrac{1}{2}(f_{i+1/2} + f_{i-1/2})$$

(B.56)

With this definition, the first averaged central difference is

$$
\begin{aligned}
\mu \delta y_i &= \tfrac{1}{2}(f_{i+1/2} + f_{i-1/2}) \\
&= \tfrac{1}{2}[(y_{i+1} - y_i) + (y_i - y_{i-1})] \\
&= \tfrac{1}{2}[y_{i+1} - y_{i-1}]
\end{aligned}
$$

(B.57)

If we make use of the operator $e^{h\mathcal{D}}$,

$$\mu \delta y_i = \tfrac{1}{2}(e^{h\mathcal{D}} - e^{-h\mathcal{D}})y_i = (\sinh h\mathcal{D})y_i$$

(B.58)

Recalling that

$$\sinh Z = Z + \frac{1}{3!}Z^3 + \frac{1}{5!}Z^5 + \cdots$$

$$\mu \delta y_i = \left(h\mathcal{D} + \frac{h^3 \mathcal{D}^3}{3!} + \frac{h^5 \mathcal{D}^5}{5!} + \cdots \right) y_i$$

(B.59)

Solving for $\mathcal{D}y_i$,

$$\mathcal{D}y_i = \frac{1}{h}(\mu \delta y_i) - \frac{h^2}{6}\mathcal{D}^3 y_i - \cdots$$

(B.60)

or

$$\mathcal{D}y_i = \frac{1}{2h}(y_{i+1} - y_{i-1}) + O(h^2)$$

(B.61)

We can write

$$
\begin{aligned}
\delta^2 y_i &= \delta(\delta y_i) = \delta y_{i+1/2} - \delta y_{i-1/2} = (y_{i+1} - y_i) - (y_i - y_{i-1}) \\
&= (y_{i+1} + y_{i-1}) - 2y_i = (e^{h\mathcal{D}} + e^{-h\mathcal{D}})y_i - 2y_i \\
&= 2(\cosh(h\mathcal{D}) - 1)y_i
\end{aligned}
$$

(B.62)

Since

$$\cosh Z = 1 + \frac{Z^2}{2!} + \frac{Z^4}{4!} + \frac{Z^6}{6!} + \cdots$$

$$\cosh(h\mathcal{D}) - 1 = \frac{h^2 \mathcal{D}^2}{2!} + \frac{h^4 \mathcal{D}^4}{4!} + \frac{h^6 \mathcal{D}^6}{6!} + \cdots = \frac{1}{2}\delta^2 y_i$$

(B.63)

Solving for $\mathscr{D}^2 y_i$,

$$\mathscr{D}^2 y_i = \frac{1}{h^2}(y_{i+1} - 2y_i + y_{i-1}) + O(h^2) \tag{B.64}$$

Comparing this expression with the expression for the second derivatives based upon unilateral differences, either forward or backward, we see that the error of the central difference formula approaches zero with decreasing h at a faster rate than the corresponding unilateral formula. This is generally true and we can state that the central difference formulas are more accurate than unilateral formulas containing the same number of terms.

We had previously shown that

$$\delta y_i = y_{i+1/2} - y_{i-1/2} \quad \text{and}$$
$$y_{i+1/2} = (e^{h\mathscr{D}/2})y_i$$
$$y_{i-1/2} = (e^{-h\mathscr{D}/2})y_i$$

Consequently

$$\delta y_i = (e^{h\mathscr{D}/2} - e^{-h\mathscr{D}/2})y_i = 2(\sinh(h\mathscr{D}/2))y_i \tag{B.65}$$

It follows, therefore that

$$\delta^n y_i = 2^n(\sinh^n(h\mathscr{D}/2))y_i \tag{B.66}$$

and

$$\delta^n y_i = 2^n\left(\frac{h\mathscr{D}}{2} + \frac{h^3\mathscr{D}^3}{8\cdot 3!} + \frac{h^5\mathscr{D}^5}{32\cdot 5!} + \cdots\right)^n y_i \tag{B.67}$$

As an example,

$$\delta^2 y_i = 2^2\left(\frac{h\mathscr{D}}{2} + \frac{h^3\mathscr{D}^3}{48} + \frac{h^5\mathscr{D}^5}{3840} + \cdots\right)^2 y_i$$

$$= 4\left[\frac{h^2\mathscr{D}^2}{4} + \frac{h^4\mathscr{D}^4}{2\cdot 28} + \left(\frac{1}{2\cdot 3840} + \frac{1}{48^2}\right)h^6\mathscr{D}^6 + \cdots\right]y_i \tag{B.68}$$

so that

$$\mathscr{D}^2 y_i = \frac{1}{h^2}(y_{i+1} - 2y_i + y_{i-1}) + O(h^2) \tag{B.69}$$

For practical reasons, this formula is useful particularly for even differences. Using the averaging operator, we can obtain

$$\mathscr{D}^3 y_i = \frac{1}{2h^3}(y_{i+2} - 2y_{i+1} + 2y_{i-1} - y_{i-2}) + O(h^2)$$

$$\tag{B.70}$$

$$\mathscr{D}^4 y_i = \frac{1}{h^3}(y_{i+2} - 4y_{i+1} + 6y_i - 4y_{i-2} + y_{i-2}) + O(h^2)$$

B.1.5 Extrapolation of finite-difference results (sometimes called 'Richardson's extrapolations')

Consider the problem of buckling of a strut hinged at its ends. The differential equation is

$$\frac{d^2y}{dx^2} + \frac{P}{EI} y = 0$$

Making use of the central difference approximation to the second derivative, this becomes the **recursion formula**,

$$\frac{1}{h^2}(y_{i+1} - 2y_i + y_{i-1}) + \frac{P}{EI} y_i = 0$$

or (B.71)

$$y_{i+1} + \left(\frac{Ph^2}{EI} - 2\right)y_i + y_{i-1} = 0$$

This formula is to be applied at all pivotal (or division) points, and in this manner, a sufficient number of algebraic equations are obtained from which an approximate value of the critical load ($P = P_{cr}$) may be deduced. Note that the central difference formula has an error of order h^2.

- The solution based on two divisions of the strut is $P_2 = 8EI/L^2$.
- The three division based solution is $P_3 = 9EI/L^2$.
- When four divisions are made, $P_4 = 9.375EI/L^2$.

For convenience, we let the coefficient of EI/L^2 be denoted by K. Recalling that the FD formula had an error of order h^2, we may postulate that the error of the solution to the complete problem is also of order h^2. Accordingly, if K_2, K_3 and K_4 are the values obtained above, and K without the subscript is the correct value, we may write for each case, since $h_n = L/n$,

$$error_n = C\left(\frac{L}{n}\right)^2 = K - K_n$$

or (B.72)

$$K - C\frac{L^2}{n^2} = K_n$$

The unknown constant C is independent of the spacing. Letting $q = CL^2$, we have

$$K - \frac{q}{(2)^2} = 8, \quad K - \frac{q}{(3)^2} = 9, \quad K - \frac{q}{(4)^2} = 9.3715 \qquad (B.73)$$

Taking any pair of these, we may solve for K.

- Using the first two, $K = 9.8$.
- Using the last two, $K = 9.849$.

Assuming that the error may be expressed more accurately, as was seen by the Taylor series result, by using a two-term error function,

$$\text{error} = C_2\left(\frac{L}{n}\right)^2 + C_4\left(\frac{L}{n}\right)^4 \tag{B.74}$$

we can obtain more accurate values of K. When five segments are used,

$$P_{cr} = \frac{9.55EI}{L^2} \tag{B.75}$$

Recalling the values obtained for two, three and four divisions

n	$P_{cr}/EI/L^2$
2	8.0
3	9.0
4	9.3715
5	9.55

one may observe that, as the number of divisions is increased, the calculated critical load tends toward the correct value. This is generally true, and use will be made subsequently of this trend to extrapolate to a better approximation.

The 'exact' value can be obtained in the following way. Let $k_2 = 8$, $k_3 = 9$, $k_4 = 9.3715$:

$$\text{error}_2 = k - k_2 = \frac{C_2L^2}{(2)^2} + \frac{C_4L^4}{(2)^4}$$

$$\text{error}_3 = k - k_3 = \frac{C_2L^2}{(3)^2} + \frac{C_4L^4}{(3)^4} \tag{B.76}$$

$$\text{error}_4 = k - k_4 = \frac{C_2L^2}{(4)^2} + \frac{C_4L^4}{(4)^4}$$

Letting $C_2L^2 = q_1$, and $C_4L^4 = q_2$

$$K - q_1/(2)^2 - q_2/(2)^4 = 8$$
$$K - q_1/(3)^2 - q_2/(3)^4 = 9 \tag{B.77}$$
$$K - q_1/(4)^2 - q_2/(4)^4 = 9.3715$$

Solving for K,

$$K = 9.865\,52$$

and

$$P_{cr} = 9.865\,52\,\frac{EI}{L^2} \tag{B.78}$$

Note that the 'exact' value is

$$P_{cr} = \frac{\pi^2 EI}{L^2} \doteqdot 9.8696 \frac{EI}{L^2} \qquad \text{(B.79)}$$

It is concluded that we can obtain a more accurate value of the solution by the use of extrapolation than by the method of increasing the number of pivotal points.

B.1.6 Integration of differential equations by finite difference methods

Boundary value problems of second order
Considering the equilibrium of a generic element (Figure B.6), the differential equation for this problem is found to be

$$\frac{d^2 y}{dx^2} + \frac{q}{T} = 0 \qquad \text{(B.80)}$$

with the boundary conditions

$$\begin{aligned} y(0) &= 0 \\ y(L) &= 0 \end{aligned} \qquad \text{(B.81)}$$

The solution to the problem by the FD method may now proceed along the following lines:
We replace the derivative in the differential equation by the appropriate FD approximation. In the present case, we replace the second derivative by the central difference formula,

$$\frac{d^2 y}{dx^2} \approx \frac{1}{h^2} \delta^2 y_i \qquad \text{(B.82)}$$

which has an error of order h^2. The differential equation is thus transformed to the difference equation,

$$y_{i-1} - 2y_i + y_{i+1} = \frac{-h^2 q_i}{T} \qquad \text{(B.83)}$$

We divide the L into n equal lengths, $h = L/n$, and denote $y(0)$ by y_0, and $y(L)$ by y_n. By virtue of the stated boundary conditions, $y_0 = y_n = 0$. We substitute

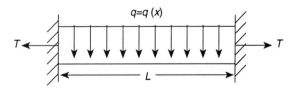

Figure B.6

the appropriate numbers into equation B.83 at each intermediate division point or pivotal point, and thus obtain $n - 1$ simultaneous algebraic equations in the values of y at $x = x_1, x_2, ..., x_{n-1}$.

If the resulting equations have sufficient regularity, they may be solved by the difference equation method. On the other hand, they may have to be solved by straightforward matrix operation. If $n = 10$, the error should be less than 1%.

One may use the extrapolation scheme for quick solution. If $y(L/2)$ is needed, we may do the following.

1. Solve the problem for $n = 4$ and $n = 6$.
2. Considering the values thus obtained at the mid-point, use the extrapolation technique to refine this value.

Partial differential equation of higher order

Three-dimensional structures such as folded plate shells can be solved with relative ease by finite differences. Such structures have complicated internal boundaries because of mixed boundary conditions. In order to reduce the number of pivotal points, the in-plane elasticity and normal to the plane problems can be separated, and furthermore, two sets of fourth-order partial differential equations can be replaced by four sets of second-order partial differential equations. By doing so, the number of necessary pivotal points inside the boundary is only three, and unidirectional differences are necessary only at the boundary. Thus the set of partial differential equations is replaced by a set of linear algebraic equations which can be solved by direct matrix operation.

B.2   Numerical procedures for calculation

B.2.1   General remarks

This section illustrates the application of numerical procedures for the calculation of shears and moments, and slopes and deflections.

1. The procedures are, in general, approximate, but lead to exact moments when the loading diagram is made up of segments of straight lines or, with the use of higher degree approximating curves, when the loading curve is made up of segments of parabolas.
2. Deflections are also exact if the M/EI curve is made up of straight line segments or appropriate parabolic segments.
3. The procedure of calculation is explained mostly by the figures. No further explanation is made unless necessary. Even though high performance computers are available, engineers must be able to anticipate the results and check the range of the computer results, since the errors made by a computer may amount to orders of magnitude. Frequently, engineers should be able to make a quick decision by simple and fast calculation for comparative designs at the planning or preliminary design stage.

B.2.2   Shears and moments

Concentrated loads

Example 1.   Cantilever beam

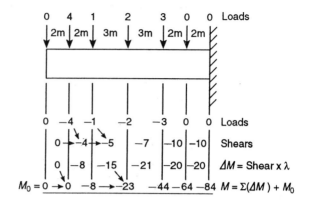

Example 2.   Simply supported beam

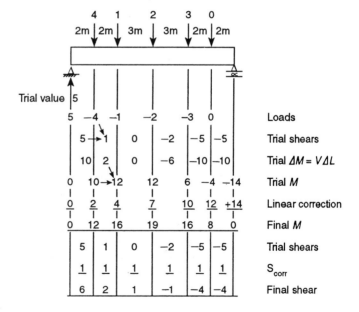

The moment at the right end must be reduced to zero by a **correction** such that all other moments are to be modified by a **linear correction**.

There is associated with the linear correction to the moments, a corresponding **constant correction** to the shears, and its value, in this example, is

$$S_{corr} = \frac{+14}{14} = 1$$

where the numerator represents the required moment correction at the end and the denominator is the span.

We may also reduce the calculated shears and moments of the cantilever beam to the proper values for a simply supported beam.

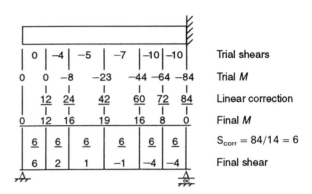

Example 3. Special case of uniform spacing of concentrated loads

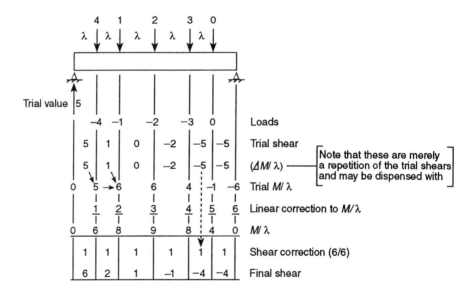

Polygonal loading curve

By dividing the elementary trapezoid (Figure B.7) into two triangular parts, it is easily seen that the trapezoid loading has equivalents, R_a and R_b located at the boundaries of the trapezoid, where

$$R_{ab} = \frac{\lambda}{2} a \frac{2}{3} + \frac{\lambda b}{2} \frac{1}{3} = \frac{\lambda}{6}(2a + b)$$

$$R_{ba} = \frac{\lambda}{6}(2b + a)$$

(B.84)

Figure B.7

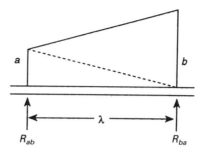

The use of equal and opposite quantities at these locations will give the correct moment at either end of the interval, λ, and at all points outside of the interval.

If two successive intervals of a continuous polygonal loading curve (Figure B.8) have the same length, λ, then the equivalent intermediate concentrated load is

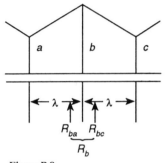

Figure B.8

$$R_b = R_{ba} + R_{bc} = \frac{\lambda}{6}(a + 4b + c) \qquad (B.85)$$

Note that the use of the equivalent concentration (having the numerical value given above but acting in the direction of the applied load) gives the correct bending moments at the division points, but **must** be resolved into its two components, R_{ba} and R_{bc}, to obtain the correct value of shear at the intermediate division point.

Application of polygonal curve equivalents to moment and shear diagrams

Example 1.   Simply supported beam with uniform load

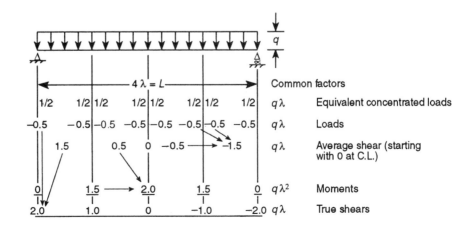

Example 2.   Cantilever beam with linearly varying load

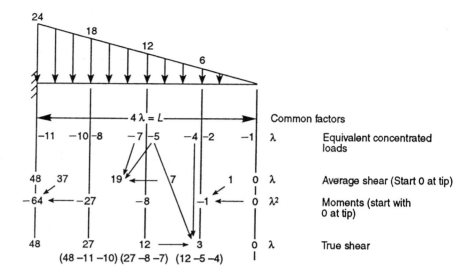

Use

$$R_{ab} = \frac{\lambda}{6}(2a + b), \quad R_{ba} = \frac{\lambda}{6}(a + 2b)$$

to obtain

$$\frac{\lambda}{6}(2 \times 24 + 18) = 11$$

$$\frac{\lambda}{6}(24 + 2 \times 18) = 10$$

$$\frac{\lambda}{6}(2 \times 18 + 12) = 8 \text{ etc.}$$

Properties of the general second-degree curve
Given three ordinates a, b, c, at $x = 0$, λ, 2λ (Figure B.9), the equation of the curve of second degree which passes through these points, i.e. $(0, a)$, (λ, b), $(2\lambda, c)$ is

$$q = \frac{a}{2}(X - 1)(X - 2) - bX(X - 2) + \frac{c}{2}X(X - 1)$$

where $X = x/\lambda$.
    The equivalent concentrated loads are determined by the following:

$$R_{ab} + R_{ba} = \lambda \int_0^1 q \, dX \tag{B.86}$$

and

$$R_{ba} = \lambda \int_0^1 Xq \, dX \tag{B.87}$$

Figure B.9

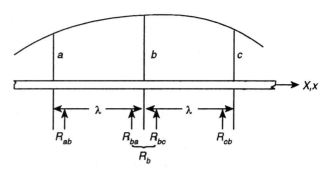

Using the above expression for q, these formulas yield the following results:

$$R_{ab} = \frac{\lambda}{24}(7a + 6b - c)$$

$$R_{ba} = \frac{\lambda}{24}(3a + 10b - c)$$

(B.88)

By analogy we obtain

$$R_{bc} = \frac{\lambda}{24}(3c + 10b - a)$$

Consequently, since $R_b = R_{ba} + R_{bc}$, we obtain

$$R_b = \frac{\lambda}{12}(a + 10b + c)$$

(B.89)

which we may also write as

$$R_b = \frac{\lambda}{12}(a - 2b + c) + \lambda b$$

(B.90)

Example 1. Reduction of the previous cantilever problem to simply supported beam

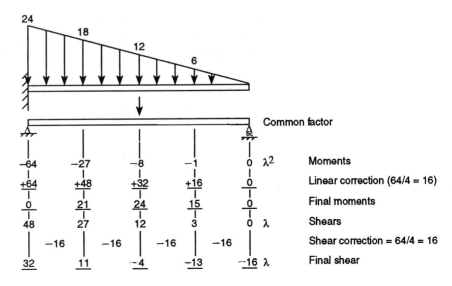

B.2.3   Slopes and deflections

Deflection of beams
We recall the analogy between

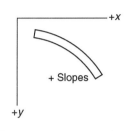

Figure B.10

- Loads and angle changes
- Shears and slopes
- Moments and deflections

The angle change is defined as $-M/EI$ and is considered an upward or positive load. Positive slope (Figure B.10) corresponds to an increase in deflection as we proceed in the direction of the increasing independent variable. Positive deflection is taken downward.

Example 1.   Simply supported beam under uniform load

The M-diagram is a parabola. Hence the formulas and resultants will be exact (if EI = const. for the time being) for any number of intervals. For the left end

$$R_{ab} = \frac{\lambda}{24}\,(7a + 6b - c) = -\frac{\lambda}{12}\ (7)$$

At the second point

$$R_b = \lambda b + \frac{\lambda}{24}\,(a - 2b + c) = -\frac{\lambda}{12}\ (34)$$

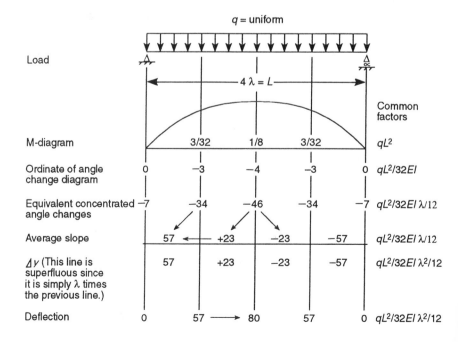

At the third (or mid-) point, one may compute the concentration either as $R_3 = R_{32} + R_{34}$ or may use

$$R_3 = \lambda b + \frac{\lambda}{12}(a - 2b + c)$$

where $b = f_3$, $a = f_2$, $c = f_4$. Using the latter form,

$$R_3 = -\frac{\lambda}{12} \quad (46)$$

$$\text{Deflection at mid point} = \frac{80}{32EI} qL^2 \frac{\lambda^2}{12} = \frac{80qL^2}{32EI} \frac{1}{12} \left(\frac{L}{4}\right)^2 = \frac{5qL^4}{384EI}$$

$$\text{Deflection at quarter point} = \left(\frac{57}{80}\right)\left(\frac{5qL^4}{384EI}\right)$$

$$\text{Slope at left end} = (57 + 7) \frac{qL^2}{32EI} \frac{\lambda}{12} = \frac{qL^3}{24EI}$$

Simplified procedure for smooth angle change curves
Recall that

$$R_b = \lambda b + \frac{\lambda}{12}(a - 2b + c)$$

Let us consider the deflections due to the second term in this formula, $(\lambda/12)(a - 2b + c)$, which we will call Part II (λb will be Part I). Suppose that we now have a smooth angle change diagram with ordinates a, b, etc. at the equally spaced division points (Figure B.11).

To obtain the required results, and recognizing that we will subsequently impose a 'linear' or 'rigid body' displacement, we can take the 'average slope' (Figure B.12) in any interval in a completely arbitrary way. Hence we will start with $(b - a)$ in the first interval.

The deflections resulting from Part II and thus are seen to be $\lambda^2/12$ times the ordinates to the angle change diagram plus a rigid body displacement. The magnitude and character of the rigid body displacement are to be determined

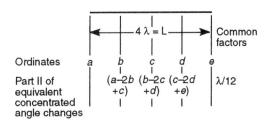

Figure B.11

Figure B.12

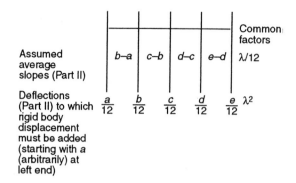

Assumed average slopes (Part II)

| b–a | c–b | d–c | e–d | λ/12 |

Deflections (Part II) to which rigid body displacement must be added (starting with *a* (arbitrarily) at left end)

| $\frac{a}{12}$ | $\frac{b}{12}$ | $\frac{c}{12}$ | $\frac{d}{12}$ | $\frac{e}{12}$ | λ^2 |

by imposing upon the deflections computed as the combination of Part I and Part II such linear variation as is necessary to satisfy the end conditions.

Example 1.   Uniform simply supported beam under uniform load

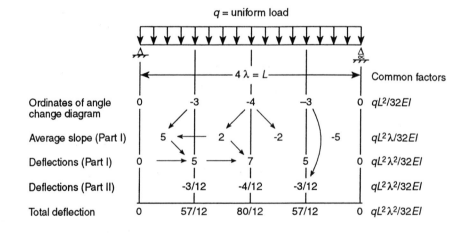

					Common factors
Ordinates of angle change diagram	0	-3	-4	-3	0	$qL^2/32EI$
Average slope (Part I)	5	2	-2	-5		$qL^2\lambda/32EI$
Deflections (Part I)	0	5	7	5	0	$qL^2\lambda^2/32EI$
Deflections (Part II)		-3/12	-4/12	-3/12		$qL^2\lambda^2/32EI$
Total deflection	0	57/12	80/12	57/12	0	$qL^2\lambda^2/32EI$

No rigid body displacement is required in this case since we observe the requirement that the slope at the actual center line is zero and the deflection at the ends is zero.

Example 2.   Simply supported beam with varying cross-section

Point No.	(0)	(1)	(2)	(3)	(4)	(5)	(6)	Common factors
Ordinates of M-diagram	0	5	8	9	8	5	0	$q\lambda^2/2$
Ordinates of angle change	0	5	8	9 ¦18	16	10	0	$-q\lambda^2/4EI_0$

Equivalent concentrated angle change
$R_{ab}=\lambda/24\,(7a+6b-c)$
$R_{ba}=\lambda/24\,(3a+10b-c)$
$R_{bc}=\lambda/24\,(3c+10b-a)$

	−11	−21 ¦−37−43¦−51 −53 ¦−106−102¦−86 −74 ¦−42 −22						$q\lambda^3/48EI_0$
Assumed average slope	(322) 311	(290) 253	(210) 159	(106) 0	(−102)−188	(−262)−304	(−326)	$q\lambda^3/48EI_0$
Trial deflection	0	311 → 564	723	723	535	231		$q\lambda^4/48EI_0$
Linear correction	0	−38.5	−77.0	−115.5	−154	−192.5	−231	$q\lambda^4/48EI_0$
Final deflection	0	272.5	487.0	607.5	569.0	342.5	0	$q\lambda^4/48EI_0$
Slope at div. pt.	322	290	210	106	−102	−262	−326	$q\lambda^3/48EI_0$
Slope correction (231/6=38.5)	−38.5	−38.5	−38.5	−38.5	−38.5	−38.5	−38.5	$q\lambda^3/48EI_0$
Final slope	283.5	251.5	171.5	67.5	−140.5	−300.5	−364.5	$q\lambda^3/4EI_0$

Buckling of bars
The buckling of uniform beams is discussed in section 4.1.5.

Example 1.   Bar with change in section

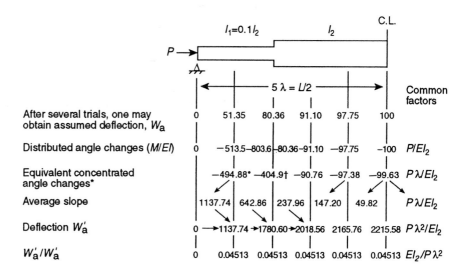

							Common factors
After several trials, one may obtain assumed deflection, W_a	0	51.35	80.36	91.10	97.75	100	
Distributed angle changes (M/EI)	0	−513.5	−803.6	−91.10	−97.75	−100	P/EI_2
Equivalent concentrated angle changes*		−494.88*	−404.9†	−90.76	−97.38	−99.63	$P\lambda/EI_2$
Average slope		1137.74	642.86	237.96	147.20	49.82	$P\lambda/EI_2$
Deflection W'_a	0 →	1137.74 →	1780.60 →	2018.56	2165.76	2215.58	$P\lambda^2/EI_2$
W'_a/W'_a	0	0.04513	0.04513	0.04513	0.04513	0.04513	$EI_2/P\lambda^2$

* At $x = \lambda$

$$\frac{\lambda}{12}(a + 10b + c) = -494.88*$$

† At $x = 2\lambda$

$$R_{ab} = \frac{\lambda}{24}(7a + 6b - c)$$

$$\Rightarrow \frac{\lambda}{24}(7 \times 80.36 + 6 \times 91.1 - 97.75) = 42.1$$

$$+ \frac{\lambda}{24}(7 \times 803.6 + 6 \times 513.5 - 0) = 362.8$$

$$= 404.9**$$

At (or near) convergence, W_a should be equal to W'_a or $W_a/W'_a = 1$. Hence $0.045\,13EI_2/P\lambda^2 = 1$, which yields

$$P_{cr} = 4.513 \frac{EI_2}{L^2}$$

Example 2. Buckling of a bar built-in at one end

Figure B.13

We must first obtain a suitably correct deflection curve for the case shown in Figure B.13. This may be done by a series of successive approximations or trials.

$$W_a/W'_a(\text{Ave}) = 0.026\,18, \quad P_{cr} = (0.026\,18)(12)(64)\frac{EI}{L^2} = 20.11\frac{EI}{L^2}$$

$$\text{Actual } P_{cr} = 20.16\frac{EI}{L^2}$$

By the use of the first iteration, we get

$$P_{cr} = \frac{12EI}{3880(L/8)^2} = \frac{19.75EI}{L^2}$$

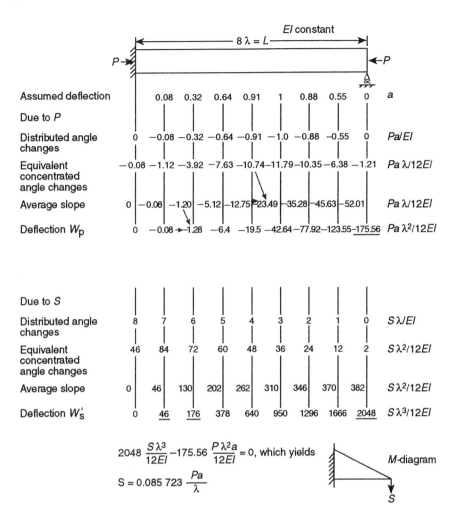

Figure B.14

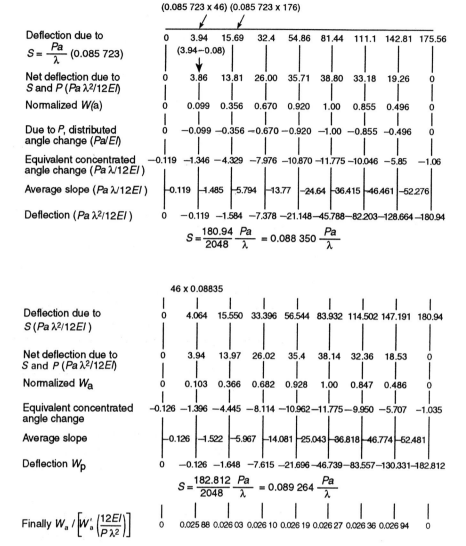

(0.085 723 x 46) (0.085 723 x 176)

Deflection due to $S = \dfrac{Pa}{\lambda}$ (0.085 723): 　0　3.94　15.69　32.4　54.86　81.44　111.1　142.81　175.56
(3.94−0.08)

Net deflection due to S and P ($Pa\lambda^2/12EI$): 　0　3.86　13.81　26.00　35.71　38.80　33.18　19.26　0

Normalized $W(a)$: 　0　0.099　0.356　0.670　0.920　1.00　0.855　0.496　0

Due to P, distributed angle change (Pa/EI): 　0　−0.099　−0.356　−0.670　−0.920　−1.00　−0.855　−0.496　0

Equivalent concentrated angle change ($Pa\lambda/12EI$): −0.119　−1.346　−4.329　−7.976　−10.870　−11.775　−10.046　−5.85　−1.06

Average slope ($Pa\lambda/12EI$): −0.119　−1.485　−5.794　−13.77　−24.64　−36.415　−46.461　−52.276

Deflection ($Pa\lambda^2/12EI$): 　0　−0.119　−1.584　−7.378　−21.148　−45.788　−82.203　−128.664　−180.94

$$S = \frac{180.94}{2048}\frac{Pa}{\lambda} = 0.088\,350\,\frac{Pa}{\lambda}$$

46 x 0.08835

Deflection due to S ($Pa\lambda^2/12EI$): 　0　4.064　15.550　33.396　56.544　83.932　114.502　147.191　180.94

Net deflection due to S and P ($Pa\lambda^2/12EI$): 　0　3.94　13.97　26.02　35.4　38.14　32.36　18.53　0

Normalized W_a: 　0　0.103　0.366　0.682　0.928　1.00　0.847　0.486　0

Equivalent concentrated angle change: −0.126　−1.396　−4.445　−8.114　−10.962　−11.775　−9.950　−5.707　−1.035

Average slope: −0.126　−1.522　−5.967　−14.081　−25.043　−36.818　−46.774　−52.481

Deflection W_p: 　0　−0.126　−1.648　−7.615　−21.696　−46.739　−83.557　−130.331　−182.812

$$S = \frac{182.812}{2048}\frac{Pa}{\lambda} = 0.089\,264\,\frac{Pa}{\lambda}$$

Finally $W_a / \left[W'_a \left(\dfrac{12EI}{P\lambda^2} \right) \right]$: 　0　0.025 88　0.026 03　0.026 10　0.026 19　0.026 27　0.026 36　0.026 94　0

B.3　Application of difference equation

B.3.1　Continuous beams

This problem may be readily formulated in terms of the rotations of the tangents to the elastic curve at each support by use of the slope–deflection formula. We thus obtain at all joints except the 'joint' at the left end (Figure B.15):

$$\theta_{n+1} + 4\theta_n + \theta_{n-1} = 0 \tag{B.91}$$

$$M_n = \frac{2EI}{L}(2\theta_n + \theta_{n+1}), \quad M_n = \frac{2EI}{L}(2\theta_n + \theta_{n-1}) \tag{B.92}$$

Figure B.15

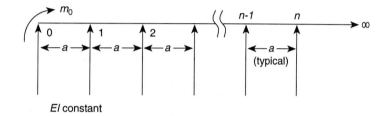

EI constant

At the first joint (joint 0), we have by the slope deflection formula

$$\frac{2EI}{a}(2\theta_0 + \theta_1) = M_0 \tag{B.93}$$

Equation B.91 generates an 'infinite' set of simultaneous equations, in tridiagonal form.

The algebraic equation having the form

$$y_{i+1} + c_1 y_i + c_2 y_{i-1} = f(x)$$

in which c_1 and c_2 are constants, y_{i-1}, y_i and y_{i+1} are the values of the dependent variables at the three successive points, usually equally spaced, $y = y(x)$, and $f(x)$ is a specified function of the independent variables, is called a nonhomogeneous, linear difference equation of second order with constant coefficients.

Consider the homogeneous difference equation of second order and with constant coefficients,

$$y_{i+1} + c_1 y_i + c_2 y_{i-1} = 0 \tag{B.94}$$

A form of solution is suggested by the success and generality of the solution used to handle linear differential equations with constant coefficients. With this motivation, we assume a solution of the form,

$$y = c \mathrm{e}^{\lambda x i}$$

In many cases, the equivalent simple form,

$$y = c\beta^{xi} \quad (\beta = \mathrm{e}^{\lambda})$$

or

$$y_i = c\beta^i \tag{B.95}$$

is found to be sufficiently general and more convenient. We seek to determine β and to evaluate c. The latter will be seen to depend upon the boundary conditions, in strict analogy to the case of differential equations. Substitution of equation B.95 into equation B.94 with x taken on a dimensionless scale yields

$$[c\beta^{i+1}] + c_1[c\beta^i] + c_2[c\beta^{i-1}] = 0 \tag{B.96}$$

We can now divide through by the factor $c\beta^{i-1}$, and we obtain

$$\beta^2 + c_1\beta + c_2 = 0 \tag{B.97}$$

Hence, equation B.95 is a solution to equation B.94, if β satisfies equation B.97.

In the case of the second-order difference equation, the resulting characteristic equation or indicial equation, equation B.97, is a quadratic equation and can be solved by the application of the quadratic formula. In the present case, this gives

$$\beta = \frac{-c_1 \pm \sqrt{c_1^2 - 4c_2}}{2} \tag{B.98}$$

which in general, yields two values or roots of the characteristic equation. Denoting these by β_1 and β_2, the solution to equation B.94 is now seen to be

$$y_i = \bar{c}_1 \beta_1^i + \bar{c}_2 \beta_2^i \tag{B.99}$$

Applying this method of solution to the semi-infinite continuous beam for which the difference equation is $\theta_{i+1} + 4\theta_i + \theta_{i-1} = 0$, we assume a solution of the form $\theta_i = \bar{c}\beta^i$ and substituting into the difference equation leads to the following algebraic characteristic equation,

$$\beta^2 + 4\beta + 1 = 0 \tag{B.100}$$

The roots of this equation are

$$\beta = -0.268, \quad -3.732 \tag{B.101}$$

We may now write the general solution of the beam problem which we are considering as

$$\theta_i = \bar{c}_1(-0.268)^i + \bar{c}_2(-3.732)^i \tag{B.102}$$

We can take the boundary condition approaching the 'far' end of the beam to be $\theta_\infty = 0$. Evaluating θ_i at $i \to \infty$ gives

$$\theta_\infty = \bar{c}_1(-0.268)^\infty + \bar{c}_2(-3.732)^\infty = \bar{c}_1(0) + \bar{c}_2(\pm\infty) \tag{B.103}$$

Since this result should be zero, i.e. the rotations at the support vanish as i becomes large without limit, it follows that $\bar{c}_2 = 0$. This leaves our solution in the form

$$\theta_i = \bar{c}_1(-0.268)^i \tag{B.104}$$

At the 'zero' end of the beam, we have from the slope deflection equation,

$$M_0 = \frac{2EI}{a}(2\theta_0 + \theta_1) \tag{B.105}$$

or

$$\frac{M_0 a}{2EI} = 2\bar{c}_1(-0.268)^0 + \bar{c}_1(-0.268)^1 = \bar{c}_1[2(1) - 0.268] = 1.732\bar{c}_1 \tag{B.106}$$

so that

$$\bar{c}_1 = \frac{M_0 a}{2EI}\frac{1}{1.732} = \frac{\sqrt{3}M_0 a}{6EI}$$

$$\theta_i = \frac{\sqrt{3}M_0 a}{6EI}(-0.268)^i \tag{B.107}$$

The same problem can be solved using the three-moment theorem, $m_{i+1} + 4m_i + m_{i-1} = 0$, with boundary conditions

$$M_0 = M_0$$
$$M_\infty = 0 \tag{B.108}$$

B.3.2 Case of multiple roots

If, for the indicial equation, β_1 is an m-fold root, then the solution corresponding to this root is

$$y_i = [c_1 + c_2 i + c_2 i^2 + \dots c_m i^{m-1}]\beta_1^i \tag{B.109}$$

To this solution must be added the solutions corresponding to the distinct roots and to other multiple roots, if any, to construct the solution of the reduced equation.

Example

Consider the difference equation,

$$y_{i+1} - 2ay_i + a^2 y_{i-1} = 0 \tag{B.110}$$

Substitution of the assumed solution $y_i = c\beta^i$ into the difference equation yields

$$c\beta^{i+1} - 2ac\beta^i + a^2 c\beta^{i-1} = 0 \tag{B.111}$$

Dividing through by $c\beta^{i-1}$ yields $\beta^2 - 2a\beta + a^2 = 0$ which can be written as $(\beta - a)^2 = 0$. Hence $\beta = a$ is a double root.

Motivated by the method used in the solution of the linear ordinary differential equation in the analogous case, we try a solution of the form $y = cia^i$. Substitution into the original equation yields

$$c(i + 1)a^{i+1} - 2acia^i + a^2 c(i - 1)a^{i-1} = 0 \tag{B.112}$$

Each term now contains the common factor ca^{i+1} and the above equation reduces to

$$ca^{i+1}[(i + 1) - 2i + (i - 1)] = ca^{i+1}(0) = 0 \tag{B.113}$$

thus providing the validity of the second solution.

B.3.3 Case of complex roots

There are occasions in which the indicial equation has complex roots. If the indicial equation has only real coefficients, then its complex roots will occur only in conjugate pairs. Thus, the roots would occur in pairs

$$\beta_1 = a + jb$$
$$\beta_2 = a - jb \tag{B.114}$$

where $j = \sqrt{-1}$. Then the solution corresponding to this conjugate complex pair will be

$$y_i = A(a + jb)^i + B(a - jb)^i \tag{B.115}$$

This can be written in the polar coordinate form,

$$y_i = A(\rho - e^{j\theta})^i + B(\rho - e^{-j\theta})^i \tag{B.116}$$

where $\rho = \sqrt{a^2 + b^2}$, and $\theta = \tan^{-1}(b/a)$. Hence, we may also write

$$y_i = \rho^i (A \, e^{ij\theta} + B \, e^{-ij\theta}) = \rho^i(c_1 \sin i\theta + c_2 \cos i\theta) \tag{B.117}$$

in which c_1 and c_2 are new arbitrary constants.

Example

$$y_{i+1} - 2y_i + 2y_{i-1} = 0 \tag{B.118}$$

Assume $y_i = \beta^i$, then substitution into the difference equation yields

$$\beta^{i+1} - 2\beta^i + 2\beta^{i-1} = 0 \tag{B.119}$$

or

$$\beta^{i-1}(\beta^2 - 2\beta + 2) = 0$$

For the assumed solution to be valid, the two roots must satisfy the condition

$$\beta^2 - 2\beta + 2 = 0 \tag{B.120}$$

and are therefore given by

$$\beta = \frac{2 \pm \sqrt{4 - 8}}{2} = 1 \pm \sqrt{j} \tag{B.121}$$

Hence $a = 1$, $b = 1$,

$$\rho = \sqrt{1^2 + 1^2} = \sqrt{2}$$
$$\theta = \tan^{-1}(1/1) = \pi/4 \tag{B.122}$$

The general solution may, therefore, be written in the form

$$(\sqrt{2})^i \left[c_1 \sin\left(\frac{i\pi}{4}\right) + c_2 \cos\left(\frac{i\pi}{4}\right) \right] \tag{B.123}$$

B.3.4 Particular solution of nonhomogeneous equations

When the homogeneous part of the difference equation has constant coefficients and when the right-hand side of the difference equation is made up of terms such as c^i, $\sin ci$, $\cos ci$, i^p with $p = 0, 1, 2, ..., N$, we may use the method of undetermined coefficients to obtain the particular solution.

Example 1   (Figure B.16)

At all joints except the left end, the three-moment equation is

$$M_{i+1} + 4M_i + M_{i-1} = ql^2/2 \tag{B.124}$$

Figure B.16

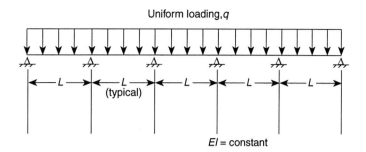

Uniform loading, q

EI = constant

We first obtain a solution to the reduced equation, $M_{i+1} + 4M_i + M_{i-1} = 0$. Assuming $M_i = c\beta^i$, substitution into the reduced difference equation now yields

$$\beta^2 + 4\beta + 1 = 0 \tag{B.125}$$

which has roots

$$\beta_{1,2} = \frac{-4 \pm \sqrt{16 - 4}}{2} = -2 \pm \sqrt{3} \tag{B.126}$$

and therefore the general solution of the reduced equation is

$$M_i = c_1(-0.268)^i + c_2(-3.732)^i \tag{B.127}$$

Since the right-hand side of the nonhomogeneous equation does not contain terms which occur in the solution of the reduced equation, and since the right-hand side is simply a constant, we try a particular solution of the form, $y_i = c_3$. Substituting into the original nonhomogeneous equation, we obtain

$$c_3 + 4c_3 + c_3 = \frac{ql^2}{2} \quad \text{or} \quad c_3 = \frac{ql^2}{12} \tag{B.128}$$

The complete general solution is the sum of the solution of the reduced equation and the particular solution:

$$M_i = c_1(-0.268)^i + c_2(-3.732)^i + \frac{ql^2}{12} \tag{B.129}$$

where the arbitrary constants are to be evaluated so that the above solution satisfies the boundary conditions.

$$M_0 = 0 \quad \text{yields} \quad c_1 = -\frac{ql^2}{12}$$

$$M_\infty = \frac{ql^2}{12} \quad \text{yields} \quad c_2 = 0 \tag{B.130}$$

Therefore

$$M_i = \frac{ql^2}{12}\left[1 - (-0.268)^i\right] \tag{B.131}$$

Example 2   (Figure B.17)

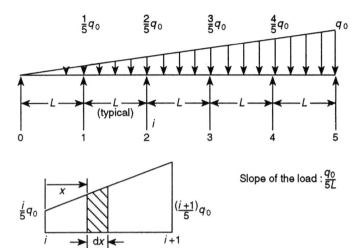

Figure B.17

Between the supports i and $i + 1$, the intensity of the load at the point at distance x from support i is

$$q = \frac{iq_0}{5} + \frac{q_0}{5L} x \tag{B.132}$$

The general three-moment equation at the ith support is

$$M_{i+1}L_{i+1} + 2M_i(L_{i+1} + L_{i-1}) + M_{i-1}L_{i-1}$$
$$= -\sum P_{i-1}L_{i-1}^2(k_1 - k_1^3) - \sum P_{i+1}L_{i+1}^2(2k_2 - 3k_2^2 + k_2^3)$$
$$+ 6EI\left(\frac{m}{L_{i-1}} + \frac{n}{L_{i+1}}\right) \tag{B.133}$$

The m/L_{i-1} and n/L_{i-1} terms are zero because there is no settlement.
  Substituting $q\,dx$ into P and x/L into k,

$$\sum P_{i+1}L_{i+1}^2(2k_2 - 3k_2^2 + k_2^3) = \int_0^L q\,dx\left(2\frac{x}{L} - \frac{3x^2}{L^2} + \frac{x^3}{L^3}\right)L^2$$
$$= \frac{q_0}{5}L^2\int_0^L\left(i + \frac{x}{L}\right)\left(2\frac{x}{L} - 3\frac{x^2}{L^2} + \frac{x^3}{L^3}\right)dx$$
$$= \frac{q_0}{5}L^3\left(\frac{i}{4} + \frac{7}{60}\right) \tag{B.134}$$

Between the supports $i - 1$ and i,

$$q = (i - 1)\frac{q_0}{5} + \frac{q_0}{5L}x$$

$$\sum P_{i-1}L_{i-1}^2(k_1 - k_1^3) = \int_0^L q\,dxL^2\left(\frac{x}{L} - \frac{x^3}{L^3}\right)dx$$

$$= \frac{q_0}{5}L^3\left(\frac{i}{4} - \frac{7}{60}\right) \tag{B.135}$$

Therefore

$$M_{i+1} + 4M_i + M_{i-1} = \frac{q_0 L^2}{5}\left(\frac{i}{4} + \frac{7}{60} + \frac{i}{4} - \frac{7}{60}\right) = \frac{q_0 L^2}{10} i \quad \text{(B.136)}$$

The homogeneous part is $M_{i+1} + 4M_i + M_{i-1} = 0$. Assuming $M_i = c\beta^i$, the three-moment equation (homogeneous part) becomes $\beta^2 + 4\beta + 1 = 0$, or

$$\beta_{1,2} = -2 \pm \sqrt{3} \quad \text{(B.137)}$$

and we obtain

$$M_{i,\text{hom}} = c_1(-0.268)^i + c_2(-3.732)^i \quad \text{(B.138)}$$

Assuming the particular solution to be

$$M_{i,p} = Ai + B \quad \text{(B.139)}$$

substitute into the three-moment equation, to obtain

$$A(i+1) + B + 4Ai + 4B + A(i-1) + B = 6(Ai + B) = \frac{q_0 L^2}{10} i \quad \text{(B.140)}$$

which implies

$$B = 0, \quad 6Ai = \frac{q_0 L^2}{10} i \quad \text{or} \quad A = \frac{q_0 L^2}{60} \quad \text{(B.141)}$$

and we have

$$M_i = c_1(-0.268)^i + c_2(-3.732)^i + \frac{q_0 L^2}{60} i \quad \text{(B.142)}$$

From the boundary conditions,

$$M_{i=0} = 0, \quad c_1 + c_2 = 0 \quad \text{(B.143)}$$

$$M_{i=5} = 0, \quad -0.001\,395 c_1 - 724 c_2 + \frac{q_0 L^2}{60}(5) = 0 \quad \text{(B.144)}$$

from which we obtain

$$M_i = \frac{q_0 L^2}{12}\left[\frac{i}{5} - \left(1 + \frac{1}{72.3}\right)(-0.268)^i + \frac{1}{72.3}(-3.732)^i\right] \quad \text{(B.145)}$$

B.3.5 Application to taut strings

In Figure B.18, F is the weight of equally spaced beads. The initial tension, T, is assumed to be sufficiently large so that the displacements of the beads are small. With this restriction it follows that we may treat the tensile force as constant, i.e., invariant with respect to the deflections. From the free body

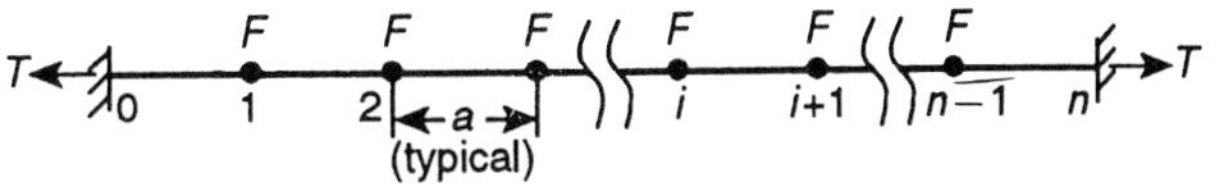

Figure B.18

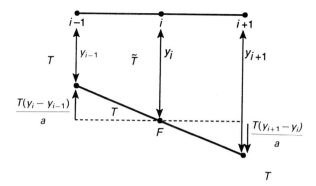

diagram (Figure B.19), summing forces in the vertical direction, we obtain

$$\frac{\uparrow T}{a}(y_i - y_{i-1}) - \frac{\downarrow T}{a}(y_{i+1} - y_i) = F\downarrow \tag{B.146}$$

or

$$y_{i+1} - 2y_i + y_{i-1} = -\frac{Fa}{T} \tag{B.147}$$

For the reduced equation, we assume

$$y_i = c\beta^i \tag{B.148}$$

and we obtain, upon substitution into the reduced equation

$$\beta^2 - 2\beta + 1 = 0 \quad \text{or} \quad (\beta - 1)(\beta - 1) = 0 \tag{B.149}$$

The homogeneous solution is

$$y_i = c_1 + c_2 i \tag{B.150}$$

For the particular solution, we may try

$$y_i = c_3 i^2 \tag{B.151}$$

since the constant and first power terms, which we would ordinarily try, have already been included in the homogeneous solution. Substituting into the complete difference equation, we obtain

$$c_3 = -\frac{Fa}{2T} \tag{B.152}$$

The complete solution is

$$y_i = c_1 + c_2 i - \frac{Fa}{2T} i^2 \tag{B.153}$$

The boundary conditions are

$$y_0 = 0, \quad y_n = 0 \tag{B.154}$$

Figure B.20

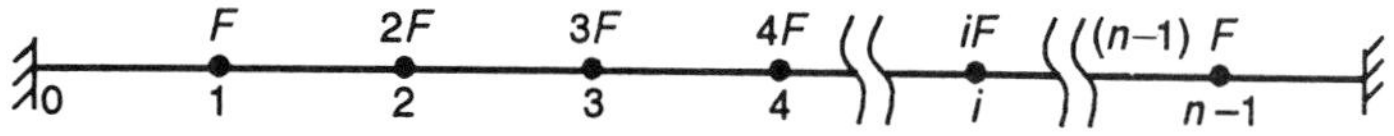

from which we obtain

$$c_1 = 0$$

$$c_2 = \frac{Fa}{2T} n \tag{B.155}$$

The complete solution to the problem, therefore, is

$$y_i = \frac{Fa}{2T}(ni - i^2) = \frac{Fa}{2T}(n - i)i \tag{B.156}$$

Consider the case when the weights of the beads increase (Figure B.20). From equilibrium of the ith bead, we obtain

$$y_{i+1} - 2y_i + y_{i-1} = -\frac{iFa}{T} \tag{B.157}$$

For the homogeneous solution, we proceed as the previous problem, and we obtain

$$y_{i,h} = c_1 + c_2 i \tag{B.158}$$

Because we have a double root for the homogeneous solution and the right-hand side of the equation has the first order of i, we must assume the particular solution in the third power of i. Let $y_{i,p} = A i^3$.

Substituting this equation into the equilibrium equation, we get $A = -Fa/6T$, and

$$y_i = c_1 + c_2 i - \frac{Fa}{6T} i^3 \tag{B.159}$$

From the boundary condition, $y_{(0)} = 0$, $y_{(n)} = 0$, we obtain

$$y_i = \frac{Fa}{6T} i(n^2 - i^2) \tag{B.160}$$

B.3.6 Analysis of symmetric two-column towers
From Figure B.21, we denote the following:

- c_i a number proportional to EI/h for the column of the ith story;
- G_i a number proportional to EI/b for the given beam at the top of the story, the factor of proportionality being the same as for the column;
- h_i the height of the ith story;
- F_i lateral force applied at the level of the top of the ith story;
- S_i shear at the ith story, given as $F_0 + \sum_{s=1}^{i} F_s$;
- $M_i = s_i h_i$;
- θ_i rotation, in consistent but possibly anomalous units, of the joints at the ith level.

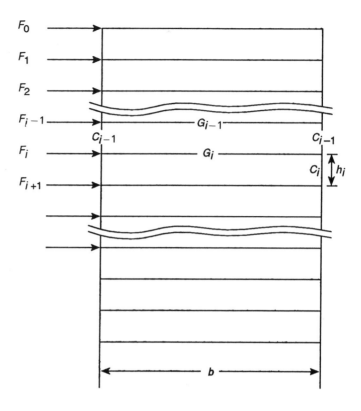

At any joint, we can obtain the following equation, from the moment equilibrium condition,

$$c_i\theta_{i+1} - (6G_i + c_i + c_{i-1})\theta_i + c_{i-1}\theta_{i-1} = -\frac{M_i + M_{i-1}}{2} \qquad \text{(B.161)}$$

If the column has appropriate regularity, the solution of the reduced equation is obtained by assuming it to be of the form

$$\theta_i = A\beta^i$$

where A is an arbitrary constant. The substitution may be made in the left-hand side of equation B.161 either after or before inserting the proper numerical values for the relative stiffness.

In the case where the columns of all stories are equal and the girders of all stories are equal, we may write equation B.161 in the form

$$\theta_{i+1} - (6r + 2)\theta_i + \theta_{i-1} = -\frac{M_i + M_{i-1}}{2c} \qquad \text{(B.162)}$$

where $r = G/c$.

Substituting $\theta_i = A\beta^i$ into the reduced equation B.162 yields

$$\beta^{i+1} - (6r + 2)\beta^i + \beta^{i-1} = 0 \qquad \text{(B.163)}$$

Upon dividing through by β^{i-1},

$$\beta^2 - (6r + 2)\beta + 1 = 0 \tag{B.164}$$

The roots of this equation are

$$\beta = 3r + 1 \pm \sqrt{3r(3r + 2)} \tag{B.165}$$

The homogeneous solution will then be

$$\theta_i = A_1\beta_1^i + A_2\beta_2^i \tag{B.166}$$

where β_1 and β_2 are the two values obtained from the quadratic formula above.

The particular solution is also readily obtained in certain cases. For example, if there is a load at the top, F_0, and the loads at the other levels are zero, then the right-hand side of equation B.161 is constant, namely, $-F_0h/c$. The corresponding particular solution is readily seen to be a constant and its value is simply

$$\theta_i = \frac{F_0 h}{6rc} \tag{B.167}$$

so that, for this case, the complete solution is

$$\theta_i = A_1\beta_1^i + A_2\beta_2^i + \frac{F_0 h}{6rc} \tag{B.168}$$

The bending moments at the ends of the girders are given by

$$M_{gi} = -3G_i\theta_i \tag{B.169}$$

The bending moments at the top and at the bottom of the ith story column may be written, respectively;

$$M_{c,i,i+1} = \frac{M_i}{4} - \frac{c_i}{2}(\theta_i - \theta_{i+1})$$

$$M_{c,i+1,i} = \frac{M_i}{4} - \frac{c_i}{2}(\theta_{i+1} - \theta_i) \tag{B.170}$$

$$= \frac{M_i}{4} + \frac{c_i}{2}(\theta_i - \theta_{i+1})$$

At the top of the frame, the values of the rotations must be such as to satisfy the equilibrium condition for the top joint, $\sum M = 0$. Note that this equation is not one of the set of difference equations for which the general solution has already been obtained. Since this is not yet satisfied (by the general solution) we may use it as a boundary condition. To apply this boundary condition, we observe that the moment at the end of the top girder is

$$M_{g,0} = -3G_0\theta_0 \tag{B.171}$$

and the moment at the top of the top story column is

$$M_{c,01} = \frac{M_0}{4} - \frac{c}{2}(\theta_0 - \theta_1) \tag{B.172}$$

The sum of the moments is

$$\frac{M_0}{4} - \frac{c}{2}(\theta_0 - \theta_1) - 3G_0\theta_0 = 0 \tag{B.173}$$

or

$$\frac{M_0}{4} - \frac{c}{2}(\theta_0 - \theta_1) - 3G_0\theta_0 = m$$

if we have m at the top.

If the general solution is now substituted into this equation for θ_0 and θ_1, we obtain one equation involving the arbitrary constants. At the lower end of the frame, if the columns may be assumed to be built in, we may take the boundary condition as

$$\theta_b = 0$$

This method can be used for several boundary condition possibilities, cases such as pin-end or partial (elastic) fixity at the base.

As an example, consider a ten-story frame, fixed at the bottom, with columns and girders with equal stiffness, acted upon by a single load at the top, F. cs and Gs may be taken as unity. Then

$$M_0 = M_1 = ..., M_{10-1} = Fh \tag{B.174}$$

and the difference equation is

$$\theta_{i+1} - 8\theta_i + \theta_{i-1} = -M_0 \tag{B.175}$$

Assuming $\theta_i = A\beta^i$, and substituting into the reduced difference equation, we obtain

$$\beta^2 - 8\beta + 1 = 0 \tag{B.176}$$

from which we get

$$\beta = 0.127, \quad 7.873$$

The particular solution is readily seen to be

$$\theta_i = \frac{M_0}{6}\left(\frac{M_0}{6c} \quad \text{with } c = 1\right) \tag{B.177}$$

The complete solution is, therefore,

$$\theta = A_1(0.127)^i + A_2(7.873)^i + \frac{M_0}{6} \tag{B.178}$$

Since $(0.127)^{10}$ is negligible, the boundary condition, $\theta_{10} = 0$, may be treated as

$$A_2(7.873)^{10} = -\frac{M_0}{6}$$

or $\tag{B.179}$

$$A_2 = \frac{-M_0}{6(7.873)^{10}}$$

For the second boundary condition, we use the equilibrium equation at the top joint (joint 0):

$$M_{c,0} + M_{g,0} = 0$$

which yields

$$3\theta_0 + \frac{1}{2}(\theta_0 - \theta_1) = \frac{M_0}{4}$$

or

$$7\theta_0 - \theta_1 = \frac{M_0}{2} \tag{B.180}$$

Using the general solution,

$$7\left(A_1 + A_2 + \frac{M_0}{6}\right) - \left[A_1(0.127) + A_2(7.873) + \frac{M_0}{6}\right] = \frac{M_0}{2} \tag{B.181}$$

From which we get

$$A_1 = -\frac{M_0}{13.746} \tag{B.182}$$

Finally, therefore,

$$\theta_i = M_0\left[\frac{1}{6} - \frac{(0.127)^i}{13.746} - \frac{(7.873)^i}{6(7.873)^{10}}\right] \tag{B.183}$$

References

1. Goldberg, J. E. and Kim, D. H. (1966) Analysis of triangularly folded plate roofs of umbrella type, in *16th General Congress of Applied Mechanics*, Tokyo, Japan, Oct., p. 280.
2. Kim, D. H. (1971) Theory of non-prismatic folded plate structures. *Proc. Korean Soc. Civ. Engrs*, Dec.
3. Goldberg, J. E. and Kim, D. H. (1967) The effect of neglecting the radial moment terms in analyzing a sectorial plate by means of finite differences, in *Proceedings 7th International Symposium on Space Technology and Science* (eds Kuroda, Y. *et al.*), Tokyo, Japan, pp. 267–78.

Appendix C. Matrices and determinants

C.1 Matrix

Let A and B be sets. A **function** with **domain** A and **range** contained in B is a rule f which assigns to each element $a \in A$ ($\in$ means 'is an element of') one and only one element $f(a)$ of B. The function T is called a linear transformation if

1. $T(x + y) = T(x) + T(y)$ for all $x, y \in V_n$ (additive)
2. $T(\alpha x) = \alpha T(x)$ for all $x \in R$ and for all $x \in V_n$ (homogeneity)

An $m \times n$ matrix performs a **linear transformation** of V_n to V_m when premultiplied, and V_m to V_n when postmultiplied. The **domain** of a transformation is defined as the set of elements which undergo transformation. The **range** of a transformation is the set of elements which is formed by the transformation operation on the elements in the domain. The range is often called the **image** of the domain under transformation. The range of a linear transformation on V_n is a **subspace** of V_m. If $[A]$ is an $m \times n$ matrix, then the set of points $[y] = [A][x]$ (for all $[x]$ in V_n) is a subspace of V_m. The range of the transformation, i.e. the subspace generated by $[y]$ is the subspace of V_m spanned by the columns of $[A]$. The dimension of the range is the maximum number of linearly independent columns in $[A]$. It may happen that the subspace of V_m is V_m itself.

A set of vectors, $[a_1], ..., [a_m]$, from V_n is said to be **linearly dependent** if there exist scalars λ_i not all zero such that

$$\lambda_1[a_1] + \lambda_2[a_2] + \cdots + \lambda_m[a_m] = 0$$

If the only set of λ_i for which this equation holds is $\lambda_1 = \lambda_2 = \cdots = \lambda_m = 0$, then the vectors are said to the **linearly independent**. If a set of vectors is linearly independent, then any subset of these vectors is also linearly independent. Suppose that $k < m$ is the maximum number of linearly independent vectors in a set of m vectors, $[a_1], [a_2], ..., [a_m]$, from V_n, then, given any linearly independent subset of k vectors in this set, every other vector in the set can be expressed as a linear combination of these k vectors.

A set of vectors, $[a_1], ..., [a_r]$ from V_n is said to **span** or generate V_n if every vector in V_n can be written as a linear combination of $[a_1], ..., [a_r]$.

A **basis** for V_n is a linearly independent subset of vectors from V_n which spans the entire space.

The **rank** of the linear transformation $T(V_n) \to V_m$ is the dimension of the linear subspace $T(V_n)$ of V_m. The rank of $[T]$ is the maximum number of linearly independent rows of $[T]$.

Given $[A] = [a_{ij}]_{m \times n}$; then

Row rank $[A]$ = the maximum number of linearly independent rows of $[A]$

Column rank $[A]$ = the maximum number of linearly independent columns of $[A]$

Given $[A]_{m \times n}$, a **minor** of order k is the matrix formed after crossing out all but k rows and k columns of $[A]$. Suppose there is a minor of order k whose determinant is not zero, but suppose that every minor of order $k + 1$ has the determinant of zero. Then the **determinant rank** of $[A]$ is k.

A **matrix** is formed with the m vectors as columns. If the rank is m, then the vectors are linearly independent. If the rank, $k < m$, the vectors are not linearly independent. If $[A]$ is a nonsingular nth-order matrix, $|A| \neq 0$. The columns (or rows) of $[A]$ are linearly independent and form a basis for V_n and the rank of $[A]$ is n. Conversely, if we have n linearly independent vectors from V_n, we can form $[A]$ with the vectors as its columns. The rank of $[A]$ is n, and $|A| \neq 0$. Since $[A]$ is nonsingular, $[A]$ has an inverse.

A matrix is defined as a rectangular array of numbers arranged into rows and columns:

$$[A] = \begin{bmatrix} a_{11} & a_{12} & \cdots & a_{1n} \\ a_{21} & a_{22} & \cdots & a_{2n} \\ \vdots & \vdots & & \vdots \\ a_{m1} & a_{m2} & \cdots & a_{mn} \end{bmatrix}$$

We call this array an m by n matrix (written $m \times n$) since it has m rows and n columns. Note that a matrix has no numerical value. It is simply a convenient way of representing arrays of numbers. Note also that a double subscript is used to represent any one of the elements (often called entries or **matrix elements**), since a matrix is generally a two-dimensional array of numbers. The first subscript refers to the **row** and the second to the **column**. a_{ij} refers to the element in the ith row, jth column. There is no relation between the number of rows and the number of columns. Any matrix with the same number of rows as columns is called a **square matrix**. A square matrix with n rows and n columns is called an nth-order matrix.

Two matrices, $[A]$ and $[B]$ are **equal**, written $[A] = [B]$, if the corresponding elements are equal. Thus, $[A] = [B]$ if and only if $a_{ij} = b_{ij}$ for all i, j. If $[A] = [B]$, then $[B] = [A]$.

Given a matrix $[A]$ and a **scalar** λ, the **product** of λ and $[A]$ is defined as

$$\lambda[A] = \begin{bmatrix} \lambda a_{11} & \lambda a_{12} & \cdots & \lambda a_{1n} \\ \lambda a_{21} & \lambda a_{22} & \cdots & \lambda a_{2n} \\ \vdots & \vdots & & \vdots \\ \lambda a_{m1} & \lambda a_{m2} & \cdots & \lambda a_{mn} \end{bmatrix}$$

Note that $\lambda[A] = [A]\lambda$.

The **sum** $[C]$ of a matrix $[A]$, with m rows and n columns, and a matrix $[B]$, with m rows and n columns, is a matrix with m rows and n columns whose elements are shown as

$$c_{ij} = a_{ij} + b_{ij}$$

Written in detail,

$$[C] = [A] + [B] = \begin{bmatrix} c_{11} & \cdots & c_{1n} \\ \vdots & & \vdots \\ c_{m1} & \cdots & c_{mn} \end{bmatrix}$$

$$= \begin{bmatrix} a_{11} & \cdots & a_{1n} \\ \vdots & & \vdots \\ a_{m1} & \cdots & a_{mn} \end{bmatrix} + \begin{bmatrix} b_{11} & \cdots & b_{1n} \\ \vdots & & \vdots \\ b_{m1} & \cdots & b_{mn} \end{bmatrix}$$

$$= \begin{bmatrix} a_{11} + b_{11} & \cdots & a_{1n} + b_{1n} \\ \vdots & & \vdots \\ a_{m1} + b_{m1} & \cdots & a_{mn} + b_{mn} \end{bmatrix}.$$

Note that $[A] + [A] = 2[A]$, and $2[A] + [A] = 3[A]$, etc. Note also that

$$[A] + [B] = [B] + [A] \quad \text{(associative law)}$$

$$[A] + ([B] + [C]) = ([A] + [B]) + [C]$$
$$= [A] + [B] + [C] \quad \text{(commutative law)}$$

Subtraction is defined as

$$[C] = [A] - [B] = [A] + (-1)[B]$$

We subtract two matrices by subtracting the corresponding elements:

$$c_{ij} = a_{ij} - b_{ij} \quad \text{for all} \quad i, j$$

Given an $m \times n$ matrix $[A]$ and an $n \times r$ matrix $[B]$, the product $[A][B]$ is defined to be an $m \times r$ matrix $[C]$. The elements of $[C]$ are computed from the elements of $[A]$ and $[B]$ according to

$$c_{ij} = \sum_{k=1}^{n} a_{ik} b_{kj}, \quad i = 1, \ldots, m, \quad j = 1, \ldots, r$$

$[A]$ is called the **premultiplier** and $[B]$ is the **postmultiplier**. From this equation, we can conclude the following rule:

The two matrices $[A]$ and $[B]$ are conformable for multiplication to yield the product $[A][B]$ if and only if the number of columns in $[A]$ is the same as the number of rows in $[B]$.

Note that

$$\sum_{k=1}^{m} \left(\sum_{j=1}^{n} a_k b_{kj} \right) = \sum_{k=1}^{m} \left(a_k \sum_{j=1}^{n} b_{kj} \right) = \sum_{j=1}^{n} \left(\sum_{k=1}^{m} a_k b_{kj} \right) = \sum_{j,k} a_k b_{jk}$$

In general, $[A][B] \neq [B][A]$. If $[A][B] = [B][A]$, the matrices are said to **commute**.

Matrices can be represented as a row of column **vectors** or as a column of row vectors. Recall that if we have m rows of 'n-space' vectors, this 'sum' of vectors can represent a second 'order' **tensor**.

An **mth order tensor** has m vector spaces. Each of these vectors has an n-dimensional subspace. Thus, an n-dimensional **vector space** is a tensor of the first order. A **scalar** is a zeroth-order tensor. All physical quantities can be expressed by a tensor. Force, displacement, velocity, etc. are vectors with three-dimensional space, i.e. $n = 3$. As explained in Chapter 3, both stress and strain are second-order tensors because of their directional dependency of area. Both stiffnesses and compliances are fourth-order tensors since they have four subscripts. Recall

$$\sigma_{ij} = C_{ijkl} \varepsilon_{kl}$$
$$\varepsilon_{ij} = S_{ijkl} \sigma_{kl} \tag{5.1}$$

The number of subscripts (or superscripts) expresses the order of the tensor. If all of m vectors of an mth order tensor have equal numbers of dimensions, n, i.e. equal n-dimensional spaces, the number of the components is

$$N = n^m.$$

The **identity matrix** of order n, is a square matrix having ones (1) along the main diagonal and zeros elsewhere:

$$[I] = \begin{bmatrix} 1 & 0 & 0 & 0 & \ldots & 0 \\ 0 & 1 & 0 & 0 & \ldots & 0 \\ \vdots & & & & & \vdots \\ 0 & 0 & 0 & 0 & \ldots & 1 \end{bmatrix}$$

or

$$[I] = \|\delta_{ij}\|, \text{ with } \delta_{ij} = \begin{cases} 1, & i = j \\ 0, & i \neq j \end{cases}$$

The symbol $\|\delta_{ij}\|$ is called the **Kronecker delta**.

Let $[A]$ be a square matrix of order n. We can construct the **matrix polynomial**:

$$\lambda_k [A]^k + \lambda_{k-1} [A]^{k-1} + \cdots + \lambda_0 [I],$$

which is also an $n \times n$ matrix.

For any scalar λ, the square matrix

$$[S] = \|\lambda \delta_{ij}\| = \lambda [I]$$

is called a **scalar matrix**.

A square matrix, $[D] = \|\lambda_i \delta_{ij}\|$ is called a **diagonal matrix**.

A matrix whose elements are all zero is called a **null** or **zero matrix** and is denoted by $[0]$. We can have

$$[A] + [0] = [A] = [0] + [A]$$

$$[A] - [A] = [0]$$

$$[A][0] = [0]$$

$$[0][A] = [0]$$

Note that $[A][B] = [0]$ does not indicate that $[A] = [0]$ or $[B] = [0]$. There are non-null matrices $[A]$ and $[B]$, whose product, $[A][B] = [0]$. As an example,

$$\begin{bmatrix} 1 & 5 \\ 0 & 0 \end{bmatrix} \begin{bmatrix} 5 & 0 \\ -1 & 0 \end{bmatrix} = \begin{bmatrix} 0 & 0 \\ 0 & 0 \end{bmatrix}$$

The **transpose** of a matrix $[A] = \|a_{ij}\|$, denoted by $[A^T]$, is a matrix formed from $[A]$ by interchanging rows and columns such that row i of $[A]$ becomes column i of the transpose matrix. Thus

$$[A^T] = \|a_{ji}\| \quad \text{when} \quad [A] = \|a_{ij}\|$$

Note that if $[A]$ is $m \times n$, $[A^T]$ is $n \times m$.

If $[C] = [A] + [B]$, then

$$[C^T] = [A^T] + [B^T]$$

since

$$c_{ij}^T = c_{ji} = a_{ji} + b_{ji} = a_{ij}^T + b_{ij}^T$$

The transpose of the product is the product of the transposes in reverse order:

$$([A][B])^T = [B^T][A^T]$$

This result can be extended to any finite number of factors:

$$([A_1][A_2] \dots [A_n])^T = [A_n^T] \dots [A_2^T][A_1^T]$$

Note that

$$[I^T] = [I]$$

$$[A^T]^T = [A]$$

since

$$(a_{ij}^T)^T = a_{ji}^T = a_{ij}$$

A **symmetric matrix** is a matrix $[A]$ with property of

$$[A] = [A^T]$$

A **skew-symmetric matrix** is a matrix $[A]$ for which

$$[A] = -[A^T]$$

and is also a square matrix, and

$$a_{ij} = -a_{ji} \quad \text{and} \quad a_{ii} = 0$$

Any square matrix can be written as the sum of a symmetric and a skew-symmetric matrix:

$$[A] = [A] + \frac{[A^T]}{2} - \frac{[A^T]}{2}$$

$$= \frac{[A] + [A^T]}{2} + \frac{[A] - [A^T]}{2}$$

$$= [A_s] + [A_a]$$

Furthermore

$$\left(\frac{[A] + [A^T]}{2}\right)^T = \frac{[A] + [A^T]}{2}$$

$$\left(\frac{[A] - [A^T]}{2}\right)^T = -\frac{[A] - [A^T]}{2}$$

$[A_s]$ is symmetric and $[A_a]$ is skew-symmetric.

If we delete all but k rows and s columns of an $m \times n$ matrix, $[A]$, the resulting $k \times s$ matrix is called a **submatrix** of $[A]$. The rule for addition of partitioned matrices is the same as the rule for addition of ordinary matrices if the submatrices are conformable for addition. The multiplication of matrices by blocks requires that the partitioning of the columns in the premultiplier be the same as the partitioning of the rows in the postmultiplier.

C.2 Determinant

The **determinant** of an nth-order matrix, $[A] = \|a_{ij}\|$, written $|A|$, is defined to be the number computed from the following sum involving n^2 elements in $[A]$:

$$|A| = \sum (\pm) a_{1i} a_{2j} \dots a_{nr}$$

The sum is taken over all permutations of the second subscripts. A term is assigned a plus sign if $(i, j, \dots, r)$ is an **even permutation** of $(1, 2, \dots, n)$, and a minus sign if it is an **odd permutation**.

Any rearrangement of the natural order of n integers is called **permutation** of these integers. The interchange of two integers is called a **transposition**. The number of **inversions** in a permutation of n integers is the number of pairs of elements (not necessarily adjacent) in which a larger integer precedes a smaller one. A permutation is **even** when the number of inversions is even, and odd when the number of inversions is odd.

The definition of a determinant indicates that only **square matrices** have determinants associated with them. One and only one element from each row and column of $[A]$ appears in every term of the summation. While a matrix has no numerical value, a **determinant is a number**.

To review: the second-order determinant is

$$\begin{vmatrix} a_{11} & a_{12} \\ a_{21} & a_{22} \end{vmatrix} = a_{11}a_{22} - a_{12}a_{21}$$

If we have two simultaneous linear equations with two unknowns,

$$a_{11}x_1 + a_{12}x_1 = b_1$$
$$a_{21}x_1 + a_{22}x_1 = b_2$$

then

$$[A][X] = [b]$$

and if $|A| \neq 0$,

$$x_1 = \frac{\begin{vmatrix} b_1 & a_{12} \\ b_2 & a_{22} \end{vmatrix}}{\begin{vmatrix} a_{11} & a_{12} \\ a_{21} & a_{22} \end{vmatrix}} \qquad x_2 = \frac{\begin{vmatrix} a_{11} & b_1 \\ a_{21} & b_2 \end{vmatrix}}{\begin{vmatrix} a_{11} & a_{12} \\ a_{21} & a_{22} \end{vmatrix}}$$

A **third-order determinant** is defined as

$$\begin{vmatrix} a_{11} & a_{12} & a_{13} \\ a_{21} & a_{22} & a_{23} \\ a_{31} & a_{32} & a_{33} \end{vmatrix} = a_{11}a_{22}a_{33} - a_{12}a_{21}a_{33} + a_{12}a_{23}a_{31} - a_{13}a_{22}a_{31}$$
$$+ a_{13}a_{21}a_{32} - a_{11}a_{23}a_{32}$$

An interchange of two columns as well as an interchange of two rows in an nth-order matrix $[A]$ changes the sign of $|A|$.

For a square matrix $[A]$, the determinant of the transposed matrix is the same as the determinant of the matrix itself. Thus

$$|A| = |A^T|$$

$$|A^T| = |A|$$

We also notice that

$$|\lambda[A]| = \lambda^n|A|$$

$$|-[A]| = (-1)^n|A|$$

Note that $|\lambda[A]| \neq \lambda|A|$.

The **cofactor** A_{ij} of the element a_{ij} of any square matrix $[A]$ is $(-1)^{i+j}$ times the determinant of the submatrix obtained from $[A]$ by deleting row i and column j. An **expansion** by row i of $|A|$ can be expressed as

$$|A| = \sum_{j=1}^{n} a_{ij}A_{ij}$$

where i can be any row. Now, we can evaluate a determinant by **expansion by cofactors**. As an example, we expand

$$|A| = \begin{vmatrix} a_{11} & a_{12} & a_{13} \\ a_{21} & a_{22} & a_{23} \\ a_{31} & a_{32} & a_{33} \end{vmatrix}$$

in cofactors by row 2 as follows:

$$|A| = a_{21}A_{21} + a_{22}A_{22} + a_{23}A_{23}$$

where

$$A_{21} = (-1)^{2+1}\begin{vmatrix} a_{12} & a_{13} \\ a_{32} & a_{33} \end{vmatrix} = a_{32}a_{13} - a_{12}a_{33}$$

$$A_{22} = (-1)^{2+2}\begin{vmatrix} a_{11} & a_{13} \\ a_{31} & a_{33} \end{vmatrix} = a_{11}a_{33} - a_{13}a_{31}$$

$$A_{23} = (-1)^{2+3}\begin{vmatrix} a_{11} & a_{12} \\ a_{31} & a_{32} \end{vmatrix} = a_{12}a_{31} - a_{11}a_{32}$$

In general

$$|[A] + [B]| \neq |A| + |B|$$

If one column (or row) of a square matrix $[A]$ is a linear combination of the other columns (or rows), $|A| = 0$. We have

$$|A| = \sum_j a_{ij}A_{ij} = \sum_i a_{ij}A_{ij}$$

$$\sum_j a_{kj}A_{ij} = \sum_j a_{ji}A_{jk} = 0 \quad (i \neq k)$$

Combining the above two equations,

$$\sum_j a_{ij}A_{kj} = \sum_j a_{ji}A_{jk} = |A|\delta_{ki}$$

Thus, we conclude that an expansion by a row or column i in terms of the cofactors of row or column k vanishes when $i \neq k$, and is equal to $|A|$ when $i = k$.

Suppose that $[A][B]$ are matrices of order n. If $[C] = [A][B]$, then $|C| = |A||B|$, that is the determinant of the product is the product of the determinants.

C.3 The matrix inverse

Given a square matrix $[A]$, if there exists a matrix $[A]^{-1}$ which satisfies the relation

$$[A]^{-1}[A] = [A][A]^{-1} = [I]$$

then $[A]^{-1}$ is called the **inverse** or **reciprocal** of $[A]$. Note that $[A]^{-1}$ does not mean $1/[A]$ or $[I]/[A]$ but is merely a notation.

The **adjoint** of matrix $[A]$, $[A^+]$, is the transpose of the matrix obtained from $[A]$ by replacing each element a_{ij} by its cofactor A_{ij}. Thus

$$[A^+] = \| a_{ij}^+ \| = \begin{bmatrix} A_{11} & A_{21} & \cdots & A_{n1} \\ A_{12} & A_{22} & \cdots & A_{n2} \\ \vdots & \vdots & & \vdots \\ A_{1n} & A_{2n} & \cdots & A_{nn} \end{bmatrix}$$

and $[A][A^+] = [A^+][A] = |A|[I]$. Therefore

$$[A]^{-1} = \frac{1}{|A|} [A^+], \quad |A| \neq 0$$

This provides us with a procedure to compute inverses.

The square matrix $[A]$ is said to be singular if $|A| = 0$, **nonsingular** if $|A| \neq 0$. Note that only nonsingular matrices have inverses and every nonsingular matrix has an inverse. As an example, consider a nonsingular matrix,

$$[A] = \begin{bmatrix} a_{11} & a_{21} \\ a_{12} & a_{22} \end{bmatrix}$$

The matrix of the cofactors is

$$\begin{bmatrix} a_{22} & -a_{21} \\ -a_{12} & a_{11} \end{bmatrix}$$

In this case, the cofactors are determinants of the first order. The adjoint (transpose of the above matrix) is

$$[A^+] = \begin{bmatrix} a_{22} & -a_{12} \\ -a_{21} & a_{11} \end{bmatrix}$$

Then

$$[A]^{-1} = \frac{1}{|A|} \begin{bmatrix} a_{22} & -a_{12} \\ -a_{21} & a_{11} \end{bmatrix}$$

If $[A] = [a_{11}]$, $|A| = a_{11}$, $[A^+] = [1]$ and $[A]^{-1} = [1/a_{11}]$, $a_{11} \neq 0$.

The product of two nth-order nonsingular matrices is nonsingular and

$$([A][B])^{-1} = [B]^{-1}[A]^{-1}$$

In fact, the product of any finite number of nonsingular matrices is nonsingular and the inverse is the product of the inverses in reverse order.

We also have that if $[A]$ is nonsingular,

$$([A]^{-1})^{-1} = [A]$$

$$[A^T]^{-1} = ([A]^{-1})^T$$

Even though $[A][B] = 0$ does not necessarily imply that $[A] = 0$ or $[B] = 0$, if either $[A]$ or $[B]$ is nonsingular, the **other** is a **null matrix**.

The inverse of a matrix can be computed by **partitioning**. Let $[A]$ be an $n \times n$ nonsingular matrix, and be partitioned as

$$[A] = \begin{bmatrix} [\alpha] & [\beta] \\ [\gamma] & [\delta] \end{bmatrix}$$

in which $[\alpha]$ is an $s \times s$ submatrix, $[\beta]$ an $s \times m$ submatrix, $[\gamma]$ an $m \times s$ submatrix and $[\delta]$ an $m \times m$ submatrix. Here, $n = m + s$. Then $[A]^{-1}$ exists and can be partitioned in the same way as $[A]$:

$$[A]^{-1} = \begin{bmatrix} [B] & [C] \\ [D] & [E] \end{bmatrix}$$

in which $[B]$ is $s \times s$, $[C]$ is $s \times m$, $[D]$ is $m \times s$ and $[E]$ is $m \times m$. The submatrices of $[A]^{-1}$ can be obtained as

$$[B] = ([\alpha] - [\beta][\delta]^{-1}[\gamma])^{-1}$$

$$[C] = -[B][\beta][\delta]^{-1}$$

$$[D] = -[\delta]^{-1}[\gamma][B]$$

$$[E] = [\delta]^{-1} - [\delta]^{-1}[\gamma][C]$$

Since $[A]^{-1}$ exists, the submatrices $[B]$, $[C]$, $[D]$ and $[E]$ exist. Since $[\alpha]$, $[\beta]$, $[\gamma]$ and $[\delta]$ are known, if $[\delta]^{-1}$ does exist, all the calculations can be carried out to compute $[B]$, $[C]$, $[D]$ and $[E]$.

If $[A]$ is any square matrix, we can have the series

$$F(A) = \sum_{n=0}^{\infty} \lambda_n [A]^n$$

This series expression holds true if and only if this converges and this series cannot converge unless

$$\lim_{n \to \infty} \lambda_n [A]^n = 0.$$

It is convenient to define $[A]^0 = [I]$. If $[A]$ satisfies

$$0 \leq a_{ij} < 1 \quad (\text{all } i, j)$$

$$\sum_{i=1}^{n} a_{ij} < 1 \quad (\text{all } i, j)$$

we can write

$$([I] - [A])^{-1} = \sum_{k=0}^{\infty} [A]^k = [I] + [A] + [A]^2 + \cdots$$

This equation is of advantage particularly when $([I] - [A])$ is of a high order. We can compute the inverse to any desired degree of accuracy by taking a sufficient number of terms in expansion.

Table index

Author index

Subject index